WDM TECHNOLOGIES: PASSIVE OPTICAL COMPONENTS

WDM TECHNOLOGIES: PASSIVE OPTICAL COMPONENTS

Volume II

Edited by

Achyut K. Dutta
Banpil Photonics, Inc.
San Jose, California, USA

Niloy K. Dutta
University of Connecticut
Storrs, Connecticut, USA

Masahiko Fujiwara
Networking Research Laboratories
NEC Corporation
Tsukuba, Ibaraki, Japan

ACADEMIC PRESS

An imprint of Elsevier Science

Amsterdam Boston London New York Oxford Paris
San Diego San Francisco Singapore Sydney Tokyo

This book is printed on acid-free paper. ∞

Academic Press
An imprint of Elsevier Science
525 B Street, Suite 1900, San Diego, CA 92101-4495, USA
http://www.academicpress.com

Academic Press
An imprint of Elsevier Science
84 Theabold's Road, London WC1X 8RR, UK
http://www.academicpress.com

Library of Congress Catalog Card Number: 2002117165

International Standard Book Number: 0-12-225262-4

PRINTED IN THE UNITED STATES OF AMERICA
03 04 05 06 MB 9 8 7 6 5 4 3 2 1

Dedicated to our parents,
Harish Chandra and Kalpana Rani Dutta,
Debakar and Madabor Datta,
and to our families,
Keiko, Jayoshree, Jaydeep,
Sudeep Hiroshi, and Cristine Dutta

Contents

Contributors

Yale Cheng (Chapters 3 & 4), Fiber Optic Product Group, JDS Uniphase Corporation, Nepean, Ontario, Canada.

Achyut K. Dutta (Chapter 1), Banpil Photonics, Inc., 1299 Parkmoor Avenue, San Jose, California, 95126, USA.

Niloy K. Dutta (Chapters 1 & 8), Department of Physics and Photonics Research Center, University of Connecticut, Storrs, CT 06269-3046, USA.

Masahiko Fujiwara (Chapter 1), Networking Research Laboratories, NEC Corporation, 34 Miyukigaoka, Tsukuba, Ibaraki 305-8501, Japan.

Yoshinori Hibino (Chapter 6), NTT Photonics Laboratories, Tokai, Ibaraki 319-1193, Japan.

Lih-Yuan Lin (Chapter 5), Tellium Inc., 2 Crescent Place, Oceanport, NJ 07757-0901, USA.

Y. Calvin Si (Chapter 3), Fiber Optic Product Group, JDS Uniphase Corporation, Nepean, Ontario, Canada.

Yan Sun (Chapter 9), Onetta Inc., 1195 Borregas Avenue, Sunnyvale, CA 94089, USA.

Atul Srivastava (Chapter 9), Onetta Inc., 1195 Borregas Avenue, Sunnyvale, CA 94089, USA.

Shu Namiki (Chapter 10), Fitel Photonics Laboratory, Furukawa Electric Co., Ltd. 6 Yawata Kaigan Dori, Ichihara, Chiba 290-8555, Japan.

Katsunari Okamoto (Chapter 2), NTT Electronics Ltd., 162 Tokai, Naka-Gun, Ibaraki-Prefecture, 319-1193, Japan.

Hideaki Okayama (Chapter 7), Oki Electric Industry Co. Ltd., 550-5, Higashiasakawa-cho, Hachioji-shi, Tokyo 193-8550, Japan.

Yuzo Yoshikuni (Chapter 11), NTT Photonics Laboratories, Nippon Telegraph and Telephone Corporation, 3-1, Morinosato Wakamiya, Atsugi-shi, Kanagawa 243-0198, Japan.

Foreword

The WDM Revolution

These volumes are about wavelength division multiplexing (WDM), the most recent technology innovation in optical fiber communications. In the past two decades optical communications has totally changed the way we communicate. It is a revolution that has fundamentally transformed the core of telecommunications, its basic science, its enabling technology, and its industry. The WDM innovation represents a revolution within the optical communications revolution, allowing the latter to continue its exponential growth.

The existence and advance of optical fiber communications is based on the invention of the laser, particularly the semiconductor junction laser, the invention of low-loss optical fibers, and related disciplines such as integrated optics. We should never forget that it took more than 25 years from the early pioneering ideas to the first large-scale commercial deployment of optical communications, the Northeast Corridor system linking Washington and New York in 1983 and New York with Boston in 1984. This is when the revolution got started in the marketplace, and when optical fiber communications began to seriously impact the way information is transmitted. The market demand for higher capacity transmission was assisted by the continuing power increase of computers and their need to be interconnected. This is a key reason for the parallel explosive growth of optical fiber transmission technology, computer processing, and other key information technologies. These technologies have combined to meet global demand for new information services including data, Internet, and broadband services—and, most likely, their rapid advance has helped fuel this demand. This demand continues its strong growth—internet traffic alone, even by reasonably conservative estimates, keeps doubling every year. (Today, we optical scientists and engineers, puzzle over the question of why this traffic growth does not appear to be matched by a corresponding growth in revenue.) Another milestone in the optical communications revolution we remember with pride is the deployment of the first transatlantic fiber system, TAT8, in 1988. (Today, of course, the map of undersea systems deployed in the oceans of the globe looks like a dense spider web.) It was around this time that researchers began exploring the next step forward, optical fiber amplifiers and WDM transmission.

WDM technology and computer architecture show an interesting parallel. Both systems trends—pulled by demand and pushed by technology advances—show their key technological figure of merit (computer processing power in one case, and fiber transmission capacity in the other) increasing by a factor 100 or more every 10 years. However, the raw speed of the IC technologies computers and fiber transmission rely on increases by about a factor of 10 only in the same time frame. The answer of computer designers is the use of parallel architectures. The answer of the designers of advanced lightwave systems is similar: the use of many parallel high-speed channels carried by different wavelengths. This is WDM or "dense WDM." The use of WDM has other advantages, such as the tolerance for WDM systems of the high dispersion present in the low loss window of embedded fibers, the fact that WDM can grow the capacity incrementally, and that WDM provides great simplicity and flexibility in the network.

WDM required the development of many new enabling technologies, including broadband optical amplifiers of high gain, integrated guided-wave wavelength filters and multiplexers, WDM laser sources such as distributed-feedback (DFB) lasers providing spectral control, high-speed modulators, etc. It also required new systems and fiber techniques to compensate for fiber dispersion and to counteract nonlinear effects caused by the large optical power due to the presence of many channels in the fiber. The dispersion management techniques invented for this purpose use system designs that avoid zero dispersion locally, but provide near-zero dispersion globally.

Vigorous R&D in WDM technologies led to another milestone in the history of optical communications, the first large-scale deployment of a commercial WDM system in 1995, the deployment of the NGLN system in the long-distance network of AT&T.

In the years that followed, WDM led the explosive growth of optical communications. In early 1996 three research laboratories reported prototype transmission systems breaking through the terabit/second barrier for the information capacity carried by a single fiber. This breakthrough launched lightwave transmission technology into the "tera-era." All three approaches used WDM techniques. Five years later, in 2001 and exactly on schedule for the factor-100-per-decade growth rate, a WDM research transmission experiment demonstrated a capacity of 10 Tb/s per fiber. This is an incredible capacity: Recall that, at the terabit/second rate, the hair-thin fiber can support a staggering 40 million 28-K baud data connections, transmit 20 million digital voice telephony channels, or a half million compressed digital TV channels. Even more importantly, we should recall that the dramatic increase in lightwave systems capacity has a very strong impact on lowering the cost of long-distance transmission. The Dixon-Clapp rule projects that the cost per voice channel reduces with the square root of the systems capacity. This allows one to estimate that this technology growth rate reduces the technology cost of transmitting one voice channel by a factor of 10 every 10 years. As a consequence of this trend, one finds that the distance of transmission plays a smaller and smaller role in the equation of telecom economics: An Internet user, for example, will click a web site regardless of its geographical distance.

WDM technology is progressing at a vigorous pace. Enabled by new high-speed electronics, the potential bit-rate per WDM channel has increased to 40 Gb/s and higher, broadband Raman fiber amplifiers are being employed in addition to the early erbium-doped fiber amplifiers, and there are new fibers and new techniques for broadband dispersion compensation and broadband dispersion management, etc. The dramatic decrease in transmission cost, combined with the unprecedented capacities appearing at a network node as well as the new traffic statistics imposed by the Internet and data transmission, have caused a rethinking of long-haul and ultra-long-haul network architectures. New designs are being explored that take advantage of the fact that WDM has opened up a new dimension in networking: It has added the dimension of wavelength to the classical networking dimensions of space and time. New architectures, transparent to bit-rate, modulation format, and protocol, are under exploration. A recent example for this are the demonstrations of bit-rate transparent fiber cross-connects based on photonic MEMS fabrics, arrays of micromirrors fabricated like integrated silicon integrated circuits.

Exactly because of this rapid pace of progress, these volumes will make a particularly important contribution. They will provide a solid assessment and teaching of the current state of the WDM art, serving as a valuable basis for further progress.

Herwig Kogelnik
Bell Labs
Lucent Technologies
Crawford Hill Laboratory
Holmdel, NJ 07733-0400

Acknowledgments

Future communication networks will require total transmission capacities of several Tb/s. Such capacities could be achieved by wavelength division multiplexing (WDM). This has resulted in an increasing demand of WDM technology in communication. With this increase in demand, many students and engineers are migrating from other engineering fields to this area. Based on our many years of experience, we felt that it is necessary to have a set of books that could help all engineers wishing to work or already working in this field. Covering a fast-growing subject such as WDM technology is a very daunting task. This work would not have been possible without the support and help of all chapter contributors. We are indebted to our current and previous employers, NEC Research Labs, Fujitsu, Bell Laboratories, and the University of Connecticut for providing an environment, that encouraged the intellectual stimulation for our research and development in the field of optical communication and its applications. We are grateful to our collaborators over the years. We would also like to convey our appreciation to our colleagues with whom we have worked for many years. Last but not least, many thanks also go to our family members for their patience and support, without which this book could not have been completed.

Achyut K. Dutta
Niloy K. Dutta
Masahiko Fujiwara

Chapter 1 | Overview

Achyut K. Dutta

Banpil Photonics, Inc., 1299 Parkmoor Avenue, San Jose, CA, 95126 USA

Niloy K. Dutta

Department of Physics and Photonics Research Center University of Connecticut, Storrs, CT 06269-3046 USA

Masahiko Fujiwara

Networking Research Laboratories, NEC Corporation 34 Miyukigaoka, Tsukuba, Ibaraki 305-8501, Japan

1.1. Prospectus

With the recent exponential growth of Internet users and the simultaneous proliferation of new Internet protocol applications such as web browsing, e-commerce, Java applications, and video conferencing, there is an acute need for increasing the bandwidth of the communications infrastructure all over the world. The bandwidth of the existing SONET and ATM networks is pervasively limited by electronic bottlenecks, and only recently was this limitation removed by the first introduction of wavelength division multiplexing (WDM) systems in the highest capacity backbone links. The capacity increase realized by the first WDM systems was quickly exhausted/utilized, and both fueled and accommodated the creation of new Internet services which, in turn, are now creating a new demand for bandwidth in more distant parts of the network. The communication industries are thus at the onset of a new expansion of WDM technology necessary to meet the new and unanticipated demand for bandwidth in elements of the telephony and cable TV infrastructure previously unconsidered for WDM deployment. The initial deployments of WDM were highly localized in parts of the communications infrastructure and supported by a relatively small group of experts. The new applications in different parts of the network must be implemented by much larger groups of workers from a tremendous diversity of technical backgrounds. To serve this community

WDM TECHNOLOGIES: PASSIVE OPTICAL COMPONENTS
$35.00

ISBN: 0-12-225262-4

involved with the optical networking, a series of volumes covering all WDM technologies (from the optical components to networks) is introduced.

Many companies and new start-ups are trying to make WDM-based products as quickly as possible to become a leader in that area. As the WDM-based products need wide knowledge, ranging from components to network architecture, it is difficult for the engineers to grasp all the related areas quickly. Today, the engineers working specifically in one area are always lacking in other areas, which impedes the development of the WDM products. The main objective of these volumes will be to give details on the WDM technology varying from the components (all types) to network architecture. We expect that this series of volumes would not only be useful for graduate students specifically in the fields of electrical engineering, electronic engineering, and computer engineering, but for class instructors as either the class textbook or as a reference.

Because the major developments in optical communication networks have started to capture the imagination of the computing, telecommunications, and opto-electronics industries, we expect that industry professionals will find this book useful as a well-rounded reference. Through our wide experience in industries involved in optical networking and optical components, we know that many engineers who are expert in the physical layer still must learn the optical system and networks and corresponding engineering problems in order to design new state-of-the-art optical networking products. It is these individuals for whom we prepared these books.

1.2. Organization and Features of the Volume

Writing books on this broad an area is not an easy task, as the volumes will cover from optical components (to be/already deployed) to the network. WDM coverage ranges from electrical engineering to computer engineering, and the field itself is still evolving. This volume is not intended to include any details about the basics of the related topics; readers much search out their own reference material on more basic issues, especially undergraduate-level books. This series can then provide a systematic in-depth understanding of multidisplinary fields to graduate students, engineers, and scientists, to increase their knowledge and potentially ability to contribute more to these WDM technologies.

In preparing this series we have attempted to combine research, development, and education on WDM technologies to allow tight coupling between the network architectures and device capabilities. Research on WDM has taught us that, without sound knowledge of device or component capabilities and limitations, you can produce architecture that could be completely unrealizable, or sophisticated technology with limited or no usefulness. We hope our approach will be helpful to professional and academic personnel, working in different area of WDM technologies, in avoiding these problems.

This series on various areas of WDM technologies is divided into four volumes, each of which is divided into a few parts to provide a clear concept among the readers or educators of the possibilities of their technologies in particular networks of interest to them. The series starts with two complete volumes on optical components. As many chapters are components related, we decided to publish two volumes for active and passive components. This format should prove more manageable and convenient for the reader. Other volumes are on the optical system and optical networks. Volume I gives a clear view of the WDM components, especially all kinds of active optical components. Volume II, covering key passive optical components, follows this. Volume III covers WDM networks and their architecture, possibly implementable in near-future networks. Finally, Volume IV will describe the WDM system, especially including a system aspects chapter implementable in the WDM equipment. All of these volumes cover not only recent but also future technologies. The contents of each volume will be explained in each book as Chapter 1, to accommodate users who choose to buy just one volume. This chapter contains only the survey of this volume.

1.3. Survey

WDM TECHNOLOGIES: PASSIVE OPTICAL COMPONENTS

Unlike most of the available textbooks on optical fiber communication, Volume II covers several key passive optical components and their key technologies, from the standpoint of WDM-based application. Based on our own hands-on experience in this area for the past 25 years, we tend to cover only those components and technologies that could be practically used in most WDM communication. This volume is divided into four parts. Part I: WDM Multiplexer/Demultiplexer Technologies, Part II: Optical Switching Technologies, Part III: Optical Amplifier Technologies, and Part IV: Critical Technologies. The chapters of each part are now briefly surveyed in an attempt to put the elements of the book into context.

Part I: WDM Multiplexer/Demultiplexer Technologies

Although optical technologies are replacing most transmission lines, the nodes of the networks, such as switching and cross-connect nodes, are still dependent on relatively slow electrical technologies. This limits the throughput all over the networks due to the limitation of electrical circuits. Making the nodes optical is important for solving these issues, and it requires multiplexing and demultiplexing by using optical technologies. Multiplexers/Demultiplexers are the key optical passive components that have revolutionized the optical network, extending its horizon from core to edge. These components play the key role in realizing all optical networks. This part covers the integrated router made from planar lightwave circuit (PLC) technology, and also discrete MUX/DEMUX and circulator, frequently used in the optical networks.

Chapter 2: Arrayed-Waveguide Grating (AWG) Router

In WDM systems, the network requires dynamic wavelength allocation because the number of wavelengths is limited. There are three types of router, based on how the data are processed to route the next hop. The Type I router, which is denoted wavelength router, has the highest feasibility at present. Arrayed-waveguide grating (AWG) is the most suitable device for the wavelength router. In Chapter 2, Katsunari Okamoto, the pioneer and inventor of many key PLC-based passive components, describes the design, fabrication, and performance of the arrayed-waveguide grating (AWG) based on the PLC technologies. He also explains the application of the AWG device to enable a class I photonic wavelength router.

Chapter 3: Optical Multiplexer/Demultiplexer: Discrete

WDM is considered to be one of the key enabling technologies in advanced optical communication networks. The key passive components in the WDM system are the wavelength division multiplexing and demultiplexing devices that combine/split lights with different wavelengths into different outputs. Several technologies are available for fabricating the MUX/DEMUX devices. Each of them has its own key significant features. The principles, structures, and performances of different MUX/DEMUX devices based on the thin film dielectric filter, fiber Bragg grating, AWG, and diffraction grating, are described in Chapter 3 by Y. C. Si and Y. Cheng. They also go on to provide the different technological advantages and issues.

Chapter 4: Circulator

The invention of the optical circulator in the 1990s can make it easier to enable many key passive components such as the erbium-doped optical amplifier, fiber Bragg grating etc., for WDM applications. The optical circulator works as the non-reciprocal polarization rotation of the Faraday effect, transmitting light one port to the next sequential port with a maximum intensity, but at the same time blocking any light transmission to the previous port. Yale Cheng describes the operational principle, design, performance, and application of the optical circulator in Chapter 4.

Part II: Optical Switching Technologies

As carriers and service providers continue their quest for profitable network solutions, they have shifted their focus from raw bandwidth to rapid provisioning, delivery, and management of revenue-generating services. Inherent transparency to the data rate of the transmission wavelength and data format are very much required in the next-generation optical network. Optical switches are the key components to provide bandwidth management, ring interconnect, protection, or mesh restoration in the optical layer. This part covers the different kinds of optical switches based on the electromechanical, thermooptic, and electrooptic effects for the applications to optical networks.

Chapter 5: Micro-Electro-Mechanical Systems (MEMS) for Optical-Fiber Communications

Micro-electro-mechanical systems (MEMS), studied since the 1980s, have very recently been considered to be powerful means of implementing various optical devices. This is possible due to the unique capability of this technology to integrate optical, mechanical, and electrical components on a single wafer. Recently, MEMS components have gone from the lab to commercial availability, and as this is being written are in carrier trials. In Chapter 5, L-Y. Lin, with many years of experience in the MEMs field, explains MEMs fabrication technology, and use of MEMs for transmission engineering. Finally, she provides the applications of MEMS as optical switches to the optical network.

Chapter 6: PLC-type Thermooptic Switches for Photonic Networks

Development of planar lightwave circuit (PLC) technologies has enabled many optical devices, including the optical switch. PLC-type thermooptic switches are a kind of optical switch, developed for a variety of applications in the optical networks. The port numbers of these switches can be increased from a small number to a medium-large number. As the matured Si-technology is being used, the suitability of its integration with other PLC devices/photonic devices will enable various functional optical components. In Chapter 6, Y. Hibino, a pioneer of thermooptic switches, reviews the principle of different kind of the thermooptic switches, their characteristics, and applications to optical networks.

Chapter 7: Lithium Niobate and Electrooptic Guided-Wave Optical Switch

Using electrooptic effects in ferro-electric crystals, such as Lithum Niobate ($LiNbO_3$), the optical switch can be fabricated. In Chapter 7, H. Okayama reviews the development of guided-wave optical switch technology using an electrooptic material, especially $LiNbO_3$, substrate. He also briefly examines the operation principle and features of almost all of the guided-wave optical switch.

Part III: Optical Amplifier Technologies

In fiber-optic communication, regenerators for amplifying the signals are used as the optical signal weakens and gets distorted as it travels through the fiber. Several orders of magnitude improvement in transmission capacity at low cost, considered to be the revolution in the telecommunication, would not have been possible without the development of the optical amplifier. This part covers different optical amplifier technologies, for optical communication application.

Chapter 8: Semiconductor Optical Amplifiers (SOA)

The operating principle of semiconductor optical amplifier is very similar to the semiconductor laser (which is covered in Chapter 2 of *WDM Technologies: Active Optical Components*). The main advantage is that it can be integrated monolithically with other optical devices on the single wafer for increasing functionality. In Chapter 8, the operating principles, design, fabrication, and characteristics of the InP-based optical amplifier are explained by N. K. Dutta, a pioneer in semiconductor laser and amplifier.

Chapter 9: Optical Amplifiers for Terabit Capacity Lightwave Systems

The economic advantages provided by lightwave systems would not have been possible without the development of the Erbium-doped fiber amplifier (EDFA). Since the first report in 1987 [1, 2], the EDFA has revolutionized optical communications. The significant performance improvements of the high-power semiconductor laser diodes (which are covered in Chapter 3 of *WDM Technologies: Active Optical Components*) helps to improve EDFA performance, and replaces the expensive electronics regenerator, which is based on the optical-electrical-optical (OEO) conversion. A. K. Srivastava and Y. Sun having many years of experience in EDFA, explain the fundamentals of EDFAs, its noise behavior, design consideration, and performance with respect to the network and system. They also briefly review the application of EDFA and the Raman cascade amplifier for high-capacity long-haul optical communications, in particular those carrying signals at 40 Gb/s line rates. The system performance of semiconductor optical (SOA) cascaded with EDFA is also demonstrated.

Chapter 10: Fiber Raman Amplifier

The frequency shift of light due to the scattering of molecules was first discovered by C. V. Raman, and this scattering was named after the discoverer. In the optical Raman amplifier, this principle is used to amplify the signals inside the fiber. The fiber Raman amplifier was extensively studied in the 1970s to 1980s before the discovery of the EDFA. As the Raman amplifier needs more than >100 mW pumped power, it was not realistic to achieve in compact LD, for application in communications. In the late 1990s, when the pumped laser technologies had matured enough to achieve a few hundred mW of power, and the reckless demands on bandwidth from Internet applications had risen, fiber Raman amplifiers were revisited for use in the real field of optical communication for better system performance. The main principle, design considerations, and important characteristics such as gain and noise performances of Raman amplifiers are given in Chapter 10, by Shu Namiki, a pioneer in this field. Shu also covers key components composing fiber Raman amplifiers and also examples of their system applications.

Part IV: Critical Technologies

The intention of this part is to include future technology, which has not yet found commercial application, being engineered for forward-looking system demonstration.

Chapter 11: Semiconductor Monolithic Circuit

Today's technology constraints make it difficult to achieve all types of optical components (active and passive devices) on a single substrate. To increase functionality, however, it is highly essential to either use hybridization techniques or to develop monolithic integration of a number of devices. Many technologies are already developed focusing on hybrid integration, and some are also underway to realize monolithic integration on the single substrate. Hybrid technology is covered in Chapter 13 of *WDM Technologies: Active Optical Components*. Chapter 11 (of this volume), by Y. Yoshikuni, who has many years of experience in monolithic integration, describes monolithic integrating technology for photonics devices with higher functionality. This chapter explains state-of-the-art future functional devices, fabricated monolithically on the single wafer, based on compound semiconductors (e.g., GaAs, InP).

References

1. R. J. Mears *et al.* "Low noise erbium-doped fiber amplifier operating at 1.54 μm," Electron. Lett., 23(1987), 1026–1028.
2. E. Desurvire, J. R. Simpson, and P. C. Becker, "High gain erbium-doped traveling wave fiber amplifier," Optics Lett., 12(1987), 888–890.

Part 1 WDM Multiplexer/ Demultiplexer Technologies

Chapter 2 | Arrayed-Waveguide (AWG) Router

Katsunari Okamoto

NTT Electronics Ltd., 162 Tokai, Naka-Gun Ibaraki-Prefecture 319-1193, Japan

Upgrading telecommunication networks to increase their capacity is becoming increasingly important due to the rapid increase in network traffic caused by multi-media communications in recent years. Although optical technologies are replacing most transmission lines, the nodes of the networks, such as switching and cross-connect nodes, still depend on relatively slow electrical technologies. This will be a serious problem because nodes in the networks will limit the throughput all over the networks, due to the limitations of the electrical circuits. Making the nodes optical, therefore, is important for solving these issues. It requires multiplexing and demultiplexing by using optical technologies. The time-division multiplexing or TDM systems that are widely used in existing optical communications systems inherently depend on electrical circuits for multiplexing and demultiplexing. The nodes in TDM systems use consequent optical–electrical conversion, electrical demulti- and multiplexing, and electrical–optical conversion. This means the throughput of the node is limited by the processing speed in the electrical circuits. Wavelength division multiplexing, or WDM, technologies, on the other hand, enable optical multi- and demultiplexing because individual signals have different light wavelengths and can be separated easily by wavelength-selective optical elements. This may enable us to construct WDM networks in which node functionality is supported by optical technologies without electrical mux/demux.

WDM TECHNOLOGIES: PASSIVE
OPTICAL COMPONENTS
$35.00

ISBN: 0-12-225262-4

Type I	Wavelength router
	Static wavelength allocation
	Dynamic wavelength allocation
Type II	Label switch router
	Optoelectronic label switch router (OLSR)
	Photonic label switch router (PLSR)
Type III	Packet switch router
	Optoelectronic packet switch router (OPSR)
	Photonic packet switch router (PPSR)

Fig. 2.1 Category of photonic routers.

There are three types of photonic routers proposed so far. Figure 2.1 categorizes these routers [1]. Type I router, which is denoted wavelength router, has the highest feasibility at present. Arrayed-waveguide grating (AWG) is the most suitable device for the wavelength router. One issue of this router is the limitation of the number of routes, because the number of wavelengths is limited. In order to solve the problem, dynamic wavelength allocation is proposed. In such a case, wavelength management with wavelength assignment algorithms and its optical hardware remain to be solved. A key device for a wavelength router is an optical switch with a large number of input/output ports.

Type II router utilizes a label to route optical data over a photonic network. Input packets are aggregated into a burst data based on the destination area, and a label is assigned to the burst data. The aggregated burst data is routed based only on the label. Two kinds of type II routers have been proposed. One is a router in which the label is O/E (optical to electrical) converted, and subsequent processes such as looking up the next hop router and controlling the optical switch for routing are done in the electronic domain. This router is called an optoelectronic label switch router (OLSR). The other router is called a photonic label switch router (PLSR), in which label processing is carried out in the optical domain. The most difficult issue in PLSR is how to generate and detect a photonic label.

Type III router is to handle packets themselves. There are also two kinds of type III routers. One is a router in which the IP address in the packet is O/E converted, and all the routing process is done in the electronic domain. This is called an optoelectronic packet switch router (OPSR). An all-optical packet handling router, which is called here a photonic packet switch router (PPSR), can also be considered. However, optical memory circuits are required to process all the routing procedures in the optical domain. Feasibility of PPSR has not been proved yet.

In this chapter, AWG devices are described in detail. AWG has a unique property of $N \times N$ signal interconnection when N input and N output ports are fully utilized. $N \times N$ interconnectivity of AWG enables us to construct Type I wavelength routers.

2.1. Waveguide Fabrication

Silica-based planar lightwave circuits (PLCs) are fabricated with various kinds of technologies. The substrate material is either silica or silicon. Several typical fabrication technologies are (1) flame hydrolysis deposition (FHD) [2], (2) low-pressure chemical vapor deposition (LPCVD), and (3) plasma-enhanced chemical vapor deposition (PECVD). FHD was originally developed for optical fiber fabrication. Fine glass particles are produced in the oxy-hydrogen flame and deposited on Si or glass substrate, which is placed on the turntable. After depositing undercladding and core glass layers, the wafer is heated to high temperature for consolidation. The circuit pattern is fabricated by photolithography and reactive ion etching (RIE). Then core ridge structures are covered with an overcladding layer and consolidated again. The thickness of under- and overcladding is about 20 μm.

LPCVD involves steps like the ones used in microelectronics silicon processing. The waveguide is formed on a substrate by low-pressure chemical vapor deposition of an undoped thick buffer layer ($\sim$15 μm), on which a core layer is deposited. PECVD is widely used in silicon integrated circuit applications. Unlike LPCVD, the gases used for depositing films are dissociated by electron impact in a glow discharge plasma. This allows deposition of relatively dense oxide films at low temperatures ($<300°$C). The mechanism of film formation in this technique is distinctly different from FHD and LPCVD, due to the highly energetic nature of the charged particles which impinge on the substrates and the growing films.

Because the typical bending radius R of silica waveguide is approximately 2–25 mm, the chip size of the large-scale integrated circuit becomes several centimeters square. Therefore, propagation loss reduction and the uniformity of refractive indices and core geometries throughout the wafer are strongly required. Propagation loss of 0.1 dB/cm was obtained in a 2-m-long waveguide with $\Delta = 2\%$ index difference (R $= 2$ mm) [3]

Table 2.1 **Waveguide Parameters and Propagation Characteristics**

	Low-Δ	*Medium-Δ*	*High-Δ*	*Super High-Δ*
Index Difference (%)	0.3	0.45	0.75	1.5∼2.0
Core Size (μm)	8 × 8	7 × 7	6 × 6	4.5 × 4.5 ∼ 3 × 3
Loss (dB/cm)	<0.01	0.02	0.04	0.07
Coupling Loss* (dB/point)	<0.1	0.1	0.4	2.0
Bending Radius** (mm)	25	15	5	2

* Coupling loss with standard single-mode fiber
** Bending radius at which bending loss in a 90-degree arc is 0.1 dB

and loss of 0.035 dB/cm was obtained in a 1.6-m-long waveguide with $\Delta = 0.75\%$ index difference (R = 5 mm) [4], respectively. Further loss reduction down to 0.017 dB/cm has been achieved in a 10-m-long waveguide with $\Delta = 0.45\%$ index difference (R = 15 mm) [5]. Table 2.1 summarizes the waveguide parameters and propagation characteristics of four kinds of waveguides. The propagation losses of low-Δ and medium-Δ waveguides are about 0.01 dB/cm, and those of high-Δ and super high-Δ waveguides are about 0.04–0.07 dB/cm, respectively. The low-Δ waveguides are superior to the high-Δ waveguides in terms of fiber coupling losses with the standard single-mode fibers. On the other hand, the minimum bending radii for high-Δ waveguides are much smaller than those of low-Δ waveguides. Therefore, high-Δ waveguides are indispensable to construct highly integrated and large-scale optical circuits such as $N \times N$ star couplers, arrayed-waveguide grating multiplexers, and dispersion equalizers.

2.2. Principle of Operation and Fundamental Characteristics of AWG

An $N \times N$ arrayed-waveguide grating (AWG) multiplexer is very attractive in optical WDM networks because it is capable of increasing the aggregate transmission capacity of single-strand optical fiber [6–9]. The arrayed-waveguide grating consists of input/output waveguides, two focusing slab regions, and a phase-array of multiple channel waveguides with the constant path length difference ΔL between neighboring waveguides (Fig. 2.2). In the first slab region, input waveguide separation is D_1, the array waveguide separation is d_1, and the radius of curvature is f_1, respectively. Generally the waveguide parameters in the first and the second slab regions may be

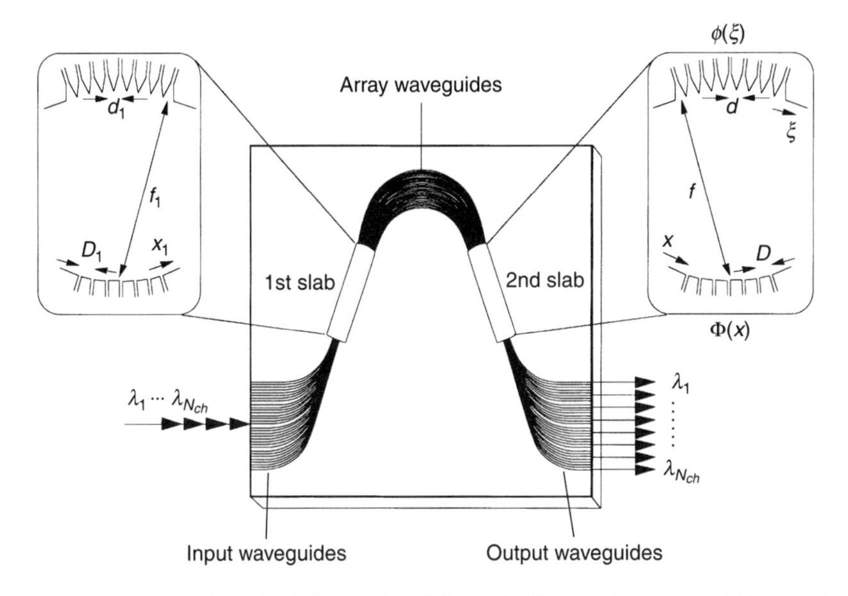

Fig. 2.2 Configuration of AWG. Reprinted from *Fundamentals of Optical Waveguides Photonic Networks*, K. Okamoto, Chapter 9 © 2000, with permission from Elsevier Science.

different. Therefore, in the second slab region the output waveguide separation is D, the array waveguide separation is d, and the radius of curvature is f, respectively. The input light at the position of x_1 (x_1 is measured in the counter-clockwise direction from the center of input waveguides) is radiated to the first slab and then excites the arrayed waveguides. The excited electric field amplitude in each array waveguide is a_i ($i = 1$–N) where N is total number of array waveguides. Amplitude profile a_i is usually a Gaussian distribution. After traveling through the arrayed waveguides, the light beams constructively interfere into one focal point x (x is measured in the counter-clockwise direction from the center of output waveguides) in the second slab. The location of this focal point depends on the signal wavelength because the relative phase delay in each waveguide is given by $\Delta L/\lambda$.

Figure 2.3 shows a schematic view of the light path, which originates from input waveguide (x_1), passes through 1st slab–array waveguide–2nd slab and converges to the output waveguide (x). Let us consider the phase retardations for the two light beams passing through the $(i-1)$-th and i-th array waveguides. The geometrical path lengths of two beams in the 1st and 2nd slab regions are approximated as shown in Fig. 2.3. The difference of the total phase retardations for the two light beams passing through the $(i-1)$-th and i-th array waveguides must be an integer multiple of 2π in

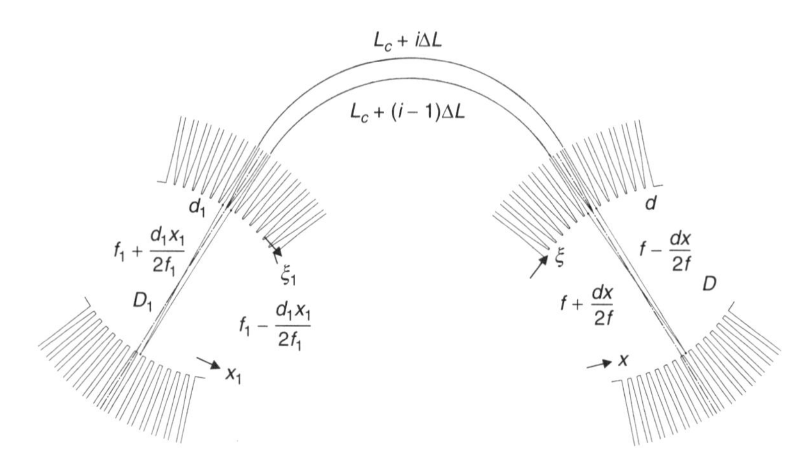

Fig. 2.3 Geometry for AWG design. Reprinted from *Fundamentals of Optical Waveguides Photonic Networks*, K. Okamoto, Chapter 9 © 2000, with permission from Elsevier Science.

order that two beams constructively interfere at the focal point x. Therefore we have the interference condition expressed by

$$\beta_s(\lambda_0)\left(f_1 - \frac{d_1x_1}{2f_1}\right) + \beta_c(\lambda_0)[L_c + (i-1)\Delta L] + \beta_s(\lambda_0)\left(f + \frac{dx}{2f}\right)$$
$$= \beta_s(\lambda_0)\left(f_1 + \frac{d_1x_1}{2f_1}\right) + \beta_c(\lambda_0)[L_c + i\Delta L]$$
$$+ \beta_s(\lambda_0)\left(f - \frac{dx}{2f}\right) - 2m\pi, \tag{2.1}$$

where β_s and β_c denote the propagation constants in slab region and array waveguide, m is an integer, λ_0 is the center wavelength of the WDM system, and L_c is the minimum array waveguide length. Subtracting common terms from Eq. (2.1), we obtain

$$\beta_s(\lambda_0)\frac{d_1x_1}{f_1} - \beta_s(\lambda_0)\frac{dx}{f} + \beta_c(\lambda_0)\Delta L = 2m\pi. \tag{2.2}$$

When the condition $\beta_c(\lambda_0)\Delta L = 2m\pi$ or

$$\lambda_0 = \frac{n_c\Delta L}{m} \tag{2.3}$$

is satisfied for λ_0, the light input position x_1 and the output position x should satisfy the condition

$$\frac{d_1x_1}{f_1} = \frac{dx}{f}. \tag{2.4}$$

In Eq. (2.3) n_c is an effective index of the array waveguide ($n_c = \beta_c/k \quad k$: wavenumber in vacuum) and m is called a diffraction order. The preceding equation means that when light is coupled into the input position x_1, the output position x is determined by Eq. (2.4). Usually the waveguide parameters in the first and second slab regions are the same. Therefore, input and output distances are equal as $x_1 = x$. The dispersion of the focal position x with respect to the wavelength λ for the fixed light input position x_1 is given by differentiating Eq. (2.2) with respect to λ as

$$\frac{\Delta x}{\Delta \lambda} = -\frac{N_c f \Delta L}{n_s d \lambda_0}, \tag{2.5}$$

where n_s is the effective index in the slab region, N_c is the group index of the effective index n_c of the array waveguide ($N_c = n_c - \lambda\, dn_c/d\lambda$), respectively. The dispersion of the focal position x with respect to the frequency ν for the fixed light input position x_1 is also given by differentiating Eq. (2.2) with respect to ν as

$$\frac{\Delta x}{\Delta v} = \frac{N_c f \Delta L}{n_s d \nu_0}, \tag{2.6}$$

where $\nu_0 = c/\lambda_0$. The output waveguide separation is $|\Delta x| = D$ when $\Delta\lambda$ is the channel wavelength spacing (Δv: frequency spacing) of the WDM signal. Generally, the waveguide parameters in the first and second slab regions are the same; they are, $D_1 = D$, $d_1 = d$, and $f_1 = f$. Putting these relations into Eqs. (2.5) or (2.6), the wavelength and frequency spacing in output side for the fixed light input position x_1 are given by

$$\Delta\lambda = \frac{n_s d D \lambda_0}{N_c f \Delta L}, \tag{2.7}$$

and

$$\Delta v = \frac{n_s d D \nu_0}{N_c f \Delta L}. \tag{2.8}$$

The spatial separation of the m-th and $(m+1)$-th focused beams for the same wavelength is given from Eq. (2.2) as

$$X_{FSR} = x_m - x_{m+1} = \frac{\lambda_0 f}{n_s d}. \tag{2.9}$$

X_{FSR} represents the free spatial range of AWG. Number of available wavelength channels N_{ch} is given by dividing X_{FSR} with the output waveguide separation D as

$$N_{ch} = \frac{X_{FSR}}{D} = \frac{\lambda_0 f}{n_s d D}. \tag{2.10}$$

Normally, center frequency ν_0 (or wavelength λ_0), channel spacing $\Delta\upsilon$ (or $\Delta\lambda$), and total number of channels N_{ch} are given from system requirements. Also input/output waveguide separation D and array-waveguide separation d are specified in accordance with waveguide geometries. First, radius of curvature f (focal length) is given by

$$f = \frac{n_s d D N_{ch}}{\lambda_0}. \tag{2.11}$$

Then, path length difference ΔL is obtained from Eq. (2.7) or (2.8) as

$$\Delta L = \frac{n_s d D \lambda_0}{N_c f \Delta\lambda} = \frac{n_s d D \nu_0}{N_c f \Delta\upsilon}. \tag{2.12}$$

Figure 2.4 shows spectral transmittance of 32-ch, 50-GHz spacing AWG. Crosstalks of less than −40 dB have been achieved. By the improvement of the fabrication technology and the optimization of the waveguide configurations, good crosstalk characteristics have also been achieved in 64-ch, 50-GHz spacing AWG and 256-ch, 25-GHz spacing AWG as shown in Figs. 2.5 and 2.6.

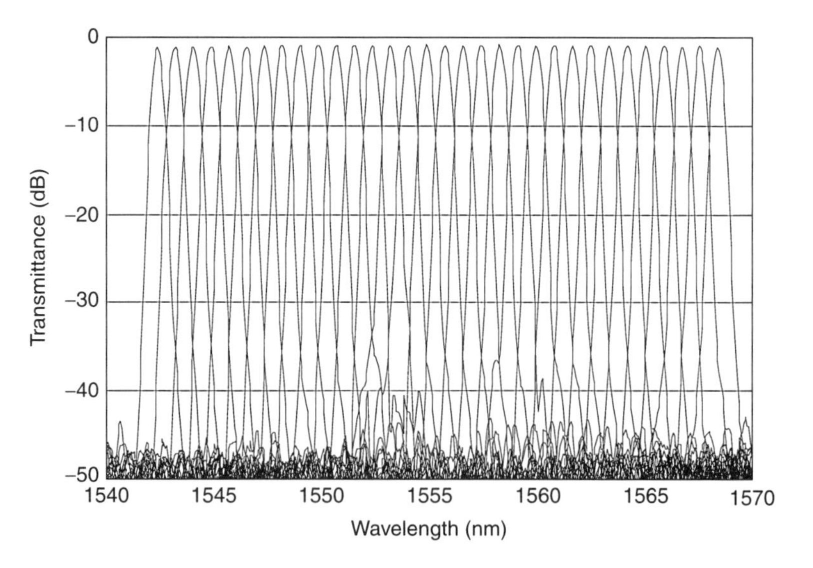

Fig. 2.4 Demultiplexing properties of a 32-ch, 100-GHz spacing AWG.

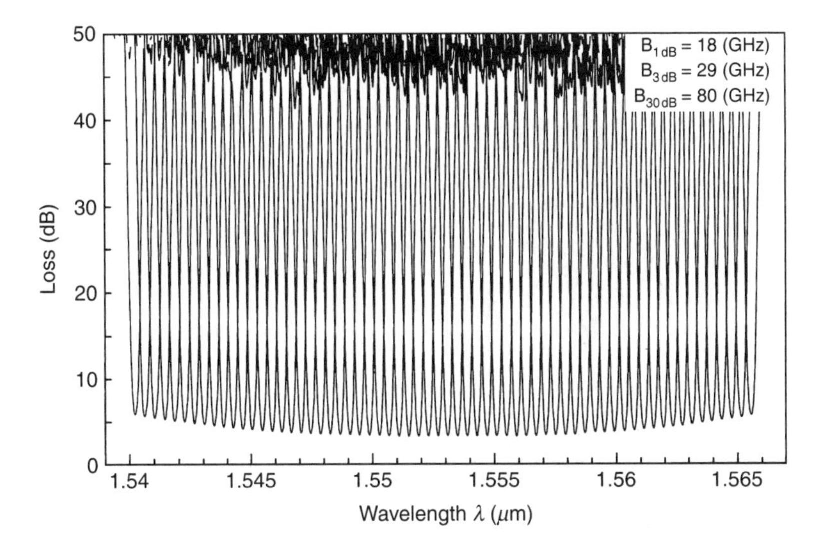

Fig. 2.5 Demultiplexing properties of a 64-ch, 50-GHz spacing AWG. Reprinted from *Fundamentals of Optical Waveguides Photonic Networks*, K. Okamoto, Chapter 9 © 2000, with permission from Elsevier Science.

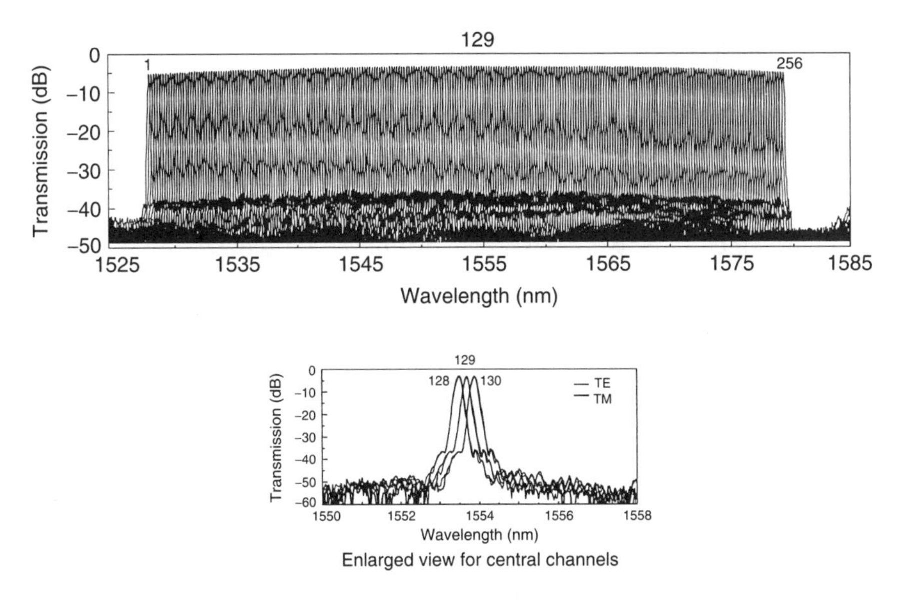

Fig. 2.6 Demultiplexing properties of a 256-ch, 25-GHz spacing AWG.

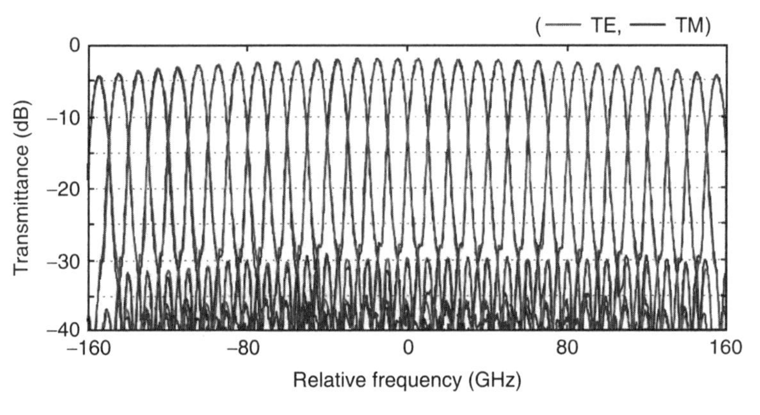

Fig. 2.7 Demultiplexing properties of a 32-ch, 10-GHz spacing AWG.

Crosstalk to other channels is caused by the sidelobes of the focused beam in the second slab region. These sidelobes are mainly attributed to the phase fluctuations of the total electric field profile at the output side array–slab interface, because the focused beam profile is the Fourier transform of the electric field in the array waveguides. The phase errors are caused by the nonuniformity of effective-index and/or core geometry in the array-waveguide region. Because path length difference ΔL is inversely proportional to channel spacing $\Delta \upsilon$ (or $\Delta \lambda$), total area of array-waveguide region becomes large in narrow channel spacing AWGs. Therefore, phase errors become much more susceptible to effective-index nonuniformities. Figure 2.7 shows spectral transmittance of 32-ch, 10-GHz spacing AWG. Crosstalks are about -25 dB. In order to get better crosstalk characteristics, further improvement in waveguide fabrication technology is required. There is also an alternative method to obtain better crosstalk characteristics in which effective-index nonuniformities are reduced by the post-processing technique.

2.3. Phase Error Compensation of AWG

Crosstalk improvement is the major concern for AWG multiplexers, especially for narrow channel spacing AWGs and $N \times N$ AWG routers. Phase errors in the AWGs can be measured by using Fourier transform spectroscopy [10]. Because phase errors with low spatial frequency generate sidelobes near the main focused beam, the slow phase variations are most harmful for crosstalk characteristics.

In order to improve the crosstalk characteristics of AWGs, a phase error compensation (trimming) experiment is carried out using 10-GHz spacing

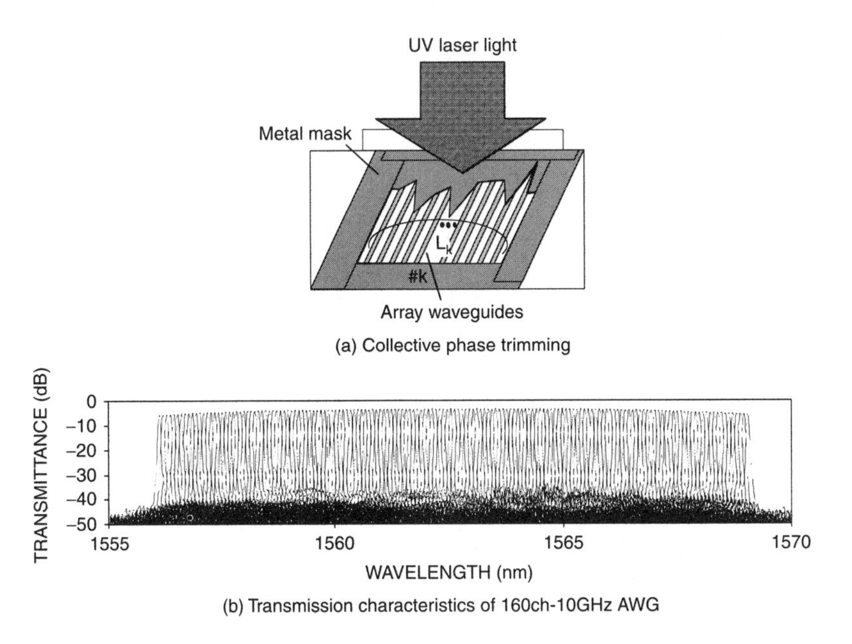

Fig. 2.8 Demultiplexing properties of a 160-ch, 10-GHz spacing AWG after phase trimming.

AWG [11]. Figure 2.8 shows the configuration of phase-compensated AWG using a UV (ultraviolet) ArF laser for phase trimming. Phase errors of the AWGs are compensated for by using the photo-induced refractive index change that occurs under UV irradiation. First, the phase error in each array waveguide is measured by using Fourier transform spectroscopy. An opening is then formed in the metal mask. UV light is irradiated through the window of the metal mask. The shape of the window is formed in such a way that the length of opening, corresponding to each array waveguide, exactly matches with the phase error value to be compensated. Figure 2.8(b) shows spectral transmittance of 160-ch, 10-GHz spacing AWG after phase-error compensation. Crosstalk of the AWG before phase trimming was −25 dB. By the phase-error compensation, crosstalk is improved to about −36 to −39 dB.

2.4. Tandem AWG Configuration

In order to realize much larger numbers of channels, for example upto 1000, tandem concatenation is extremely important. Figure 2.9 shows configuration of a 10 GHz-spaced 1010-channel WDM filter that covers both the C (conventional) and L (long) fiber amplifier bands [12]. It consists of a

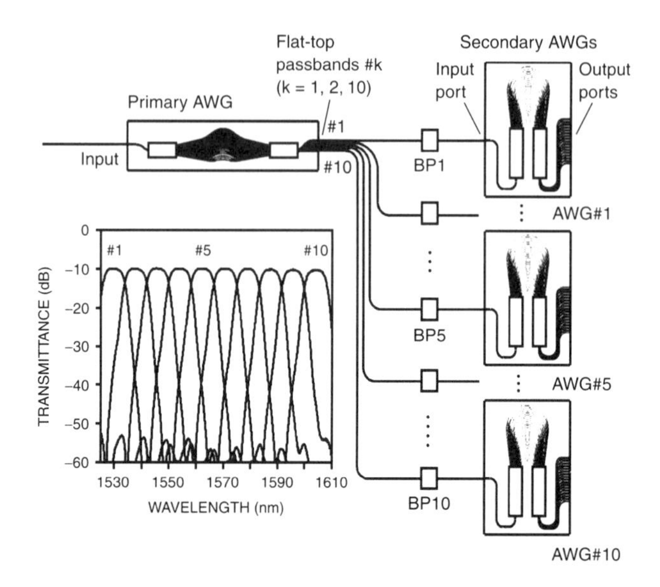

Fig. 2.9 Configuration of tandem AWG filter. Inset shows transmittances of primary AWG when combined with BP #k (k = 1, 2, 10). K. Takada, et al., 10 "GHz-spaced 1010-channel AWG filter achieved by tandem connection of primary and secondary AWGs," Photon. Tech. Lett. © 2001 IEEE.

primary 1 × 10 flat-top AWG with a 1-THz channel spacing and 10 secondary 1 × 160 AWGs (AWG #k with k = 1, 2, 10) with 10-GHz spacing. Phase errors of the latter AWGs were compensated for by using the photo-induced refractive index change as shown in Fig. 2.8. Crosstalk of secondary AWGs is around −32 dB and the sidelobe levels in these passbands were less than −35 dB. This tandem configuration enables us to construct flexible WDM systems; that is, secondary AWGs can be added when bandwidth demand increases. Also, this configuration will be essential for the construction of hierarchical crossconnect (XC) systems such as fiber XC, band XC, and wavelength XC.

Output port #k (k = 1, 2, 10) of the primary AWG was connected to the input port of AWG #k through an optical fiber and an interference bandpass filter BP #k. Two conditions were imposed on these AWGs. First, the center wavelength of 160 channels of AWG #k was designed to coincide with that of the flat-top passband #k from primary AWG output #k. Then passband #k was sliced with the AWG #k without any noticeable loss. Second, the sidelobe components of flat-top passband #k were removed by BP #k as shown in the Fig. 2.9 inset. Therefore, one passband within the FSR was obtained from one output port of AWG #k. 101 wavelengths

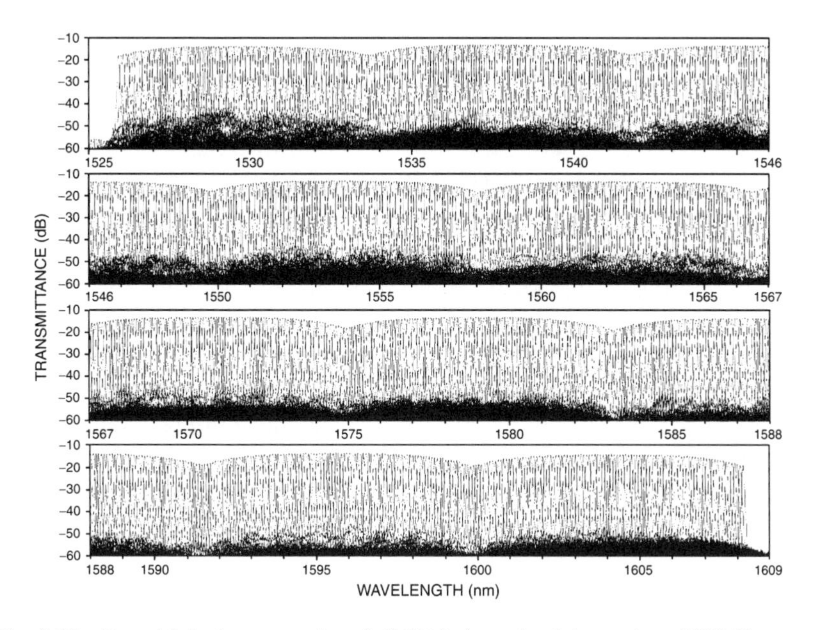

Fig. 2.10 Demultiplexing properties of all 1010 channels of the tandem AWG filter measured when unpolarized light was launched into the input port of the primary AWG.

were selected from every AWG #k. Figure 2.10 shows the demultiplexing properties of all the channels of the tandem AWG filter. There were a total of 1010 channels and they were all aligned at 10 GHz intervals with no channel missing in the 1526 to 1608 nm wavelength range. The loss values ranged from 13 to 19 dB. The main origin of the 13 dB loss is the 10 dB intrinsic loss of the primary AWG. This can be reduced by about 9 dB when a 1×10 interference filter is used instead of flat-top AWG.

2.5. $N \times N$ AWG Router

In principle, $N \times N$ signal interconnection can be achieved in AWG when free spectral range (FSR) of AWG is N times the channel spacing. Here FSR is given from Eq. (2.2) by

$$FSR = \frac{n_c \nu_0}{N_c m}. \tag{2.13}$$

Generally, light beams with three different diffraction orders of $m-1$, m, and $m+1$ are utilized to achieve $N \times N$ interconnections [9]. The cyclic property provides an important additional functionality as compared to simple multiplexers or demultiplexers and plays a key role in more complex devices as add/drop multiplexers and wavelength switches.

However, such interconnectivity cannot always be realized with the conventional AWGs. The typical diffraction order of 32-channel AWG with 100-GHz channel spacing is $m = 60$ as calculated by Eq. (2.2). Because FSR is inversely proportional to m, substantial pass frequency mismatch is brought by the difference between three FSRs. Also, insertion losses of AWG for peripheral input and output ports are 2–3 dB higher than those for central ports, as shown in Fig. 2.5. These noncyclic frequency characteristics and loss nonuniformity in conventional AWGs are the main obstacles preventing the development of practical $N \times N$ routing networks.

A novel $N \times N$ ($N = 4$–32) arrayed-waveguide grating having uniform-loss and cyclic-frequency characteristics is proposed and fabricated to solve the problems in conventional AWGs [13]. Figure 2.11 shows the

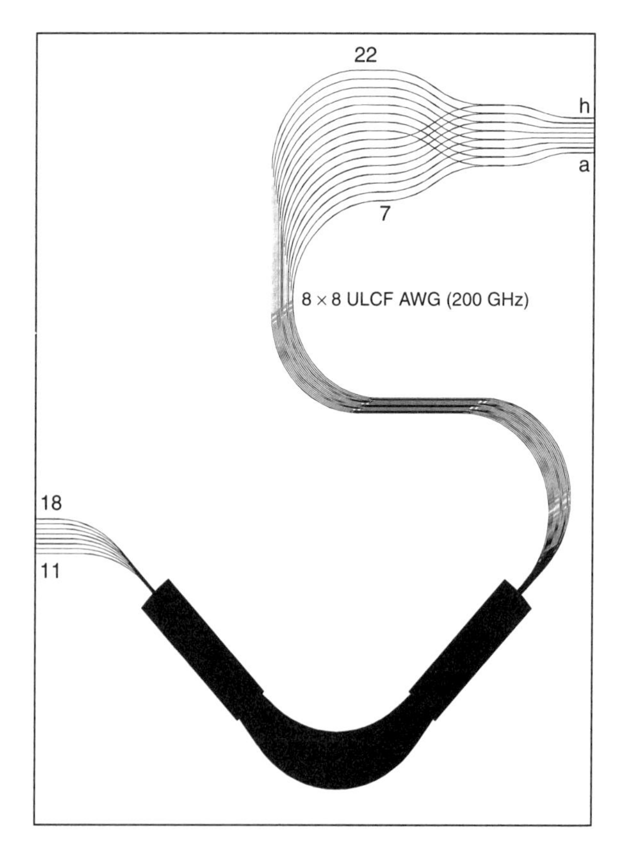

Fig. 2.11 Schematic configuration of uniform-loss and cyclic-frequency (ULCF) AWG.

schematic configuration of an 8×8 uniform-loss and cyclic-frequency (ULCF) arrayed-waveguide grating. It consists of a 28-channel AWG multiplexer with 200-GHz spacing and 8 optical combiners, which are connected to 16 output waveguides of the multiplexer. The arc length of the slab is f = 8.07 mm and the number of array waveguides is 120, having the constant path length difference $\Delta L = 35.3\ \mu$m between neighboring waveguides. The diffraction order is $m = 33$, which gives a free spectral range of $FSR = 5.6$ THz. In the input side, 8 waveguides ranging from #11 to #18 are used for the input waveguides so as to secure the uniform loss characteristics. In the output side, two waveguides ($(i+6)$-th and $(i+14)$-th waveguide for i = 1–8) are combined through waveguide intersection and a multimode interference (MMI) coupler [14] to make one output port. Because the peripheral output ports are not used, uniform loss characteristics are obtained. Figure 2.12 shows the principle of how the ULCF arrayed-waveguide grating is constructed. Each of the 8 input ports (#11–#18) can carry 8 different wavelengths $\lambda_{11}, \lambda_{12}, \lambda_{13}, \lambda_{14}, \lambda_{15}, \lambda_{16}, \lambda_{17}$, and λ_{18}. The 8 wavelengths coupled into, for example, input port #18 are distributed among output ports #a to #h. The 8 wavelengths carried by other input ports are distributed in the same way, but cyclically rotated. In this way each output port receives 8 different wavelengths, one from each input port.

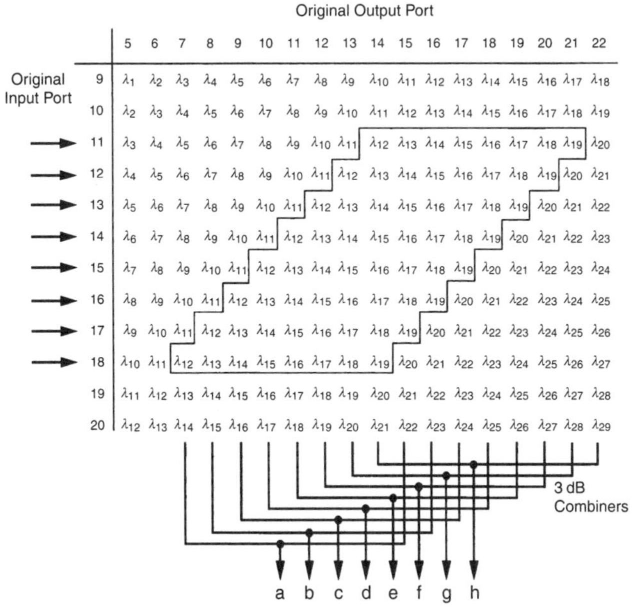

Original Output Port

Original Input Port	5	6	7	8	9	10	11	12	13	14	15	16	17	18	19	20	21	22
9	λ_1	λ_2	λ_3	λ_4	λ_5	λ_6	λ_7	λ_8	λ_9	λ_{10}	λ_{11}	λ_{12}	λ_{13}	λ_{14}	λ_{15}	λ_{16}	λ_{17}	λ_{18}
10	λ_2	λ_3	λ_4	λ_5	λ_6	λ_7	λ_8	λ_9	λ_{10}	λ_{11}	λ_{12}	λ_{13}	λ_{14}	λ_{15}	λ_{16}	λ_{17}	λ_{18}	λ_{19}
→ 11	λ_3	λ_4	λ_5	λ_6	λ_7	λ_8	λ_9	λ_{10}	λ_{11}	λ_{12}	λ_{13}	λ_{14}	λ_{15}	λ_{16}	λ_{17}	λ_{18}	λ_{19}	λ_{20}
→ 12	λ_4	λ_5	λ_6	λ_7	λ_8	λ_9	λ_{10}	λ_{11}	λ_{12}	λ_{13}	λ_{14}	λ_{15}	λ_{16}	λ_{17}	λ_{18}	λ_{19}	λ_{20}	λ_{21}
→ 13	λ_5	λ_6	λ_7	λ_8	λ_9	λ_{10}	λ_{11}	λ_{12}	λ_{13}	λ_{14}	λ_{15}	λ_{16}	λ_{17}	λ_{18}	λ_{19}	λ_{20}	λ_{21}	λ_{22}
→ 14	λ_6	λ_7	λ_8	λ_9	λ_{10}	λ_{11}	λ_{12}	λ_{13}	λ_{14}	λ_{15}	λ_{16}	λ_{17}	λ_{18}	λ_{19}	λ_{20}	λ_{21}	λ_{22}	λ_{23}
→ 15	λ_7	λ_8	λ_9	λ_{10}	λ_{11}	λ_{12}	λ_{13}	λ_{14}	λ_{15}	λ_{16}	λ_{17}	λ_{18}	λ_{19}	λ_{20}	λ_{21}	λ_{22}	λ_{23}	λ_{24}
→ 16	λ_8	λ_9	λ_{10}	λ_{11}	λ_{12}	λ_{13}	λ_{14}	λ_{15}	λ_{16}	λ_{17}	λ_{18}	λ_{19}	λ_{20}	λ_{21}	λ_{22}	λ_{23}	λ_{24}	λ_{25}
→ 17	λ_9	λ_{10}	λ_{11}	λ_{12}	λ_{13}	λ_{14}	λ_{15}	λ_{16}	λ_{17}	λ_{18}	λ_{19}	λ_{20}	λ_{21}	λ_{22}	λ_{23}	λ_{24}	λ_{25}	λ_{26}
→ 18	λ_{10}	λ_{11}	λ_{12}	λ_{13}	λ_{14}	λ_{15}	λ_{16}	λ_{17}	λ_{18}	λ_{19}	λ_{20}	λ_{21}	λ_{22}	λ_{23}	λ_{24}	λ_{25}	λ_{26}	λ_{27}
19	λ_{11}	λ_{12}	λ_{13}	λ_{14}	λ_{15}	λ_{16}	λ_{17}	λ_{18}	λ_{19}	λ_{20}	λ_{21}	λ_{22}	λ_{23}	λ_{24}	λ_{25}	λ_{26}	λ_{27}	λ_{28}
20	λ_{12}	λ_{13}	λ_{14}	λ_{15}	λ_{16}	λ_{17}	λ_{18}	λ_{19}	λ_{20}	λ_{21}	λ_{22}	λ_{23}	λ_{24}	λ_{25}	λ_{26}	λ_{27}	λ_{28}	λ_{29}

3 dB Combiners

a b c d e f g h

Fig. 2.12 Principle of operation in ULCF AWG.

Table 2.2 **Wavelength Interconnectivity Pattern of 8 × 8 ULCF AWG**

	λ12	*λ13*	*λ14*	*λ15*	*λ16*	*λ17*	*λ18*	*λ19*
Wavelength (nm)	193.5	193.3	193.1	192.9	192.7	192.5	192.3	192.1
Frequency (THz)	1549.315	1550.918	1552.524	1554.134	1555.747	1557.363	1558.983	1560.604

	Out a	*b*	*c*	*d*	*e*	*f*	*g*	*h*
Input 18	λ12	λ13	λ14	λ15	λ16	λ17	λ18	λ19
11	λ13	λ14	λ15	λ16	λ17	λ18	λ19	λ12
12	λ14	λ15	λ16	λ17	λ18	λ19	λ12	λ13
13	λ15	λ16	λ17	λ18	λ19	λ12	λ13	λ14
14	λ16	λ17	λ18	λ19	λ12	λ13	λ14	λ15
15	λ17	λ18	λ19	λ12	λ13	λ14	λ15	λ16
16	λ18	λ19	λ12	λ13	λ14	λ15	λ16	λ17
17	λ19	λ12	λ13	λ14	λ15	λ16	λ17	λ18

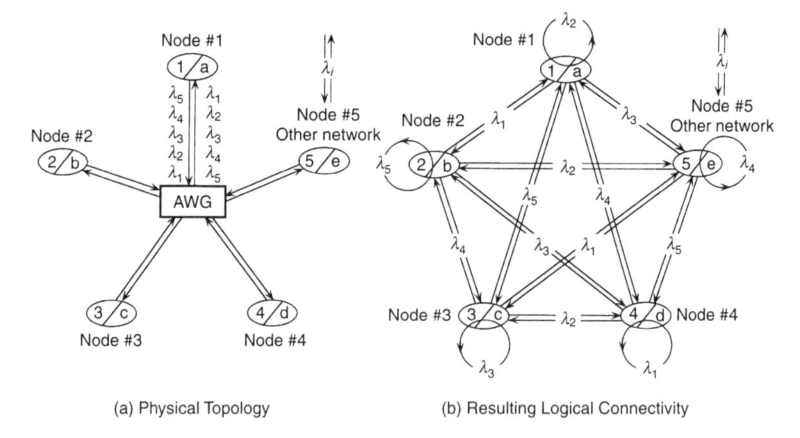

Fig. 2.13 Principle of operation in ULCF AWG. Reprinted from *Fundamentals of Optical Waveguides Photonic Networks*, K. Okamoto, Chapter 9 © 2000, with permission from Elsevier Science.

Table 2.2 summarizes the wavelength interconnectivity pattern of an 8 × 8 ULCF AWG. To realize such an interconnectivity scheme in a strictly nonblocking way using a single wavelength, a huge number of switches would be required. Using a cyclic property of ULCF AWG, this functionality can be achieved with only one AWG. An example of an all-optical $N \times N$ ($N = 5$) interconnection system using AWG as a router is shown in Fig. 2.13. Figure 2.13(a) shows the physical topology between the AWG

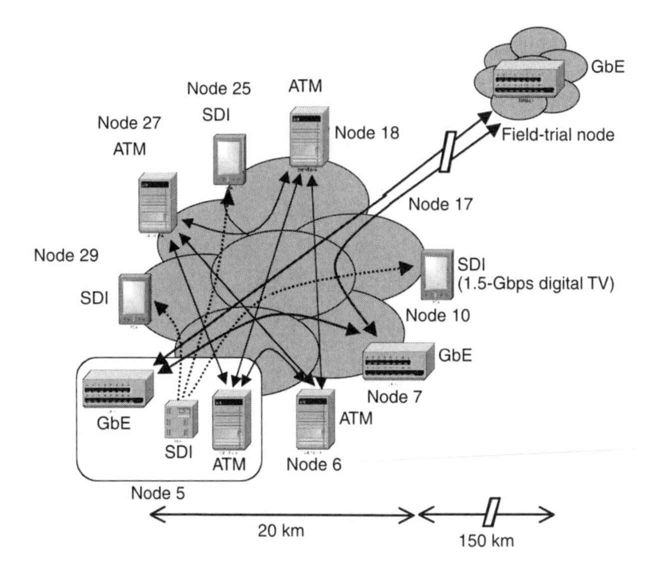

Fig. 2.14 Experimental configuration for multi-protocol networking. K. Kato et al., "32 × 32 full-mesh (1024 path) wavelength–routing WDM network based on uniform-loss cyclic-frequency arrayed-waveguide grating," Electron Lett. © 2000 IEEE.

router and N nodes. Some routes may be used for connection to other networks. Based on the interconnectivity of AWG in Fig. 2.13, the resulting logical connectivity patterns become $N \times N$ star network as shown in Fig. 2.13(b). All the nodes can communicate with each other at the same time, thus enabling N^2 optical connections simultaneously. Signals can be freely routed by changing the carrier wavelength of the signal. When combined with the wavelength conversion lasers, this AWG router can construct signal routing networks without using space-division optical switches.

A 32 × 32 full-mesh (1024 path) wavelength-routing WDM network experiment using 32 × 32 ULCF AWG has been demonstrated [15]. Figure 2.14 shows an experimental configuration for multi-protocol networking. The key characteristics of the $N \times N$ network are low coherent crosstalk, low loss, loss uniformity, and a wide passband throughout all the $N \times N$ paths. The power crosstalk of 32 × 32 ULCF AWG was reduced to −40 dB. Bit error rates at 10 Gbps NRZ were measured at randomly chosen paths as well as the worst passband path. The sensitivity was as low as −36 dBm at a bit error rate of 10^{-12}. Thus, the system has a throughput of 10 Tbps (32 × 32 × 10 Gbit/s). The performance is well suited for a scalable metropolitan area, campus backbone, and heavy traffic local area networks where several different protocols are co-implemented.

This full-mesh WDM interconnection system is also applicable as a platform for wavelength-routing networks.

2.6. Optical Add/Drop Multiplexer

An optical add/drop multiplexer (ADM) is a device that gives simultaneous access to all wavelength channels in a WDM communication system. A novel integrated-optic ADM is fabricated, and basic functions of individually routing 16 different wavelength channels with 100-GHz channel spacing has been demonstrated [16]. The waveguide configuration of 16-ch optical ADM is shown in Fig. 2.15. It consists of 4 arrayed-waveguide gratings and 16 double-gate thermooptic (TO) switches. Four AWGs are allocated with their slab regions crossing each other. These AWGs have the same grating parameters, which are channel spacing of 100 GHz and free spectral range of 3300 GHz (26.4 nm) at the 1.55 μm region. Equally spaced WDM signals, $\lambda_1, \lambda_2, \ldots, \lambda_{16}$, which are coupled to the main input port (add port) in Fig. 2.15, are first demultiplexed by the AWG_1 (AWG_2) and then 16 signals are introduced into the lefthand-side arms (righthand-side arms) of double-gate TO switches. The cross angle of the intersecting waveguides is designed to be larger than 30 degrees so as to make the

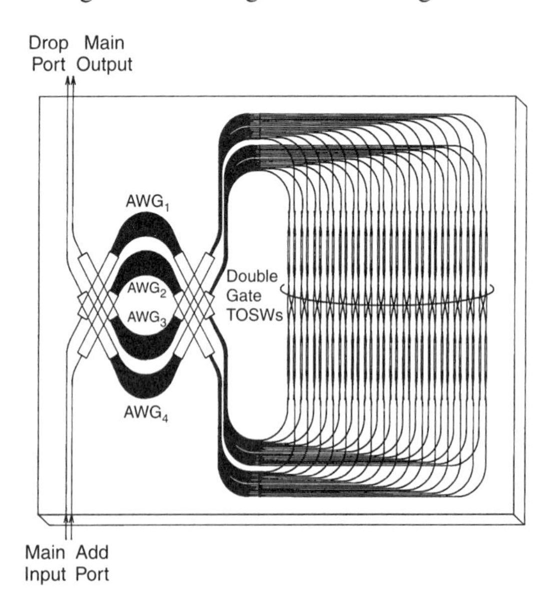

Fig. 2.15 Configuration of 16-ch optical. Reprinted from *Fundamentals of Optical Waveguides Photonic Networks*, K. Okamoto, Chapter 9 © 2000, with permission from Elsevier Science.

crosstalk and insertion loss negligible. Here the "off" state of the double-gate switch is defined as the switching condition where a signal from the left input port (right input port) goes to the right output port (left output port) in Fig. 2.15. The "on" state is then defined as the condition where a signal from the left input port (right input port) goes to the left output port (right output port). When the double-gate switch is "off," the demultiplexed light by AWG_1 (AWG_2) goes to the cross arm and is multiplexed again by the AWG_3 (AWG_4). On the other hand, if the double-gate switch is "on," the demultiplexed light by AWG_1 (AWG_2) goes to the through arm and is multiplexed by the AWG_4 (AWG_3). Therefore, any specific wavelength signal can be extracted from the main output port and led to the drop port by changing the corresponding switch condition. A signal at the same wavelength as that of the dropped component can be added to the main output port when it is coupled into an add port in Fig. 2.15. Figure 2.16 shows light transmission characteristics from main input port to main output port (solid line) and drop port (dotted line) when TO switches SW_2, SW_4, SW_6, SW_7, SW_9, SW_{12}, SW_{13}, and SW_{15}, for example, are turned to "on." The on–off crosstalk is smaller than -30 dB with the on-chip losses of 8–10 dB. Selected signals λ_2, λ_4, λ_6, λ_7, λ_9, λ_{12}, λ_{13} and λ_{15} are extracted from main output port (solid line) and led to the drop port (dotted line). The present optical ADM can transport all input signals to the succeeding stages without inherent power losses. Therefore, these ADMs are very attractive for

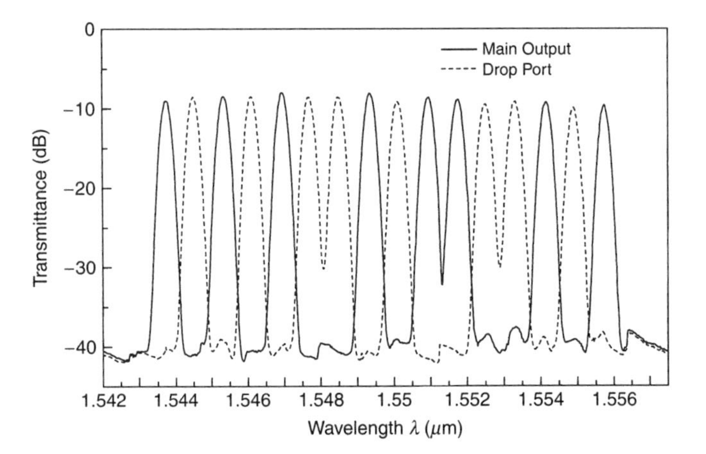

Fig. 2.16 Light transmission characteristics from main input port to main output port (solid line) and drop port (dotted line) when TO switches SW2, SW4, SW6, SW7, SW9, SW12, SW13, and SW15 are turned to "on." Reprinted from *Fundamentals of Optical Waveguides Photonic Networks*, K. Okamoto, Chapter 9 © 2000, with permission from Elsevier Science.

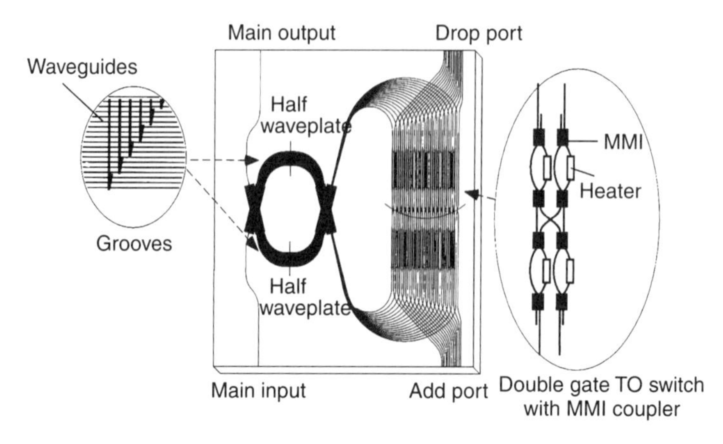

Fig. 2.17 Configuration of athermal 16-channel optical ADM.

all optical WDM routing systems and allow the network to be transparent to signal formats and bit rates.

In order to make PLC devices temperature insensitive, silica-based AWGs incorporating silicone adhesive in the arryed-waveguide region have been developed [17, 18]. Because silicone adhesive has a negative refractive index change with temperature, it can cancel the positive thermal coefficient of silica waveguides. Figure 2.17 shows the configuration of a 16-channel ADM with two AWGs and 16 double-gate TO switches [19]. Channel spacing of the ADM is 100 GHz. Segmented-type trapezoidal grooves were fabricated in the arrayed waveguides and then filled with a silicone adhesive. Figure 2.18 shows the transmission spectra for the light from main input port to main output port over the 10–80°C range when the 2nd, 5th, and 10th TO switches are activated. All spectra almost coincide with each other; that is, the center wavelength, losses, and crosstalk of all channels are almost insensitive to temperature changes. The center wavelength variations of all channels are within –0.06 nm.

It is widely recognized that optical hybrid integration is potentially a key technology for fabricating advanced integrated optical devices [20]. A silica-based waveguide on a Si substrate is a promising candidate for the hybrid integration platform, as high-performance PLCs have already been fabricated using silica-based waveguides and Si has highly stable mechanical and thermal properties that make it suitable as an optical bench. Figure 2.19 shows a schematic configuration of a hybrid-integrated optical wavelength selector (OWS) [21]. It consists of two AWGs and a 16-ch semiconductor optical amplifier (SOA) gate array, which is hybrid integrated on PLC. Sixteen main channels are connected to SOA gates and two monitor

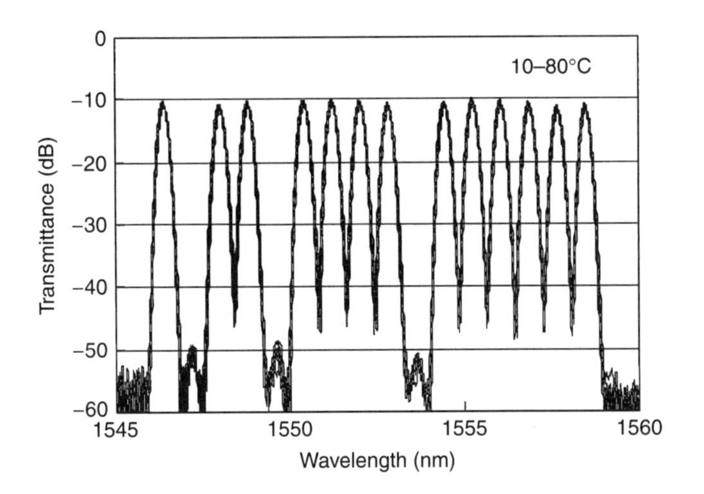

Fig. 2.18 Characteristics of athermal 16-channel optical ADM over 10 to 80°C.

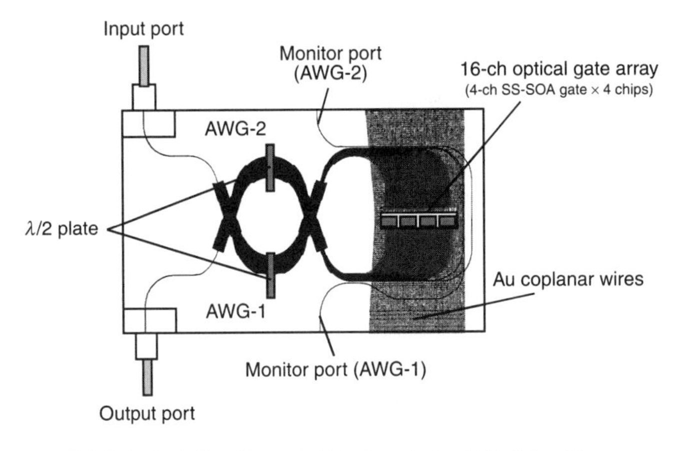

Fig. 2.19 Schematic configuration of optical wavelength selector. R. Kasahara et al., "A compact optical wavelength selector composed of arrayed-waveguide gratings and an optical gate array integrated on a single PLC platform," Photon. Tech. Lett., 12 © 2000 IEEE.

ports are attached to the AWGs so as to measure each AWG performance independently. Sixteen SOA gates are grouped into four 4-ch gate arrays. A spot-size converter is equipped at both ends of each SOA to achieve high coupling efficiency under butt coupling conditions with silica waveguides. Spot-size converted SOAs are precisely mounted on the terraced silicon substrate using a passive alignment technique. Au coplanar wiring is used to achieve high-speed operation over 1 GHz response time. Figure 2.20 shows

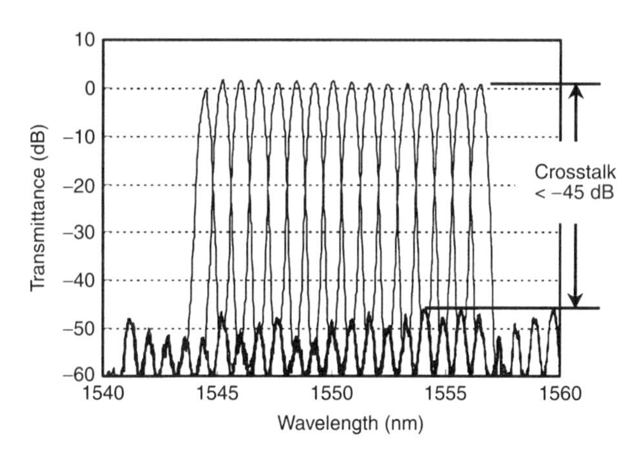

Fig. 2.20 Transmission spectra of OWS at an optical input power of −10 dBm and a gate injection current of 35–50 mA. R. Kasahara et al., "A compact optical wavelength selector composed of arrayed-waveguide gratings and an optical gate array integrated on a single PLC platform," Photon. Tech. Lett., 12 © 2000 IEEE.

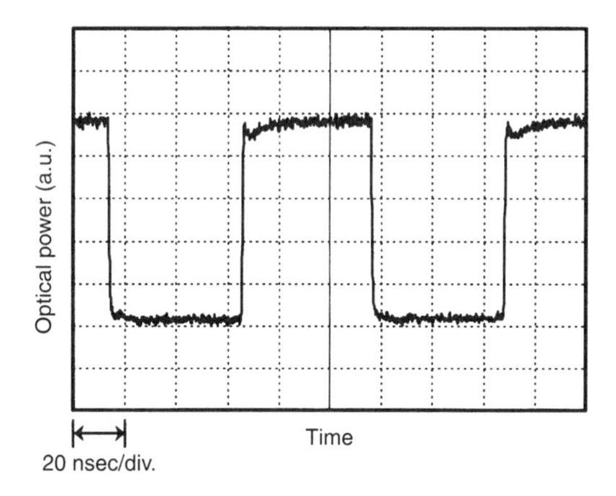

Fig. 2.21 Switching characteristics measured at port 8. R. Kasahara et al., "A compact optical wavelength selector composed of arrayed-waveguide gratings and an optical gate array integrated on a single PLC platform," Photon. Tech. Lett., 12 © 2000 IEEE.

the transmission spectra of OWS at an input optical power of −10 dBm and at a gate injection current of 35–50 mA. In the measurement, each SOA gate was successively opened and transmission spectra was measured by using a tunable laser source. Transmittance of 0–2 dB was obtained for every channel; that is, the entire loss caused by fiber coupling losses, AWG losses, and SOA-to-waveguide coupling losses is completely compensated by the gain of SOA. An average polarization dependent loss is 0.5 dB and an extinction ratio is better than −45 dB, respectively. Figure 2.21 shows the switching characteristics of the OWS at port 8. Rise and fall times are both less than 1 nsec.

2.7. Optical Label Recognition Circuit for Photonic Label Switch Router

Optical address signal recognition is a key technology for future packet-switched networks, where each router has to provide an ultrahigh throughput exceeding the electronic speed limits. Several all-optical address recognition schemes have been proposed. They include autocorrelation with matched filters [22] and a bit-wise AND operation with stored reference pulse patterns [23]. However, these approaches need numerous matched filters or reference patterns to distinguish the address, because they determine the proper address pattern by comparing the degree of matching between the incoming address and reference patterns. Here, we propose a novel optical circuit for recognizing an optical pulse pattern, based on an optical digital-to-analog (D/A) converter fabricated on a silica-based planar lightwave circuit (PLC) [25]. The circuit converts the optical pulse pattern to analog optical amplitude, and enables us to recognize the address. Figure 2.22 shows the operational principle of the optical D/A converter. An incoming 4-bit optical pulse train "$C_0C_1C_2C_3$" is first split into four duplicates. Each duplicate is relatively delayed by 0, $\Delta\tau$, $2\Delta\tau$, and $3\Delta\tau$, where $\Delta\tau$ is the time interval of the incoming pulse, and weighted with the

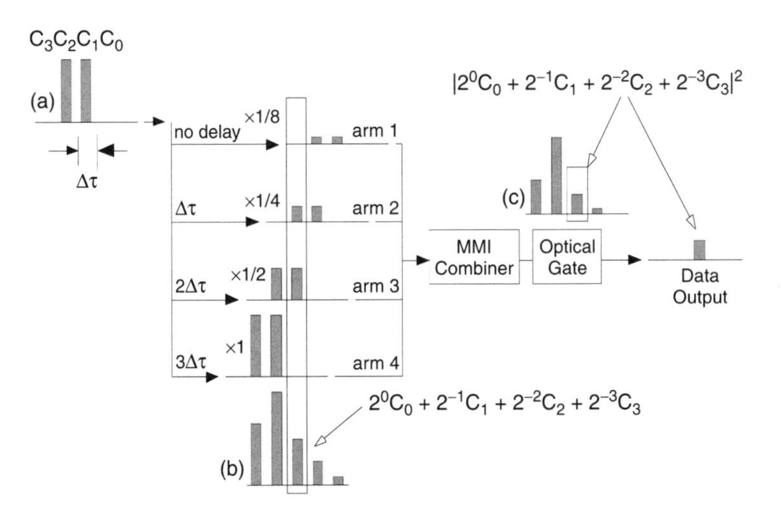

Fig. 2.22 Operational principle of optical D/A converter.

coefficients 2^0, 2^{-1}, 2^{-2}, and 2^{-3}, respectively. The weighted pulses are then recombined and one of the output pulses is extracted with an optical gate. As shown in Fig. 2.22, the intensity of the gated pulse is given by the square of the D/A conversion of the input pulse sequence, as

$$I = \frac{1}{16}|2^0 C_0 + 2^{-1} C_1 + 2^{-2} C_2 + 2^{-3} C_3|^2. \tag{2.14}$$

Then, the incoming pulse pattern can be recognized by the intensity of the output data I.

Figure 2.23 shows the configuration of a fabricated 4-bit optical D/A converter, which features a coherent optical transversal filter [24]. It consists of a 1×4 multimode interference (MMI) splitter, delay lines with a relative delay time $\Delta\tau$, thermooptic phase controllers, thermooptic switches as amplitude controllers, and a 4×1 MMI combiner. The time delay $\Delta\tau$ is set at 100 psec to deal with a 10 Gb/s pulse train. Weighting coefficients in the D/A converter were adjusted by supplying electric power to thermooptic heaters while monitoring them using Fourier transform spectroscopy [10]. The adjustment errors for phase and amplitude were 0.03 rad ($\lambda/200$) and 2%, respectively. The total device loss was about 9.9 dB, including a 2^{-m} weighting loss of 6.6 dB and a fiber coupling loss of 1 dB. 10-Gb/s RZ pulse sequences, generated by an electrooptic modulator, were coupled into the D/A converter to confirm its operation. The source wavelength of the laser was 1550.320 nm. Figures 2.24(a) and (b) show an input pulse

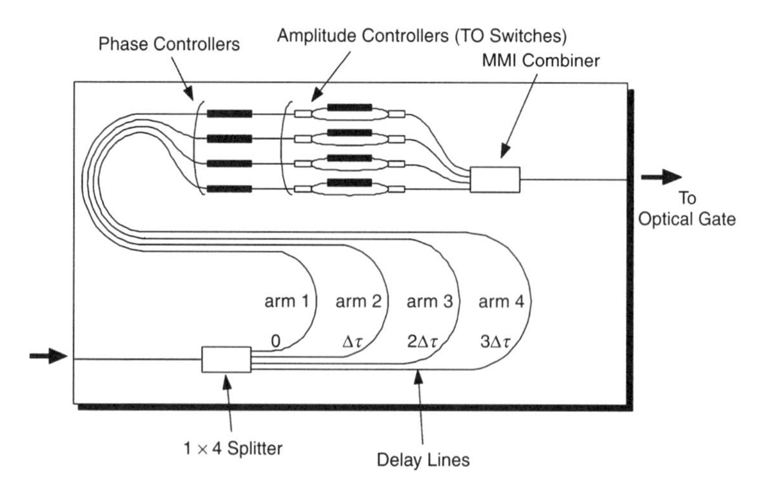

Fig. 2.23 Schematic configuration of 4-bit optical D/A converter. T. Saida et al., "Integrated optical digital-to-analogue converter and its application to pulse pattern recognition," Electron Lett. © 2001 IEEE.

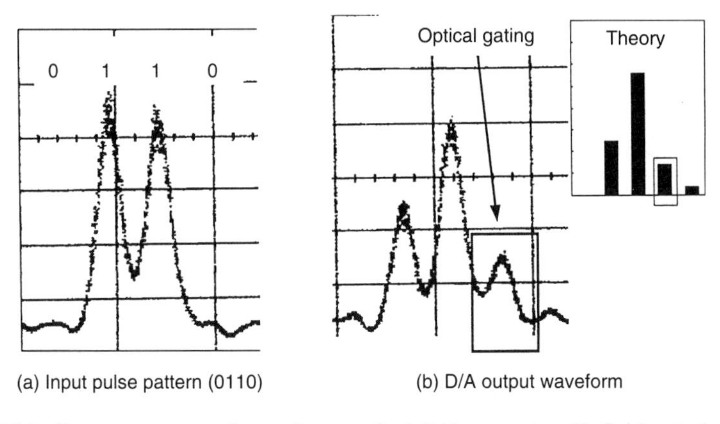

Fig. 2.24 Data output waveforms from optical D/A converter. T. Saida et al., "Integrated optical digital-to-analogue converter and its application to pulse pattern recognition," Electron Lett. © 2001 IEEE.

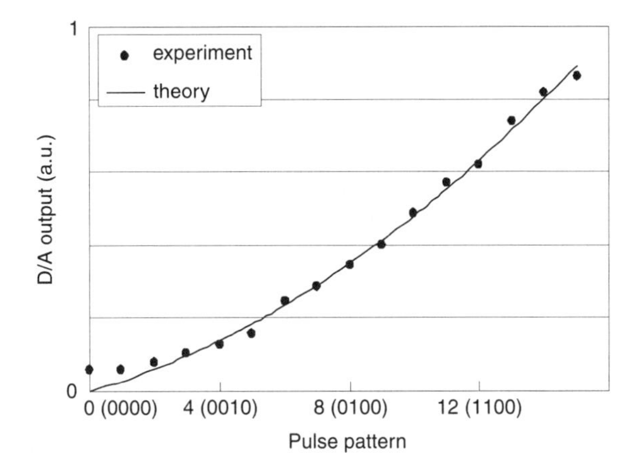

Fig. 2.25 Relationship between incoming pulse patterns and D/A output. T. Saida et al., "Integrated optical digital-to-analogue converter and its application to pulse pattern recognition," Electron Lett. © 2001 IEEE.

pattern ("0110") and an output waveform from the D/A converter, respectively, which are observed on a sampling oscilloscope. The light of the peak marked by the solid square in Fig. 2.24(b) corresponds to the digital-to-analog converted "0110" patterns. The inset in Fig. 2.24(b) shows the theoretical output waveform, which agrees well with the measured waveform. Figure 2.25 shows the measured relationship between incoming pulse patterns and D/A output. The solid line is a theoretical curve. Although, with the current measurement system, the D/A output has a minimum detection limit due to noise, the experimental and theoretical data agree very well.

This optical label recognition circuit will be a key element in future photonic packet routing networks.

2.8. Summary

Although silica-based waveguides are simple circuit elements, various functional devices are fabricated by utilizing spatial multi-beam or temporal multi-stage interference effects such as arrayed-waveguide grating multiplexers and lattice-form programmable filters. Hybrid integration technologies will further enable us to realize much more functional and high-speed devices. The PLC technologies supported by continuous improvements in waveguide fabrication, circuit design, and device packaging will further proceed to a higher level of integration of optics and electronics aiming at the next generation of telecommunication systems.

References

1. T. Aoyama, "Photonic Networking—Only way to cope with IP traffic explosion," 4th International Topical Workshop on Contemporary Photonic Technologies CPT 2001, (Jan. 15–17, 2001), Tokyo Japan, MB-1, 5–10.
2. M. Kawachi, "Silica waveguide on silicon and their application to integrated-optic components," Opt. and Quantum Electron., 22 (1990), 391–416.
3. S. Suzuki, K. Shuto, H. Takahashi, and Y. Hibino, "Large-scale and high-density planar lightwave circuits with high-Δ GeO_2-doped silica waveguides," Electron. Lett., 28 (1992), 1863–1864.
4. Y. Hibino, H. Okazaki, Y. Hida, and Y. Ohmori, "Propagation loss characteristics of long silica-based optical waveguides on 5 inch Si wafers," Electron. Lett., 29 (1993), 1847–1848.
5. Y. Hida, Y. Hibino, H. Okazaki, and Y. Ohmori, "10-m long silica-based waveguide with a loss of 1.7 dB/m," IPR'95 IThC6 (1995), Dana Point, CA.
6. M. K. Smit, "New focusing and dispersive planar component based on an optical phased array," Electron. Lett., 24 (1988), 385–386.
7. H. Takahashi, S. Suzuki, K. Kato, and I. Nishi, "Arrayed-waveguide grating for wavelength division multi/demultiplexer with nanometer resolution," Electron. Lett., 26 (1990), 87–88.
8. C. Dragone, C. A. Edwards, and R. C. Kistler, "Integrated optics $N \times N$ multiplexer on silicon," Photon. Tech. Lett., 3 (1991), 896–899.
9. K. Okamoto, Fundamentals of Optical Waveguides Photonic Networks (Academic Press, London, 2000), chapter 9.

10. K. Takada, Y. Inoue, H. Yamada, and M. Horiguchi, "Measurement of phase error distributions in silica-based arrayed-waveguide grating multiplexers by using Fourier transform spectroscopy," Electron. Lett., 30 (1994), 1671–1672.
11. K. Takada, T. Tanaka, M. Abe, T. Yanagisawa, M. Ishii, and K. Okamoto, "Beam adjustment free crosstalk reduction in 10 GHz-spaced arrayed-waveguide grating via photosensitivity under UV laser irradiation through metal mask," Electron. Lett., 36 (2000), 60–61.
12. K. Takada, M. Abe, T. Shibata, M. Ishii, Y. Inoue, H. Yamada, Y. Hibino, and K. Okamoto, "10 GHz-spaced 1010-channel AWG filter achieved by tandem connection of primary and secondary AWGs," Photon. Tech. Lett., 13 (2001), 577–578.
13. K. Okamoto, H. Hasegawa, O. Ishida, A. Himeno, and Y. Ohmori, "32 × 32 arrayed-waveguide grating multiplexer with uniform loss and cyclic frequency characteristics," Electron. Lett., 33 (1997), 1865–1866.
14. F. B. Veerman, P. J. Schalkwijk, E. C. M. Pennings, M. K. Smit, and B. H. Verbeek, "An optical passive 3-dB TMI-coupler with reduced fabrication tolerance sensitivity," Jour. Lightwave Tech., 10 (1992), 306–311.
15. K. Kato *et al.*, "32 × 32 full-mesh (1024 path) wavelength-routing WDM network based on uniform-loss cyclic-frequency arrayed-waveguide grating," Electron. Lett., 36:15 (2000), 1294–1296.
16. K. Okamoto, M. Okuno, A. Himeno, and Y. Ohmori, "16-channel optical Add/Drop multiplexer consisting of arrayed-waveguide gratings and double-gate switches," Electron. Lett., 32 (1996), 1471–1472.
17. Y. Inoue, A. Kaneko, F. Hanawa, H. Takahashi, K. Hattori, and S. Sumida, "Athermal silica-based arrayed-waveguide grating multiplexer," Electron. Lett., 33 (1997), 1945–1946.
18. A. Kaneko, S. Kamei, Y. Inoue, H. Takahashi, and A. Sugita, "Athermal silica-based arrayed-waveguide grating (AWG) multiplexers with new low loss groove design," Paper TuO1-1, Proc. OFC '99 (1999), 204–206.
19. T. Saida, A. Kaneko, T. Goh, M. Okuno, A. Himeno, K. Takiguchi, and K. Okamoto, "Athermal silica-based optical add/drop multiplexer consisting of arrayed-waveguide gratings and double gate thermo-optical switches," Electron. Lett., 36 (2000), 528–529.
20. Y. Yamada, S. Suzuki, K. Moriwaki, Y. Hibino, Y. Tohmori, Y. Akatsu, Y. Nakasuga, T. Hashimoto, H. Terui, M. Yanagisawa, Y. Inoue, Y. Akahori, and R. Nagase, "Application of planar lightwave circuit platform to hybrid integrated optical WDM transmitter/receiver module," Electron. Lett., 31 (1995), 1366–1367.
21. R. Kasahara, M. Yanagisawa, A. Sugita, I. Ogawa, T. Hashimoto, Y. Suzaki, and K. Magari, "A compact optical wavelength selector composed of arrayed-waveguide gratings and an optical gate array integrated on a single PLC platform," Photon. Tech. Lett., 12 (2000), 34–36.

22. K. Kitayama and N. Wada, "Photonic IP Routing," Photon. Tech. Lett., 11:12 (1999), 1689–1691.
23. D. Cotter, J. K. Lucek, M. Shabeer, K. Smith, D. C. Rogers, D. Nesset, and P. Gunning, "Self-routing of 100 Gbit/s packets using 6 bit 'keyword' address recognition," Electron. Lett., 31:17 (1995), 1475–1476.
24. K. Okamoto, H. Yamada, and T. Goh, "Fabrication of coherent optical transversal filter consisting of MMI splitter/combiner and thermo-optic amplitude and phase controllers," Electron. Lett., 35:16 (1999), 1331–1332.
25. T. Saida, K. Okamoto, K. Uchiyama, K. Takiguchi, T. Shibata, and A. Sugita, "Integrated optical digital-to-analogue converter and its application to pulse pattern recognition," Electron. Lett., 37:20 (2001), 1237–1238.

Chapter 3 | Optical Multiplexer/Demultiplexer: Discrete

Y. Calvin Si
Yale Cheng

Fiber Optic Product Group, JDS Uniphase Corporation, Nepean, Ontario, Canada

3.1. Introduction

Wavelength division multiplexing (WDM) has emerged as a key enabling technology in advanced optical communication networks. The use of WDM technology not only significantly increases the capacity of the existing fiber optic networks without an increase in fiber counts, but also provides advantages in network management, network provision, and flexible services. The vast progress in advanced WDM networks has been the main driver of the recent rapid growth of the telecommunications industry. The key passive component in the WDM system is the wavelength division multiplexing and demultiplexing (MUX/DEMUX) device, which combines/splits lights with different wavelengths into different outputs. Several technologies have been used to fabricate MUX/DEMUX devices. Each of them has some distinguishing key features and is suitable for different applications. In this chapter, the principles, structures, and performances of the thin film dielectric filter, fiber Bragg grating, arrayed-waveguide grating (AWG), and diffraction grating based MUX/DEMUX devices are described.

3.2. Parameters of MUX/DEMUX Devices

Components used as MUX/DEMUX elements in WDM optical systems have several distinct features and stringent performance requirements.

WDM TECHNOLOGIES: PASSIVE OPTICAL COMPONENTS
$35.00

ISBN: 0-12-225262-4

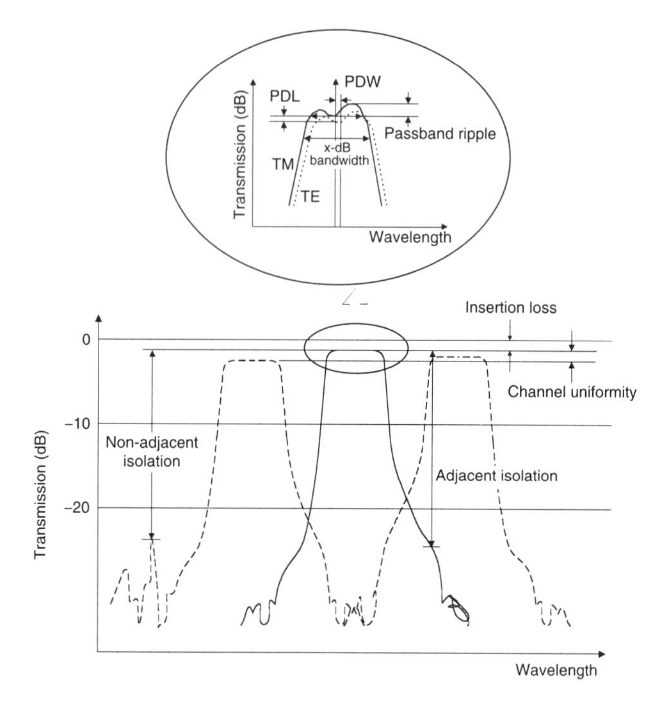

Fig. 3.1 Definition of various performance parameters of a MUX/DEMUX device.

The use of commonly defined terms (parameters) helps in better understanding and comparison of technology differences and advantages. International standards have been established in order to provide guidelines in component and system developments. One of the standards is the frequency grid used a as carrier of optical information in dense WDM (DWDM) systems, defined by the International Telecommunication Union (ITU-T recommendation G.692), where equally spaced optical frequencies (100 GHz spacing) are defined with the reference at 193.10 THz. The nominal center frequency and corresponding nominal center wavelength for the 100 GHz spacing is shown in Table 3.1. The various performance parameters of a MUX/DEMUX device are illustrated in Fig. 3.1 and described in the following text.

Insertion loss is one of the most important parameters and can be defined as the loss level from the peak point of the channel (peak loss) or within a certain passband range (loss over passband).

Channel uniformity (or insertion loss variation) is the maximum insertion loss difference among all channels.

Passband ripple is defined as the maximum peak-to-peak loss variation within the passband of one channel.

Table 3.1 **ITU Frequency Grid and Corresponding Wavelength**

Frequency (THz)	*Wavelength (nm)*	*Frequency (THz)*	*Wavelength (nm)*	*Frequency (THz)*	*Wavelength (nm)*	*Frequency (THz)*	*Wavelength (nm)*
197.20	1520.25	194.20	1543.73	191.20	1567.95	188.20	1592.95
197.10	1521.02	194.10	1544.53	191.10	1568.77	188.10	1593.79
197.00	1521.79	194.00	1545.32	191.00	1569.59	188.00	1594.64
196.90	1522.56	193.90	1546.12	190.90	1570.42	187.90	1595.49
196.80	1523.34	193.80	1546.92	190.80	1571.24	187.80	1596.34
196.70	1524.11	193.70	1547.72	190.70	1572.06	187.70	1597.19
196.60	1524.89	193.60	1548.51	190.60	1572.89	187.60	1598.04
196.50	1525.66	193.50	1549.32	190.50	1573.71	187.50	1598.89
196.40	1526.44	193.40	1550.12	190.40	1574.54	187.40	1599.75
196.30	1527.22	193.30	1550.92	190.30	1575.37	187.30	1600.60
196.20	1527.99	193.20	1551.72	190.20	1576.20	187.20	1601.46
196.10	1528.77	193.10	1552.52	190.10	1577.03	187.10	1602.31
196.00	1529.55	193.00	1553.33	190.00	1577.86	187.00	1603.17
195.90	1530.33	192.90	1554.13	189.90	1578.69	186.90	1604.03
195.80	1531.12	192.80	1554.94	189.80	1579.52	186.80	1604.88
195.70	1531.90	192.70	1555.75	189.70	1580.35	186.70	1605.74
195.60	1532.68	192.60	1556.55	189.60	1581.18	186.60	1606.60
195.50	1533.47	192.50	1557.36	189.50	1582.02	186.50	1607.47
195.40	1534.25	192.40	1558.17	189.40	1582.85	186.40	1608.33
195.30	1535.04	192.30	1558.98	189.30	1583.69	186.30	1609.19
195.20	1535.82	192.20	1559.79	189.20	1584.53	186.20	1610.06
195.10	1536.61	192.10	1560.61	189.10	1585.36	186.10	1610.92
195.00	1537.40	192.00	1561.42	189.00	1586.20	186.00	1611.79
194.90	1538.19	191.90	1562.23	188.90	1587.04	185.90	1612.65
194.80	1538.98	191.80	1563.05	188.80	1587.88	185.80	1613.52
194.70	1539.77	191.70	1563.86	188.70	1588.73	185.70	1614.39
194.60	1540.56	191.60	1564.68	188.60	1589.57	185.60	1615.26
194.50	1541.35	191.50	1565.50	188.50	1590.41	185.50	1616.13
194.40	1542.14	191.40	1566.31	188.40	1591.26	185.40	1617.00
194.30	1542.94	191.30	1567.13	188.30	1592.10	185.30	1617.88

Insertion loss of a MUX/DEMUX device typically changes with the state of polarization of the light input into the device and the change is also wavelength dependent. Polarization-dependent loss (PDL) is the loss variation over all polarization states at a given wavelength.

Channel spacing is the center wavelength (frequency) difference between channels and typically has a fixed number (equal channel spacing) in dense WDM systems.

Channel offset is the center wavelength (frequency) difference between the measured center wavelength (frequency) and that defined by the ITU standard.

Passband is the usable wavelength (frequency) range for a given spectral flatness and loss level. Commonly used passbands are 0.5 dB bandwidth (spectral width where the filter response passes through the level 0.5 dB down from the peak), 1 dB bandwidth, and 3 dB bandwidth.

Temperature-dependent wavelength shift (TDλ) is a measure of the temperature stability of a MUX/DEMUX device and is defined as the wavelength (frequency) shift over the operating temperature range of the device and typically described in nm/°C or GHz/°C units.

Polarization-dependent wavelength shift (PDW) is the center wavelength drift caused by the variation in polarization state of the incoming light.

Adjacent channel isolation is the worst case optical power leakage from adjacent channels at a given bandwidth.

Non-adjacent channel isolation is the worst case optical power leakage from all non-adjacent channels over the entire spectrum of the device. In some cases it may also be called noise floor.

30 dB figure of merit is a measure to evaluate the relation between the bandwidth and isolation of a filter and is defined as the ratio of the 0.5 dB bandwidth to that of 30 dB bandwidth. The 30 dB figure of merit of an ideal square-shaped filter is 1, and the higher the figure of merit is, the more difficult the filter is to make.

3.3. Dielectric Thin-film Interference Filter-based MUX/DEMUX

One of the most widely used devices for multiplexing/demultiplexing multi-channel optical signals in the DWDM optical communication systems is the dielectric thin-film interference filter-based device due to its simplicity, technological maturity, and design flexibility for achieving low loss, high isolation, and wide bandwidth performances. Dielectric thin-film filters have been used for filtering a light with a particular wavelength (such as color filters in the visible wavelength) for a long time, and the principle of the dielectric thin-film filter has been long understood. The thin-film filters typically have three types—long pass, short pass, and bandpass filters. However, the filters used in the DWDM optical communication systems are completely different from the color filters in terms of structure, performance requirement, and degree of process control requirement. For MUX/DEMUX of DWDM signals, high transmission (typically more than 95%), wide bandwidth (typically more than 50% of the channel spacing), and high isolation (more than 25 dB at adjacent channel) are required. To achieve these requirements, bandpass filters are primarily used due to the narrow spacing of channels (typically in the order of nanometers).

Dielectric thin-film bandpass filters consist of cavities and quarter wavelength layers deposited on a substrate glass using difference techniques. MUX/DEMUX devices for WDM applications have to pass very stringent environmental standards such as high temperature, high humidity, and various mechanical test conditions, and the life span of the devices is expected to be at least 25 years as specified by Telcordia standards (Telcordia GR-1209 and GR-1221). Thin-film filters produced by using conventional deposition methods such as e-beam evaporation and ion-assisted deposition can no longer satisfy the temperature and long-term stability requirements. Different techniques such as plasma-assisted e-beam deposition and sputtering have been developed for producing high-density thin-film filters for DWDM applications.

3.3.1. BASIC STRUCTURE OF THE DIELECTRIC THIN-FILM INTERFERENCE FILTERS

Narrow bandpass dielectric thin-film interference filters are basically Fabry-Perot etalons with distributed multi-layer mirrors. As shown in Fig. 3.2, a typical single-cavity thin-film filter structure consists of a cavity layer sandwiched by alternating quarter-wavelength-thick layers of high and low refractive index materials. SiO_2 is the most common material used as the low index material and TiO_2 and Ta_2O_5 are typically used as high index materials due to their high index, low absorption, and stability at the 1550 nm range. The thickness of the cavity layer determines the center wavelength of the filter, while alternating quarter wavelength layers determine the reflectivity. The numbers of layers vary from design to design depending on the particular performance requirements.

When a light beam with different wavelength (frequency) components is launched onto the filter, part of the light with the frequency component matching the resonant frequency of the cavity is transmitted and the rest of the light beam is reflected by the filter.

Properties of the single-cavity thin-film filter are similar to that of a single-cavity etalon filter with narrow bandwidth and limited isolation. However, in the WDM systems wide bandwidth and high isolation are desired due to the signal modulation requirement, center wavelength drift of laser sources, etc. It is known that the passband of a thin-film filter can be

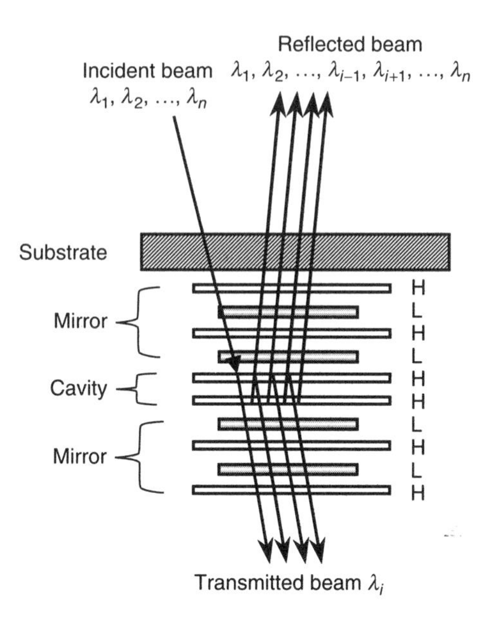

Fig. 3.2 Structure of a typical single-cavity thin-film filter.

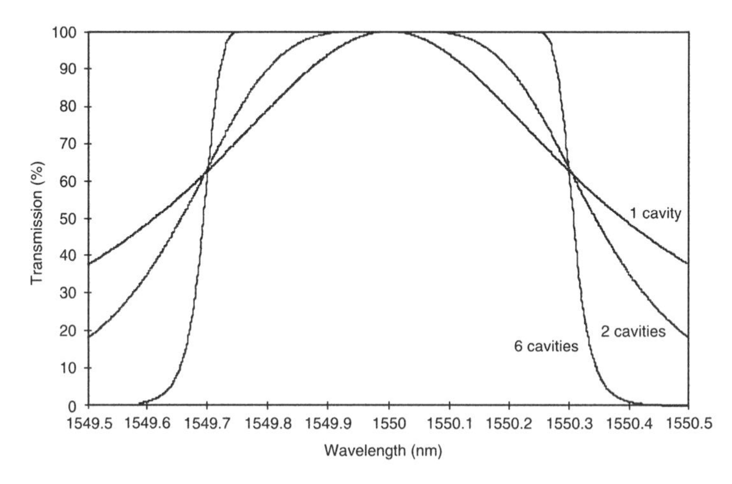

Fig. 3.3 Comparison of passband and isolation of thin-film filters with different cavities.

flattened to yield a square-top passband by cascading multi-cavities in the structure [1–3]. The comparison of passband and isolation for filters with different cavities is shown in Fig. 3.3, where the transmission shapes of single-cavity, two-cavity, and six-cavity filters for applications in DWDM systems with a 100 GHz spacing are shown. It is clear that better filtering performance can be obtained by increasing cavities in the filter design. However, in practice, increasing cavity will result in the increase of layers and increased difficulties in film deposition. With current deposition technologies, thin-film filters with up to 300 layers are being produced with good yield.

3.3.2. DESIGN AND ANALYSIS OF THE THIN-FILM FILTER

Many methods, such as graphical vector method, admittance diagrams, analytical synthesis, and numerical methods, have been developed for designing and analyzing the thin-film filters [4–6]. Among them, the matrix theory of multi-layer systems-based numerical method is the most flexible method for designing and analyzing complex thin-film filters involving a large number of layers, especially thin-film filters for DWDM applications where close to 100 layers are normally required [7].

For simplicity, we consider a single-layer thin film bounded by two infinite mediums. Parameters that have to be considered are refractive index n_1 and thickness d of the layer, the refractive indexes of the substrate n_s and the medium n_m, the beam incident angle θ, the wavelength of the light λ, and the polarization plane of the beam.

The transmittance T and the reflectance R of the multi-layer thin-film filter can be calculated from a matrix formulation derived from Maxwell's equations. From the boundary conditions for the electric and magnetic vectors, the following equations can be obtained;

$$E_m + E'_m = E_1 + E'_1 \tag{3.1}$$

$$H_m - H'_m = H_1 - H'_1 \tag{3.2}$$

$$E_1 e^{i\delta} + E'_1 e^{-i\delta} = E_s \tag{3.3}$$

$$H_1 e^{i\delta} - H'_1 e^{-i\delta} = H_s \tag{3.4}$$

$$n_m E_m - n_m E'_m = n_1 E_1 - n_1 E'_1 \tag{3.5}$$

$$n_1 E_1 e^{i\delta} - n_1 E'_1 e^{-i\delta} = n_s E_s \tag{3.6}$$

where E_m, H_m, E'_m, and H'_m are the incident and reflected electric and magnetic vectors in the medium, E_1, E'_1, H_1, and H'_1 are the transmitted and reflected electric and magnetic vectors in the thin film, and E_s and H_s are the transmitted electric and magnetic vectors in the substrate, respectively. δ is the phase introduced by the thin-film layer and is given as

$$\delta = \frac{2\pi}{\lambda}(n_1 d \cos\theta) \tag{3.7}$$

where λ is the wavelength of the light and $(n_1 d \cos\theta)$ is the effective optical thickness for a given incident angle of θ.

The following relations can be obtained from Eqs. (3.1–3.6);

$$1 + \frac{E'_m}{E_m} = \left(\cos\delta - i\frac{n_s}{n_1}\sin\delta\right)\frac{E_s}{E_m} \tag{3.8}$$

$$n_m - n_m\frac{E'_m}{E_m} = (-in_1 \sin\delta + n_s\cos\delta)\frac{E_s}{E_m} \tag{3.9}$$

By using a 2×2 matrix M_1 and the amplitude reflection r and transmission t coefficients, Eqs. (3.7) and (3.8) can be expressed as

$$\begin{pmatrix} 1 \\ n_m \end{pmatrix} + \begin{pmatrix} 1 \\ -n_m \end{pmatrix} r = M_1 \begin{pmatrix} 1 \\ n_s \end{pmatrix} t \tag{3.10}$$

where

$$M_1 = \begin{pmatrix} \cos\delta & -\frac{i}{n_1}\sin\delta \\ -in_1\sin\delta & \cos\delta \end{pmatrix} \tag{3.11}$$

$$t = \frac{E_s}{E_m} \quad \text{and} \quad r = \frac{E'_m}{E_m}$$

For a thin-film filter consisting of N layers, the following equation can be derived using a similar approach:

$$\begin{pmatrix} 1 \\ n_m \end{pmatrix} + \begin{pmatrix} 1 \\ -n_m \end{pmatrix} r = M_1 M_2 \ldots M_N \begin{pmatrix} 1 \\ n_s \end{pmatrix} t \tag{3.12}$$

where $M_1, M_2, \ldots, M_N$ are matrixes representing each layer, and can be expressed as Eq. (3.11) with n_1 and δ replaced by those of each layer. From Eq. (3.12), the amplitude reflection r and transmission t coefficients of the multi-layer filter can be expressed as

$$r = \frac{n_m A + n_s n_m B - C - n_s D}{n_m A + n_s n_m B + C + n_s D} \tag{3.13}$$

$$t = \frac{2n_m}{n_m A + n_s n_m B + C + n_s D} \tag{3.14}$$

where A, B, C, and D are components of the matrix M, and are expressed as

$$M = M_1 M_2 ... M_N = \begin{pmatrix} A & B \\ C & D \end{pmatrix}$$

The intensity transmittance T and reflectance R can be obtained as

$$T = |t|^2 \tag{3.15}$$

$$R = |r|^2 \tag{3.16}$$

The phase change on the transmission and reflection are given by

$$\sigma_t = \arg t \tag{3.17}$$

$$\sigma_r = \arg r \tag{3.18}$$

From Eq. (3.7), it is clear that the center wavelength of a bandpass filter is incident angle dependent. The center wavelength shifts toward the shorter wavelength with increased incident angle. This property is currently being used in MUX/DEMUX manufacturing to accurately tune the center wavelength to match the ITU frequency grid. However, there is a limit in terms of incident angle due to polarization dependence of the filter. The polarization property of the filter can be analyzed using the effective refractive index instead of the pure material refractive index in the M matrix (Eq. (3.11)) [8]. The effective refractive indexes for the p- and s-polarization are given as

$$n_p = \frac{n_1}{\cos\theta} \quad \text{p-polarization} \tag{3.19}$$

$$n_s = n_1 \cos\theta \quad \text{s-polarization} \tag{3.20}$$

The effective refractive index for p-polarization increases with the incident angle while that for s-polarization decreases with the incident angle.

Therefore, with a large incident angle, center wavelengths of the filter for p- and s-polarization no longer overlap each other, resulting in a narrower net passband.

Temperature and environmental stability of the filter center wavelength is a very important factor in MUX/DEMUX devices, especially in narrow channel spacing applications. It is understood that the temperature coefficient of the center wavelength drift is caused by the internal stress build-up during the deposition of the thin-films and is largely affected by the thermal expansion coefficient of the substrate. It has been demonstrated that by using high thermal expansion glass as the substrate, the temperature coefficient of the center wavelength drift can be reduced to nearly zero [9].

In high bit-rate transmission, chromatic dispersion is one of the limiting factors in achievable distance. The thin-film filter is basically a resonant device; therefore, its chromatic dispersion is relatively high, especially for square-top-shaped filters. For example, the chromatic dispersion of a typical 4-cavity filter for 100 GHz application is in the range of 20–30 ps/nm. However, recent design has shown that the chromatic dispersion of the thin-film filters can be minimized through optimal design of the cavities and reflecting layers. Reduction of a factor of three has been achieved.

3.3.3. MUX/DEMUX DEVICES USING DIELECTRIC THIN-FILM FILTERS

MUX/DEMUX devices can be formed by placing a thin-film interference bandpass filter between two lenses and launching a collimated beam onto the filter. An optical signal with a particular wavelength matching the passband of the filter will pass through the filter and the rest of the signals will be reflected by the filter.

Figure 3.4 shows the schematic diagram of a typical 3-port DEMUX device for extracting one channel from a number of channels using graded index (GRIN) lenses [10]. In this particular case, an optical signal light with N channels (wavelengths) is launched into the device through an

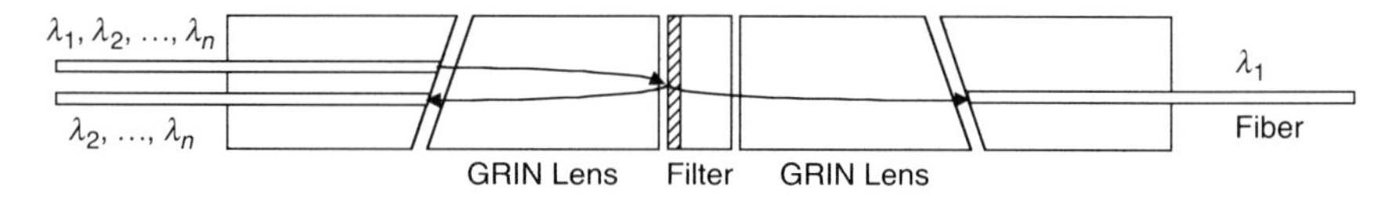

Fig. 3.4 Schematic diagram of a typical 3-port thin-film filter-based DEMUX device.

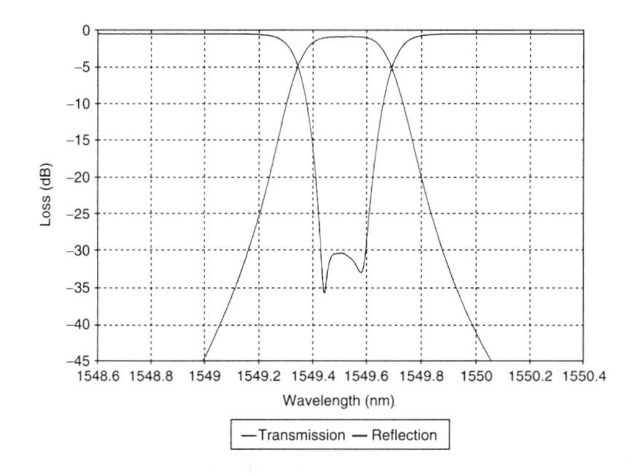

Fig. 3.5 Typical transmission and reflection spectrum of a thin-film filter-based DEMUX device for 50 GHz applications.

input fiber. The signal with the wavelength of λ_1 is coupled to an output optical fiber after passing through the filter and signals with wavelengths of λ_2 to λ_n are reflected by the filter and coupled into another output fiber. With this structure, the center wavelength of the filter can be easily tuned to a specific wavelength (i.e., ITU grid) by varying the beam angle through the filter, without the need for tight control on the filter or active adjustment of the filter. The beam angle can be adjusted by simply changing the core distance between two fibers [11]. Figure 3.5 shows the typical transmission and reflection spectrum of the DEMUX device shown in Fig. 3.4 for use in a 50 GHz-spaced DWDM systems using a 5-cavity bandpass filter. A wide square-top pass-band and high isolation have been achieved with very good loss.

Three methods have been used to form a MUX/DEMUX device for separating multiple channels with different wavelengths using the thin-film filters. The first is to use a wavelength-independent star coupler to equally split the power of an incoming light into *N* branches and then place thin-film filters corresponding to the desired wavelength at each branch as shown in Fig. 3.6. MUX/DEMUX devices using this method are very easy to make; however, they suffer from high insertion loss due to the intrinsic power splitting (for example, at least 9 dB loss for 8 channels) and are only suitable for demultiplexing a small number of channels (typically less than 4 channels). The second method is to sequentially cascade multiple 3-port devices as shown in Fig. 3.7. This method offers good flexibility in channel configuration and improved insertion loss while maintaining

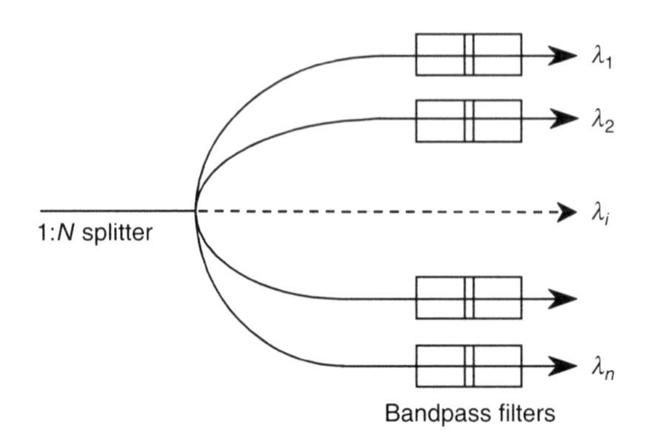

Fig. 3.6 Schematic diagram of a multi-channel DEMUX device using $1 \times N$ splitter and thin-film filters.

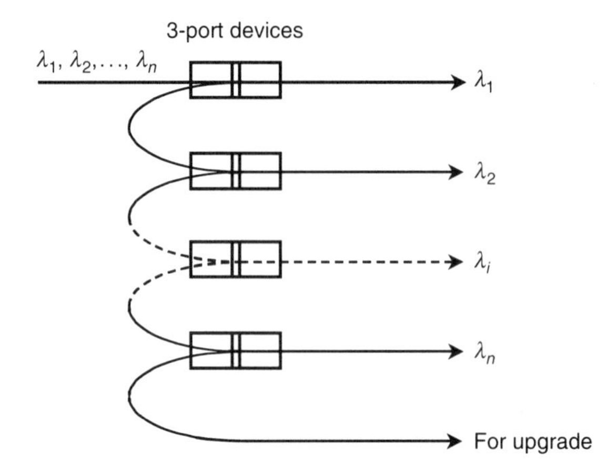

Fig. 3.7 Schematic diagram of a multi-channel DEMUX using cascaded thin-film filter-based 3-pot couplers.

manufacturability similar to the 3-port devices shown in Fig. 3.4. Another advantage of this method is that an upgradable modular approach can be used to reduce the initial deployment cost. 40-channel DEMUX devices are commercially available using this method and a typical performance of a 40-channel 100 GHz spacing DEMUX is shown in Fig. 3.8(a) with very low insertion loss by using the modular design shown in Fig. 3.8(b). The third method is to cascade multiple thin-film filters in a collimated beam as shown in Fig. 3.9 [12]. This method has the potential for providing the lowest insertion loss because light is only coupled into fiber once

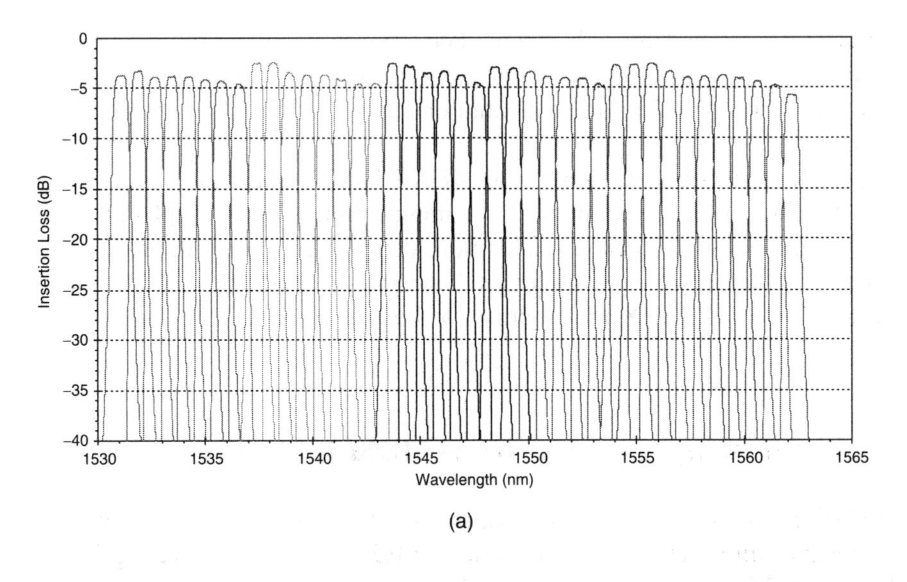

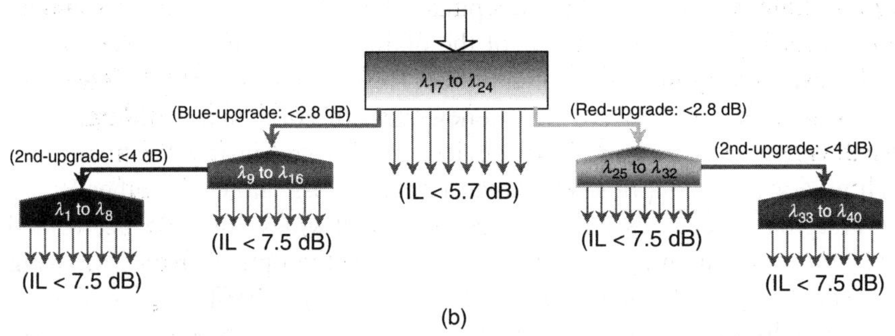

Fig. 3.8 Typical spectrum of a 40-channel thin-film filter-based DEMUX for 100 GHz applications (a) and the corresponding modular configuration (b).

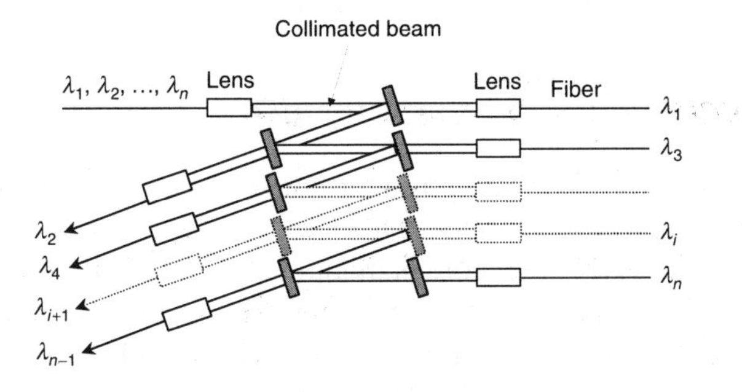

Fig. 3.9 Schematic diagram of a multi-channel DEMUX using cascaded thin-film filters in collimated beams.

for each channel. However, this method requires the highest control on filters and manufacturing processes due to the cascading configuration; that is, the performance of all sequential channels is affected by the previous channels.

Currently, thin-film filter-based MUX/DEMUX is the most widely deployed technology in commercial communication systems of up to 10 Gb/s. With the continued design innovation, deposition process improvement, and optical processing technology advance in device design, thin-film filter-based MUX/DEMUX will continue to be one of the key technologies in dense WDM systems.

3.4. Fiber Bragg Grating-based MUX/DEMUX

Most of the passive components used in optical communication systems, such as thin-film filter-based devices, are based on bulk optics and require coupling light in and out of the optical fibers, leading to certain loss and the need for active positioning of the optical fiber. With the discovery of the photosensitivity in the optical fiber [13, 14], fiber Bragg grating-based devices have been developed as a new class of components providing in-fiber devices where functional devices are formed inside the fiber core, eliminating the coupling-associated insertion loss and costs. Fiber Bragg grating-based devices are relatively simple devices, and can be formed by simply modulating the refractive index of the optical fiber core using an ultraviolet (UV) light source. Fiber Bragg grating-based optical filters and dispersion compensators have become commonly used devices in the advanced optical communication systems. Fiber Bragg grating-based filters are typically rejection filters and are commonly used with optical circulators to form MUX/DEMUX devices.

3.4.1. PRINCIPLE OF THE FIBER BRAGG GRATING

The simplest fiber Bragg grating consists of a periodic refractive index variation in the core of an optical fiber, as shown in Fig. 3.10, where the refractive index of the fiber core is modulated with a period of Λ. When a light with a broad spectrum is launched into one end of a fiber containing a fiber Bragg grating, the part of the light with wavelength matching the Bragg grating wavelength will be reflected back to the input end, with the rest of the light passing through to the other end. From the momentum

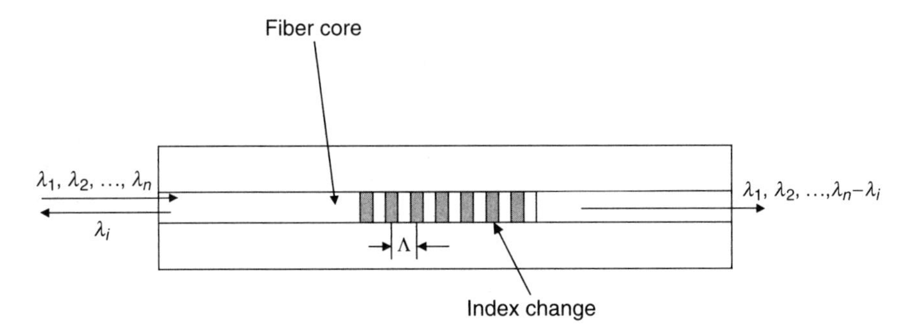

Fig. 3.10 Schematic diagram of a fiber Bragg grating.

conservation requirement of the Bragg grating condition, the following equation can be obtained:

$$2\left(\frac{2\pi n_{\mathrm{eff}}}{\lambda_B}\right) = \frac{2\pi}{\Lambda} \tag{3.21}$$

where n_{eff} is the effective refractive index of the fiber core and λ_B the wavelength of the light reflected by the Bragg grating. Therefore, The Bragg grating wavelength can be expressed as

$$\lambda_B = 2n_{\mathrm{eff}}\Lambda \tag{3.22}$$

Note that the Bragg grating wavelength is the function of the effective index and the period of the grating. Therefore, the fiber Bragg grating can be used as a MUX/DEMUX device in WDM systems for extracting a signal (channel) with a particular wavelength from a stream of signals (channels).

Several methods have been developed to fabricate the fiber Bragg grating. Among them, interferometric [15], phase mask [16] and point-by-point techniques [17] using UV light are the most common ways of making fiber Bragg gratings for optical communication applications.

In the interferometric method, a UV light beam is split into two beams and recombined to form an interference pattern on the fiber. The interference pattern can be changed by adjusting the phase relation between the two beams. The main advantage of this method is the flexibility for producing completely different gratings without any major change of the set-up. However, because free space optics is used in this method, it requires extremely stable optical set-up, and any environmental change (for example, vibration, temperature, and airflow variations) could disturb the interference pattern, resulting in imperfect writing. Therefore, this method is mainly suitable for experimental and prototyping purposes.

In the phase mask method, a collimated UV beam is spatially modulated by a diffractive phase mask in front of a fiber, and the fringe pattern produced by the interference of the diffracted beams is used for writing of the grating. The advantage of the phase mask method is the simplicity of the optics and reduced sensitivity to the environment. However, it reduces the flexibility because a phase mask has to be made for almost every type of grating. Therefore, this method is suitable for large-volume production.

The point-by-point method uses a focused UV beam directly launched onto the fiber to change the refractive index of the fiber core and is mainly used to write the long period gratings.

The standard fibers for telecommunication applications are not photosensitive enough to the UV light, so a technique called hydrogen-loading has been used to increase the UV sensitivity [18]. In the hydrogen-loading process, a fiber is soaked in high-pressure hydrogen for a period of time so that the hydrogen molecules are diffused into the core of the fiber, resulting in increased photosensitivity.

3.4.2. PROPERTIES OF THE FIBER BRAGG GRATING

As mentioned in section 3.2, for applications in WDM systems, insertion loss, isolation (crosstalk), passband bandwidth, wavelength stability, polarization effect, etc. are very important parameters.

Insertion loss of the fiber Bragg grating-based devices is mainly determined by the reflectivity of the grating. The reflectivity at the Bragg grating wavelength is given as [19]

$$R = \tanh^2\left[\frac{\pi \Delta n L}{\lambda}\left(1 - \frac{1}{V^2}\right)\right] \tag{3.23}$$

where Δn is the refractive index change, L the grating length, and V the normalized frequency of the fiber. It is noted that the reflectivity increases with the increase of the refractive index change and the grating length. High reflectivity (strong grating with large refractive index change) is desired for DWDM applications and more than 95% reflectivity has been regularly achieved.

In DWDM applications, it is very important to have high rejection at adjacent channels to reduce the crosstalk. Isolation (or crosstalk) of the grating is determined by the sidelobes of the grating. Figure 3.11 shows a

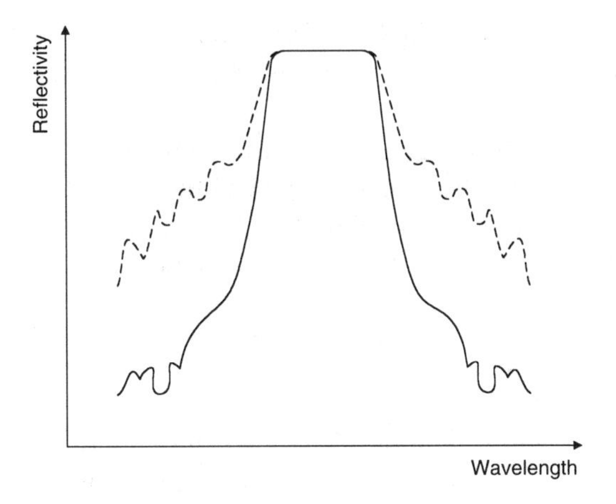

Fig. 3.11 Typical reflection spectrum of a fiber Bragg grating for WDM applications, where broken and solid curves correspond to spectrum of a uniform grating and that of an apodized grating, respectively.

typical reflection spectrum of a grating for WDM applications, where the broken curve corresponds to that of a uniform grating. It is noticed that besides the main reflection peak, there are a lot of sidelobes at the adjacent wavelengths caused by the multiple reflections from opposite ends of the grating region. Techniques have been developed to reduce the sidelobes and an apodization technique [20] is currently been used, where instead of using a uniform refractive index change across the whole region of the grating, refractive index change is gradually reduced toward the end of the grating, resulting in reduced multiple reflections from the end. As shown in Fig. 3.11 (solid curve), the sidelobes can be significantly reduced (up to 20 dB reduction).

The full-width half maximum bandwidth for a strong grating can be approximated as [21]

$$BW = \lambda_B \sqrt{\left(\frac{\Delta n}{2n_0}\right)^2 + \left(\frac{1}{m}\right)^2} \tag{3.24}$$

where Δn is the refractive index variation, n_0 the average refractive index, and m the number of grating planes. The bandwidth increases with the refractive index change, again suggesting that the strong grating is desirable. It is known that the spectrum of the fiber Bragg grating can be broadened to near square shape from the sync or Gaussian shape by deep exposure to the UV light during the formation of the grating, resulting in saturation of the refractive index change and reduction of the effective length of the grating (thus reducing m) because the transmitted light is depleted by reflection [22].

It is known that the Bragg grating wavelength changes with the strain and temperature of the fiber because both parameters affect the effective refractive index and the grating period (Eq. (3.22)) [23]. The relation between the wavelength shift and the strain ε can be expressed as

$$\Delta\lambda_S = \lambda_B(1 - p)\varepsilon \tag{3.25}$$

where p is the effective strain-optic constant and a function of the effective refractive index, strain-optic tensor, and the Poisson's ratio. For a typical germanium-doped fiber, the strain-induced wavelength change at wavelength range around 1550 nm is estimated to be about 1.2 pm/$\mu\varepsilon$.

The temperature-induced wavelength change is related to the thermal expansion of the fiber and the thermal coefficient of the refractive index of the fiber core. Because the thermal expansion of the fiber (silica) is about 5.5×10^{-7}, while the thermal coefficient of the refractive index of a typical germanium-doped silica fiber is about 8.6×10^{-6}, the temperature effect of the refractive index change is clearly the dominant cause of the wavelength shift. Therefore, the temperature-induced wavelength shift can be approximated as

$$\Delta\lambda_T = \lambda_B \frac{1}{n_{\text{eff}}} \Delta T \frac{dn}{dT} \tag{3.26}$$

It is found that the typical temperature drift of the center wavelength is about 13 pm/°C at the 1550 nm range. In MUX/DEMUX applications, strain-induced wavelength drift is used to compensate the temperature-induced wavelength drift in packaging the fiber Bragg gratings (athermal packing) [24, 25]. The athermal package is designed such that the fiber Bragg grating is compressed with an increase in temperature, while the grating is stretched, with a decrease in temperature, so that the strain- and temperature-induced wavelength drifts compensate each other, resulting in a temperature-insensitive device.

Stability and reliability are also important issues for field application of the fiber Bragg grating. Because the grating is formed with the permanent refractive index change in the fiber core, the stability of the index change is one of the key reliability concerns. Thermal decay is one of the factors affecting the stability of the grating [26, 27]. Annealing the grating at a temperature much higher (typically over 300°C) than the operating temperature has been found to be effective in stabilizing the grating.

Radiation mode coupling of the fiber Bragg grating is one of the limiting factors in WDM applications, because it limits the usable overall wavelength range, hence reducing the total number of channels. Cladding and radiation mode coupling cause lower transmission of the grating at the wavelength shorter than the reflection wavelength and multiple sharp valleys in transmittance due to higher order mode coupling. The common way of reducing this effect is to use a higher-NA fiber [28], and more than 10 nm usable wavelength range has been achieved by using special high-NA fibers.

Similar to the thin-film filter, the fiber Bragg grating has relatively large negative chromatic dispersion, therefore, is not suitable for high bit-rate long-distance transmission systems. However, on the other hand the chromatic dispersion property of the fiber Bragg grating has been actively used to compensate for the chromatic dispersion in standard optical fibers.

3.4.3. FIBER BRAGG GRATING-BASED MUX/DEMUX DEVICES

Single-channel Add/Drop and MUX/DEMUX modules can be formed using the fiber Bragg grating. Because the fiber Bragg grating is a reflective device, circulators are typically used together with the fiber Bragg grating for MUX/DEMUX in WDM applications. A typical schematic diagram of the fiber Bragg grating-based single-channel MUX/DEMUX is show in Fig. 3.12, where a fiber Bragg grating-based band rejection filter is connected to port 2 of a 3-port optical circulator. In operation as a demultiplexer, an optical signal containing N wavelengths is launched into port 1 of the circulator. After reaching the grating, one of the wavelengths (λ_1) is

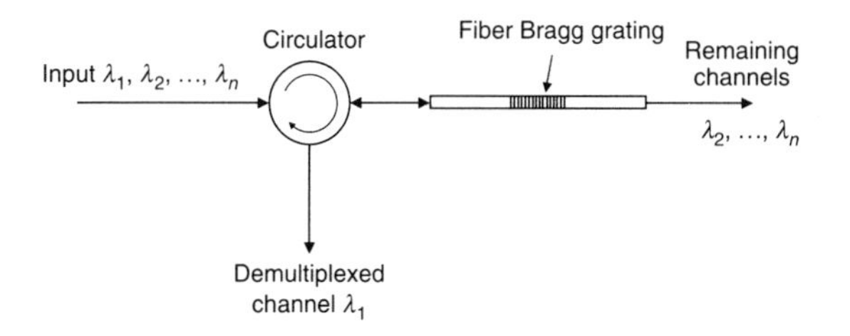

Fig. 3.12 Schematic diagram of a fiber Bragg grating-based single-channel DEMUX device.

reflected by the grating and extracted through port 3 of the circulator, and the rest of the wavelengths are passed through the grating without being affected.

To separate several wavelengths, the configuration shown in Fig. 3.12 can be cascaded in series. However, this is not a desired solution for demultiplexing large numbers of channels because the insertion loss and cost increase linearly with the increase of the channels due to the use of optical circulators. Several methods have been proposed to demultiplex multiple channels using hybrid configurations consisting of fiber Bragg grating and other forms of MUX/DEMUX devices. Two examples are shown in Fig. 3.13 and Fig. 3.14, respectively. In the configuration shown in Fig. 3.13 a 1×4 splitter and 8 grating-based filters are used to form a 4-channel

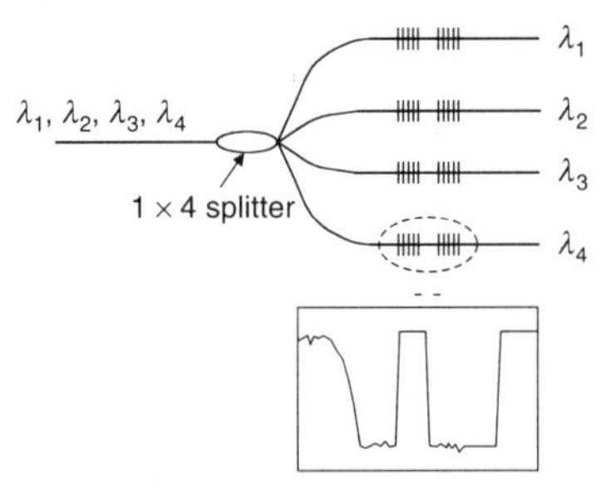

Fig. 3.13 Schematic diagram of 4-channel DEMUX device using 1×4 splitter and transmissive fiber Bragg gratings.

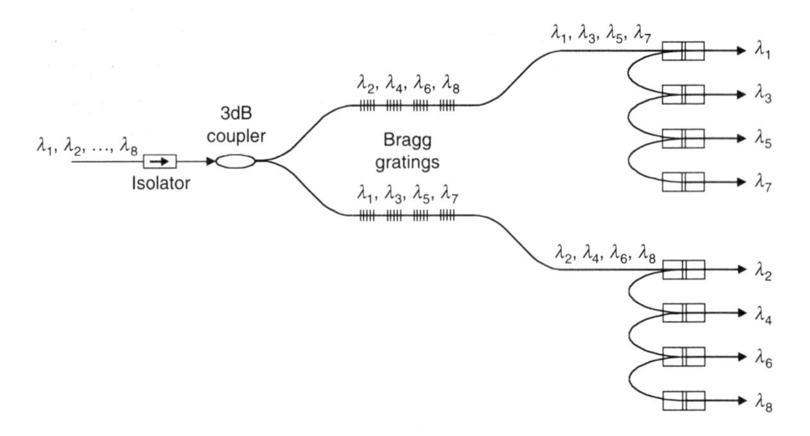

Fig. 3.14 Schematic diagram of an 8-channel DEMUX using combination of fiber Bragg gratings and thin-film filters.

DEMUX device [29]. The gratings are used in pairs to form a transmission filter for passing a particular wavelength while rejecting the other wavelengths, and square-top spectrum with very high isolation is achieved. However, the main drawback of this method is the high loss associated with the use of splitters, especially with a high number of channels. A hybrid approach for reducing the insertion loss is shown in Fig. 3.14, where fiber Bragg gratings and thin-film filters are used to form a DEMUX device for separating an 8-channel signal [30]. In this configuration, a 1×2 (3 dB) coupler is used to split the input into 2 output arms, and 4 reflection gratings are used in each arm to block alternative channels. The transmitted channels in each arm are then separated by using thin-film-based DEMUX devices. This configuration reduces the loss by eliminating the use of optical circulators and at the same time reduces the requirement on thin-film filters because only alternate channels are de-multiplexed by the filters. This structure can be expanded to more than 8 channels.

Similar to thin-film filter-based MUX/DEMUX devices, the fiber Bragg grating can only process one wavelength at a time. Therefore, sequential cascading is typically required for multiple wavelengths, resulting in relatively high loss and cost for a large number of wavelengths (wide wavelength ranges). The other limitation factor is the radiation mode coupling, which limits the usable bandwidth of the short wavelength range (from the rejection band edge) to about 12 nm. Currently, FBG is mainly used in niche MUX/DEMUX applications, such as add/drop, that require very high figure-of-merit filters and very dense spacing (50 GHz and 25 GHz).

3.5. Planar Lightwave Circuits-based MUX/DEMUX

Planar lightwave circuits (PLC) have been attracting attention in the optical research community for more than 10 years due to their potential for integration of multiple optical functions on a single substrate to form integrated optical circuits, and compatibility with the semiconductor manufacturing technology, leading to potential lower cost. With the increased demand for information bandwidth and the emerging of new technologies, more and more different types of components and different functions are required in the advanced optical networks, resulting in increased interest in the planar lightwave circuits. One of the widely deployed PLC devices is the AWG-based MUX/DEMUX device (also called phased-array gratings or waveguide grating routers) for DWDM optical system applications.

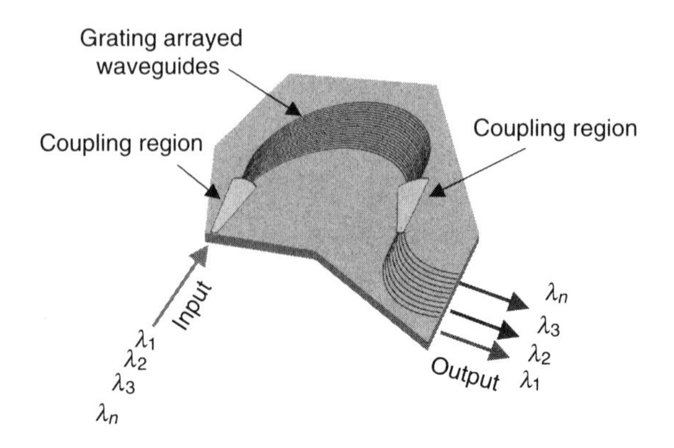

Fig. 3.15 Schematic diagram of an AWG-based MUX/DEMUX device.

3.5.1. PRINCIPLE OF THE AWG

The principle of an AWG-based MUX/DEMUX is similar to that of diffraction grating-based devices, and is based on multi-beam interference [31–33]. As shown in Fig. 3.15, a typical AWG MUX/DEMUX device consists of N numbers of input and output waveguides, two coupling regions (similar to the function of a lens in a diffraction grating-based device), and a phase array of multiple-channel waveguides (similar to the function of the diffraction grating) with a constant path length difference ΔL between adjacent waveguides.

In operation, the input light from an input waveguide is diffracted in the coupling region and excites all the arrayed channel waveguides. After traveling through the channel waveguides, light beams from the channel waveguides are diffracted again and constructively interfere with each other at different focal points in the second coupling region, then couple into the output waveguides. Because the relative phase delay among the channel waveguides is wavelength dependent, the location of the focal point depends on the wavelength of the input light as, light beams with different wavelengths are focused into different output waveguides resulting in the demultiplexing of optical signals.

Similar to that of a diffraction grating, the grating equation for the AWG can be expressed as

$$n_s d(\sin\theta_\mathrm{i} + \sin\theta_\mathrm{o}) + n_0 \Delta L = m\lambda \tag{3.27}$$

where θ_i and θ_o are input and output beam angle, n_0 and n_s the effective refractive indexes of the channel waveguides at the center wavelength and the coupling region, respectively, d the pitch of the channel waveguides, and m the grating diffraction order. The first item in Eq. (3.27) represents the phase difference in the input and output coupling regions, and the second item represents the phase difference from the channel waveguides. If the phase difference is only generated by the length difference in the channel waveguides, then the diffraction order is given as

$$m = n_0 \frac{\Delta L}{\lambda_c} \tag{3.28}$$

where λ_c is the center wavelength of the intended operating wavelength range.

The wavelength dispersion of the focal position can be expressed as

$$\frac{dp}{d\lambda} = -\frac{n_w f \Delta L}{n_s d \lambda_c} \tag{3.29}$$

where n_w is the group effective refractive index of the channel waveguides, and f the focal length of the coupling region. If the material used as the channel waveguides has a wavelength-dependent refractive index, then n_w should be

$$n_w = n_0 - \lambda \frac{dn_0}{d\lambda} \tag{3.30}$$

where $\frac{dn_0}{d\lambda}$ is the wavelength dispersion of the refractive index of the channel waveguides. Therefore, for given channel spacing $\Delta\lambda$, diffraction order m, and output waveguides spacing D, the required focal length can be obtained as

$$f = \frac{D}{\Delta\lambda} \frac{n_s d}{m} \frac{n_0}{n_w} \tag{3.31}$$

The path length difference can be obtained from Eq. (3.29)

$$\Delta L = \frac{n_s d D \lambda_c}{n_w f \Delta\lambda} \tag{3.32}$$

The free spectral range (FSR) of the AWG can be calculated from Eq. (3.28) as the following:

$$FSR = \lambda_c \left(\frac{m}{m-1} \frac{n_f}{n_0} - 1 \right) \tag{3.33}$$

where n_f is the refractive index of the channel waveguide at the wavelength of one FSR off from the center wavelength. If the material dispersion of the channel waveguide is negligible ($n_f = n_0$), then the FSR can be approximated as

$$FSR = \lambda_c \left(\frac{1}{m - 1} \right) \tag{3.34}$$

3.5.2. *PROPERTIES OF AWG-BASED MUX/DEMUX*

Because in the AWG-based MUX/DEMUX devices the all-functional elements are integrated into a single chip on a substrate, it is relatively simple to manufacture the device once the chip is fabricated. The only remaining process is the attachment of fibers to the waveguide chip. The fabrication of the AWG chip is based on batch processing using the techniques developed in the semiconductor industry, therefore, it is relatively independent of the number of channels. In the AWG device all wavelengths are processed at the same time (parallel processing), so the insertion loss is relatively uniform over all wavelengths and does not increase linearly with an increase in channel counts due to the non-cascading feature. The contribution of the insertion loss is mainly from the coupling loss between the waveguides and fibers and the propagating loss in the waveguides. Typically an insertion loss of 4–5 dB is achieved for a 40-channel AWG device. The other features of the AWG devices include compact size and flexibility for integration with other components such as detectors and attenuators. A typical spectrum and the packaged device of a 32-channel AWG MUX/DEMUX device are shown in Fig. 3.16 and Fig. 3.17, respectively. The spectrum

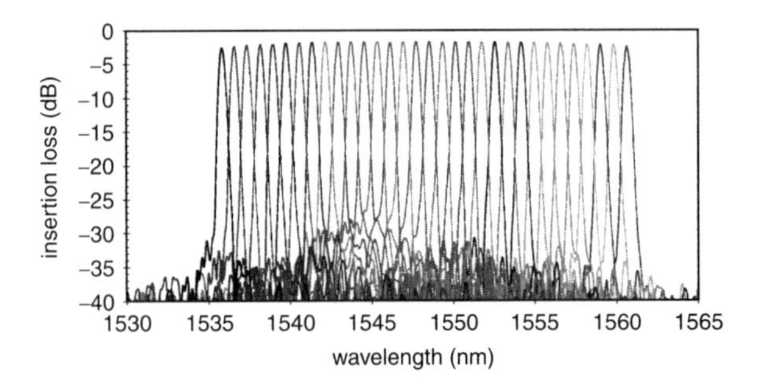

Fig. 3.16 Typical spectrum of a narrow-band 32-channel AWG DEMUX for 100 GHz applications.

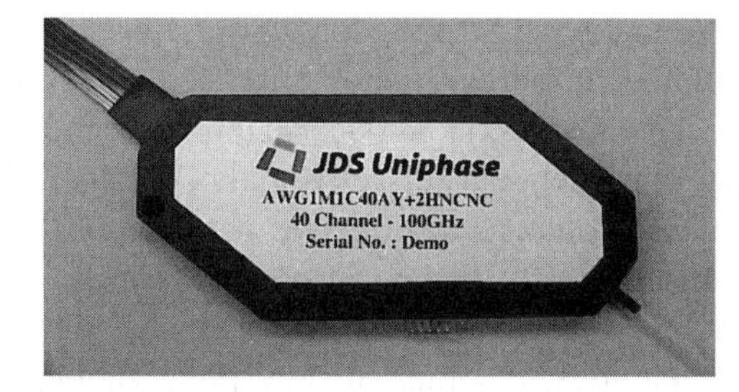

Fig. 3.17 Example of a packaged 40-channel AWG MUX/DEMUX device.

shape of the AWG device is intrinsically a Gaussian shape, which gives the lowest loss but may not be the best choice for some applications because the narrow passband requires tight wavelength control on both the AWG device and the optical source. Some research has shown that by altering the input and output waveguides or coupling region designs, the passband of the AWG devices can be broadened to yield a wide-band spectrum shape [34–36]. However, broadening of the spectrum shape comes with the price of increased insertion loss of 2–3 dB.

Compared to other technologies, AWG-based MUX/DEMUX devices have some distinguishing features, such as temperature- and polarization-dependent wavelength shift and cumulative crosstalk, etc. As shown in the grating equation (Eq. 3.27), at a fixed grating order, the center wavelength of the AWG changes with the variation in the path length difference and refractive index. Because the path length difference is realized through the physical length difference among channel waveguides, the performance of the AWG device is determined by the properties of the materials used as the channel waveguides. For the common silica-on-silicon AWG, the temperature behavior of the device is dominated by the temperature coefficient of the refractive index of the silica glass ($dn/dT = 1.1 \times 10^{-5}/^{\circ}C$), resulting in a wavelength drift of about 12 pm/°C. Therefore, temperature control is normally required in the AWG-based MUX/DEMUX devices. This is the main disadvantage of the AWG devices, because it requires not only constant power consumption and electronic driving circuits, but also monitoring of the system. Achieving athermal (temperature-insensitive) AWG MUX/DEMUX devices has been the recent focus of AWG development. Several methods have been reported, such as the use of silicone trench (because the refractive index of the silicone has a negative temperature coefficient of $-37 \times 10^{-5}/^{\circ}C$) and temperature-compensated packaging techniques [37, 38].

Because the AWG device is formed in waveguides, birefringence in the waveguides is another important factor. Birefringence in the waveguides will not only affect the insertion loss but also introduce the wavelength shift through polarization-dependent refractive index change. Birefringence in the waveguide causes change in effective path length between TE and TM modes, resulting in different focal position for TE and TM modes. Although advancement in fabrication techniques has minimized birefringence in the waveguides, resulting in low polarization-dependent loss, polarization-dependent wavelength drift in the AWG devices is still an issue. Several compensation methods have been proposed, including inversion of the polarization state in the middle of the channel waveguides using a polyimide half-waveplate [39] and the focal point correction using a birefringent crystal.

Unlike the thin-film filter-based device where the transmission outside the passband decreases almost infinitely with the wavelength, as can be seen from Fig. 3.16, the filtering in the AWG-based device levels out to a certain level and forms a noise floor. Therefore, cumulative crosstalk becomes an issue in the actual applications. It has been shown that the noise floor is caused by the scattering due to phase errors in the waveguide fabrication, resulting in imperfect focusing of the light [40]. The phase errors are caused by the refractive index variation within the waveguides and dimension fluctuation in the channel waveguides. Techniques have been developed to reduce or correct the phase error, such as the use of post UV trimming [41].

AWG-based MUX/DEMUX devices are linear phase filters and have intrinsically very small chromatic dispersion. This provides a significant advantage in high bit-rate long haul transmissions. Currently, AWG is mainly used in very high channel count (>32 channel) MUX/DEMUX applications, and is also a suitable solution for achieving a higher level of component integration in WDM systems.

3.6. Diffraction Grating-based MUX/DEMUX

Diffraction gratings have been used in spectroscopy to separate polychromatic light into monochromatic components. A diffraction grating is an array of reflecting (or transmitting) elements separated by a distance comparable to the wavelength of the light. When a light with different wavelengths is incident on a grating, the grating diffracts each wavelength component to different directions according to the incident angle, wavelength of the light, and the grating structure. Based on this property, diffraction gratings have been studied for multiplexing/demultiplexing multiple channels of different wavelengths in optical communication systems since the late 1970s [42, 43].

3.6.1. PRINCIPLE OF DIFFRACTION GRATINGS [44]

We consider first the case when a monochromatic light is launched onto a grating surface as shown in Fig. 3.18. Each grating groove can be considered as being a very small, slit-shaped diffracted source, and the diffracted light forms a diffracted wavefront. Under certain conditions, the diffracted lights from all facets of the grooves are in phase with each other, so they combine constructively and can be imaged to a single spot.

As shown in Fig. 3.18, when a light of wavelength λ is launched to a grating with an angle α, the light is diffracted toward different directions along angle β_{i}, where i is the diffraction order. The angles are measured from the grating normal and have positive or negative signs depending on whether the diffracted light is on the same side or opposite side as the incident light. The particular case when the light is diffracted back into the same direction as the incident light (i.e., $\alpha = \beta$) is called Littrow configuration. The geometrical path difference between light from adjacent grooves is given as

$$\Delta p = d(\sin\alpha + \sin\beta)\cos\phi \tag{3.35}$$

where ϕ is the angle between the incident beam and the plane perpendicular to the groove facet. In most of the applications $\phi = 0$. From the principle

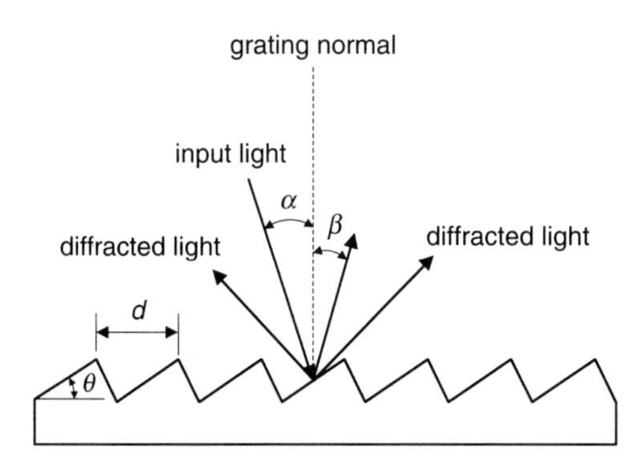

Fig. 3.18 Schematic diagram of a diffraction grating.

of interference, only when the path difference equals the integral of the wavelength, will the diffracted light be in phase and form constructive interference. Therefore, the following grating equation can be obtained assuming $\phi = 0$:

$$d(\sin\alpha + \sin\beta) = m\lambda \tag{3.36}$$

where m is the diffraction order and an integer. Equation (3.36) can be written as the following using the grating pitch G ($=1/d$), which is the number of grooves per millimeter:

$$(\sin\alpha + \sin\beta) = m\lambda G \tag{3.37}$$

When a light of multiple wavelengths is incident on the grating, the diffraction angle for each wavelength can be expressed as the following from Eq. (3.37):

$$\beta_\lambda = \arcsin(m\lambda G - \sin\alpha) \tag{3.38}$$

It can be seen from Eq. (3.38) that when $m = 0$, β_λ equals α and is independent of the wavelength. Therefore, the grating acts as a plane mirror. The feature of the wavelength-dependent diffraction angle of the diffraction grating is the basic foundation for forming MUX/DEMUX devices.

Angular dispersion is one of the important parameters because the larger the angular dispersion, the easier it is to separate different wavelength components. The angular dispersion can be obtained by differentiating the grating equation as

$$\frac{\partial\beta}{\partial\lambda} = \frac{m}{d\cos\beta} = \frac{Gm}{\cos\beta} = \frac{\sin\alpha + \sin\beta}{\lambda\cos\beta} \tag{3.39}$$

It can be seen that the angular dispersion increases with the grating pitch G, and for a given wavelength the angular dispersion can be considered as a function of incident and diffraction angles.

There are many ways to make the diffraction gratings. One of the most common and traditional ways is mechanical ruling, where grooves of the grating are cut using machines. Ruled gratings are very flexible in terms of the groove shape and variations. Gratings can also be replicated from the ruled master gratings where the grooves are formed in a very thin layer of resin and the grooved layer is then cemented to a substrate. Holographic interference has been used to make gratings since the late 1960s where the interference pattern is formed using collimated light beams and recorded in photoresist, and grooves are formed after chemical developing of the photoresist. One of the features of the holographic grating is its sinusoidal-shaped grooves. Other shapes are possible by post-processing using ion etching. Diffraction gratings have also been produced in semiconductor materials using etching technology mainly to produce Echelle gratings [45]. Echelle grating is a high-order and high-angle grating and gives higher resolution and dispersion than ordinary gratings.

Because the diffraction grating is a linear phase filter, its chromatic dispersion is typically very small. However, techniques used to broaden the passband may increase the dispersion.

3.6.2. MUX/DEMUX DEVICES USING THE DIFFRACTION GRATINGS

MUX/DEMUX devices can be fabricated by using the diffraction grating in combination with imaging optics such as a lens or by using a concave-shaped grating as a DEMUX device as shown in Figs. 3.19 and 3.20. The DEMUX device shown in Fig. 3.19 consists of a fiber array, a lens, and

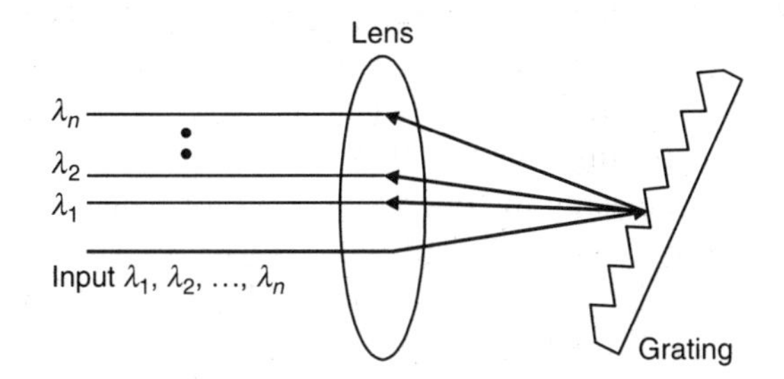

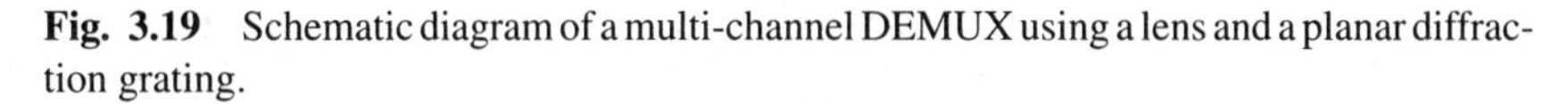

Fig. 3.19 Schematic diagram of a multi-channel DEMUX using a lens and a planar diffraction grating.

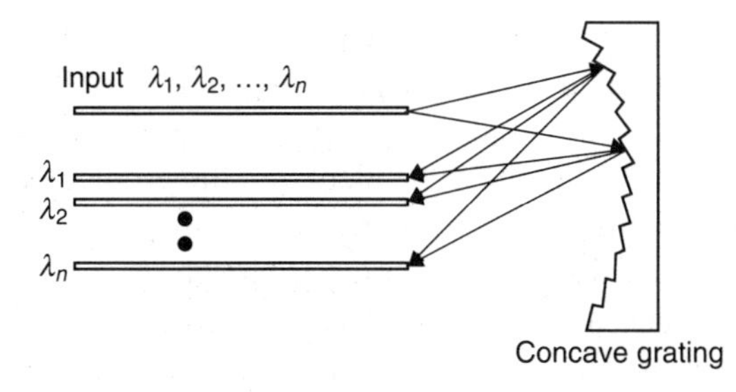

Fig. 3.20 Schematic diagram of a multi-channel DEMUX using concave diffraction grating.

a planer grating. The input beam from an input optical fiber is collimated by the lens, and the collimated beam with multiple wavelengths is incident on the grating. The input beam is diffracted by the grating with different beam angles for the different wavelengths. After passing through the lens, the diffracted beams with different wavelengths are focused into different locations and coupled into output fibers. In the configuration shown in Fig. 3.20, a fiber array and a concave grating are used [46].

The reverse linear dispersion, in units of nm/mm, is one of the important parameters in designing the diffraction grating-based MUX/DEMUX devices, and is a measure of a change in wavelength (in nm) corresponding to a change in location of the image at the focal point (core separation of the fiber array). The reverse linear dispersion can be expressed as

$$P = \frac{d \cos \beta}{mf} = \frac{\textit{channel spacing}}{\textit{fiber core spacing}} \tag{3.40}$$

where f is the effective focal length.

With a given grating order, grating pitch, incident and diffraction angles, required fiber array spacing and focal length can be determined for a given operating wavelength and channel spacing. It can be seen from Eq. (3.40) that for a given channel spacing, the required focal length is proportional to the fiber core spacing of the fiber array. Using a grating device for narrower channel spacing thus involves increasing the focal length of the imaging system, increasing the operating order, or reducing the fiber core spacing. Increasing the focal length leads to increased overall size and environmental sensitivity, which is undesirable in the telecommunication field. Therefore, in diffraction grating-based MUX/DEMUX devices for DWDM applications, efforts have been focused on reduction of the fiber core spacing using different techniques such as etching of the fiber cladding and the use of waveguide concentrators.

Loss and isolation performances of the diffraction grating-based MUX/DEMUX devices are mainly determined by the quality of the gratings. Loss has a direct relation to the grating efficiency. The efficiency of a grating is wavelength and incident angle dependent and the efficiency can be varied for a given wavelength by changing the groove shape (blazing). A typical efficiency curve for a diffraction grating is shown in Fig. 3.21. It is noted that the efficiency has a peak at the blaze wavelength and the efficiency is highly dependent on the polarization of the incident beam relative to the grooves. Typically, diffraction efficiency for s-polarized light (perpendicular to the grating grooves) is higher than that for p-polarized light (parallel

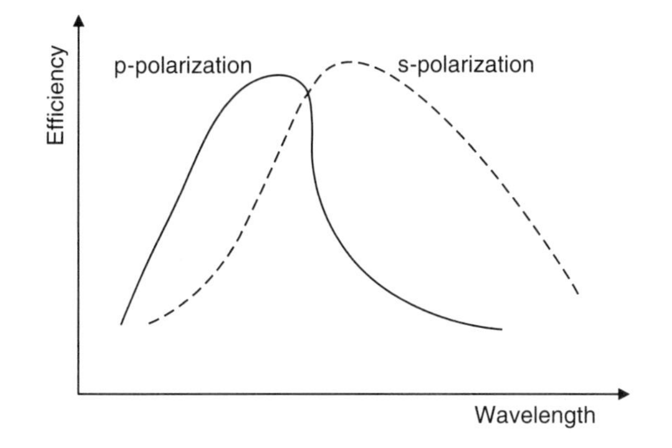

Fig. 3.21 Typical efficiency curve of a diffraction grating for different polarization states.

to the grating grooves) when the diffraction angle is above the blazing angle. Polarization-dependent loss caused by the polarization-dependent efficiency is a general problem associated with the diffraction grating-based MUX/DEMUX devices and compensation is generally required.

Isolation of the diffraction grating-based MUX/DEMUX devices is mainly determined by the resolution of the grating, which depends on the optical quality of the grating. The ultimate limit of the resolution is set by a dimensionless parameter, the resolving power R. The resolving power is defined as

$$R = \frac{\lambda}{\Delta\lambda} = mN = \frac{Nd(\sin\alpha + \sin\beta)}{\lambda} \tag{3.41}$$

where $\Delta\lambda$ is the maximum resolvable wavelength separation and N the total number of grooves illuminated on the surface of the grating. However, in practice it is very difficult to reach the limit of the resolving power due to imperfection in grating quality. The resolving power is affected not only by the incident and diffraction angles, but also by the surface quality of the grooves, uniformity of the groove spacing, and imaging optics.

Unlike the thin-film filter-based MUX/DEMUX devices, where each channel is sequentially multiplexed/demultiplexed one at a time (sequential process), resulting in large variations in the insertion loss for each channel, one of the advantages of the diffraction grating-based MUX/DEMUX device is that all channels are multiplexed/demultiplexed at the same time (parallel process), resulting in a very uniform loss performance across the whole wavelength range, and simplified processes. Therefore, the diffraction grating is more suitable for making MUX/DEMUX devices where a large number of channels are involved. On the other hand, compared to the square-top shape of the thin-film filter, the typical passband of the grating-based MUX/DEMUX is Gaussian shaped, which is less desirable in optical systems. There are different ways to flatten the passband, but they normally involve an increase of loss or increased complexity.

Diffraction grating-based MUX/DEMUX devices for DWDM applications are typically bulky due to the long focal length required to achieve narrow spacing, and their performances are largely package dependent. Thermal and mechanical management of the overall packaging is a challenge, and long-term stability is yet to be proven for MUX/DEMUX application.

3.7. Interleaver Technology

With rapid growth in Internet traffic, there is a constant demand for more communication bandwidth. In the optical communication systems, there are three ways to increase the bandwidth: the first is to lay more optical fibers, the second is to increase transmission speed, and the third is to use more wavelengths. Currently, all three options have been aggressively used. In the WDM optical system, increased channel numbers result in the decreasing of the channel spacing because the wavelength window for transmission is limited. Currently, 50 GHz-spaced DWDM systems have been deployed in large scale and 25 GHz-spaced systems are being actively developed. On the other hand, increased data rate requires wider passband of the MUX/DEMUX devices. Therefore, performance requirements have come close to the limit of the existing MUX/DEMUX technologies.

The optical interleaver has been developed to meet the ever-increasing demands on MUX/DEMUX devices. The function of an optical interleaver is to separate an incoming wavelength streams of channels into two separate wavelength streams with complimentary channels (Fig. 3.22). For example, the most commonly used interleaver can separate an optical signal with N channels of spacing D (ex. 50 GHz) into two optical signals (odd and even channels) with $N/2$ channels of spacing $2D$ (ex. 100 GHz) in each arm, thus reducing the performance requirement in each arm. Non-symmetric interleaving can also be realized.

Optical interleavers can be constructed using two principles: one is the using of resonant cavities such as multi-cavity etalons, fiber Bragg gratings, Michelson Gires-Tournois interferometer [47] and nonlinear GT; the

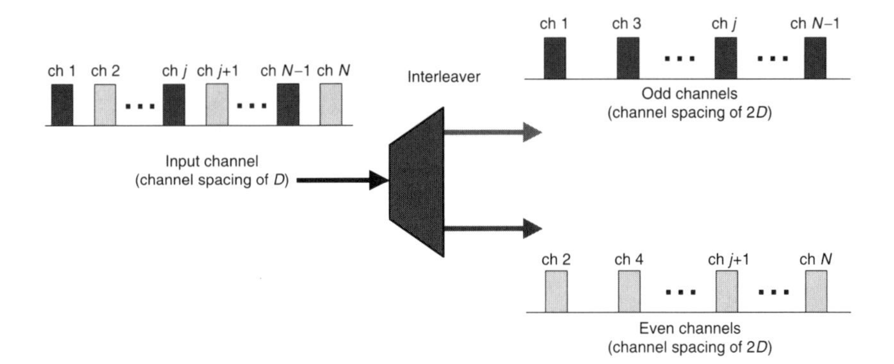

Fig. 3.22 Schematic diagram of an optical frequency interleaver.

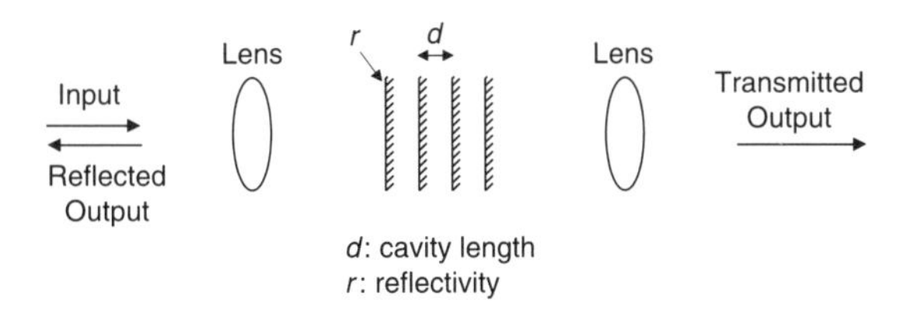

Fig. 3.23 Schematic diagram of a multi-cavity etalon-based optical interleaver.

other is using finite impulse response (FIR) filters such as polarization interference filters [48] and cascaded Mach-Zehnder filters [49]. The use of the resonant cavities can provide wider passband but with relatively higher dispersion due to the cavities. On the other hand, the use of FIR filters can provide dispersion-free performance, which is very important for high data rate transmission, but often requires more cascaded stages to achieve a flat-top filter shape.

Coherent superposition of multiple Fabry-Perot etalons has the advantage of near square shape transmission and better sidelobe suppression and is suitable for use in narrow spacing DWDM systems. Schematic diagram of a multi-cavity etalon-based optical interleaver is shown in Fig. 3.23, where the number of cavities, the reflectivity of each cavity surface, and length of each cavity can be selected to provide different FSRs and flexible transmission and reflection passbands. A typical transmission spectrum for a 4-cavity-based optical interleaver for 50 GHz DWDM applications with a uniform cavity length is shown in Fig. 3.24. The coherent superposed multi-cavity etalons can be designed and analyzed using the traditional matrix method [50].

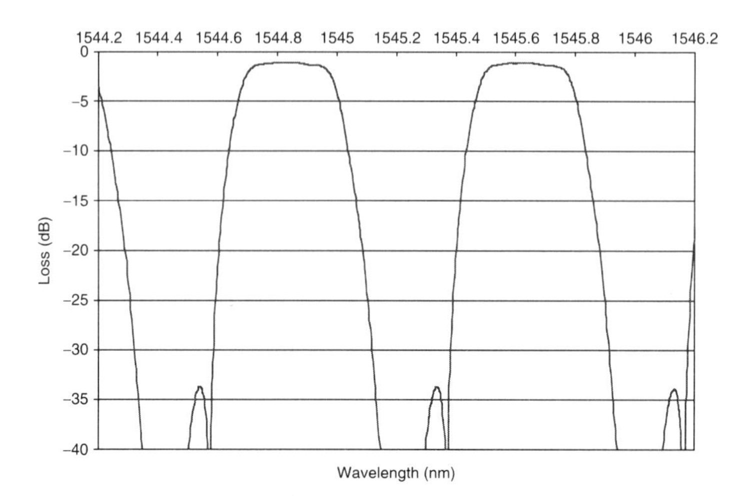

Fig. 3.24 Typical spectrum of a 4-cavity etalon optical interleaver for 50 GHz DWDM applications with a uniform cavity length.

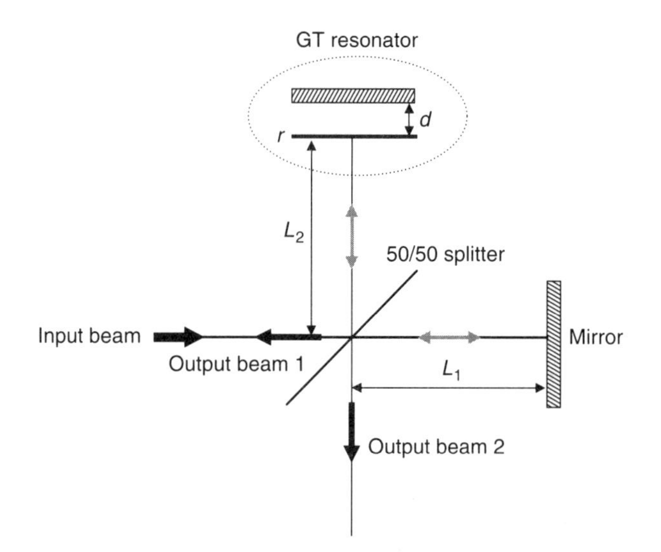

Fig. 3.25 Schematic diagram of a Michelson GT interferometer-based optical interleaver.

Gires-Tournois (GT) resonator is an asymmetric Fabry-Perot etalon with a partially reflecting front mirror and a 100% reflecting back mirror. Therefore, the GT resonator is an all-pass filter with unit amplitude and a periodic phase. An optical interleaver can be constructed by incorporating the GT resonator into a Michelson inteferometer as shown in Fig. 3.25.

In operation, an incoming optical beam is split by a 50/50 splitter into two beams with equal intensity, and one of the beams is reflected by a mirror and the other by a GT resonator. The two reflected beams form constructive interference at the beam splitter depending on the phase relation between the two arms. The FSR of the interleaver is determined by cavity length of the GT, and the passband shape can be varied by selecting the reflectivity of the GT and phase difference ($L_1 - L_2$) between the two arms of the Michelson interferometer. One drawback of this configuration is that one of the output beams overlaps with the input beam, therefore, proper arrangement such as an optical circulator, has to be used in order to gain access to both outputs.

Multi-order half-waveplate-based polarization interference filters have been used in solar physics for a long time [51, 52]. Recently, polarization interference filters have been investigated for applications in WDM systems as a MUX/DEMUX device [48]. A polarization interference filter can be constructed using birefringent crystal-based multi-order half-waveplate and different shapes of filters can be designed by cascading more than one waveplate (see Chapter 3 for details on birefringent crystals). When a linearly polarized light is launched into a half-waveplate oriented 45° to the optical axis of the waveplate, the phase retardation φ at wavelength λ_0 is given as

$$\varphi = \frac{2\pi}{\lambda_0} \Delta n_0 t \tag{3.42}$$

where Δn_0 is the birefringence of the waveplate at λ_0 and t the thickness of the waveplate. For use as an interleaver, the crystal thickness is selected such that the phase φ for one set of the wavelength equals $p\pi$ (p is an integer and an odd number) so that the polarization state of light is rotated by 90°, and at the same time the phase φ for other wavelengths equals $q\pi$ (q is an integer and an even number) so that the polarization state of light is maintained. By placing a polarization splitter after the waveplate, the two sets of the wavelengths can be separated and directed to different ports as shown in Fig. 3.26. The required crystal length can be calculated using

$$t = \frac{c}{\Delta\nu(2\Delta n)} \tag{3.43}$$

where c is the light speed in vacuum and $\Delta\nu$ is the frequency spacing (channel spacing in frequency) of the two wavelength sets. The spectrum of the single waveplate optical interleaver is sinusoidal and it has been

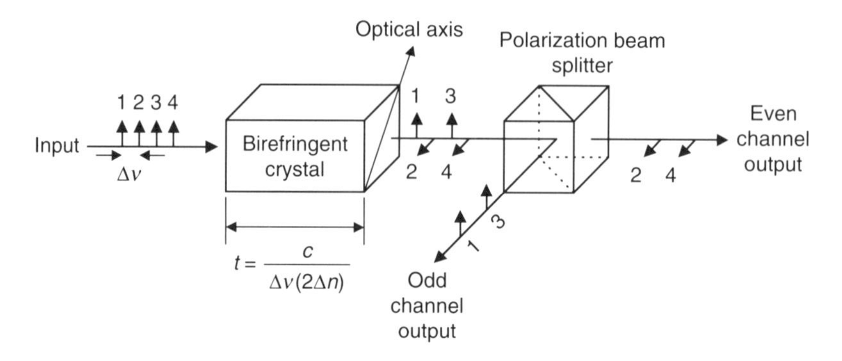

Fig. 3.26 Schematic diagram of the polarization interference filter-based optical interleaver.

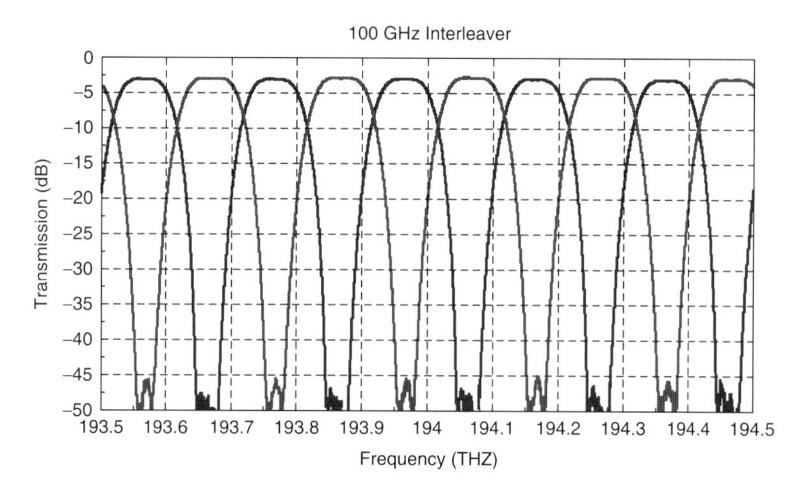

Fig. 3.27 Spectrum of a polarization interference filter-based optical interleaver for 100 GHz spacing applications with cascaded multi-order waveplates.

demonstrated that a square-top spectrum can be obtained by cascading more waveplates in series as shown in Fig. 3.27, where two waveplates with the length of t and $2t$ are cascaded in series.

The principle of cascaded Mach-Zenhder filter-based optical interleavers is similar to that of the polarization interference filters; however, the phase difference is generated by the physical path length difference instead of the refractive index difference of the birefringent crystal. Fused fiber and waveguide technologies are the best choice for fabricating these kind of filters. Similar to Eq. (3.43), the required path length difference is given as

$$\Delta L = \frac{c}{2n_m \Delta \nu} \tag{3.44}$$

where n_m is the refractive index of the light guiding material. The performance of these kind of interleavers is very similar to that of the polarization interference filters.

As an emerging technology, optical frequency interleaver has enabled various new system configurations and component applications. The hybridization of interleaver technology with other MUX/DEMUX technologies has opened a new dimension for component designers to achieve cost-effective high-performance MUX/DEMUX solutions with more flexibility.

References

1. J. S. Seeley, "Resolving power of multilayer filters," J. Opt. Soc. Am., 54 (1964) 342–346.
2. S. W. Warren, "Properties and performance of basic designs of infrared interference filters," Infrared Phys., 8 (1968) 65–78.
3. A. Thelen, in Physics of Thin Films, G. Hass and R. E. Thun, eds. (Academic Press, New York and London, 1969) 47–86.
4. H. A. Macleod, Thin Film Optical Filters (McGraw-Hill, New York, 1986).
5. Z. Knittl, Optics of Thin Films (Wiley & Sons, London, 1976).
6. S. A. Furman and A. V. Tikhonravov, Optics of Multilayer Systems, Editions Frontieres (Gif-Sur-Yvette, 1992).
7. W. Weinstein, Computations in Thin Film Optics (E. T. Heron & Co. Ltd., London, 1954).
8. K. Rabinovitch and A. Pagis, "Polarization effects in multilayer dielectric thin films," Opt. Acta., 21 (1974) 963–980.
9. H. Takashashi, "Temperature stability of thin-film narrow-bandpass filters produced by ion-assisted deposition," Appl. Opt., 34 (1995) 667.
10. S. Sugimoto, K. Minemura, K. Kobayashi, M. Seki, M. Shikada, A. Ueki, and T. Yanase, "High speed digital-signal transmission experiments by optical wavelength-division multiplexing," Electron. Lett., 13 (1977) 680.
11. Y. C. Si, G. Duck, J. Ip, and N. Teitelbaum, US patent No. 5,612,824 (1997).
12. K. Nosu, H. Ishio, and K. Hashimoto, "Multireflection optical multi/demultiplexer using interference filters," Electron. Lett., 15 (1979) 414.
13. K. O. Hill, Y. Fujii, D. C. Johnson, and B. S. Kawasaki, "Photosensitivity in optical fiber waveguides: Application to reflection filter fabrication," Appl. Phys. Lett., 32 (1978) 647–649.

14. B. S. Kawasaki, K. O. Hill, D. C. Johnson, and Y. Fujii, "Narrow-band Bragg reflectors in optical fibers," Opt. Lett., 3 (1978) 66–68.
15. G. Meltz, W. W. Morey, and W. H. Glenn, "Formation of Bragg gratings in optical fibers by a transverse holographic method," Opt. Lett., 14 (1989) 823–825.
16. K. O. Hill, B. Malo, F. Bilodeau, D. C. Johnson, and J. Albert, "Bragg gratings fabricated in monomode photosensitive optical fiber by UV exposure through a phase mask," Appl. Phys. Lett., 62 (1993) 1035–1037.
17. K. O. Hill, B. Malo, K. A. Vineberg, B. F. Bilodeau, D. C. Johnson, and I. Skinner, "Efficient mode conversion in telecommunication fiber using externally written gratings," Electron. Lett., 26 (1990) 1270–1272.
18. P. J. Lemaire, R. M. Atkins, V. Mizrahi, and W. A. Reed, "High pressure H_2 loading as a technique for achieving ultrahigh UV photosensitivity and thermal sensitivity in GeO_2 doped optical fibers," Electron. Lett., 29 (1993) 1191–1193.
19. D. K. Lam and B. K. Garside, "Characterization of single-mode optical fiber filters," Appl. Opt., 20 (1981) 440–445.
20. J. Albert, K. O. Hill, B. Malo, S. Thériault, F. Bilodeau, D. C. Johnson, and L. E. Erickson, "Apodisation of the spectral response of fibre Bragg gratings using a phase mask with variable diffraction efficiency," Electron. Lett., 31 (1995) 222–223.
21. Russel St. J., P. J. L. Archambault, and L. Reekie, "Fibre gratings," Physics World (1993) 41–46.
22. K. O. Hill and G. Meltz, "Fiber Bragg grating technology fundamentals and overview," IEEE J. Lightwave Technol, 15 (1997) 1263–1276.
23. G. Meltz and W. W. Morey, "Bragg grating formation and germanosilicate fiber photosensitivity," Proceedings SPIE (Quebec) 1516 (1991) 185–199.
24. T. E. Hammon, J. Bulman, F. Ouelette, and S. B. Poole, "A temperature compensated optical fibre Bragg grating band rejection filter and wavelength reference," 1st OECC Tech. Dig. 18C1-2, (1996).
25. T. Iwashima, A. Inoue, M. Shigematsu, M. Nishimura, and Y. Hattori, "Temperature compensation technique for fibre Bragg gratings using liquid crystalline polymer tubes," Electron. Lett., 33 (1997) 417–419.
26. T. Erdogan, V. Mizrahi, P. J. Lemaire, and D. Monroe, "Decay of UV-induced fiber Bragg gratings," J. Appl. Phys., 76 (1994) 73–80.
27. S. Kannan, J. Z. Y. Guo, and P. J. Lemaire, "Thermal stability analysis of UV-induced fiber Bragg gratings," IEEE J. Lightwave Technol, 15 (1997) 1478–1483.
28. T. Komukai and M. Nakazawa, "Efficient fiber gratings formed on high NA dispersion-shifted fiber and dispersion-flattened fiber," Japanese J. Appl. Phys., 34 (1995) L1286–L1287.

29. V. Mizrahi, T. Erdogan, D. J. DiGiovanni, P. J. Lemaire, W. M. MacDonald, S. G. Kosinski, S. Cabot, and J. E. Sipe, "Four channel fibre grating demultiplxer," Electron. Lett., 30 (1994) 780–781.
30. J. J. Pan and Y. Shi, "Steep skirt fibre Bragg grating fabrication using a new apodised phase mask," Electron. Lett., 33 (1997) 1895–1896.
31. M. K. Smit, "New focusing and dispersive component based on an optical phased array," Electron. Lett., 24 (1988) 385–386.
32. C. Dragone, "An N×N optical multiplexer using a planar arrangement of two star couplers," IEEE Photon. Technol. Lett., 3 (1991) 812–815.
33. H. Takahashi, S. Suzuki, K. Kato, and I. Nishi, "Arrayed waveguide grating for wavelength division multi/demultiplexer with nanometer resolution," Electron. Lett., 26 (1990) 87–88.
34. K. Okamoto and H. Yamada, "Arrayed-waveguide grating multiplexer with flat spectral response," Opt. Lett., 20 (1995) 43–45.
35. M. R. Amersfoort, J. B. D. Soole, H. P. LeBlanc, N. C. Andreadakis, A. Rajhel, and C. Caneau, "Passband broadening of integrated arrayed waveguide filters using multimode interference couplers," Electron. Lett., 32 (1996) 449–451.
36. D. Trouchet, A. Beguin, C. Prel, C. Lerminiaux, H. Boek, and R. O. Maschmeyer, "Passband flattening of PHASAR WDM using input and output star couplers designed with two focal points," Proc. OFC'97 (Dallas) ThM7, (1997).
37. Y. Inoue, A. Kaneko, F. Hanawa, H. Takahashi, K. Hattori, and S. Sumida, "Athermal silica-based arrayed-waveguide grating multiplexer," Electron. Lett., 33 (1997) 1945–1946.
38. G. Heise, H. W. Schneider, and P. C. Clemens, "Optical phased array filter module with passively compensated temperature dependence," ECOC'98 (Madrid), (1998) 20–24.
39. Y. Inoue, Y. Ohmori, M. Kawachi, S. Ando, T. Sawada, and H. Takahashi, "Polarization mode converter with polyimide half waveplate in silica-based planar lightwave circuits," IEEE Photon. Technol. Lett., 6 (1994) 626–628.
40. T. Goh, S. Suzuki, and A. Sugita, "Estimation of waveguide phase error in silica-based waveguides," J. Lightwave Technol., 15 (1997) 2107.
41. J. Gehler and F. Knappe, "Crosstalk reduction of arrayed waveguide gratings by UV trimming of individual waveguides without H_2-loading," OFC'2000 (Baltimore) (2000) WM9.
42. W. J. Tomlinson, "Wavelength multiplexing in multimode optical fibers," Appl. Opt., 16 (1977) 2180–2194.
43. J. Hegarty, S. D. Poulsen, K. A. Jackson, and I. P. Kaminow, "Low-loss singlemode wavelength-division multiplexing with etched fibre arrays," Electron. Lett., 20 (1984) 685–686.
44. M. C. Hutley, Diffraction Gratings (Academic Press, New York, 1982).

45. P. C. Clemens, R. März, A. Reichelt, and H. W. Schnerder, "Flat-field spectrograph in SiO_2/Si," IEEE Photon. Technol. Lett., 4 (1992) 886–887.
46. F. N. Timofeev, P. Bayvel, E.G. Churin, P. Gambini, and J. E. Midwinter, "Penalty-free operation of a concave free-space grating demultiplexer at 2.5 Gbit/s with 0.2–0.6 nm channel spacing," Proc. ECOC'96 (Oslo) 2 (1996) 321–324.
47. B. B. Dingel and M. Izutsu, "Multifunction Optical filter with a Michelson-Gires-Tournois interferometer for wavelength-division-multiplexed network system applications," Opt. Lett., 23 (1998) 1099–1101.
48. S. Pietralunga, F. Breviario, M. Martinelli, and D. Di Rocco, "Clacite frequency splitter for dense WDM transmitter," IEEE Photon. Technol. Lett., 8 (1996) 1659–1661.
49. M. Kuznetsov, "Cascaded coupler Mach-Zehnder channel dropping filters for wavelength-division-Multiplexed optical systems," J. Lightwave Technol., 12 (1994) 226–230.
50. H. Van de Stadt and J. M. Muller, "Multimirror Fabry-Perot interferometers," J. Opt. Am. A., 2 (1985) 1363–1370.
51. B. Lyot, "Optical apparatus with wide field using interference of polarized light," C. R. Acad. Sci. (Paris), 197 (1933) 1593.
52. Y. Öhman, "On some new birefringent filter for solar research," Ark. Astron., 2 (1958) 165.

Chapter 4 | Circulator

Yale Cheng

Fiber Optic Product Group, JDS Uniphase Corporation,
Nepean, Ontario, Canada

ABSTRACT

The optical circulator has become an indispensable passive component in advanced optical communication systems, especially WDM optical systems. In this chapter, the operating principles, analysis methods, and design considerations of optical circulators are described in detail, together with their typical applications in WDM systems.

4.1. Introduction

The study of optical circulators started in the early 1960s with the synergy from the microwave circulator. In the late 1970s, with the advancement of optical communications, design of optical circulators operating in optical communication wavelength windows was increasingly explored [1–3]. However, field deployment of optical circulators in optical communication systems remained very limited, due to their complexity, limited performance advantage and applications, and high cost. It is only in the 1990s that optical circulators became one of the indispensable elements in advanced optical communication systems, especially WDM systems [4, 5]. The applications of the optical circulator expanded within the telecommunications industry (together with erbium-doped fiber amplifiers and fiber Bragg gratings), but also expanded into the medical and imaging fields.

An optical circulator is a multi-port (minimum three ports) nonreciprocal passive component.

The function of an optical circulator is similar to that of a microwave circulator—to transmit a lightwave from one port to the next sequential port with a maximum intensity, but at the same time to block any light transmission from one port to the previous port. Optical circulators are based on the nonreciprocal polarization rotation of the Faraday effect. In this chapter,

WDM TECHNOLOGIES: PASSIVE
OPTICAL COMPONENTS
$35.00

ISBN: 0-12-225262-4

the operating principles, design and analysis methods, and applications of optical circulators are described in detail.

4.2. Operating Principle

4.2.1. FARADAY EFFECT

The Faraday effect is a magnetooptic effect discovered by Michael Faraday in 1845. It is a phenomenon in which the polarization plane of an electromagnetic (light) wave is rotated in a material under a magnetic field applied parallel to the propagation direction of the lightwave. A unique feature of the Faraday effect is that the direction of the rotation is independent of the propagation direction of the light, that is, the rotation is nonreciprocal. The angle of the rotation θ is a function of the type of Faraday material, the magnetic field strength, and the length of the Faraday material, and can be expressed as

$$\theta = \mathrm{VBL} \tag{4.1}$$

where V is the Verdet constant of a Faraday material, B the magnetic field strength parallel to the propagation direction of the lightwave, and L the length of the Faraday material.

The Verdet constant is a measure of the strength of the Faraday effect in a particular material, and a large Verdet constant indicates that the material has a strong Faraday effect. The Verdet constant normally varies with wavelength and temperature. Therefore, an optical circulator is typically only functional within a specific wavelength band and its performance typically varies with temperature. Depending on the operating wavelength range, different Faraday materials are used in the optical circulator.

Rare-earth-doped glasses and garnet crystals are the common Faraday materials used in optical circulators for optical communication applications due to their large Verdet constant at 1310 nm and 1550 nm wavelength windows. Yttrium Iron Garnet (YIG, $Y_3Fe_5O_{12}$) and Bismuth-substituted Iron Garnets (BIG, i.e., $Gd_{3-x}Bi_xFe_5O_{12}$, $(BiYbTb)_3Fe_5O_{12}$, $(HoTbBi)_3Fe_6O_{12}$, $(BiTb)_3(FeGa)_5O_{12}$, etc.) are the most common materials. The Verdet constant of the BIG is typically more than 5 times larger the YIG, so a compact device can be made using the BIG crystals. All these materials usually need an external magnet to be functional as a Faraday rotator. Recently, however, a pre-magnetized garnet (also call latching garnet, $(BiRE)_3(FeGa)_5O_{12}$) crystal has been developed that eliminates the use of an external magnet, providing further potential benefit in reducing overall size. Faraday rotators in optical circulators are mostly used under a saturated magnetic field, and the rotation angle increases almost linearly with the thickness of the rotator in a given wavelength (typically 40 nm) range. The temperature and wavelength dependence of the Faraday rotation angle of the typical BIG crystals at wavelength of 1550 nm is 0.04–0.07 deg/°C and 0.04–0.06 deg/nm, respectively.

4.2.2. PROPAGATION IN BIREFRINGENT CRYSTALS

Another common material used in the construction of optical circulators is the birefringent crystal. Birefringent crystals used in optical circulators are typically anisotropic uniaxial crystals (having two refractive indices with one optical axis). In an anisotropic medium, the phase velocity of the light depends on the direction of the propagation in the medium and the polarization state of the light. Therefore, depending on the polarization state of the light beam and the relative orientation of the crystal, the polarization of the beam can be changed or the beam can be split into two beams with orthogonal polarization states.

The refractive index ellipsoid for a uniaxial crystal is shown in Fig. 4.1. When the direction of the propagation is along the z-axis (optic axis), the

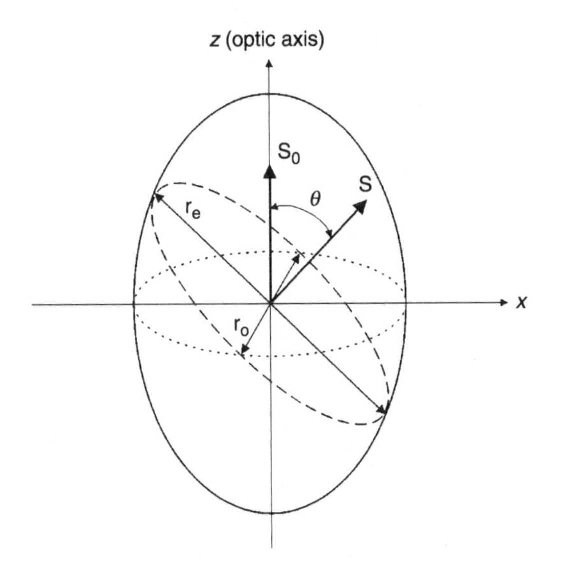

Fig. 4.1 Refractive index ellipsoid of a uniaxial crystal.

intersection of the plane through the origin and normal to the propagation direction S_0 is a circle; therefore, the refractive index is a constant and independent of the polarization of the light. When the direction of the propagation S forms an angle θ with the optic axis, the intersection of the plane through the origin and normal to S becomes an ellipse. In this case, for the light with the polarization direction perpendicular to the plane defined by the optic axis and S, the refractive index, is called the ordinary refractive index n_o, is given by the radius r_o and independent of the angle θ. This light is called ordinary ray and it propagates in the birefringent material as if in an isotropic medium and follows the Snell's law at the boundary. On the other hand, for light with the polarization direction along the plane defined by the optic axis and S, the refractive index is determined by the radius r_e and varies with the angle θ. This light is called the extraordinary ray and the corresponding refractive index is called the extraordinary refractive index n_e. In this case n_e is a function of θ and can be expressed as [6]

$$\frac{1}{n_e^2(\theta)} = \frac{\cos^2\theta}{n_o^2} + \frac{\sin^2\theta}{n_e^2} \tag{4.2}$$

The n_e varies from n_o to n_e depending on the direction of propagation. A birefringent crystal with $n_o < n_e$ is called a positive crystal, and one with $n_o > n_e$ is called a negative crystal.

Therefore, the function of a birefringent crystal depends on its optic axis orientation (crystal cutting) and the direction of the propagation of a light. Birefringent crystals commonly used in optical circulators are quartz, rutile, calcite, and YVO_4.

4.2.3. *WAVEPLATES*

One of the applications of the birefringent crystal is the waveplate (also called retardation plate). A waveplate can be made by cutting a birefringent crystal to a particular orientation such that the optic axis of the crystal is in the incident plane and is parallel to the crystal boundary (zx-plane in Fig. 4.1). When a plane wave is perpendicularly incident onto the incident plane (zx-plane), the refractive index for the polarization component parallel to the x-axis equals n_o and that parallel to the z-axis equals n_e. Therefore, when a linearly polarized light with the polarization direction parallel to the z- or x-axis is incident to the waveplate, the light beam experiences no effect of the waveplate except for the propagation time delay due to the refractive index. However, when the polarization direction of the incident light is at an angle to the optic axis, the components parallel to the x- and z-axes travel at difference velocities due to the refractive index difference. Therefore, after passing through the waveplate, a phase difference exists between these two components, and the resulting polarization of the output beam depends on the phase difference. The phase difference can be expressed as

$$\delta = \frac{2\pi}{\lambda} \Delta n t \tag{4.3}$$

where δ is the wavelength of the light, Δn the refractive index difference between the ordinary and extraordinary refractive indices, and t the thickness of the crystal. When the thickness of the crystal is selected such that the phase difference equals to $\mathrm{m} \cdot (\pi/2)$ (quarter of the wave), the waveplate is called a quarter-waveplate, and similarly the phase difference in a half-waveplate is $\mathrm{m} \cdot \pi$ (where m is called the order of the waveplate, and is an integer and odd number).

The quarter-waveplate is best known for converting a linearly polarized light into a circularly polarized light or vice versa, when a light beam is passed through the quarter-waveplate with the polarization direction at 45° to the optic axis (Fig. 4.2(a)). The half-waveplate is used most frequently to rotate the polarization direction of a linearly polarized light.

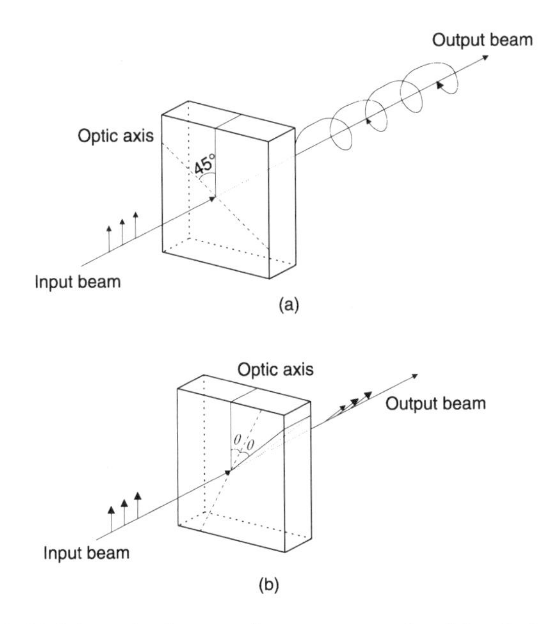

Fig. 4.2 (a) & (b) Schematic diagram of a quarter-waveplate (a) and a half-waveplate (b).

When a linearly polarized light beam is launched into a half-waveplate with an angle θ against the optic axis of the waveplate, the polarization direction of the output beam is rotated and the rotation angle equals to 2θ (Fig. 4.2(b)). Crystal quartz is widely used for making waveplates, due to its small birefringence.

4.2.4. *BEAM DISPLACER*

Another commonly used form of the birefringent crystal is the beam displacer, which is used to split an incoming beam into two beams with orthogonal polarization states, the intensity of each beam dependent on the polarization direction of the incoming beam. The birefringent crystal-based beam displacer is made by cutting a birefringent crystal in a specific orientation such that the optic axis of the crystal is in a plane parallel to the propagation direction and having an angle α to the propagation direction (Fig. 4.3). The separation d between the two output beams depends on the thickness of the crystal and the angle between the optic axis and the propagation direction, and can be expressed as

$$d = \frac{(n_e^2 - n_o^2)\tan\alpha}{n_e^2 + n_o^2 \tan^2\alpha} t \tag{4.4}$$

where t is the thickness of the crystal. The optic axis angle to yield a maximum separation is given as

$$\alpha_{\max} = \tan^{-1}\left(\frac{n_e}{n_o}\right) \tag{4.5}$$

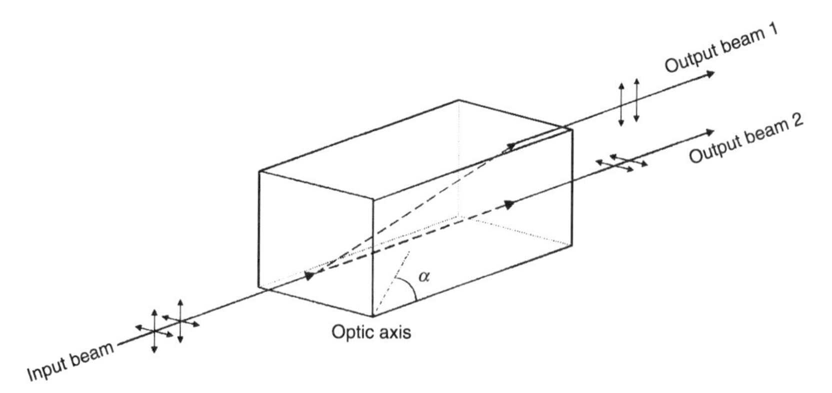

Fig. 4.3 Schematic diagram of a birefringent beam displacer.

Rutile, calcite, and YVO_4 are common birefringent materials for the beam dispacer due to their large birefringence (Δn of more than 0.2 at 1550 nm wavelength). For rutile crystal, the n_e and n_o at a wavelength of 1550 nm are 2.709 and 2.453, respectively, resulting in the α_{max} of 47.8°. For YVO_4 crystal, the n_e and n_o at a wavelength of 1550 nm are 2.149 and 1.945, respectively. Calcite is rarely used in optical circulators due to its softness and instability in a damp heat environment.

4.2.5. OPERATING PRINCIPLE OF OPTICAL CIRCULATORS

Optical circulators can be divided into two categories—the polarization-dependent optical circulator, which is only functional for a light with a particular polarization state, and the polarization-independent optical circulator, which is functional independent of the polarization state of a light. It is known that the state of polarization of a light is not maintained and varies during the propagation in a standard optical fiber due to the birefringence caused by the imperfection of the fiber. Therefore, the majority of optical circulators used in fiber optic communication systems are designed for polarization-independent operation. The polarization-dependent circulators are only used in limited applications such as free-space communications between satellites, and optical sensing.

Optical circulators can be divided into two groups based on their functionality. One is the full circulator, in which light passes through all ports in a complete circle (i.e., light from the last port is transmitted back to the first port). The other is the quasi-circulator, in which light passes through all ports sequentially but light from the last port is lost and cannot be transmitted back to the first port. For example, in the case of a full three-port circulator, light passes through from port 1 to port 2, port 2 to port 3, and port 3 back to port 1. However, in a quasi-three-port circulator, light passes through from port 1 to port 2 and port 2 to port 3, but any light from port 3 is lost and cannot be propagated back to port 1. In most applications only a quasi-circulator is required.

The operation of optical circulators is based on two main principles; polarization splitting and recombining together with nonreciprocal polarization rotation, and asymmetric field conversion with nonreciprocal phase shift.

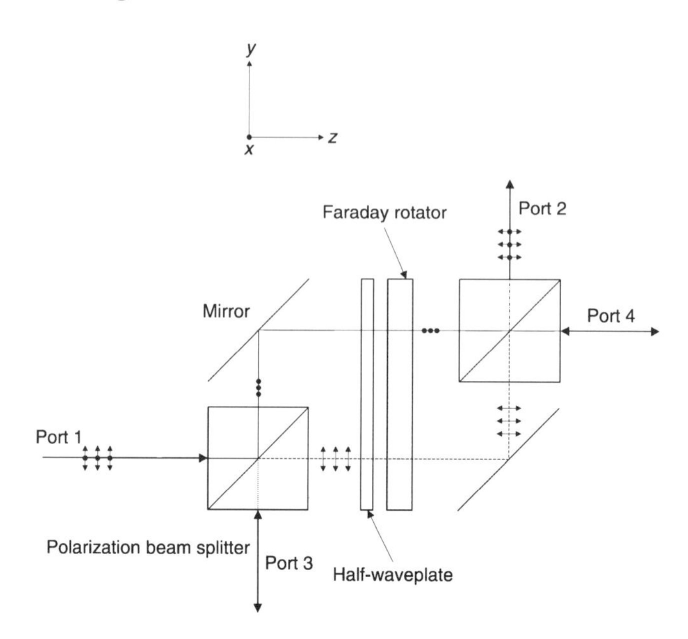

Fig. 4.4 Schematic diagram of polarization beam splitter-based circulator.

4.2.5.1. Operation Principle of Nonreciprocal Polarization Rotation-based Circulators

Dielectric coatings-based polarization beam splitters were used to construct optical circulators in the early stage of circulator development. A schematic diagram of a 4-port circulator is shown in Fig. 4.4, where two dielectric coating-based polarization beam splitter cubes were used to split the incoming beam into two beams with orthogonal polarization.

In operation, a light beam launched into port 1 is split into two beams by the polarization beam splitter that transmits the light with horizontal polarization (along the y-axis) and reflects the light with vertical polarization (along the x-axis). The two beams are then passed through a half-waveplate and a Faraday rotator. The optic axis of the half-waveplate is arranged at 22.5° to the x-axis so that the vertically polarized light is rotated by +45°. The thickness of the Faraday rotator is selected for providing 45°-polarization rotation and the rotation direction is selected to be counter-clockwise when light propagates along the z-axis direction. Therefore, the polarization of the two beams is unchanged after passing through the half-waveplate and Faraday rotator because the polarization rotation introduced by the half-waveplate (+45°) is cancelled by that of the

Faraday rotator (−45°). The two beams are recombined by the second polarization splitter and coupled into port 2.

Similarly, when a light beam is launched into port 2, it is split into two beams with orthogonal polarization by the second polarization beam splitter. Due to the non-reciprocal rotation of the Faraday rotator, in this direction the polarization rotations introduced by both the half-waveplate and Faraday rotator are in the same direction, resulting in a total rotation of 90°. Therefore, the two beams are combined by the first polarization splitter in a direction orthogonal to port 1 and coupled into port 3. The operation from port 3 to port 4 is the same as that from port 1 to port 2.

However, the isolation of this type of optical circulator was relatively low due to limited extinction ratio (around 20 dB) of the polarization beam splitters. Various designs using birefringent crystals have been proposed to increase the isolation by utilizing the high extinction ratio property of the crystal. One of the designs is shown in Fig. 4.5(a), where birefringent beam

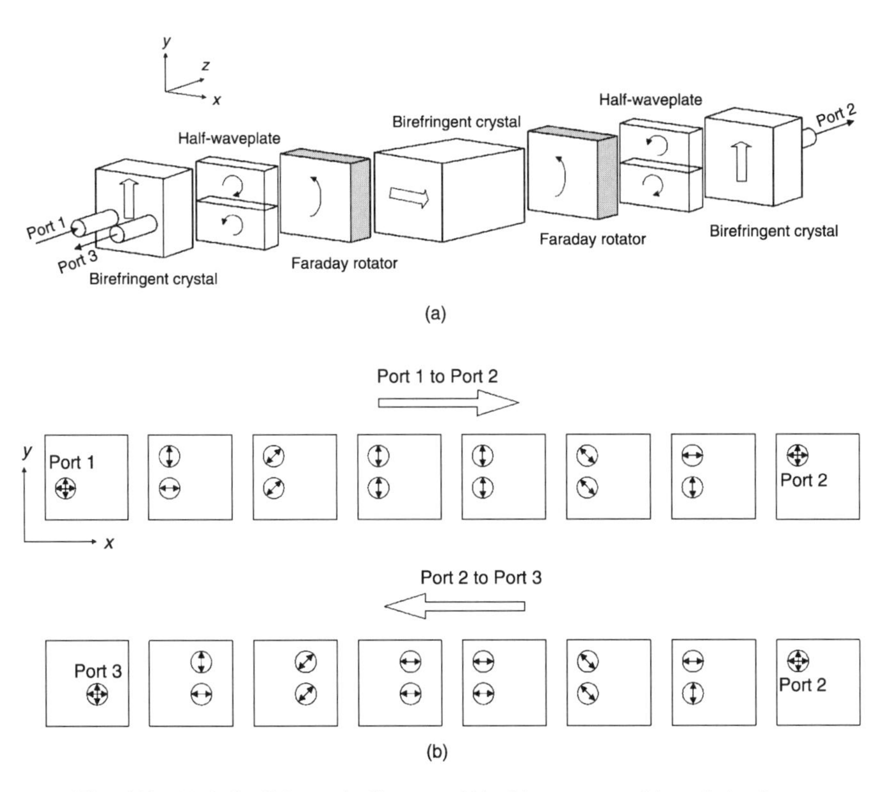

Fig. 4.5 (a) & (b) Schematic diagram of birefringent crystal-based circulator.

dispacers are used for splitting and combining of the orthogonally polarized light beams [7]. As shown in Fig. 4.5(b), where each circle indicates the beam position and the arrow inside the circle indicates the polarization direction of the beam, a light beam launched into port 1 is split into two beams with orthogonal polarization states along the y-axis. Two half-waveplates, one (upper) with its optic axis oriented at 22.5° and the other (lower) at −22.5°, are used to rotate the two beams so that their polarization direction becomes the same. The Faraday rotator rotates the polarization of both beams 45° counter-clockwise, and the two beams are vertically polarized (along the y-axis) and passed through the second birefringent crystal without any spatial position change because the polarization directions of the two beams match the ordinary ray direction of the crystal. After passing through another Faraday rotator and half-waveplate set, the two beams are recombined by the third birefringent crystal, which is identical to the first one. Similarly, a light beam launched into port 2 is split into two beams and passed through the half-waveplate set and the Faraday rotator. Due to the nonreciprocal rotation of the Faraday rotator, the two beams become horizontally polarized (along the x-axis), and are therefore spatially shifted along the x-axis by the second birefringent crystal because they match the extraordinary ray direction of the crystal. The two beams are recombined by the first crystal at a location different from port 1 after passing through the Faraday rotator and half-waveplates. The distance between port 1 and port 3 is determined by the length of the second birefringent crystal.

The use of the birefringent crystals generally results in an increase in size and cost of the circulator due to the cost of crystal fabrication. Extensive development efforts have been concentrated on improvement of various designs. Due to the performance advantages, currently all commercially deployed optical circulators are based on the use of birefringent crystals.

4.2.5.2. Operating Principle of Asymmetric Field Conversion-based Circulators

An optical circulator can be constructed using two-beam interference with nonreciprocal phase shifting [8] without the need for polarization beam splitting. One example of this kind of optical circulator is shown in Fig. 4.6, where a four-port circulator is constructed using two power splitters and nonreciprocal phase shifters. In operation, a light beam launched into port 1 is split into two beams with equal intensity by the first power splitter. The two beams are then passed through two sets of phase-shifting elements

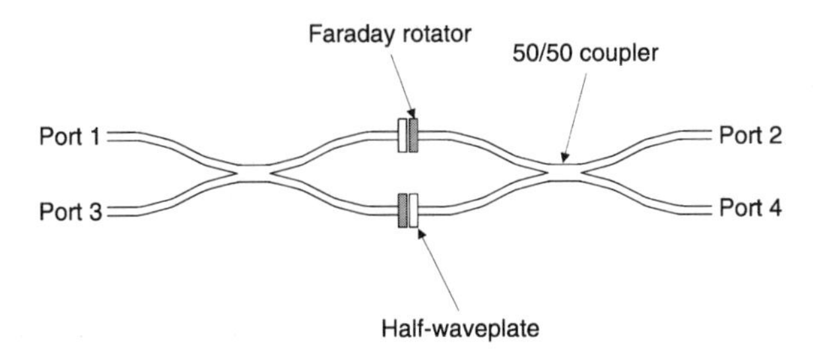

Fig. 4.6 Schematic diagram of asymmetric field conversion-based circulator.

(half-waveplate and Faraday rotator) that are selected such that they provide no phase shift between the two beams in one direction, but in the reverse direction a phase shift of π is introduced between the two beams. Therefore, from port 1 to port 2 the two beams are in phase and will be constructively recombined by the second splitter and coupled into port 2. Similarly, a light beam launched into port 2 is split into two beams by the second splitter, and passed through the phase shifter set. Because a phase shift of π is introduced this time, the two beams are out of phase and no longer will be coupled into port 1 but will be coupled into port 3 due to the out-of-phase relation between port 1 and port 3. The structure of this type of circulator is very simple and potentially could lead to lower cost. However, because phase information is used for the circulator function, control of the phase in each element and control of path length difference between beams are very critical for the performance. Currently, circulators based on this principle have only been investigated in waveguide devices and no commercial products are available due to manufacturing challenges and performance disadvantages.

4.3. Analysis of Optical Circulators

Important performance parameters of optical circulators include insertion loss, isolation, polarization-dependent loss (PDL), directivity, polarization mode dispersion (PMD), and return loss. Optical performance of polarization beam splitting-based optical circulators can be theoretically calculated using a Jones matrix. Jones matrices for various components used in the circulator are summarized in Table 4.1. The circulator shown in Fig. 4.5 can be analyzed by using these Jones matrices.

When a light beam is launched into port 1, the output from port 2 is the sum of the ordinary and extraordinary beams from the third birefringent

Table 4.1 **Jones Matrix for Various Optical Elements**

Optical Elements	*Jones Matrix*	*Notes*
Linear polarizer	$P = \begin{pmatrix} \cos^2\theta & \sin\theta\cos\theta \\ \sin\theta\cos\theta & \sin^2\theta \end{pmatrix}$	θ: azimuth angle of polarizer
Half-waveplate	$H = \begin{pmatrix} \cos(2\rho) & \sin(2\rho) \\ \sin(2\rho) & -\cos(2\rho) \end{pmatrix}$	ρ: azimuth angle of optical axis
Quarter-waveplate	$Q = \begin{pmatrix} e^{i\pi/4}\cos^2\rho + e^{-i\pi/4}\sin^2\rho & \sqrt{2}i\sin\rho\cos\rho \\ \sqrt{2}i\sin\rho\cos\rho & e^{-i\pi/4}\cos^2\rho + e^{i\pi/4}\sin^2\rho \end{pmatrix}$	ρ: azimuth angle of optical axis
General waveplate	$W = \begin{pmatrix} e^{i\delta/2}\cos^2\rho + e^{-i\delta/2}\sin^2\rho & 2i\sin\rho\cos\rho\sin(\delta/2) \\ 2i\sin\rho\cos\rho\sin(\delta/2) & e^{-i\delta/2}\cos^2\rho + e^{i\delta/2}\sin^2\rho \end{pmatrix}$	δ: retardence ρ: azimuth angle of optical axis
Faraday rotator	$F = \begin{pmatrix} \cos\beta & -\sin\beta \\ \sin\beta & \cos\beta \end{pmatrix}\begin{pmatrix} 1 & -ie_F \\ ie_F & 1 \end{pmatrix}$	β: rotation angle of Faraday rotator e_F: extinction ration of Faraday rotator

crystal and can be expressed as

$$E_{\text{out}} = E_{2\text{o}} + E_{2\text{e}} \tag{4.6}$$

where $E_{2\text{o}}$ and $E_{2\text{e}}$ can be expressed as

$$E_{2\text{e}} = \begin{pmatrix} e_\text{c} & 0 \\ 0 & 1 \end{pmatrix}\begin{pmatrix} \cos(-2\rho) & \sin(-2\rho) \\ \sin(-2\rho) & -\cos(-2\rho) \end{pmatrix}\begin{pmatrix} \cos(-\beta) & -\sin(-\beta) \\ \sin(-\beta) & \cos(-\beta) \end{pmatrix}\begin{pmatrix} e_\text{c} & 0 \\ 0 & 1 \end{pmatrix}$$
$$\bullet\begin{pmatrix} \cos(-\beta) & -\sin(-\beta) \\ \sin(-\beta) & \cos(-\beta) \end{pmatrix}\begin{pmatrix} \cos(-2\rho) & \sin(-2\rho) \\ \sin(-2\rho) & -\cos(-2\rho) \end{pmatrix}\begin{pmatrix} 1 & 0 \\ 0 & e_\text{c} \end{pmatrix}E_{\text{in}} \tag{4.7}$$

$$E_{2\text{o}} = \begin{pmatrix} 1 & 0 \\ 0 & e_\text{c} \end{pmatrix}\begin{pmatrix} \cos(2\rho) & \sin(2\rho) \\ \sin(2\rho) & -\cos(2\rho) \end{pmatrix}\begin{pmatrix} \cos(-\beta) & -\sin(-\beta) \\ \sin(-\beta) & \cos(-\beta) \end{pmatrix}\begin{pmatrix} e_\text{c} & 0 \\ 0 & 1 \end{pmatrix}$$
$$\bullet\begin{pmatrix} \cos(-\beta) & -\sin(-\beta) \\ \sin(-\beta) & \cos(-\beta) \end{pmatrix}\begin{pmatrix} \cos(2\rho) & \sin(2\rho) \\ \sin(2\rho) & -\cos(2\rho) \end{pmatrix}\begin{pmatrix} e_\text{c} & 0 \\ 0 & 1 \end{pmatrix}E_{\text{in}} \tag{4.8}$$

where e_c is the extinction ratio of the birefringent crystals, ρ and β are azimuth angle of the optic axis of the half-waveplates and rotation angle of the Faraday rotators, respectively. Signs of ρ and β represent the rotation direction and a minus sign indicates the counter-clockwise rotation. E_{in} is the Jones vector of the input beam, and for a randomly polarized light, E_{in} can be expressed as

$$E_{in} = \begin{pmatrix} aE_0 \\ bE_0 \end{pmatrix} \tag{4.9}$$

where E_0 is the amplitude of the input electric field and $a^2 + b^2 = 1$. Therefore, the insertion loss from port 1 to port 2 of the circulator can be calculated as

$$\text{Loss} = -10 \log \frac{E_{out} \cdot E_{out}^*}{E_{in} \cdot E_{in}^*} \tag{4.10}$$

where E_{in}^* and E_{out}^* are the conjugate matrices of E_{in} and E_{out}, respectively.

Similarly, the isolation of the circulator from port 2 to port 1 can be calculated as

$$\text{Iso} = -10 \log \frac{E'_{out} \cdot E'^*_{out}}{E'_{in} \cdot E'^*_{in}} \tag{4.11}$$

where E'_{out} is the output from port 2 when an input light of E'_{in} is launched into port 1 and $E'_{out} = E'_{1o} + E'_{1e} \cdot E'_{1o}$ and E'_{1e} can be expressed as

$$E'_{1e} = \begin{pmatrix} e_c & 0 \\ 0 & 1 \end{pmatrix} \begin{pmatrix} \cos(2\rho) & \sin(2\rho) \\ \sin(2\rho) & -\cos(2\rho) \end{pmatrix} \begin{pmatrix} \cos(-\beta) & -\sin(-\beta) \\ \sin(-\beta) & \cos(-\beta) \end{pmatrix} \begin{pmatrix} e_c & 0 \\ 0 & 1 \end{pmatrix}$$
$$\bullet \begin{pmatrix} \cos(-\beta) & -\sin(-\beta) \\ \sin(-\beta) & \cos(-\beta) \end{pmatrix} \begin{pmatrix} \cos(2\rho) & \sin(2\rho) \\ \sin(2\rho) & -\cos(2\rho) \end{pmatrix} \begin{pmatrix} 1 & 0 \\ 0 & e_c \end{pmatrix} E'_{in} \tag{4.12}$$

$$E'_{1o} = \begin{pmatrix} 1 & 0 \\ 0 & e_c \end{pmatrix} \begin{pmatrix} \cos(-2\rho) & \sin(-2\rho) \\ \sin(-2\rho) & -\cos(-2\rho) \end{pmatrix} \begin{pmatrix} \cos(-\beta) & -\sin(-\beta) \\ \sin(-\beta) & \cos(-\beta) \end{pmatrix} \begin{pmatrix} e_c & 0 \\ 0 & 1 \end{pmatrix}$$
$$\bullet \begin{pmatrix} \cos(-\beta) & -\sin(-\beta) \\ \sin(-\beta) & \cos(-\beta) \end{pmatrix} \begin{pmatrix} \cos(-2\rho) & \sin(-2\rho) \\ \sin(-2\rho) & -\cos(-2\rho) \end{pmatrix} \begin{pmatrix} e_c & 0 \\ 0 & 1 \end{pmatrix} E'_{in} \tag{4.13}$$

Effects of various parameters on the insertion loss and isolation can be calculated using Eqs. (4.6) to (4.13). For example, the effect of the rotation angle error of the Faraday rotators on insertion loss and isolation can be obtained using Eqs. (4.7) to (4.13), assuming $e_c = 0$ and $\rho = 22.5°$, and are given as

$$\text{Loss} = (\cos(45 + \Delta\beta) + \sin(45 + \Delta\beta))^4/4 \tag{4.14}$$

$$\text{Iso} = (\cos(45 + \Delta\beta) - \sin(45 + \Delta\beta))^4/4 \tag{4.15}$$

where $\Delta\beta$ is the rotation angle error of the Faraday rotator. The effect of retardence error can also be calculated by replacing the Jones matrix for the half-waveplate with that for the general waveplate. Temperature and wavelength behaviors of the circulator can be estimated by incorporating the temperature and wavelength dependences of the Faraday rotator and waveplates into the corresponding Jones matrices.

Although performances of optical circulators depend on the particular design and materials, typically an insertion loss of 1 dB, an isolation of more than 40 dB, a PDL of less than 0.1 dB, and a PMD of less than 0.1 ps are achievable regardless of the design. The main differences among various designs are performance stability, size, and cost.

4.4. Designs of Optical Circulators

Cost and stability have been the main limiting factors in expanding the applications of optical circulators. Recently, several designs [9–11] have been developed in an effort to reduce the cost and realize high reliability. In the design shown in Fig. 4.5, the circulator is used in a collimated beam and each port is collimated using a lens; therefore, relatively large size elements have to be used in order to construct the design due to the beam size. In recent designs, efforts have been concentrated on reducing the use of materials and size.

4.4.1. OPTICAL CIRCULATOR DESIGN USING DIVERGING BEAM

A compact low-cost circulator design has been proposed, placing optical elements in a diverging beam instead of in a collimating beam to reduce the overall use of expensive materials. As shown in Fig. 4.7, in this design, all optical elements are placed in a diverging beam between the input/output

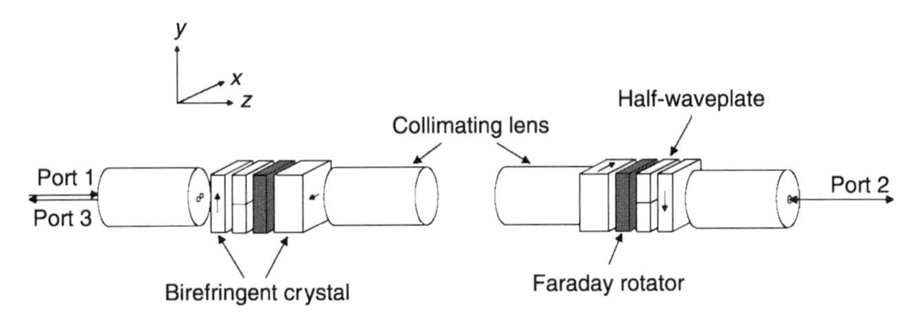

Fig. 4.7 Schematic diagram of a compact circulator based on diverging beam shifting.

ports and lenses. Two identical groups of elements are placed near the focal point of the lens, resulting in reduced size and manufacturing complexity. Each group of elements consists of two birefringent crystals, one Faraday rotator with 45° rotation angle, and two half-waveplates with their optic axes oriented in opposite directions (22.5° and −22.5°).

In operation, a light beam from port 1 is split into two orthogonally polarized beams in the y-axis by the first birefringent crystal. The two half-waveplates and the Faraday rotator are arranged such that after passing through the rotators the polarization directions of the two beams are the same and match the ordinary ray direction of the second birefringent crystal. Therefore, the two beams pass through the second birefringent crystal without any displacement. Two lenses are used for providing a one-to-one imaging system. Because the second group of the element is the same as the first one, the two beams are recombined and launched into port 2. Similarly, a light beam launched into port 2 is split and passed through to the rotators. Due to the nonreciprocal rotation of the Faraday rotator, the polarization directions of the two beams are rotated matching the extraordinary ray direction of the second birefringent crystal. Therefore, the two beams are shifted a certain amount along the x-axis and shifted again the same amount by the second birefringent crystal in the other group. If the sum of beam shifting by the two birefringent crystals is designed such that it is the same as the distance between the first and third ports, the two beams will be recombined and coupled into port 3. Because port 1 and port 3 share a single lens and the beam shifting is done at the diverging beam, the required beam shifting in this case is very small and typically equal to the fiber diameter of 125 μm. On the other hand, the required beam shifting in the design shown in Fig. 4.5 is determined by the diameter of a lens due to the use of collimated beams and is typically in the order of millimeters. To further reduce the required thickness of the birefringent crystal, mode-field

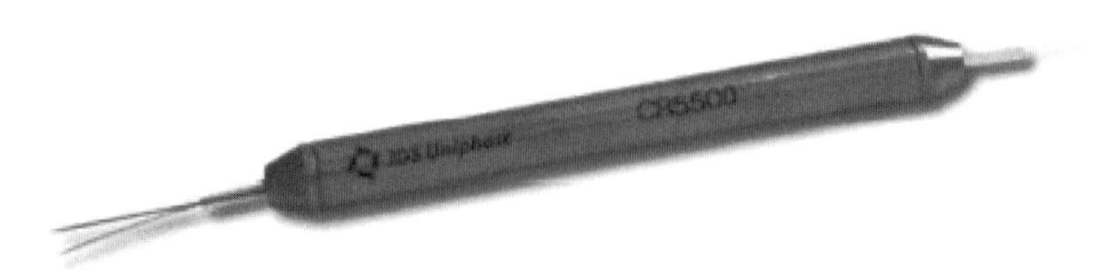

Fig. 4.8 Photograph of a packaged compact circulator using diverging beam shifting.

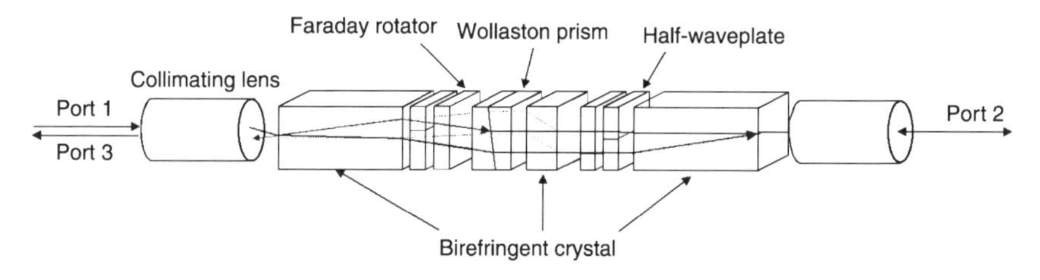

Fig. 4.9 Schematic diagram of a compact circulator based on collimated beam deflection.

diameter of the input and output fiber is expanded to reduce the divergence angle of the beam. With this compact design, a circulator with a size of 5.5 mm in diameter and less than 60 mm in length has been developed, as shown in Fig. 4.8, compared to a typical size of over 25 mm in cross-section and over 90 mm in length for the design shown in Fig. 4.5.

4.4.2. OPTICAL CIRCULATOR DESIGN USING BEAM DEFLECTION

A compact circulator using collimated beam deflection is also proposed and demonstrated. In the design, polarization-dependent angle deflection is used instead of the polarization-dependent position shift used in the design of Fig. 4.5. As shown in Fig. 4.9, a single lens is used to collimate the light for both port 1 and 3 and all elements of the circulator are positioned in the collimated beam. The main difference between the design shown in Fig. 4.5 and this design is that a Wollaston prism is used in place of a birefringent beam displacer and a single lens is used for collimating two beams.

In operation, a light beam launched into port 1 is collimated and split into two beams with orthogonal polarization by the first birefringent crystal. The polarization directions of the two beams are rotated by the half-waveplates and Faraday rotator so that they become the same. Because port 1 is off-axis of the lens, the resulting collimated beam from the lens forms an angle θ to the propagation axis. This angle is corrected by the Wollaston prism and the two beams are propagated straight to the second Faraday rotator (solid lines in Fig. 4.9). After passing through the half-waveplates and being recombined by the third birefringent crystal, the combined beam is focused by the second lens into port 2. Similarly, light launched into port 2 is collimated and split into two beams with their polarization direction rotated. Due to the nonreciprocal rotation of the Faraday rotator, the two beams from port 2 are deflected to a direction opposite to the angle θ by the Wollaston prism (dotted lines in Fig. 4.9). Therefore, after passing through the polarization rotators and the first birefringent crystal, the combined beam is focused by the first lens to a position different from that of port 1. The required deflecting angle of the Wollaston prism can be determined by the position distance between port 1 and port 3 and the focal length of the lens. This design reduces the size of materials considerably. However, because the beam splitting and recombining is still performed in the collimated beam, it still requires relatively long crystals compared to the design shown in Fig. 4.7.

4.4.3. REFLECTIVE OPTICAL CIRCULATOR

As shown in design examples described in this section, most optical circulators have a symmetric structure in terms of element materials and their relative positions. Therefore, a proposed design concept using imaging folding to redirect the light beam and reuse the common elements has advantages in reducing the overall device size and cost [4, 10]. A schematic diagram of one of the compact reflective circulator designs is shown in Fig. 4.10, where a single lens and a mirror are used to couple lights between all ports that are at the same side of the circulator. In this design, all elements are passed through twice to reduce the element account to half while maintaining the same performance as a conventional circulator.

In operation, a light beam launched into port 1 is split into two beams by the first birefringent crystal, and passed through the second crystal without any lateral position change, because the rotation angles of the polarization rotators ($+45^\circ$ or -45° rotation) are designed such that the polarization directions of the two beams match the ordinary ray direction of the second birefringent crystal. After being collimated by the lens and reflected by the mirror, the two beams are passed through the same elements again except for half-waveplates and recombined into port 2.

Similarly, a light beam launched into port 2 is split into two beams with orthogonal polarization directions. After passing through the polarization

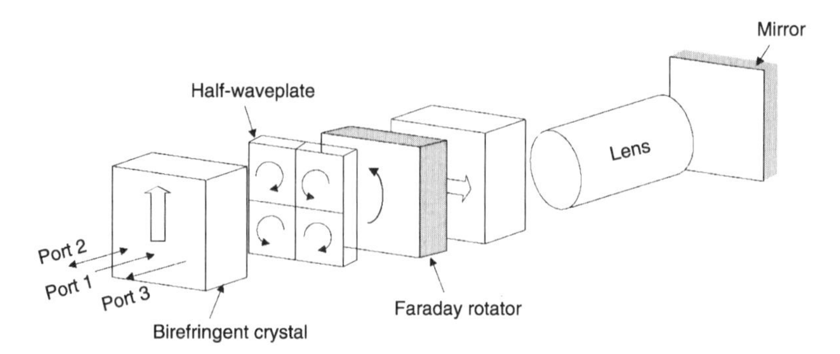

Fig. 4.10 Schematic diagram of a compact reflective circulator.

rotators, the polarization directions of both beams are aligned with the extraordinary ray direction of the second birefringent crystal due to the nonreciprocal rotation of the Faraday rotator, and the physical locations of the two beams are shifted after passing through the crystal. The two beams receive the location shift again after being reflected by the mirror and passed through the crystal. Therefore, after the proper polarization rotation the two beams are recombined at a location different from port 1 and will be coupled into port 3 if the distance between port 1 and port 3 matches two times the beam shift introduced by the second birefringent crystal. Multi-port circulators can be made by adding more ports into the design. With the reflective design, the size and required optical elements can be significantly reduced, resulting in overall cost savings.

There are many variations in the circulator design, however, all nonreciprocal polarization rotation-based designs share a common structure with a minimum of three functional elements; polarization splitting and recombining elements, nonreciprocal polarization rotation elements, and polarization-dependent beam steering (angular or positional) elements.

4.5. Applications of Optical Circulators

Optical circulators were originally used in telecommunication systems for increasing transmission capacity of existing networks. By using optical circulators in a bi-directional transmission system, the transmission capacity of the network can be easily doubled without the need for deploying additional fibers, which has become increasingly expensive. However, with the rapid advancement in optical communication technologies and the ready

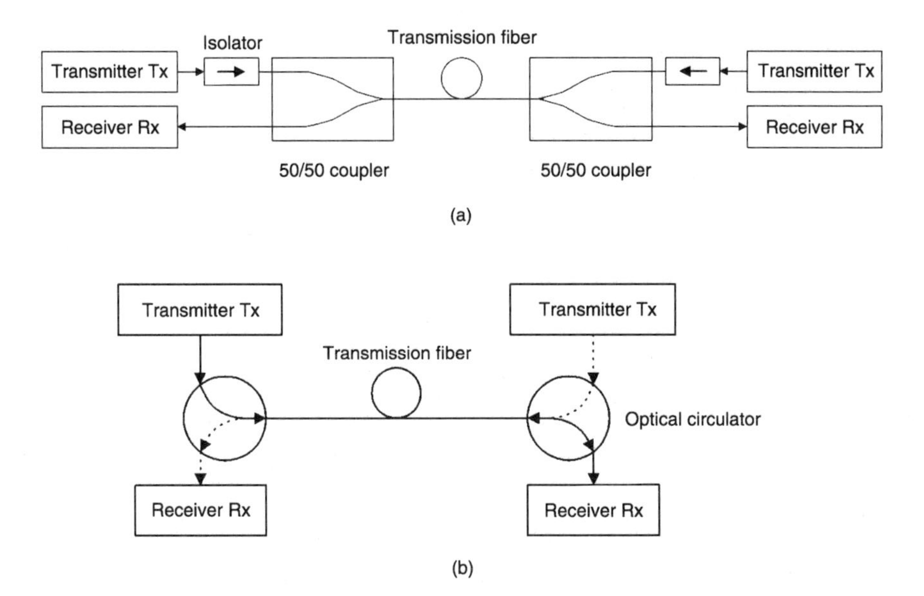

Fig. 4.11 (a) & (b) Schematic diagram of bi-directional transmission system (a) configuration with traditional 50/50 couplers, (b) configuration using optical circulators.

availability of low-cost and high-performance circulators, the applications of optical circulators have drastically expanded into not only the telecommunication industries but also the sensing and imaging fields. Optical circulators have become an especially important element in advanced optical networks such as DWDM networks.

In the traditional bi-directional optical communication system, a 50/50 (3 dB) coupler, which splits a light beam into two beams with equal intensity, was used to couple the transmitters and receivers as shown in Fig. 4.11(a). However, there are two main problems with this kind of structure. One is the need for an optical isolator in the transmitters to prevent light crosstalk between the transmitters, and the other is the high insertion loss associated with the use of the 50/50 coupler, because two couplers have to be used and each has a minimum loss of 3 dB, which results in a minimum 6 dB reduction of the link budget from the system. The use of an optical circulator can solve both of the problems by providing the isolation function as well as a loss of less than 3 dB (Fig. 4.11(b)).

Optical circulators are powerful devices for extracting optical signals from a reflective device. Therefore, optical circulators are often used in conjunction with the fiber Bragg gratings that are typically reflective devices.

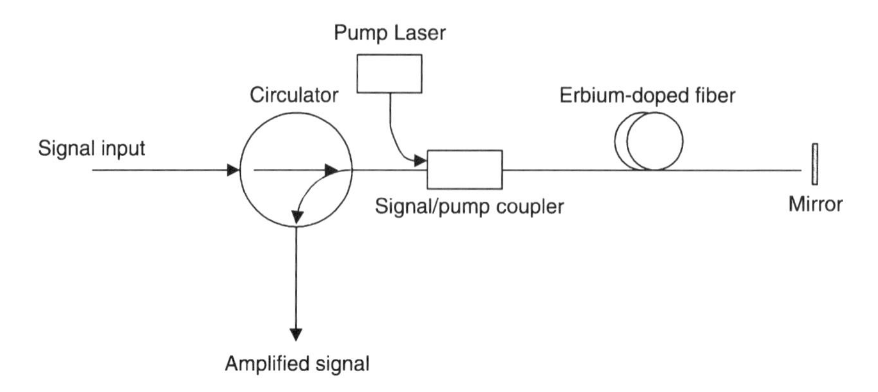

Fig. 4.12 Schematic diagram of a reflective erbium-doped fiber amplifier using optical circulator.

As described in Chapter 3, together with fiber Bragg gratings, optical circulators have become one of the dispensable elements in advanced DWDM optical networks. Circulators are used as MUX/DEMUX devices, but are also used with the fiber Bragg grating in dispersion compensation [12], tunable optical Add/Drop [13], and other applications.

Another application of the circulators is use with a mirror for double passing an optical element to increase efficiency. One example is the reflective erbium-doped fiber amplifier shown in Fig. 4.12 [14]. In operation, signal light is launched into port 1 of a circulator and passed through port 2 with minimum loss. The signal is combined with the pump light from a pump laser by a WDM coupler, and both lights are launched into an erbium-doped fiber. The amplified signal and residual pump lights are reflected by the mirror and passed through the erbium-doped fiber again so that the signal is amplified twice by the erbium-doped fiber, reducing the required length of the fiber, and the residual pump power is also re-used to increase the pump efficiency. The idea has been adapted into different devices, such as replacing the coupler and erbium-doped fiber with a dispersion compensation fiber to reduce the required fiber length and adding a Faraday rotator between the mirror and fiber to reduce the polarization-induced effects [15]. Bi-directional fiber amplifiers are also proposed for taking full advantage of the circulator [16].

With the development of advanced optical networks, applications of optical circulators are expanding rapidly and new functionality and applications are emerging quickly. For example, recently it has been reported that by adding wavelength-selective functions into circulators, a bi-directional wavelength-dependent circulator can be configured, which opens a new dimension of applications in advanced DWDM optical networks [17].

References

1. A. Shibukawa and M. Kobayashi, "Compact optical circulator for optical fiber transmission," Appl. Opt., 18 (1979) 3700–3703.
2. T. Matsumoto and K. Sato, "Polarization-independent optical circulator: an experiment," Appl. Opt., 19 (1980) 108–112.
3. M. Shirasaki, H. Kuwahara, and T. Obokata, "Compact polarization-independent optical circulator," Appl. Opt., 20 (1981) 2683–2687.
4. Y. Cheng and G. Duck, "High performance optical circulators for fiber optical communication applications," OPTO95, Paris (1995).
5. Y. Cheng, "Nonreciprocal and hybrid passive components," OFC2000, Baltimore, WF1-1 (2000).
6. M. Born and E. Wolf, Principles of Optics. (Pergamon Press, New York, 1980).
7. M. Koga and T. Matsumoto, "High-isolation polarization-insensitive optical circulator for advanced optical communication systems," J. Lightwave Technol., 10 (1992) 1210–1217.
8. Y. Okamura, T. Negami, and S. Yamamoto, "Integrated optical isolator and circulator using nonreciprocal phase shifters: a proposal," Appl. Opt., 23 (1984) 1886–1889.
9. Y. Cheng, "Optical circulator," US Patent #5991076 (1999).
10. Y. Cheng, "Optical circulator," US Patent #5930422 (1999).
11. W. Li, V. Au-Yeung, and Q. D. Guo, "Optical circulator," US Patent #5930039 (1999).
12. K. O. Hill, F. Bilodeau, B. Malo, T. Kitagawa, S. Theriault, D. C. Johnson, J. Albert, and K. Takiguchi, "Aperiodic in-fiber bragg gratings for optical fiber dispersion compensation," OFC'94, San Jose, PD17 (1994) 17–20.
13. P. Leishching, H. Bock, A. Richter, C. Glingener, P. Pace, B. Keyworth, J. Philipson, M. Farries, D. Stoll, and G. Fischer, "All-optical-networking at 0.8 Tbit/s using reconfigurable optical add/drop multiplexers," J. Photonics Technol. Lett., 12 (2000) 918–920.
14. S. Nishi, K. Aida, and K. Nakagawa, "Highly efficient configuration of erbium-doped fiber amplifier," ECOC'90, Amsterdam, 1 (1990) 99–102.
15. C. R. Giles, "Suppression of polarisation holeburning-induced gain anisotropy in reflective EDFAs," Electron. Lett., 30 (1994) 976–977.
16. Y. Cheng, N. Kagi, A. Oyobe, and K. Nakamura, "Novel fibre amplifier configuration suitable for bidirectional system," Electron. Lett., 28 (1992) 559–561.
17. T. Ducellier, K. Tai, B. Chang, J. Xie, J. Chen, L. Mao, H. Mao, and J. Wheeldon, "The 'Bidirectional Circulator': an enabling technology for wavelength interleaved bidirectional networks," ECOC2000, Munich, PD3.9 (2000).

Part 2 Optical Switching Technologies

Chapter 5 | Micro-Electro-Mechanical Systems (MEMS) for Optical-Fiber Communications

Lih-Yuan Lin

Tellium Inc., 2 Crescent Place, Oceanpart, NJ 07757-0901, USA

Fuelled by rapid growth in demand for optical network capacity and the sudden maturation of wavelength division multiplexing (WDM) technologies, the globe's long-haul optical networks are transforming themselves into systems that transport tens to hundreds of wavelengths per fiber, with each wavelength modulated at 10 Gb/s or more. Consequently, new ways of surmounting the transmission obstacles associated with hundreds of closely spaced wavelength channels becomes imperative, as does finding new ways of provisioning and restoring network traffic in units at roughly the wavelength level.

These needs have stimulated a storm of technological advancement on various fronts, involving a broad range of network elements such as tunable lasers and filters, dynamic gain-equalizers, reconfigurable wavelength add/drops, tunable dispersion-compensators, and core-transport-network optical crossconnects. Micro-electro-mechanical systems (MEMS), studied since the 1980s, have recently emerged as a powerful means of implementing such functions in compact and low-cost form, owing to the unique capability of this technology to integrate optical, mechanical, and electrical components on a single wafer. Various MEMS components and subsystems for optical-fiber communications—tunable lasers and filters, high-speed optical modulators, reconfigurable wavelength-add/drop multiplexers, dynamically adjustable gain-equalizers, tunable chromatic dispersion-compensators, polarization controllers, and optical crossconnects—have

WDM TECHNOLOGIES: PASSIVE OPTICAL COMPONENTS
$35.00

ISBN: 0-12-225262-4

been demonstrated. Some of the early resulting devices have already moved, on exceptionally short time scales, to the brink of commercial realization.

This chapter focuses on the applications of MEMS in various optical-fiber communication components and subsystems. The chapter starts with an introduction to MEMS fabrication technologies, followed by MEMS for transmission engineering, then finishes with MEMS for optical-layer networking.

5.1. MEMS Fabrication Technologies

Various fabrication technologies for making micromachining structures have been studied. Bulk micromachining creates robust mechanical structures by machining into the semiconductor substrates. LIGA technology, on the other hand, uses electroplating to create high-aspect-ratio structures with high mechanical strength above the semiconductor substrates. Surface micromachining has the advantages of integrating versatile mechanical structures, sensors, and actuators on the same substrate by epitaxial growth of structural and sacrificial layers on the substrate. Lately, silicon-on-insulator (SOI) technologies have attracted vast attention in optical-fiber communications, as they achieve the feature of bulk silicon micromachining that gives a robust, flat, and smooth optical surface, while to certain extent preserving the flexibility of surface micromachining in the integration of actuators and sensors. More recently, integration of bulk micromachining with surface micromachining by wafer bonding has also been proposed to combine the advantages of both.

5.1.1. BULK MICROMACHINING

Since the 1970s, people have realized that silicon, in addition to being the main material for IC fabrication, is also a superior platform for making mechanical structures because of its great mechanical strength, surprisingly close to steel's [1]. Various methods have since been investigated to machine mechanical structures in the silicon substrates. They include, for example, wet chemical etching, dry etching such as reactive ion-beam etching (RIE), laser and focused ion-beam machining.

In wet chemical etching, the etching shapes can be defined by the chemicals used, the crystal plane of the silicon substrate, and the dopants in the silicon. The silicon substrate is first patterned using photolithography with a masking material that has a substantially lower etching rate than silicon.

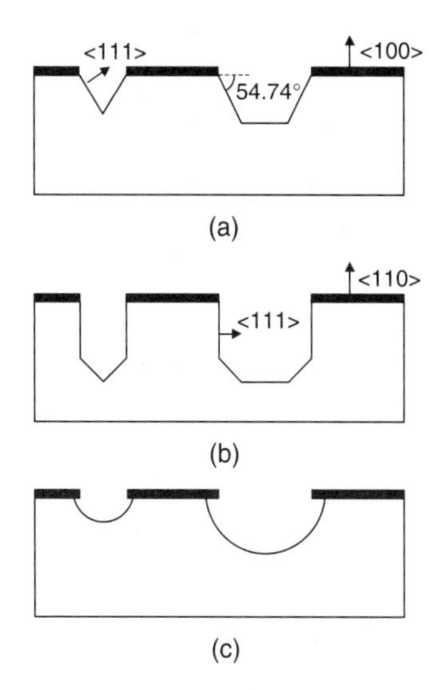

Fig. 5.1 Etching geometries by wet chemical etching in silicon bulk micromachining. (a) Anisotropic etching on <100> substrate. (b) Anisotropic etching on <110> substrate. (c) Isotropic etching.

For example, Table 5.1 summarizes the three major chemical etching processes, their etchants and masking materials, and the dopant dependence [1]. Depending on the crystal plane of the silicon substrate, the etched shape varies. Figure 5.1 illustrates the common etching geometries obtained from different crystal planes in the bulk silicon micromachining processes.

One of the earliest applications of micromachining in optical fiber communications is silicon V-grooves, which mostly utilize anisotropic etching on the <100> silicon substrates, as shown in Fig. 5.2. Using a highly anisotropic etching process, such as the KOH etching shown in Table 5.1, the etching shape of the V-grooves can be precisely defined by the photolithographically patterned masks and etching time. The resultant V-grooves are particularly suitable for highly accurate fibers or fiber arrays alignment to active or passive optoelectronic components.

Deep reactive ion-beam etching (DRIE) has often been employed to create tall structures with a high aspect ratio. Figure 5.3 shows an example of actuated micro-mirrors for 2×2 switches [2]. Photolithographically defined masks are used to pattern the mechanical structures, and the unwanted areas are removed by the ion beam. This process is often done on silicon-on-insulator (SOI) wafers, so that actuatable structures can be created after the SiO_2 layer is removed. (A more elaborate SOI process will be described

Table 5.1 **Typical Bulk Micromachining Etching Process by Chemicals [1]**

Etchant	Mixing Ratio	Temp. (°C)	Etch Rate (μm/min)	Anisotropic <100>:<111> Etch Ratio	Dopant Dependence	Masking Films (Etch Rate of Mask)
HF + HNO_3 + Water or CH_3COOH	1:3:8	22	0.7–3.0	1:1	$\leq 10^{17}$ cm^{-3} n or p reduces etch rate by ~150	SiO_2 (300 Å/min)
	1:2:1	22	40	1:1	None	Si_3N_4
Ethylene diamine + Pyrocatechol + water (EDP)	750 ml:120 gr: 100 ml	115	0.75	35:1	$\geq 7 \times 10^{19}$ cm^{-3} boron reduces etch rate by ~50	SiO_2 (2 Å/min) Si_3N_4 (1 Å/min) Au, Cr, Ag, Cu, Ta
KOH + water or isopropyl	44 gr:100 ml	85	1.4	400:1	$\geq 10^{20}$ cm^{-3} boron reduces etch rate by ~20	Si_3N_4 SiO_2 (14 Å/min)
	50 gr:100 ml	50	1.0	400:1		

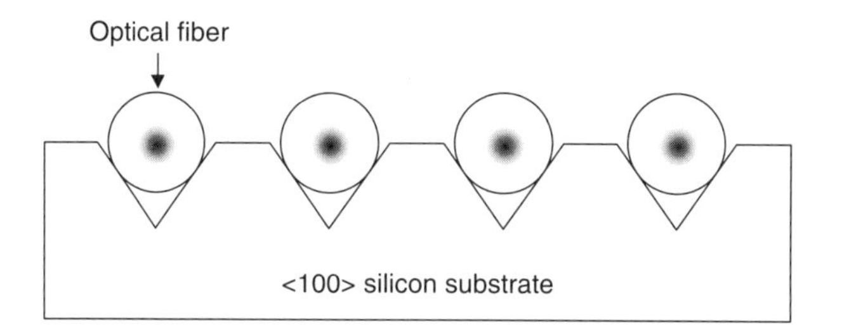

Fig. 5.2 Silicon V-grooves for optical fiber alignment.

Fig. 5.3 An actuated micro-mirror for 2 × 2 switches [2].

in Section 5.1.3.) In laser and focused ion-beam machining, high-energy lasers or ion beams are focused into a tiny spot and mechanical structures are sculptured out of the substrate by these high-energy chisels. The process is normally highly directional, requires no masks, and precise movements need to be coordinated between the high-energy beams and the substrates in order to create complicated mechanical structures.

5.1.2. SURFACE MICROMACHINING

In general, bulk micromachining has the advantage of being able to achieve high-quality mechanical and optical structures, due to its robust fabrication process. However, its simplicity in fabrication also limits its flexibility in achieving versatile mechanical structures. The fabrication process that offers complementary advantages is surface micromachining. In surface micromachining, alternating structural and sacrificial materials are deposited epitaxially and patterned photolithographically on the semiconductor substrate. Vias and connection holes between structural layers are included in the process as needed. The wafer is then subjected to release etching to selectively remove sacrificial materials. After the whole process, complicated movable mechanical structures for sensors or actuators are created. The choice of structural and sacrificial materials is determined by the etchant used in the release etching. The sacrificial material should have a very high etching rate in response to the etchant, while the structural material should have a negligible etching rate in release etching. Commonly used structural/sacrificial materials in surface-micromachined MEMS components for optical-fiber communications are polysilicon/SiO_2, Si_3N_4/SiO_2, GaAs/AlAs, and InP/InGaAs. All of these have very high differential etching rates in hydrofluoric (HF) acid.

Among the earlier demonstrated surface-micromachined structures are micro-motors [3] and comb-drive actuators [4]. Both are good examples of the surface micromachining process. Figure 5.4(a) shows the top view of a micro-wobble motor. It consists of a rotor with a bearing, and stators. The fabrication process is illustrated in Fig. 5.4(b), where the cross-section along the dotted line in Fig. 5.4(a) before release etching is shown. The fabrication process starts with the deposition of a layer of Si_3N_4 for electrical isolation. The first layer of polysilicon (POLY0) is deposited and patterned for electrical shielding. The first sacrificial layer, PSG1, consisting

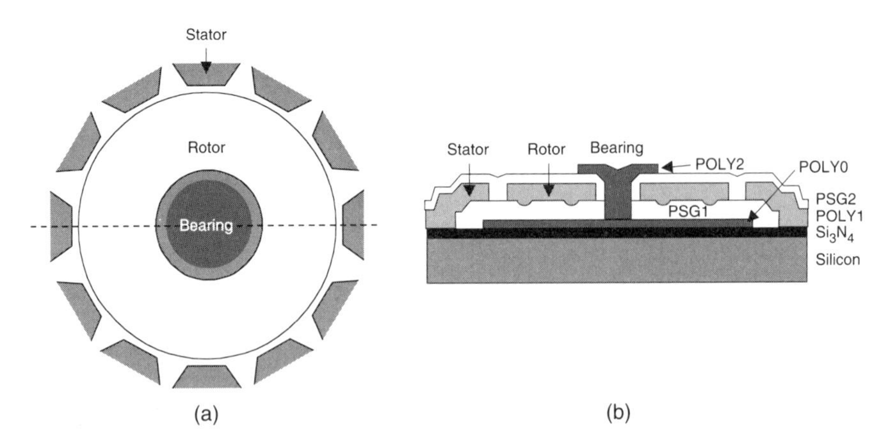

Fig. 5.4 (a) Top view of a micro-wobble motor. (b) Cross-section along the dotted line in (a), showing an example of the surface-micromachining fabrication process.

of phosphosilicate glass (PSG, whose property is similar to SiO_2), is then deposited. The next step is patterning and etching via holes for the first structural polysilicon layer (POLY1), which is deposited and patterned next. Before the deposition of the POLY1 layer, shallow dimple etching in the PSG1 layer is often performed to obtain dimple structures under the polysilicon layer to prevent stiction between the polysilicon layer and the substrate. After the deposition and patterning of the POLY1 layer, the second sacrificial PSG2 layer is deposited. PSG layers normally conform well to the topology of the POLY layers underneath. Before the deposition of the next structural polysilicon layer (POLY2), via holes are created in the PSG2 or both the PSG1 and PSG2 layers. In the former case, the POLY2 layer can be connected to the POLY1 layer. In the latter case, the POLY2 layer can be connected to the POLY0 or Si_3N_4 layer. Similarly, dimples can be created on the bottom of the POLY2 layer by shallow dimple etching in the PSG2 layer before the deposition of POLY2.

The top view of a comb-drive actuator is shown in Fig. 5.5. Both the moving combs and the static combs can be fabricated on either the POLY1 or POLY2 layer, with anchors obtained through PSG via holes. By applying a bias between the moving part and the static part, the combs of the moving part are attracted to the combs of the static part. If this bias is a periodic function with frequency equal to the resonant frequency of the moving part, resonant motion is then achieved. The fabrication process is similar to the process of micro-motors. Because the capacitance between the moving combs and the static combs is a function of the relative position

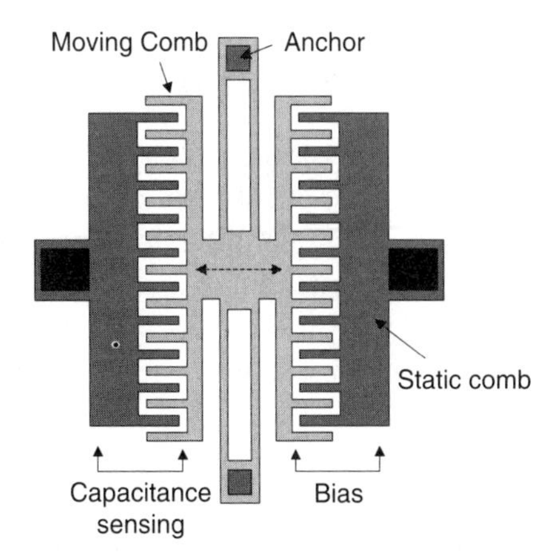

Fig. 5.5 Top view drawing of a comb-drive actuator.

between the combs, such a structure can also be used as a sensor, such as an accelerometer. Lately, vertical comb-drives have been studied for optical scanners [5] and switches. These structures, though bearing a similarity to the surface-micromachined lateral comb-drives, are in general processed by bulk micromachining on SOI wafers (whose general fabrication process will be described in the next section). In these devices, the scanning mirrors are anchored at two ends by torsion beams, as shown in Fig. 5.6(a). A set of comb structures is attached to the torsion beams. The position of these combs is vertically offset from the static combs. The bias between the combs creates vertical motion of the moving combs, and therefore a torque for the micro-mirrors scanning motion. Figure 5.6(b) shows the SEM photograph of a vertical comb-drive actuated scanning mirror.

Most of the surface-micromachined structures are limited to two-dimensional planes due to the in-plane geometry of the thin epitaxial layers. The surface-micromachined micro-hinges [6] allow three-dimensional structures to be created out of in-plane epitaxial growth. Figure 5.7(a) shows a SEM photograph of a micro-hinge. Its fabrication process is shown in Fig. 5.7(b). Similar to the fabrication of micro-motors, the process also consists of two structural polysilicon layers and two sacrificial PSG layers. After the release etching, with POLY2 hinge staples, the polysilicon plate patterned on the POLY1 layer can be rotated out of the substrate, forming a three-dimensional structure. Such an approach has been widely employed to implement three-dimensional integrated micro-optics, such as micro-mirrors, micro-lenses, and micro-gratings, on a silicon substrate, thus creating a "micro-optical bench" [7].

5.1.3. SILICON-ON-INSULATOR (SOI) TECHNOLOGY

Recently, silicon-on-insulator (SOI) technology has attracted vast interest in the community of MEMS for optical-fiber communications, especially for applications that require high-quality optical surfaces. The basic SOI wafer consists of a silicon substrate, a thin layer of silicon dioxide on top of the silicon substrate, and a layer of active silicon on top of the silicon dioxide. Multiple stacks of silicon dioxide and silicon can also be obtained by wafer bonding if desired.

The processing of SOI technology is mostly similar to the bulk micromachining process. One unique feature of SOI technology is the creation of a well-defined "active" top silicon layer (shown in Fig. 5.8) due to the sandwiched silicon dioxide layer, which can function as both an insulating

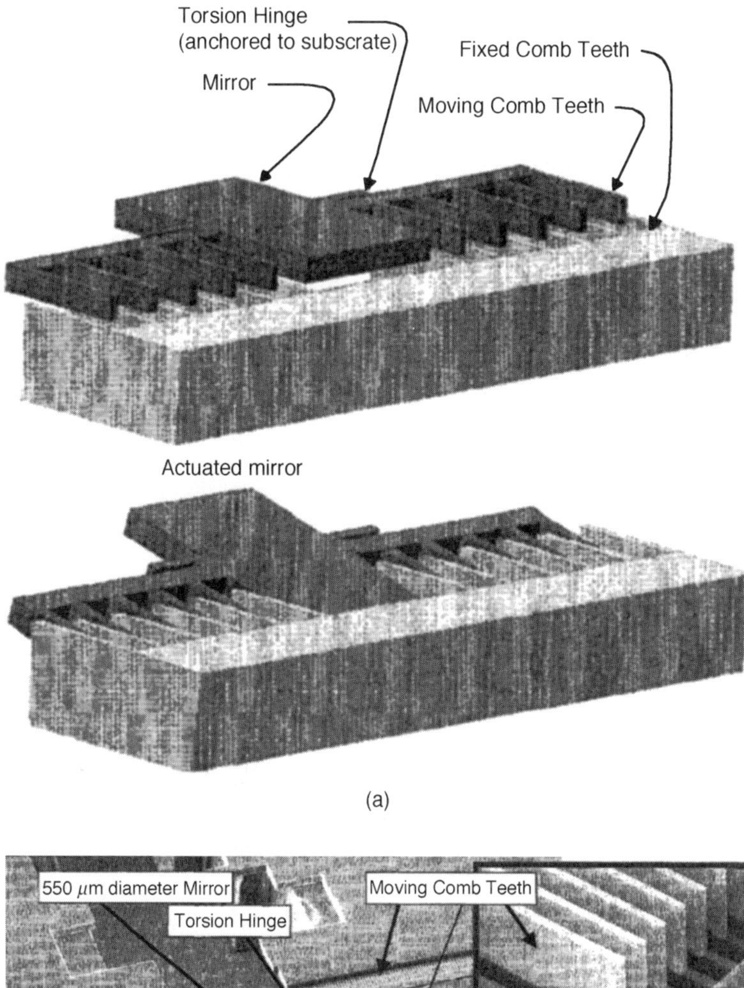

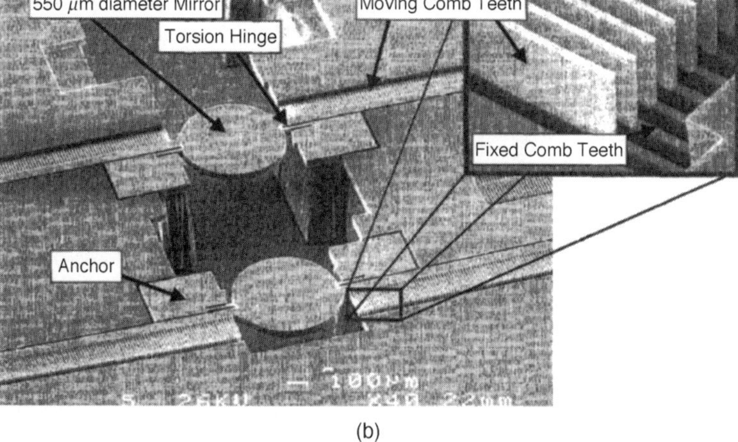

Fig. 5.6 Schematic diagram and SEM photograph of a vertical comb-drive actuated scanning mirror [5].

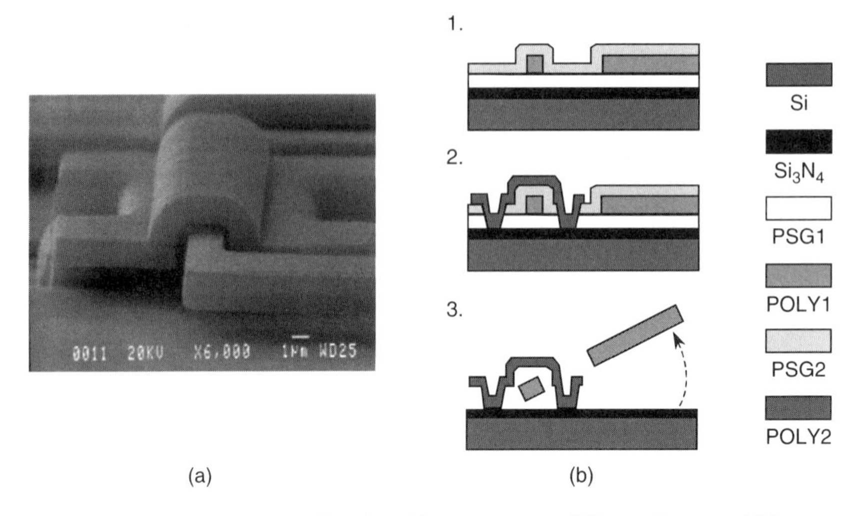

Fig. 5.7 (a) SEM photograph of a micro-hinge (courtesy of Cronos Integrated Microsystems), and (b) its fabrication process.

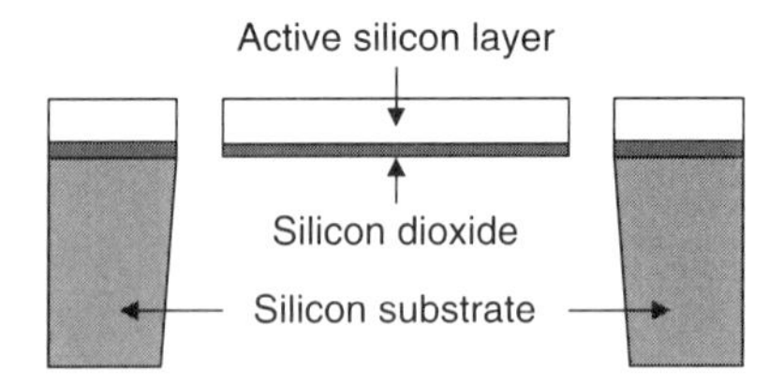

Fig. 5.8 Cross-section of a silicon-on-insulator (SOI) wafer, with part of the top active layer defined by photolithography and bulk micromachining etching. A released structural layer is formed after the bottom silicon material is etched away.

layer and an etching stop. After the bottom silicon material is etched away, the active layer can then function like the released structural layer in surface-micromachining, and be used as a sensor or actuator. Therefore, SOI technology combines the advantage of robustness in bulk micromachining with the benefit of versatility in surface micromachining.

The thickness of the active silicon layer normally ranges from a few microns to a few tens of microns. Its thickness and surface quality can be well controlled by chemical-mechanical polishing. The resulting active layer can in general achieve a much larger radius of curvature than the thin films used in surface micromachining, due to its substantial thickness. With these advantages, SOI technology has been widely explored

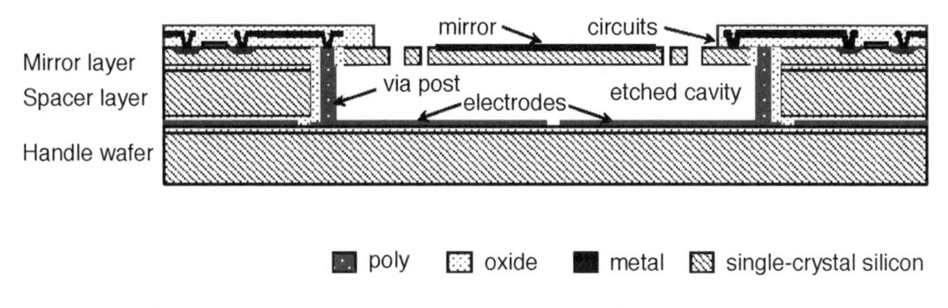

Fig. 5.9 Cross-section of a gimbal mirror by SOI technology [8].

for micro-mirrors in optical scanning and optical switching. Figure 5.9 shows the cross-section of a gimbal mirror made by SOI technology [8]. The substrate of the top SOI wafer is lapped down to the desired thickness for the spacer layer. The bottom silicon handle wafer, with electrodes patterned on the top, is then wafer-bonded to the top SOI wafer and forms a triple-stack SOI wafer. Integrated circuits for actuation and sensing can be patterned on top of the active layer, and connected to the bottom electrodes through polysilicon-filled vias. The micro-mirror and associated structures on the active silicon layer are patterned and defined by RIE. The SiO_2 and the spacer silicon material between the micro-mirror and the electrodes are subsequently removed by selective etching. The released micro-mirror can then be actuated by applying bias between the mirror and the bottom electrodes. Another way of fabricating gimbal mirrors is patterning and releasing the micro-mirror before wafer-bonding the top SOI wafer to the handle wafer.

5.1.4. OTHER FABRICATION TECHNOLOGIES

In addition to the above-mentioned fabrication technologies, various other micromachining technologies have been studied over the years. Here, two examples are described.

5.1.4.1. LIGA Process

LIGA is an abbreviation of the German words Lithographic, Galvanoformung, and Abformung, which stand for lithography, electroplating, and molding. It has been widely studied since the early days of MEMS. The LIGA process uses electroplating to create tall and robust mechanical structures with high aspect ratio above the substrate. The process is illustrated in Fig. 5.10. It starts with photolithographically patterning a thick layer of resist on top of a thin seed layer, such as titanium, which is also used as a sacrificial layer. The structural material is then electroplated into the patterned holes in the resist, followed by planarization. After the resist is dissolved, the formed structure can be used as a mold for mass production. It can also be released and formed into an actuatable structure after etching the sacrificial metal layer. Figure 5.11 shows an SEM photograph of micro-gear structures, fabricated by LIGA process.

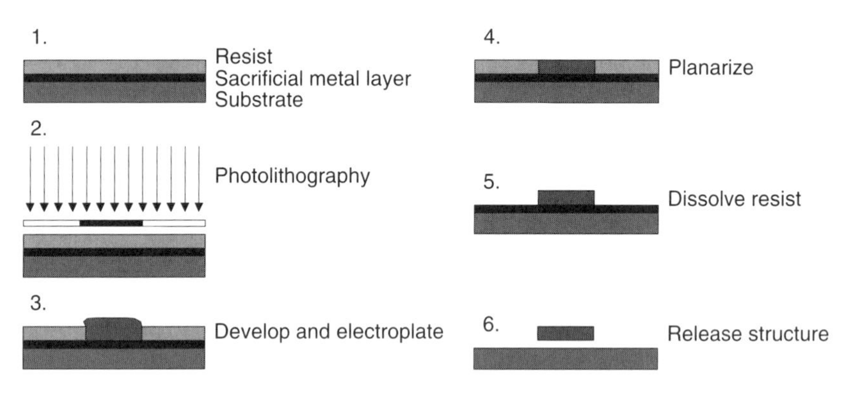

Fig. 5.10 Fabrication process of LIGA.

Fig. 5.11 SEM photograph of micro-gear structures fabricated by LIGA process (courtesy of Cronos Integrated Microsystems).

5.1.4.2. Hybrid-integrated Surface/Bulk Micromachining

Surface micromachining and bulk micromachining offer complementary advantages—surface micromachining is flexible and versatile in processing and can achieve various actuator and sensor structures with high functionality, while bulk micromachining features robust and high-quality mechanical structures. This provides a strong incentive to integrate surface micromachining and bulk micromachining processes. But as sections 5.1.1 and 5.1.2 showed, bulk micromachining and surface micromachining use quite different material systems and processing principles, so it is difficult to integrate these two monolithically.

Recently, hybrid integration of structures made by surface micromachining and bulk micromachining has been demonstrated using wafer bonding [9]. The outline of the process flow is shown in Fig. 5.12. Two wafers are first processed separately by surface micromachining and bulk micromachining. An SOI wafer with thinned substrate is used for the bulk micromachining process. It is then flip-chip bonded to the surface-micromachined wafer. In [9], photoresist is used for wafer-bonding as a demonstration. After the two wafers are bonded together, the thinned substrate of the SOI wafer is removed by DRIE with the SiO_2 layer as an etching stop. After removing the SiO_2 layer using HF, the active single-crystalline silicon layer on top can be patterned photolithographically. The remaining photoresist is then cleaned by oxygen plasma, and the whole wafer is subjected to release etching after the bonding photoresist is hard baked and cured. The etchant removes the sacrificial layer in the surface-micromachined wafer and releases

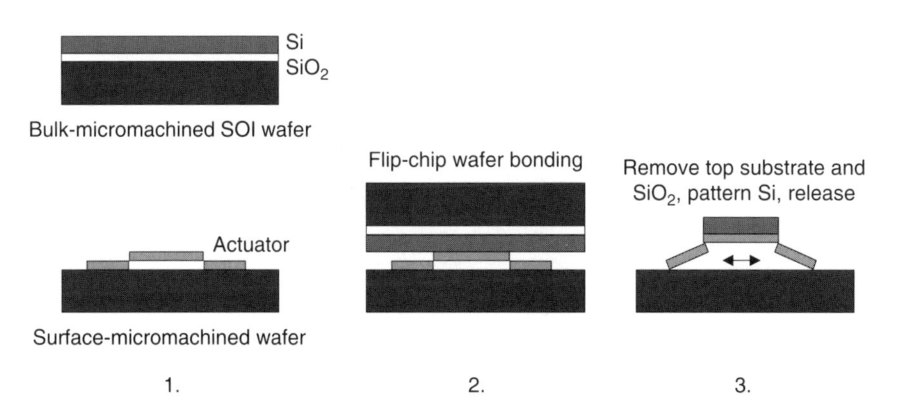

Fig. 5.12 Process flow of hybrid integration of surface micromachining and bulk micromachining.

the mechanical structures. This process integrates high-quality mechanical structures, such as flat and smooth mirrors made of single crystal silicon, with surface micromachined high-functional actuators and/or sensors.

5.2. MEMS for Transmission Engineering

Due to the large tunability that can be achieved via mechanical movement in the MEMS devices, MEMS technologies have been widely explored to provide various tuning, modulation, equalization, and compensation functionalities in WDM transmission systems. Examples include tunable lasers and receivers, tunable filters, data modulators, variable attenuators, dynamic gain equalizers, chromatic dispersion compensators, polarization-state controllers, and polarization-mode dispersion (PMD) compensators. The following sections will describe these applications of MEMS in WDM transmission engineering.

5.2.1. TUNABLE LASERS, DETECTORS, AND FILTERS

Most of the MEMS tunable filters employ the principle of tunable Febry-Perot etalons. Tunable lasers and receivers are then obtained by integrating tunable filters with lasers or receivers via MEMS processing. A Fabry-Perot etalon consists of two parallel partial reflectors, as shown in Fig. 5.13. When the incident wavelength λ and the cavity length d satisfy the following

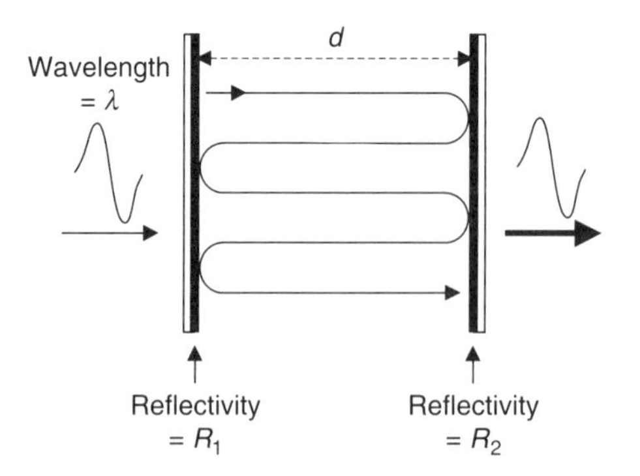

Fig. 5.13 Fabry-Perot etalon.

relation:

$$\frac{2nd}{\lambda} = q, \tag{5.1}$$

where q is an integer and n is the refractive index of the media in the cavity, resonance in the Fabry-Perot etalon is formed. The power transmission coefficient T and reflection coefficient R through the etalon are related to the reflectivity of both mirrors (R_1 and R_2), the cavity length (d), and the wavelength (λ), by [37]

$$T = \frac{(1 - R_1) \cdot (1 - R_2)}{\left(1 - \sqrt{R_1 R_2}\right)^2 + 4\sqrt{R_1 R_2}\sin^2 \frac{\omega n d}{c}}, \quad \text{and} \tag{5.2}$$

$$R = \frac{\left(\sqrt{R_1} - \sqrt{R_2}\right)^2 + 4\sqrt{R_1 R_2}\sin^2 \frac{\omega n d}{c}}{\left(1 - \sqrt{R_1 R_2}\right)^2 + 4\sqrt{R_1 R_2}\sin^2 \frac{\omega n d}{c}}, \tag{5.3}$$

where $\omega = 2\pi c/\lambda$. The transmission and reflection spectra are shown in Fig. 5.14. By changing the cavity length, the spectra, and therefore the peak transmission wavelength, can be tuned. The other parameters that characterize the Fabry-Perot etalons are free-spectra range (FSR), full-width at half-maximum (FWHM) of the transmission peak, and the finesse. The FSR of the etalon is defined as the separation of two adjacent

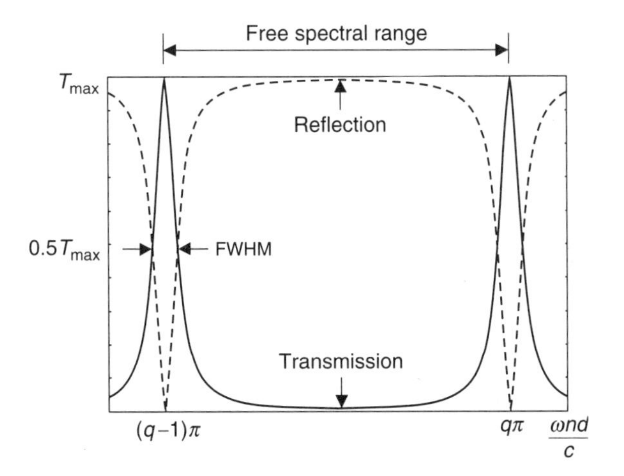

Fig. 5.14 Transmission and reflection spectra of the Fabry-Perot etalon.

transmission peaks (in frequency domain) as

$$FSR = c/2nd. \tag{5.4}$$

The FWHM is equal to $\frac{c}{2nd} \cdot \frac{1-(R_1 R_2)^{1/2}}{\pi (R_1 R_2)^{1/4}}$. The finesse F is then defined as

$$F = \frac{FSR}{FWHM} = \frac{\pi (R_1 R_2)^{1/4}}{1 - (R_1 R_2)^{1/2}}. \tag{5.5}$$

In general, it is desirable to design a Fabry-Perot etalon with high finesse and narrow FWHM. This requires the two parallel reflectors with high reflectivity (R_1, $R_2 \to 1$).

Figure 5.15(a) shows an example of MEMS tunable filters [10]. The suspended reflector consists of an AlO_x/GaAs diffractive Bragg reflector (DBR). Tuning is achieved by applying variable bias between the suspended reflector and the substrate to change the cavity length of the Fabry-Perot etalon. The tuning characteristic of this MEMS tunable filter is shown in Fig. 5.15(b).

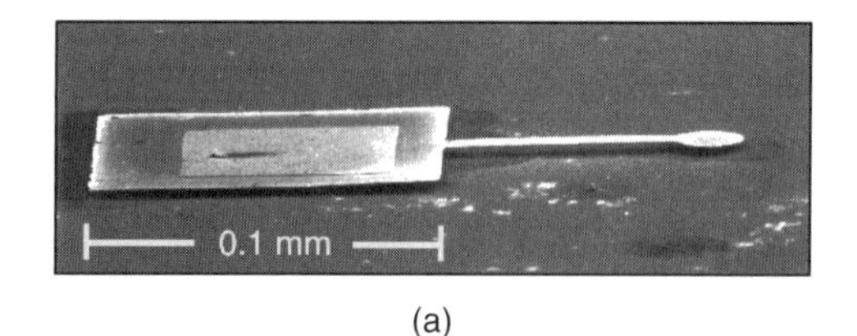

(a)

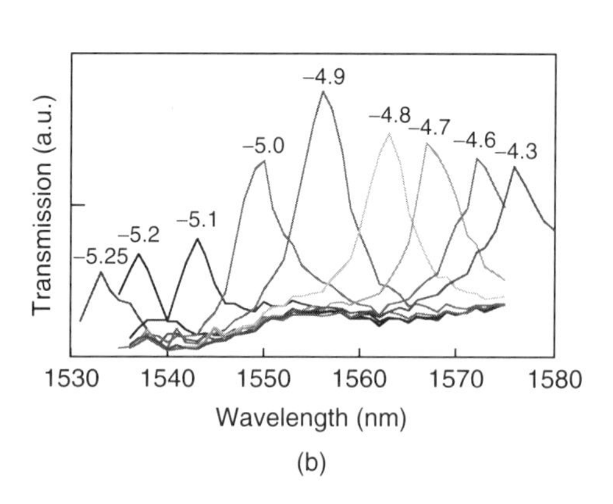

(b)

Fig. 5.15 (a) An example MEMS tunable filter, consisting of a suspended diffractive Bragg reflector. (b) Tuning of the transmission peak wavelength. The labels in the figure are the various applied biases [10].

By integrating the tunable filter with a vertical-cavity surface-emitting laser (VCSEL) or a photodetector, a tunable laser or photodetector can then be achieved. Various MEMS tunable photodetectors, light-emitting diodes (LEDs), and VCSELs have been demonstrated [11–14]. Figure 5.16 is an example of a MEMS tunable LED/photodetector [14]. A micromachined Si/SiO_2 DBR mirror membrane is integrated on top of the strain-compensated multiple quantum-wells (MQWs) that work as an LED when forward-biased and as a photodetector when reverse-biased. A p-type GaAs/AlAs Bragg mirror forms the bottom fixed mirror. Figure 5.16(a) shows the cross-section of the LED/photodetector structure and its SEM photograph. The tuning characteristics upon applying various biases between the top DBR mirror and the substrate are shown in Fig. 5.16(b). The inset in this figure shows its emission profile with no applied voltage. Recently, MEMS external cavity tunable diode lasers that employ external gratings and rotary mirrors as tuning mechanism have also been demonstrated [15].

5.2.2. DYNAMIC GAIN-EQUALIZERS AND VARIABLE OPTICAL ATTENUATORS

The emergence of erbium-doped fiber amplifiers has made long-haul WDM transmission economically irresistible. Fiber amplifiers permit multichannel transmission unencumbered by traditional constraints arising from thermal and shot noise at the receiver; however, the amplifiers contribute noise of their own, whose associated signal-to-noise ratio is proportional to the per-channel signal power levels that enter the amplifiers. Thus, to maintain signal integrity for all channels after cascading a number of fiber amplifiers, it is important to have strong and uniform signal power levels. The large nonuniformity of the erbium atom's gain spectrum makes this an intrinsically difficult problem (see Fig. 5.17) [16]. External gain-equalizers are thus required in order to equalize the power levels of WDM channels traversing a fiber-amplifier gain module.

Long-period fiber Bragg gratings and thin-film filters have for several years been successfully employed to achieve static gain-equalizers that have now attained a high state of maturity. However, such static gain-equalizers must be tailored with precision to a specified amplifier operating point. At the same time, this operating point, and thus the amplifier gain spectrum, is quite sensitively influenced by essentially all system parameters: the number of channels present in the WDM system, the power per channel, the pump power and wavelength, and the loss between amplifiers.

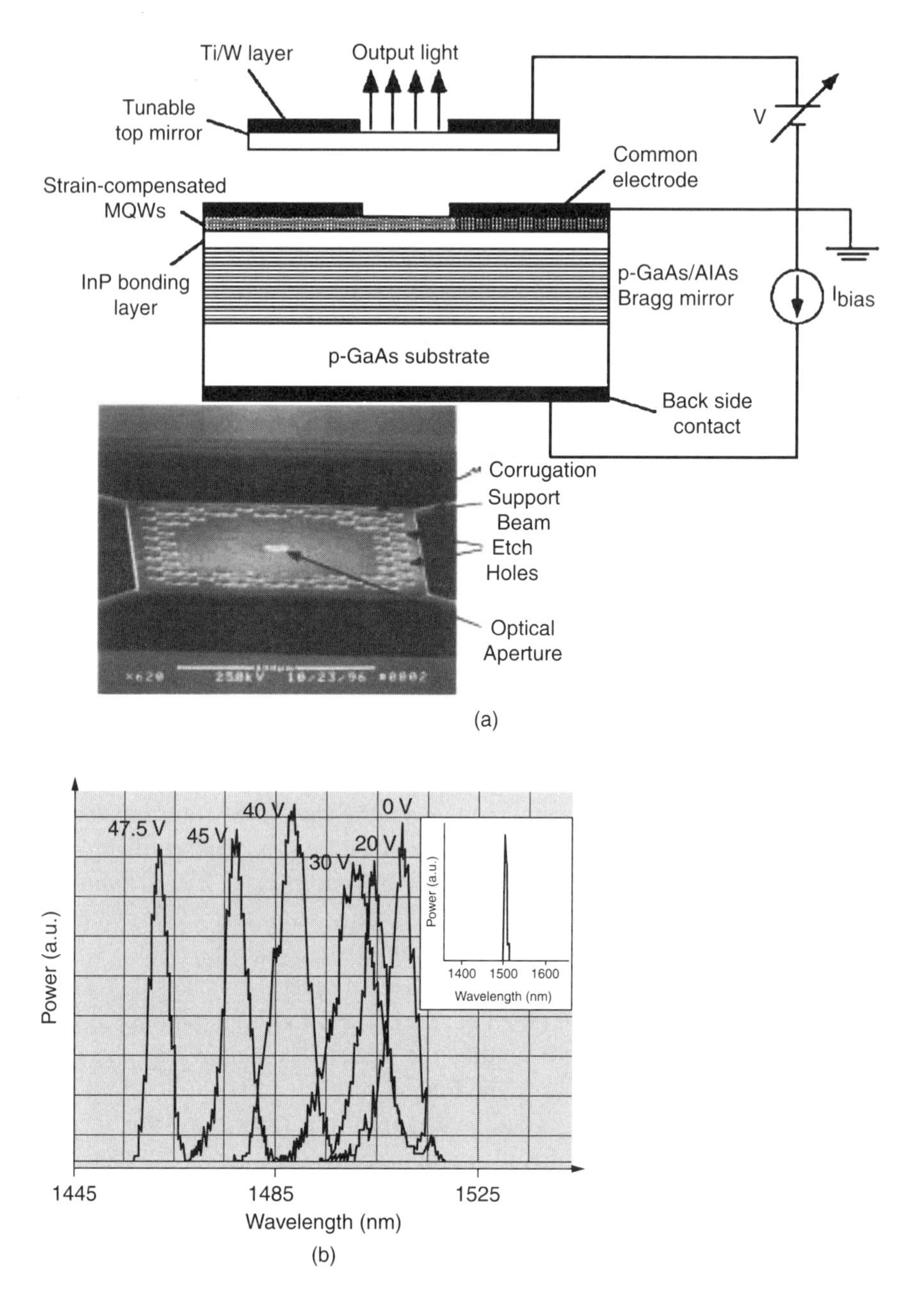

Fig. 5.16 (a) Cross-section of a MEMS tunable LED/photodetector and its SEM photograph. (b) The tuning characteristics upon applying various biases between the top DBR mirror and the substrate. The inset shows the emission profile with no applied voltage [14].

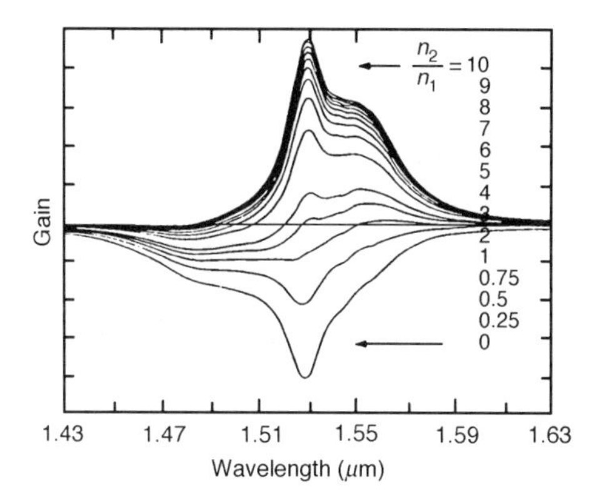

Fig. 5.17 Gain spectrum of erbium-doped fiber amplifier. n_1: ground-state population of erbium atoms. n_2: excited-state population of erbium atoms [16].

Moreover, apart from the strongly-coupled nature of WDM links, the increasing presence of WADMs in the middle of WDM links, and the parasitic Raman gain incurred along the system [17], both have the effect of making the gain spectrum of individual fiber amplifiers and their associated links fundamentally unpredictable at the level required by low-error-rate transmission engineering. Dynamic gain-equalizers consequently become essential in systems with reconfigurable wavelength add/drop that supports more than about 100 WDM channels.

MEMS technologies have been employed to implement dynamic gain equalizers [18, 19] with various approaches. One example is to utilize the "mechanical anti-reflection switch (MARS)" [20]. The MARS consists of a thin Si_3N_4 membrane fabricated by LPCVD, suspended over a silicon substrate with an air gap in between, as shown in Fig. 5.18(a). By applying a voltage between the Si_3N_4 membrane and the substrate, the thickness of the air gap can be adjusted. This results in continuous variation in the reflection coefficient (see Fig. 5.18(b)).

Such a device is an example of adjustable multi-dielectric layer systems, whose optical property can be formulated by 2×2 matrix approach [21]. Figure 5.19 shows a schematic drawing of a multi-dielectric layer system. A plane wave $E = E(x)e^{i(\omega t - \beta z)}$ incidents from the left. The electric field in the various layer of the system can be represented by

$$E(x) = \begin{cases} A_0 e^{-ik_{0x}(x-x_0)} + B_0 e^{ik_{0x}(x-x_0)}, & x < x_0 \\ A_l e^{-ik_{lx}(x-x_{l-1})} + B_l e^{ik_{lx}(x-x_{l-1})}, & x_{l-1} < x < x_l, \\ A_s e^{-ik_{sx}(x-x_N)} + B_s e^{ik_{sx}(x-x_N)}, & x_N < x \end{cases} \quad (5.6)$$

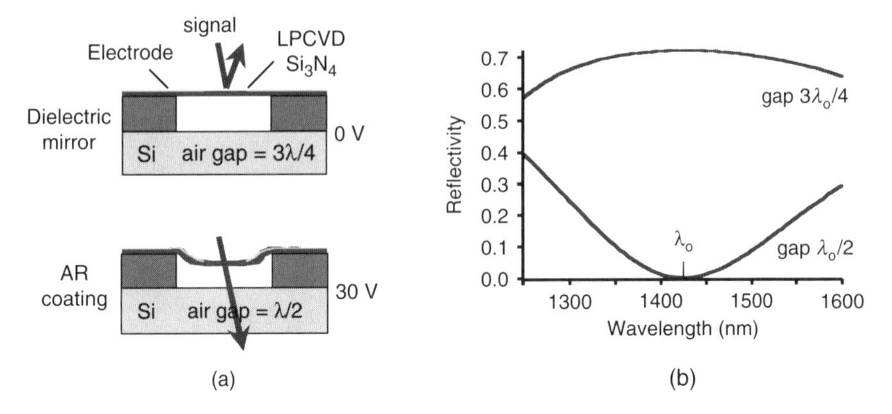

Fig. 5.18 (a) Cross-section of a mechanical anti-reflection switch (MARS). (b) Reflectivity of MARS versus wavelength with various gap thicknesses adjusted by voltage bias [20].

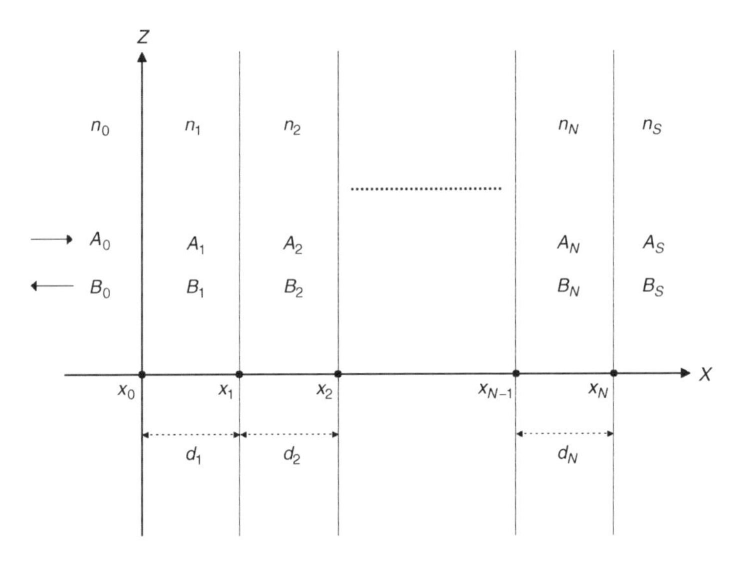

Fig. 5.19 A multi-dielectric layer system.

where $k_{lx} = n_l \frac{\omega}{c} \cos\theta_l$ is the x component of the wave factor. The incident angle of the optical wave at each interface is θ_l. For this system, the amplitudes of the forward and reflected fields are related to the adjacent ones by

$$\begin{pmatrix} A_0 \\ B_0 \end{pmatrix} = D_0^{-1} D_1 \begin{pmatrix} A_1 \\ B_1 \end{pmatrix}$$
$$\begin{pmatrix} A_l \\ B_l \end{pmatrix} = P_l D_l^{-1} D_{l+1} \begin{pmatrix} A_{l+1} \\ B_{l+1} \end{pmatrix}, \quad l = 1, 2, \ldots, N, \tag{5.7}$$

where

$$D_l = \begin{pmatrix} 1 & 1 \\ n_l \cos\theta_l & -n_l \cos\theta_l \end{pmatrix} \quad \text{for TE wave}$$
$$D_l = \begin{pmatrix} \cos\theta_l & \cos\theta_l \\ n_l & -n_l \end{pmatrix} \quad \text{for TM wave} \tag{5.8}$$

are the matrices representing the phase transformation across the interface, and

$$P_l = \begin{pmatrix} e^{i\phi_l} & 0 \\ 0 & e^{-i\phi_l} \end{pmatrix}, \quad \phi_l = k_{lx} d_l \tag{5.9}$$

accounts for the propagation of the optical wave. The transmission and reflection coefficients can then be determined from the corresponding amplitudes of the electric fields:

$$\begin{pmatrix} A_0 \\ B_0 \end{pmatrix} = \begin{pmatrix} M_{11} & M_{12} \\ M_{21} & M_{22} \end{pmatrix} \begin{pmatrix} A_S \\ B_S \end{pmatrix}$$
$$\begin{pmatrix} M_{11} & M_{12} \\ M_{21} & M_{22} \end{pmatrix} = D_0^{-1} \left[\prod_{l=1}^{N} D_l P_l D_l^{-1} \right] D. \tag{5.10}$$

The results are dependent of wavelength and the thickness of the dielectric layers. Figure 5.18(b) is an example of such results.

A dynamic gain-equalizer can then be obtained by integrating a continuous array of the mechanical anti-reflection switch, as shown in Fig. 5.20(a) [18]. A wavelength demultiplexer, e.g., a grating, projects different

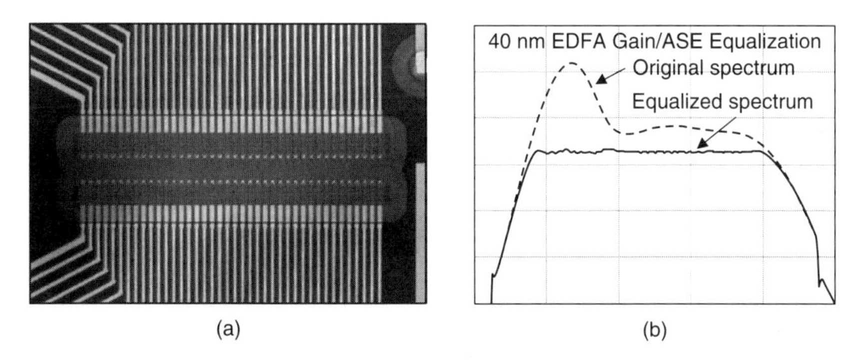

Fig. 5.20 (a) Top view photograph of a MEMS dynamic gain equalizer. (b) Flattened erbium-doped fiber amplifier gain spectrum after the equalizer [18].

wavelengths along different sections of the gain-equalizer. Various voltage bias is then applied to these different sections to achieve dynamic gain-equalization. Figure 5.20(b) shows the application of this device as a means of flattening a 40-nm fiber-amplifier gain spectrum. The device is extremely effective, is reasonably low in excess loss, and offers good prospects for scaling to large channel counts.

5.2.3. *CHROMATIC DISPERSION COMPENSATION*

As per-channel bit rate increases, so too do impairments due to chromatic dispersion. Static dispersion compensation is now well established using dispersion-compensating fibers and chirped fiber gratings. Tunable dispersion-compensation technology, on the other hand, is still in its infancy, though it will in general become necessary as systems move from 10 Gb/s to 40 Gb/s per wavelength.

MEMS offer one reasonably promising means of achieving such tunability in group-velocity dispersion [22]. The chief device structure thus far used to achieve this consists of a MARS all-pass filter—a device that is obtained by coating a high-reflective layer (metal or dielectric mirror) at the bottom of the silicon substrate of a MARS device, as shown in Fig. 5.21(a). The silicon substrate now functions as a Fabry-Perot cavity. The effective photon lifetime in the cavity can be tuned by adjusting the reflectivity of the top MARS device with voltage tuning. The reflectivity

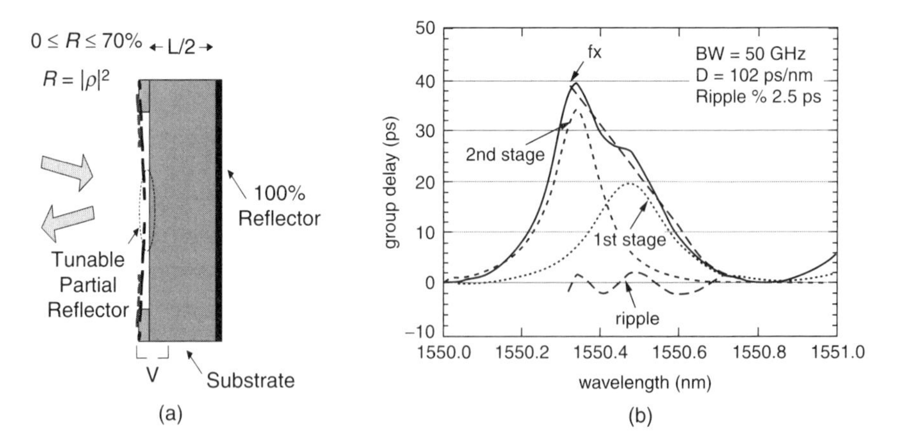

Fig. 5.21 (a) MARS all-pass filter for chromatic dispersion compensation. (b) Group delay obtained from a two-stage device with a dispersion of −102 ps/nm [22].

is wavelength dependent, and therefore results in a difference in group-delay between different wavelengths. In addition, the refractive index of the silicon substrate can be adjusted by temperature tuning, which is also wavelength dependent. The refractive index affects the effective optical path length in the cavity. This provides another degree of freedom for adjusting group-delay between different wavelengths. By cascading multiple such devices, and using the combination of voltage and temperature tuning, a variety of dispersion values and slopes can be achieved. Figure 5.21(b) shows measured results obtained using a two-stage device, with dispersion equal to -102 ps/nm. Positive dispersion slopes and different dispersion values have also been demonstrated [22].

5.2.4. POLARIZATION ENGINEERING

Because signal polarization states in transmission fiber are in practice stochastic, but certain components and subsystems exhibit polarization dependence, practical polarization controllers have long been desired. These become particularly important at 10 Gb/s and beyond, where polarization-mode dispersion (PMD) in older transmission fiber becomes a substantial impairment [23]. Such PMD-mitigation must be done channel-by-channel. Thus, in view of the increasing number of channels in lightwave communication systems, it is critical that any practical polarization controller deployed for PMD-compensation be both compact and low in cost.

The basic building block of a full polarization controller is polarization rotators. A compact MEMS polarization rotator can be achieved by surface-micromachining technology [24]. Figure 5.22 shows a schematic drawing and an SEM photograph of such a device, which rotates an input state of polarization (SOP) along its principal axis on the Poincaré sphere (shown in Fig. 5.23). The principal axis is defined by the two orthogonal $|TE\rangle$ and $|TM\rangle$ eigenstates of the polarization beam-splitter (PBS). The PBS consists of a polysilicon plate that is oriented at the Brewster angle (74° for polysilicon) to the incident optical beam. The fixed mirror and the phase-shifter are both parallel to the PBS. The phase-shifter consists of a micro-mirror and an electrode plate standing in front of the micro-mirror. Phase-shifting is achieved by fine-tuning the angle of the micro-mirror to create a phase difference between the two optical paths, which is realized by applying a bias voltage between the micro-mirror and the electrode plate.

The Poincaré sphere representation of the SOP propagating through the system is shown in Fig. 5.23. The TE and TM modes are antipodal and their

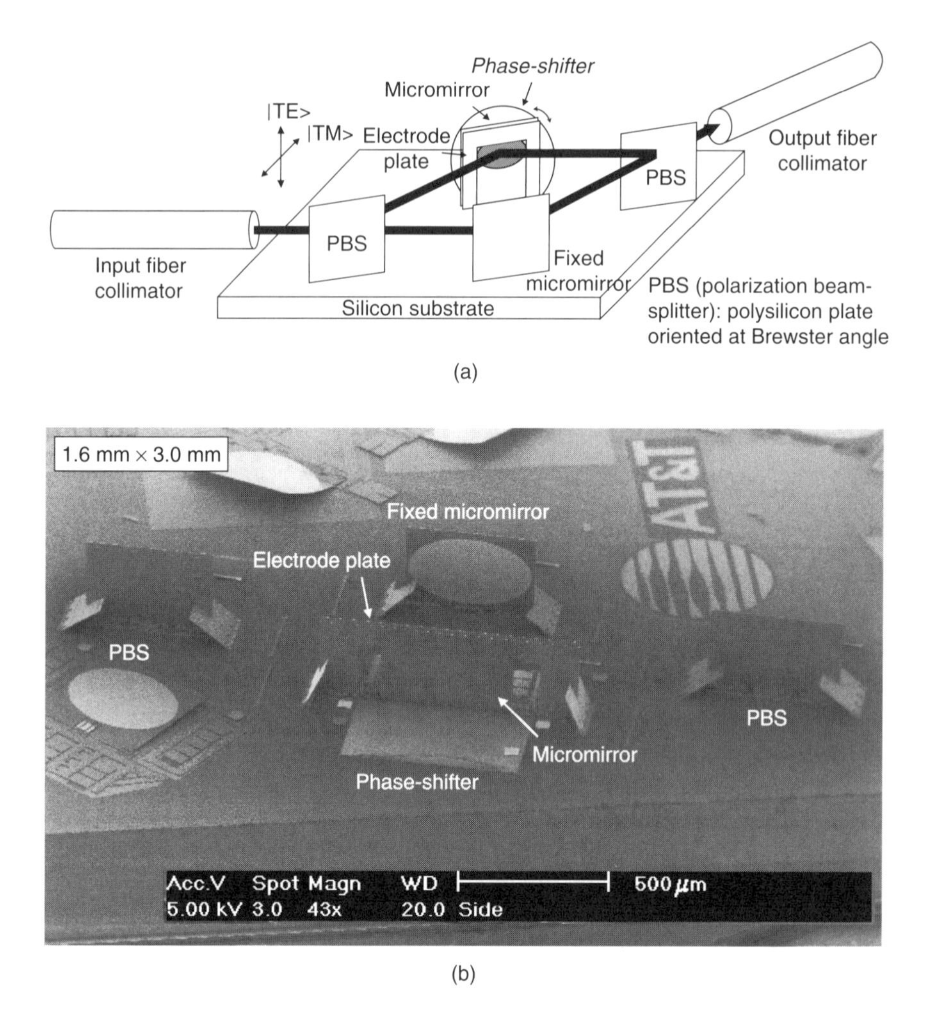

Fig. 5.22 (a) Schematic drawing, and (b) SEM photograph of a MEMS polarization rotator. (*See color plate*).

relative amplitudes and phases are described through polar and azimuthal angles, θ and φ, respectively, via

$$|SOP\rangle = \cos\theta/2 \cdot e^{-i\varphi/2}|TE\rangle + \sin\theta/2 \cdot e^{i\varphi/2}|TM\rangle. \quad (5.11)$$

Tuning the phase-shifter changes φ, and imposing a one-wavelength difference in the optical path moves the SOP in a full circle of constant latitude about the TE/TM axis, as shown by the dashed circle on the Poincaré sphere. Figure 5.24 shows the experimental results obtained using the MEMS polarization rotator. The SOP trajectory evolves on the Poincaré sphere, as

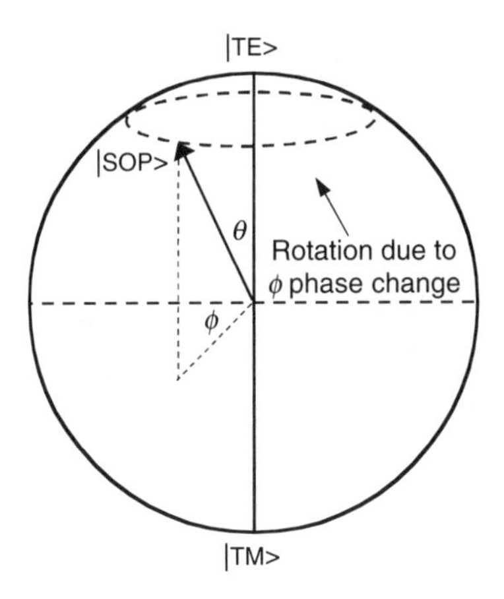

Fig. 5.23 Poincaré sphere representation of the state-of-polarization.

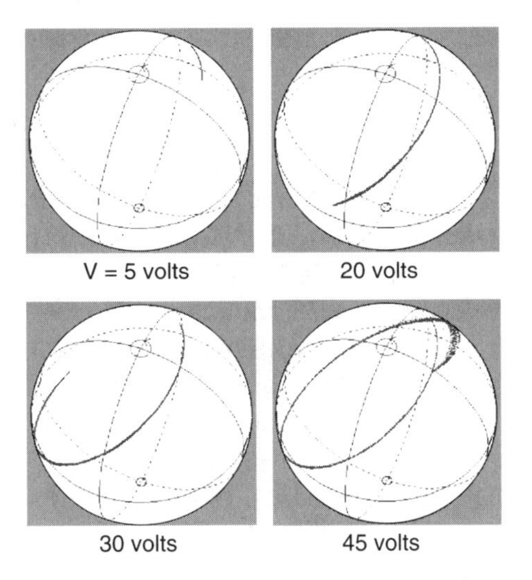

Fig. 5.24 Polarization rotation on the Poincaré sphere by applying bias to the phase shifter in the MEMS polarization rotator. (*See color plate*).

the voltage bias on the phase shifter increases. At $\sim$ 45-V bias, a full circle is achieved. The angular tuning of the phase shifter results in the slight imperfect enclosure of the circle, as this also changes the angle of the propagating optical beam. This can be improved by employing translational motion, such as a suspended membrane as the micro-mirror that can be modulated, for the phase shifter.

A full polarization controller can then be obtained by cascading three such polarization rotators, with the center stage rotated by 45° with respect

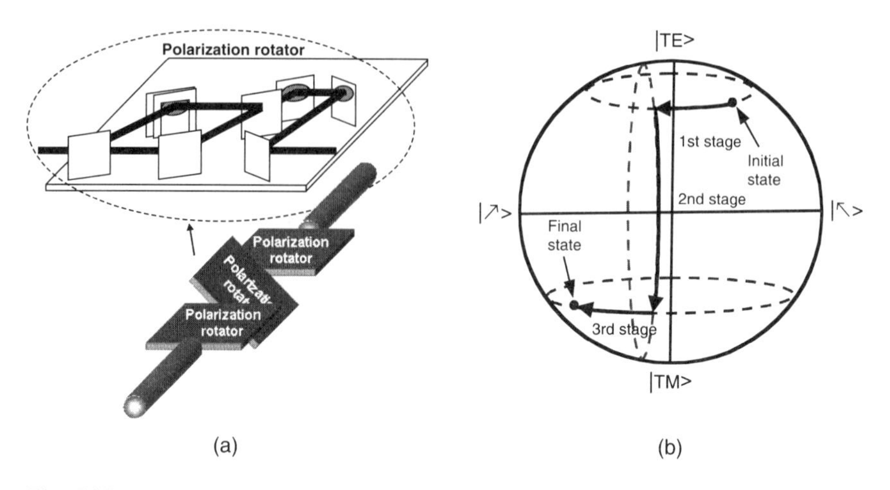

Fig. 5.25 (a) A full polarization controller obtained by cascading three stages of MEMS polarization rotators. (b) Action on the Poincaré sphere by the proposed MEMS polarization controller. (*See color plate*).

to the first and third stages, so that its principal axis is orthogonal to that of the first and third stages on the Poincaré sphere. Figure 5.25(a) shows a schematic drawing of a polarization controller. Its action on the Poincaré sphere is shown in Fig. 5.25(b). With this approach, an SOP can be moved to any final destination on the Poincaré sphere. By cascading a MEMS polarization controller with a PBS and a tunable delay line, the slow and fast axes of the optical wave can be aligned to the two principal axes of the PBS. These two components are split by the PBS, and the fast-traveling wave can then be sent to the tunable delay line. This compensates the group velocity difference between the slow-traveling and fast-traveling components, and achieves a first-order PMD compensator.

5.3. MEMS for Optical-layer Networking

With the maturation of WDM technologies, including various component technologies such as sources, amplifiers, compensators, filters, and high-speed electronics that have resulted in terabit-per-second WDM links, WDM has so far been able to keep up with the increasing demand of optical network capacity. The challenges for core networking are now swiftly shifting from transmission to switching. It is no longer problematic to transport enormous information capacity from point to point; the difficulty now resides in parting it out and managing it.

In addition to being able to access a portion of the WDM signals, there are two sources of acute switching pressure in the core long-haul network. The first is restoration—traffic must be automatically rerouted in the event of failures over time intervals on the order of $\sim$100 ms. The second is provisioning—new circuits must be established in response to service-layer requests over time intervals on the order of a few minutes. Both functions are currently carried out using electronic switches and add/drop multiplexers that operate typically on signals at 45–155-Mb/s data rates—an arrangement that becomes both unmanageable and unaffordable in core networks, just now emerging, whose larger nodes transport hundreds of Gb/s of traffic. Moreover, service-layer vehicles—IP routers—are now sprouting interfaces, at 2.5 to 10 Gb/s, that are well-matched to the per-wavelength bit rates of transport systems. Thus, there is an emerging need to provision and restore traffic at much coarser granularity, at or approaching the wavelength. What is needed, then, is a switch scalable to thousands of ports, with each port transporting 2.5–10-Gb/s signals, scalable to 40 Gb/s, and with switching times $<$10 ms. This challenge greatly outstrips the capabilities of embedded networking technology. The result has been a frenzied drive to develop coarse-granularity circuit switches with capacities that stretch into the terabits per second and beyond.

5.3.1. RECONFIGURABLE WAVELENGTH ADD/DROP MULTIPLEXERS (WADMs)

As WDM transmission system channel count increases, and unregenerated long-haul point-to-point transmission links grow longer, it becomes important to find a means of adding and dropping a fraction of the WDM channels in the middle of a transmission system, so as to avoid stranding capacity in the system. Fixed wavelength add/drop multiplexers (WADM) fashioned from thin-film filters and Bragg gratings [25] have for some years been available in mature form. MEMS has thus far had little to offer here in the way of improvements. The contribution of MEMS has typically been that it offers a means of incorporating configurability or switching into wavelength-selective structures.

Thus far, switchable WADMs have been made using planar waveguide switches [26], micro-machined tilting or on/off mirrors [27, 28], and liquid-crystal switches [29]. An example of MEMS reconfigurable WADM is shown in Fig. 5.26(a). The input optical beam is first collimated by a lens. A grating demultiplexes the multi-wavelength optical signal, and casts optical

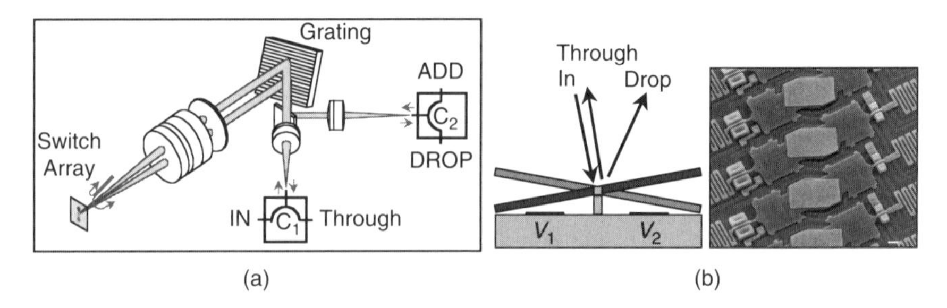

Fig. 5.26 (a) A WADM utilizing an array of tilting MEMS mirrors. (b) SEM and working principle of the MEMS mirror array [27]. (*See color plate*).

beams with different wavelengths onto different micro-mirrors in the micro-mirror array, shown in Fig. 5.26(b). The micro-mirror can be tilted in one direction, to retro-reflect the optical signal to the input fiber. A circulator then separates the reflected signal from the input signal, and sends the reflected signal to the through port. If the micro-mirror is tilted in the other direction, the optical signal is reflected to another direction, collimated by the lens, then reflected by a fixed mirror to the drop port. This free-space optical interconnection works symmetrically; therefore, by combining a circulator with the drop port, an add port can be included, which brings the signal to the through port by retracing the optical path.

In general, most of the WADMs demonstrated do not offer the feature of "client configurability." That is, a drop port (a client) is in general rigidly associated by the device structure with only one particular wavelength. As the manageability and flexibility of optical networks become increasingly critical concerns that begin to eclipse installed first cost, client configurability has become an increasingly important feature to incorporate in WADMs.

Client-configurable WADMs can be achieved using 2-D MEMS crossbar matrix switches and tunable lasers, as proposed in [30]. A suggested client-configurable WADM structure is illustrated in Fig. 5.27. The matrix switch connects the demultiplexed input channels, add ports, drop ports, and output channels to be multiplexed for transmission. The add ports use tunable lasers to select which wavelengths are to be added. Both sides of the micro-mirrors are used simultaneously for add and drop optical paths. This operation can be carried out independently for all input wavelengths, and therefore offers client configurability. This architecture is particularly useful for unidirectional networks where the add/drop ports associated with a certain input channel (wavelength) are connected to the same user. In more

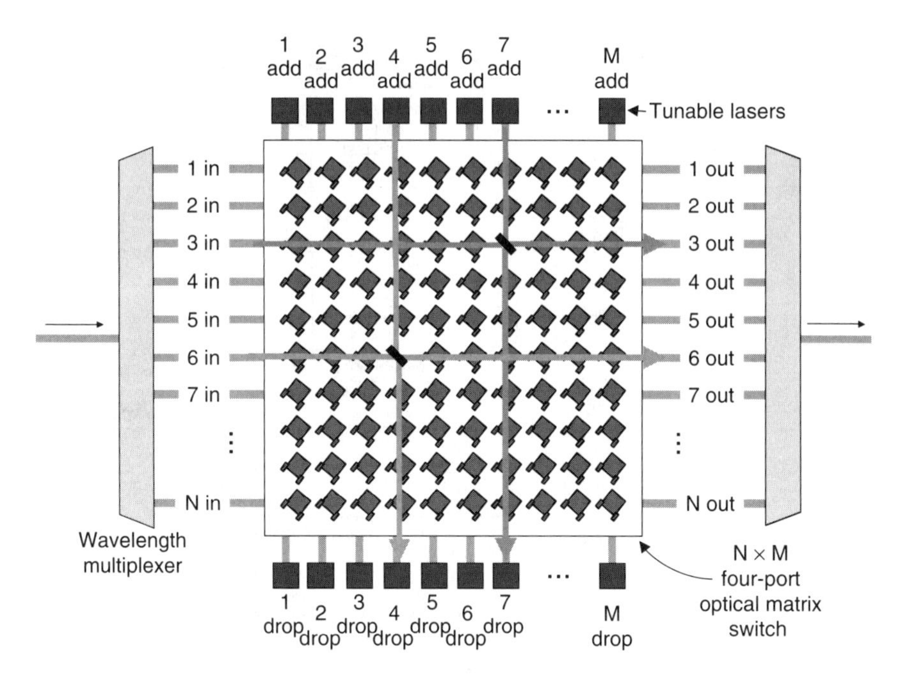

Fig. 5.27 2-D MEMS crossbar matrix switch for WADM with client configurability. (*See color plate*).

general add/drop applications where network traffic is bidirectional, the add port and the drop port associated with the same channel (wavelength) are independent of each other, and therefore the client configurability for each needs to be addressed separately. Various architectures using two crossbar matrix switches to achieve bidirectional WADM have been proposed [30]. The two matrix switches can either be fabricated on a single chip, or can be two separate chips interconnected together.

An eight-channel client-configurable WADM has been demonstrated using a 2-D MEMS crossbar switch together with eight tunable lasers [30]. The channel wavelengths in the 1550-nm band were used, with channel spacings of 100 GHz. Figure 5.28(a–c) shows the spectrum at the through port with Channels 3, 5, and 7 dropped. Also shown are the drop-port spectrum and the spectrum at the through port with Channels 3 and 5 added. The chief virtues of WADMs implemented via this approach are that they tend to offer low crosstalk between adjacent channels and very high extinction ratios, as do most of the MEMS WADMs. The loss performance at low channel counts tends to be limited by the quality of the mux/demux and fiber collimator array. Such structures utilizing 2-D crossbar matrix

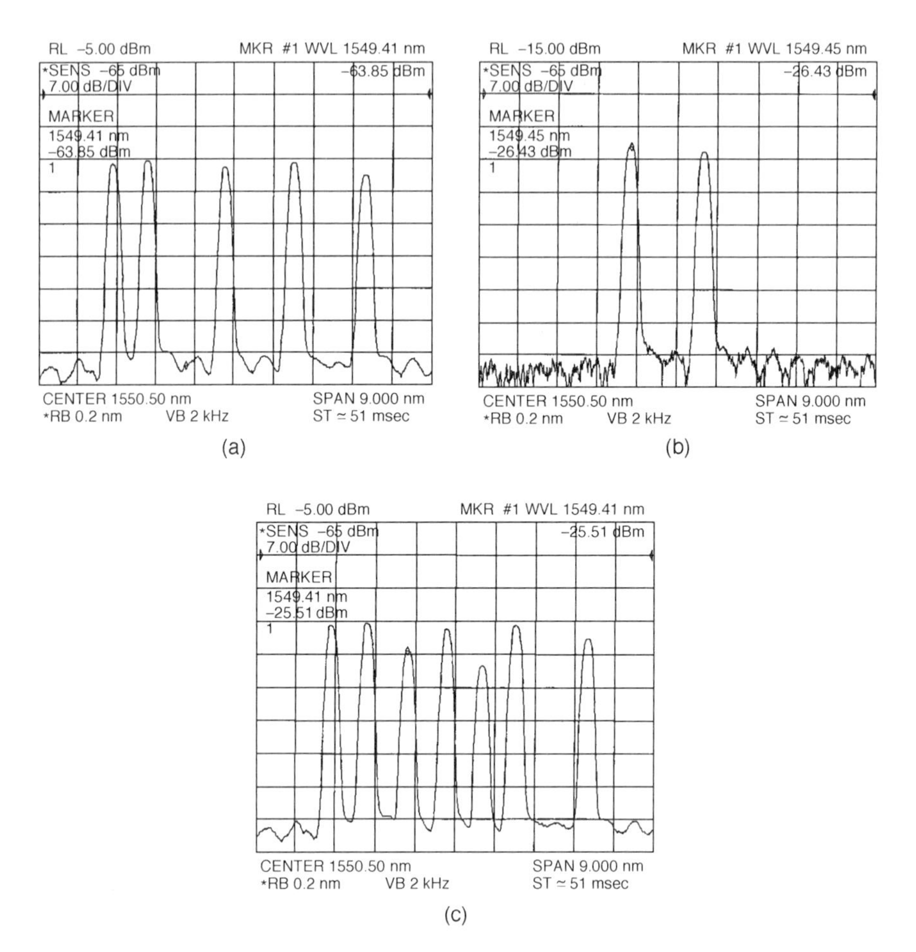

Fig. 5.28 WADM spectra with client configurability: (a) Through spectrum with Channel 3, 5, and 7 dropped. (b) Dropped Channel 3 and 5. (c) Through spectrum with Channel 3 and 5 added.

switches tend to have difficulty moving to channel counts beyond 16 or so, however, due to the loss and path-length considerations (which will be summarized in Section 5.3.2.3).

5.3.2. *OPTICAL CROSSCONNECTS (OXCS)*

Optical crossconnects (OXC) have emerged as a critical network element for constructing next-generation provisioning and restoration vehicles for emerging mesh-based optical networks. The core optical switching technology required by such network elements, however, has not progressed at a speed matching the needs of optical networks, and has so far lagged behind rival electronic-fabric-based technology.

Nevertheless, optical switching has been an area of active research for two decades. Various optical switch technologies, employing both free-space interconnection and integrated optics, have been demonstrated. However, all such technologies prior to the advent of MEMS have fallen well short of the 256–1000-port systems that are needed in core networks. Furthermore, optical switch fabrics will need to achieve low insertion loss, low switching time (on the order of a few milliseconds), low crosstalk, low polarization dependence, and low wavelength dependence. It is moreover desired that they be transparent to bit-rate in the interval of roughly 2.5 Gb/s to 40 Gb/s. These requirements impose stringent challenges on all optical-switching technologies.

MEMS technology has recently proven itself as a promising option for building such high-port-count optical switches. Optical MEMS have the fundamental advantage of being able to exploit the benefits of free-space interconnection (including low loss and crosstalk, and low polarization and wavelength dependence), together with the advantages of integrated optics (including compactness, optical pre-alignment, and low cost). Therefore, they offer the possibility of achieving high port count in a small, low-cost system with excellent optical quality. In this section, we review various optical MEMS switching technologies and their performance. We then discuss the challenges these technologies face.

5.3.2.1. 2-D Digital Crossbar Switches

2-D MEMS digital crossbar switches feature integration of a matrix of micro-mirrors on a silicon substrate, with light propagating parallel to the surface of the substrate, as shown in Fig. 5.29. The micro-mirrors are rotated, slid, or lifted into and out of the optical path to change the propagation direction of the optical beam between through state and reflection state. Figure 5.29 shows free-rotating micro-mirrors as an example. An SEM of the free-rotating micro-mirror with scratch-drive actuators is shown in Fig. 5.30 [31]. The translation motion of the actuators is converted to mirror rotation via the mechanical design. Other types of micro-mirrors for 2-D digital crossbar switches have also been demonstrated. Shown in Fig. 5.31, for example, is the stress-bending switch [32]. Here, a polysilicon cantilever is coated with a layer of gold. The stress difference between

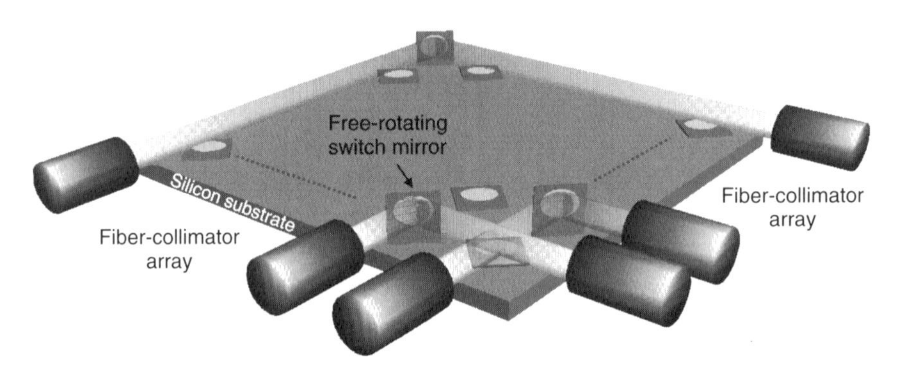

Fig. 5.29 Schematic drawing of a 2-D MEMS digital crossbar switch, with free-rotating micro-mirrors as an example.

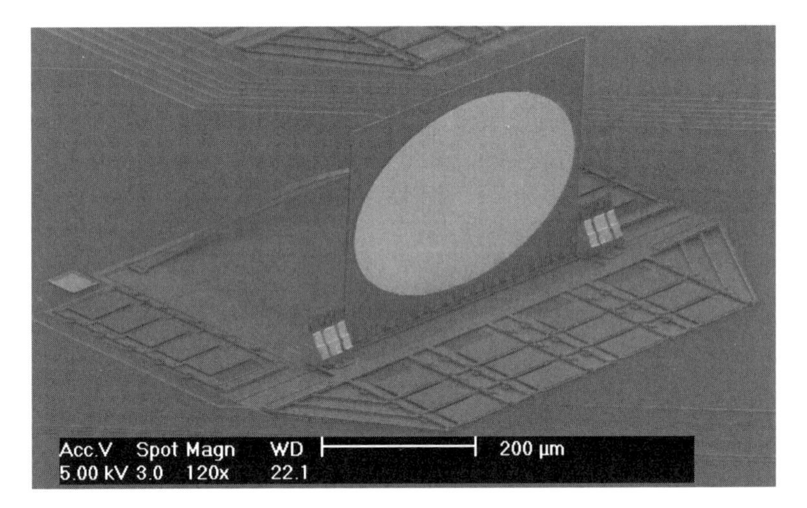

Fig. 5.30 SEM of the free-rotation micro-actuated switch mirror with scratch-drive actuators [31].

gold and polysilicon then causes the cantilever to bend upward after release etching, thus lifting the micro-mirror and resulting in through state for the switch element. The micro-mirror can be attracted to the substrate, thus placing the switch element in the reflection state, by applying a voltage between the cantilever and the substrate. Magnetically actuated torsion mirrors have also been demonstrated for 2-D digital crossbar switches [33], as shown in Fig. 5.32. The micro-mirror is rotated out of the substrate when a magnetic field is applied. A vertical stop helps to align the angle

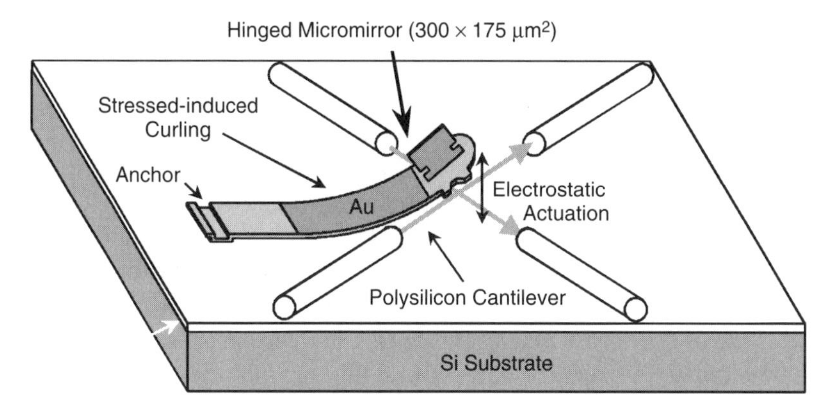

Fig. 5.31 Schematic drawing of a stress-induced bending switch [32].

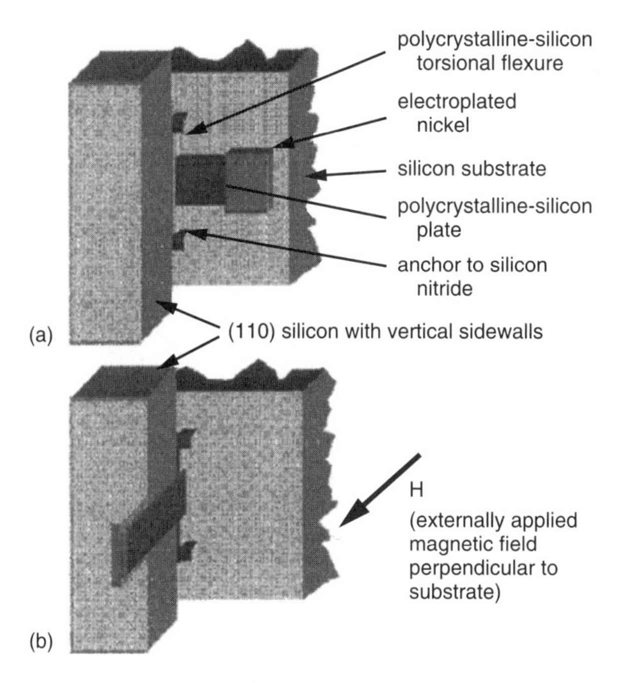

Fig. 5.32 Schematic drawing of a magnetic-actuated micro-mirror [33].

of the micro-mirror at 90° after it is rotated up. A bias is then applied between the vertical stop and the micro-mirror so that no further power is required to keep the micro-mirror in position. Torsion mirrors actuated by electro-static force have also been explored [34].

Such 2-D digital crossbar switches have been shown capable of achieving quite good optical quality, particularly in the areas of crosstalk, polarization and wavelength dependence, and bit-rate transparency; these are immediate rewards of the device's free-space nature. An 8 × 8 research prototype switch with insertion loss as low as 1.7 dB has been demonstrated [35].

A significant advantage of this technology accrues from its use of simple binary control of the micro-mirror position; this results in simple control algorithms and short switching times on the order of one millisecond. Due both to the limited wafer area and to the insertion loss arising from divergence of the optical beam propagating in free space, a 32 × 32 single-stage switch is perhaps the largest that this technology will in practice achieve. To achieve higher port-count, multi-stage nonblocking switch architectures will be needed. Figure 5.33 shows a representation of a 3-stage Clos architecture. The first and third stages consist of N layers of M × N switches, while the second stage consist of N layers of N × N switches. The port-count of the resulting 3-stage switch is M × N. To achieve strictly nonblocking performance, N needs to be greater than or equal to 2M−1. In principle, then, large strictly nonblocking switches can be made from relatively small monolithic switch elements. Although this approach has had great commercial impact in the world of electronic switches, its future is more problematic in optical systems. This is due chiefly to the rapid accumulation

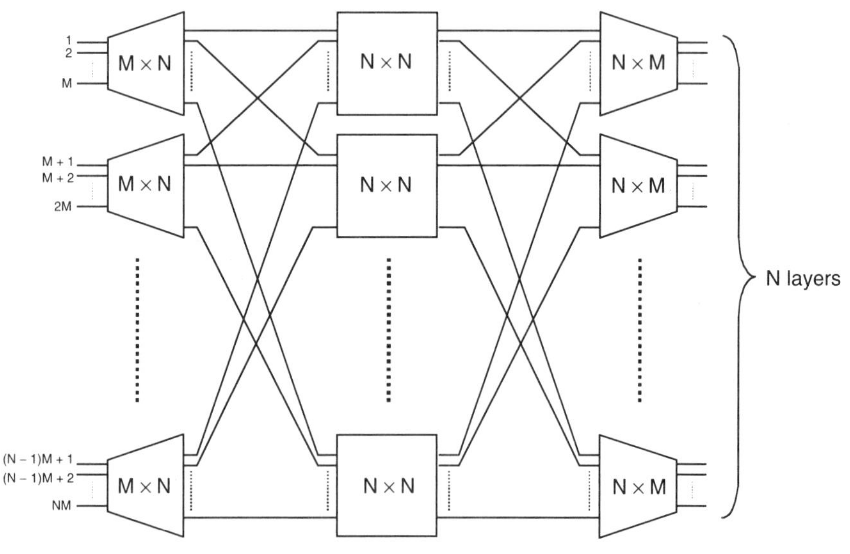

Fig. 5.33 3-stage Clos architecture.

of loss through three optical-switch stages, and to the cost and volume of the large number of fiber interconnects that are needed between stages.

5.3.2.2. 3-D Analog Beam-steering Switches

To break through the port-count barriers previously described, alternative switch architectures are needed. The most promising and widely pursued of these is the 3-D analog beam-steering structure, as shown in Fig. 5.34. The switch fabric consists of two MEMS mirror arrays, an input fiber/lens array, and an output fiber/lens array. Each fiber and lens constitute one port, associated with one mirror in the MEMS mirror array. The micro-mirrors are two-axis gimbal mirrors, as shown in Fig. 5.35. Each mirror can be tilted in both axes in analog fashion. The optical beam from an input port shines on the gimbal mirror that is associated with this port. This gimbal mirror then adjusts its angles in both axes, and redirects the optical beam to the

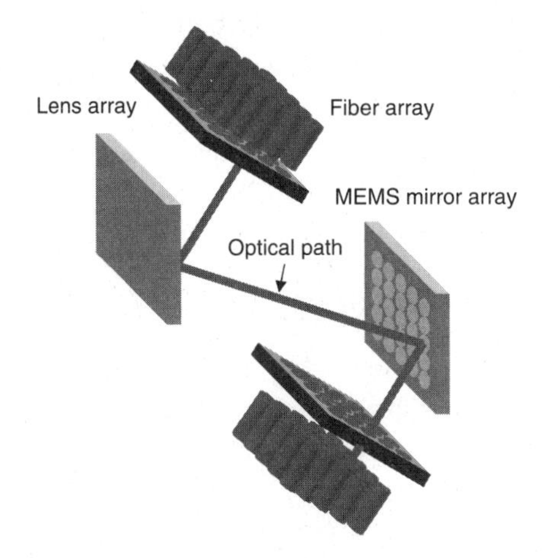

Fig. 5.34 Schematic drawing of a 3-D analog beam-steering switch. (*See color plate*).

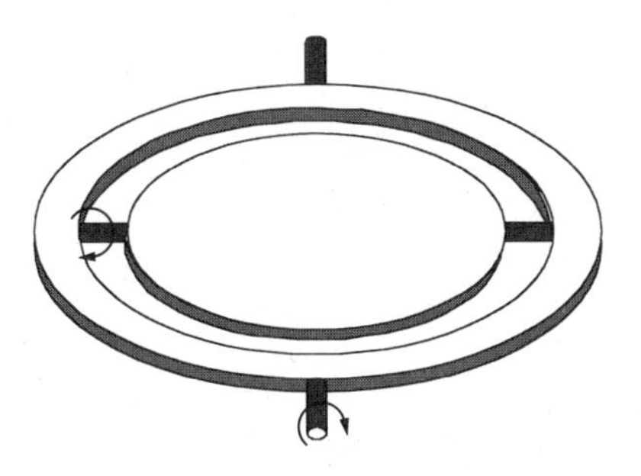

Fig. 5.35 A two-axis gimbal mirror.

mirror in the second MEMS mirror array, associated with the desired output port. The second mirror also adjusts its angles in both axes, and redirects the optical beam to the output lens/fiber with required angular precision. Two gimbal mirrors are required to perform switching because high-precision angular alignment between the optical beam and the lens/fiber, well under one milli-radian, is required to achieve low coupling loss.

This approach trades off the simplicity of digital control in the 2-D MEMS crossbar switches with the possibility of achieving 1000-port switches in a single-stage structure. The chief challenge for this technology is that of controlling the micro-mirror with the required angular precision, as will be discussed in more detail in the next section. The technology was first demonstrated [36] using switch arrays containing the individual optical switch modules that include optical fiber, collimating lens, servo LEDs, and the MEMS mirror. A quad-detector, also integrated into the optical switch module monitors the images of the LEDs on the opposite switch array and provides continuous tracking of mirror rotation angle for use in a closed-loop feedback system. Such closed-loop feedback control is crucial for achieving the required angular alignment precision.

5.3.2.3. Challenges for MEMS Optical Switching Technologies

Divergence of the Optical Beam

Although free-space optical switches in general offer impressive optical quality in the form of low crosstalk, low dispersion, low polarization dependence and low wavelength dependence, their loss performance tends to be slightly more challenging. This is because the loss in such systems is fundamentally limited by the divergence of fundamental-mode Gaussian optical beams propagating in free space [37]. The basic theoretical and experimental description of coupling loss in 2-D crossbar switches can be found in [38]. The results, summarized in Fig. 5.36, can also be applied to 3-D analog beam-steering switches when the horizontal axis is converted to corresponding propagation distances.

The divergence of a fundamental-mode Gaussian beam is described by the following two equations:

$$w(z) = w_0\sqrt{1 + (z/z_0)^2} \quad 1/e \text{ half-width versus propagation}$$
$$R(z) = z(1 + (z_0/z)^2) \quad \text{Radius of curvature of the wave-front versus propagation} \tag{5.12}$$

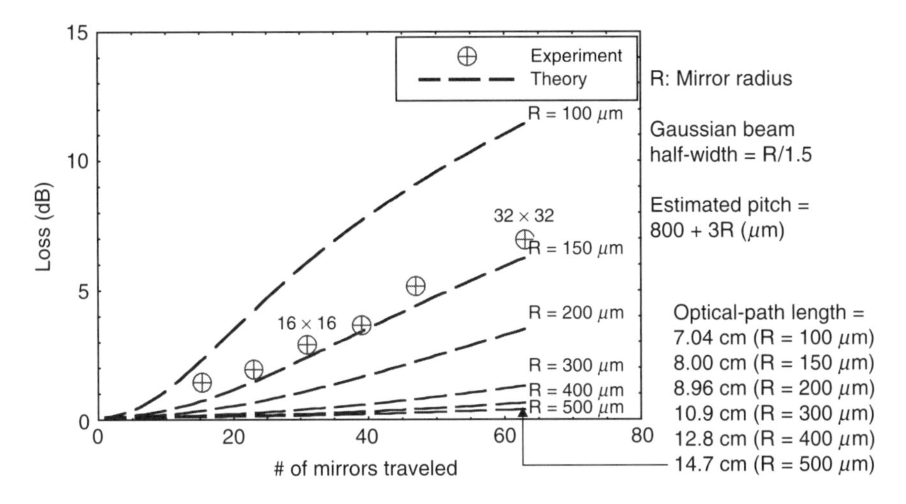

Fig. 5.36 Loss versus propagation distance in free-space optical switches, due to divergence of optical beam.

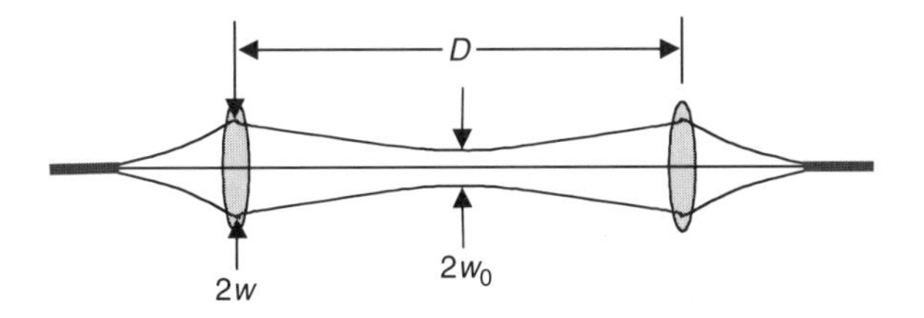

Fig. 5.37 Mode-matching in free-space optical coupling.

where w_0 is the $1/e$ half-width at the beam waist, and z is the propagation distance. $z_0 = \pi w_0^2/\lambda$ is the propagation distance when $w(z) = \sqrt{2}w_0$. Although large optical-beam diameter reduces divergence, it requires large mirrors that are difficult to control and to fabricate in practice. The number of mirrors that can be integrated on a single wafer is also correspondingly reduced. Therefore, there exists an optimal beam size after optimizing the above-mentioned considerations.

To compensate for the effects of optical-beam divergence, mode-matching, as shown in Fig. 5.37, has been very widely employed. To achieve mode-matching, the optical fiber is retracted slightly from the focal point of the lens, and slight focusing of the optical beam is achieved after the lens, so that the minimum beam-waist is located at the midpoint of the optical path. Because the mode of the optical beam matches the mode of the receiving fiber at the receiving plane, in principle, zero coupling loss can be achieved. This approach, however, is still limited by beam diffraction in free space. The diffraction limit, which is equivalent to the effect of optical-beam divergence, can be derived from Eq. (5.12) and results in

$$w = w_0\sqrt{1+\left(\frac{D\lambda}{2\pi w_0^2}\right)^2} \sim w_0\left(\frac{D\lambda}{2\pi w_0^2}\right) = \frac{D\lambda}{2\pi w_0}. \qquad (5.13)$$

For 2-D crossbar switches, the minimum distance between input and output is geometrically determined by port count and the size of the micro-mirror, which also determines the upper-limit of the beam size. The fabrication limit favors small mirrors while the diffraction limit prefers large beam size for high port count, and therefore large mirrors. The combination of the fabrication limit in mirror size and the diffraction limit results in an optimum beam width, which is in general fairly large, and therefore limits the port count in single-stage switch even when mode-matching is employed.

For 3-D analog beam-steering switches, the minimum distance between input and output is geometrically determined by the number of ports to be supported (equal to the number of micro-mirrors on the wafer) the size of the micro-mirrors, and by the maximum tilt angle of the gimbal mirror. At the same time, the optical beam width also decides the size of the micro-mirror, and therefore the size of the wafer and minimum propagation distance. These facts, in combination with the fact that long propagation distance requires a large optical beam (shown by Eq. 5.13) as well as the fabrication limit for mirror size, result in an optimal optical beam width. As a result, complete mode-matching can be achieved only if the mirror tilt angle is sufficiently large for reasonable mirror size.

Angular Control of the Micro-mirror

In free-space optical switches, the insertion loss is highly sensitive to angular misalignment, as discussed in [38]. The theoretical simulation and experimental verification of loss versus angular misalignment in a 2-D crossbar switch is shown in Fig. 5.38. The theory applies to 3-D analog beam-steering switches as well. The optical path in 3-D analog beam-steering switches is in general longer than that in 2-D digital crossbar switches; therefore, the requirement for angular alignment precision is also more stringent. Furthermore, in 2-D digital crossbar switches, the switching motion is digital, and integrated or external angular alignment

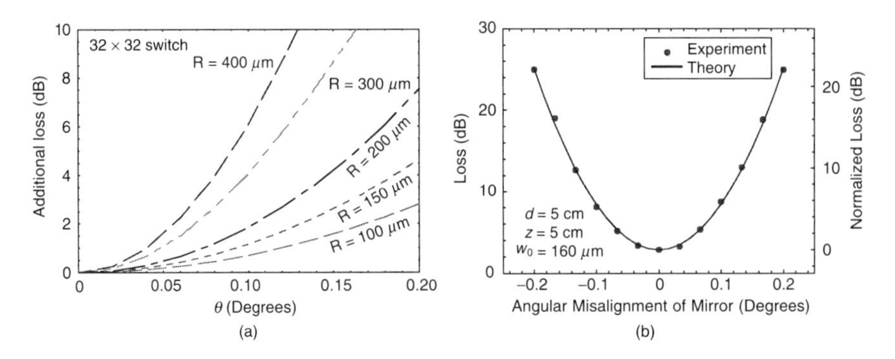

Fig. 5.38 (a) Theoretical simulation, and (b) experimental verification for loss versus angular misalignment in a 2-D crossbar switch. R is the mirror radius, which is related to the $1/e$ half-width of the optical beam w_0 by $R = 1.5\ w_0$. d is the distance between the input fiber and the micro-mirror, and z is the distance between the micro mirror and the output fiber.

structures can be incorporated to help define the angle of the switch mirrors. On the other hand, such approaches cannot be implemented in 3-D analog beam-steering switches, and controlling the angular precision of the gimbal mirrors in analog fashion becomes the chief challenge for this technology.

MEMS gimbal mirrors, being simple harmonic oscillators, have fundamental modes that exhibit a Lorentzian response to angular drive excitation given by

$$P(w) = \frac{w_0^2}{(iw)^2 + 2i\zeta w w_0 + w_0^2}, \tag{5.14}$$

where ζ, the damping coefficient of the mirror, is related to the mirror Q-factor by $Q = 1/2\zeta$. w_0 is the resonant angular frequency of the fundamental mode. Figure 5.39 contains the Bode plot (amplitude and phase responses in frequency domain) of a gimbal mirror together with its amplitude response in time when driven by a step function. The mirror has a resonant frequency of 500 Hz, and a damping coefficient of 0.1. The amplitude response settles to $1/e$ in a damping time $\tau = Q/2\pi f_0$. MEMS gimbal mirrors fabricated in single-crystal silicon, if without special fabrication design, commonly exhibit rather high Q-factors in the neighborhood of 150, and resonant frequencies on the order or 500 Hz [39]. This results in a damping time constant of ~50 ms, well in excess of the settling times required by core-network switching applications.

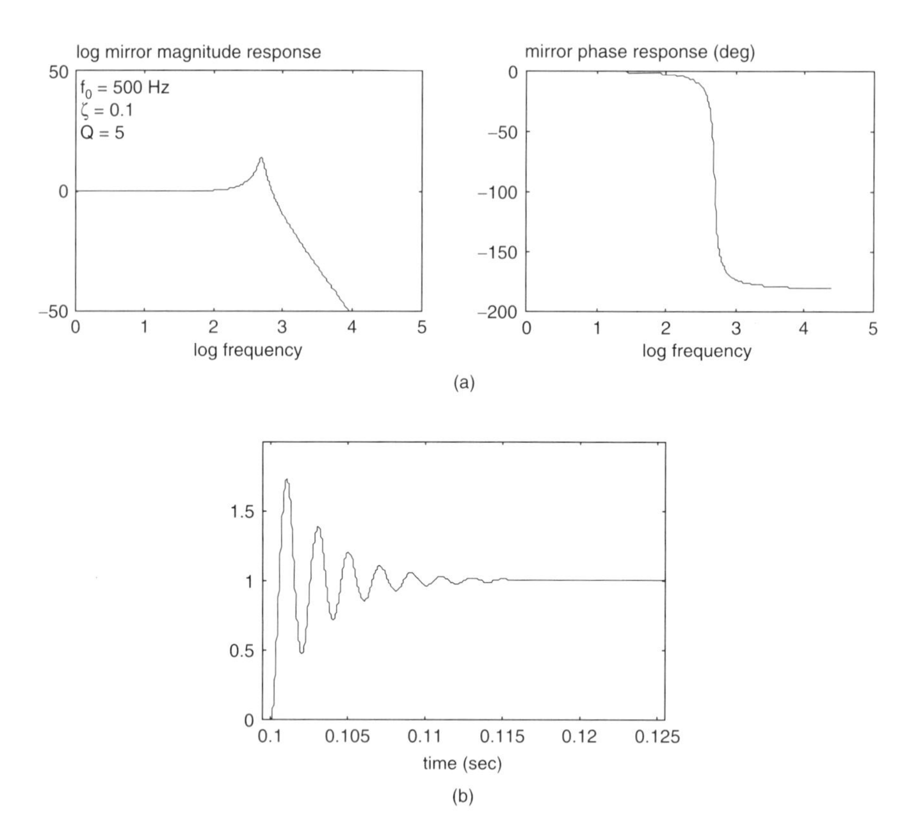

Fig. 5.39 (a) Bode plot, and (b) amplitude response in time domain, of a gimbal mirror.

If the mirror properties can be characterized precisely, one can solve the problem by tailoring the drive pulses to inverse-match the Q-factor and resonant frequency of the mirrors. This in principle offers a route to achieving short switching times, as shown in Fig. 5.40(a). However, such approaches exhibit extremely strong sensitivity to modeling error and fabrication uniformity of devices over the surface of a wafer. Figure 5.40(b) shows, for example, a 10% error in the estimation of damping coefficient, and 1% error in the estimation of resonant frequency. Both result in significant mirror overshoot and ringing. In addition, open-loop approaches confer no immunity whatever to stochastic perturbations due to shock and vibration. Therefore, closed-loop control has been actively pursued by various groups working on 3-D analog beam-steering switches, to achieve short switching time and stochastic immunity.

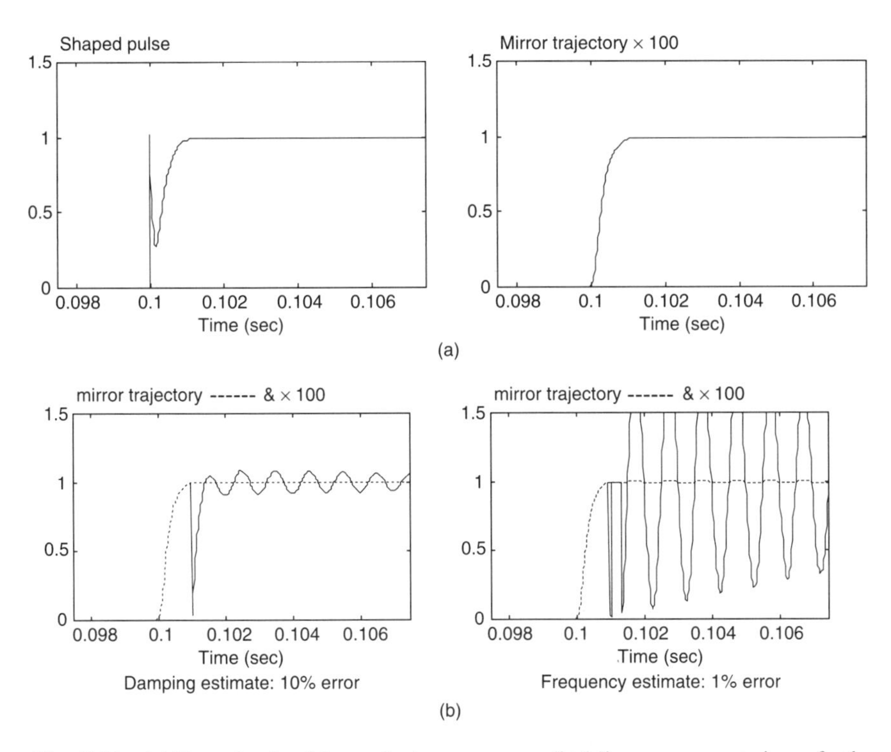

Fig. 5.40 (a) Shaped pulse drive and mirror response. (b) Mirror response to imperfectly modeled shaped pulse drive.

5.4. Conclusion

Over the past few years, MEMS technology has been widely and actively probed by the research community for possible applications in WDM communication systems. The technology offers virtues and limitations that are substantially different from those of the conventional technology it seeks to displace. Owing to its free-space nature, and the unique IC-compatible fabrication process of integrating optical, mechnical, and electrical structures on a single chip, MEMS offer compact and potentially low-cost means of implementing reconfigurable or switchable devices with substantial functionality. In this chapter, the designs and applications of MEMS in optical crossconnects, reconfigurable wavelength add/drops, dynamic gain-equalizers, chromatic dispersion-compensators, polarization controllers, tunable lasers, detectors, and filters are discussed.

References

1. K. E. Petersen, "Silicon as a mechanical material," Proc. IEEE, 70 (1982) 420–457.
2. C. Marxer and N. F. d. Rooij, "Micro-opto-mechanical 2 × 2 switch for single-mode fibers based on plasma-etched silicon mirror and electrostatic actuation," J. Lightwave Technol., 17 (1999) 2–6.
3. L. S. Fan, Y. C. Tai, and R. S. Muller, "Pin-joints, springs, cranks, gears, and other novel micromechanical structures," presented at 4th Int. Conf. Solid-State Sensors and Actuators, Tokyo (1987).
4. W. C. Tang, H. Nguyen, M. W. Judy, and R. T. Howe, "Electrostatic-comb drive of lateral polysilicon resonators," Sensors and Actuators, A21 (1990) 328–331.
5. R. A. Conant, J. T. Nee, K. Y. Lau, and R. S. Muller, "A flat high-frequency scanning micromirror," presented at Solid-State Sensor and Actuator Workshop, Hilton Head Island, SC (June 4–8, 2000).
6. K. S. J. Pister, M. W. Judy, S. R. Burgett, and R. S. Fearing, "Microfabricated hinges," Sensors and Actuators A, 33 (1992) 249–256.
7. M. C. Wu, L. Y. Lin, S. S. Lee, and K. S. J. Pister, "Micromachined free-space integrated micro-optics," Sensors and Actuators A, 50 (1995) 127–134.
8. S. Blackstone and T. Brosnihan, "SOI MEMS technologies for optical switching," presented at International Optical MEMS Conference, Okinawa, Japan (Sep. 25–28, 2001).
9. G.-D. J. Su, H. Toshiyoshi, and M. C. Wu, "Surface-micromachined 2-D optical scanners with high-performance single-crystalline silicon micromirrors," IEEE Photonics Technology Letters, 13 (2001) 606–608.
10. E. C. Vail, M. S. Wu, G. S. Li, L. Eng, and C. J. Chang-Hasnain, "Widely tunable micromachined GaAs Fabry-Perot filters," Electron. Lett., 31 (1995) 228–229.
11. M. S. Wu, E. C. Vail, G. S. Li, W. Yuen, and C. J. Chang-Hasnain, "Tunable micromachined vertical cavity surface emitting lasers," Electron. Lett., 31 (1995) 1671–1672.
12. M. S. Wu, E. C. Vail, G. S. Li, W. Yuen, and C. J. Chang-Hasnain, "Widely and continuously tunable micromachined resonant cavity detector with wavelength tracking," IEEE Photonics Technol. Lett., 8 (1996) 98–100.
13. M. C. Larson and J. J. S. Harris, "Wide and continuous wavelength tuning in a vertical-cavity surface-emitting laser using a micromachined deformable-membrane mirror," Applied Physics Letters, 68 (1996) 891–893.
14. G. L. Christenson, A. T. T. D. Tran, Z. H. Zhu, Y. H. Lo, M. Hong, J. P. Mannaerts, and R. Bhatt, "Long-wavelength resonant vertical-cavity LED/photodetector with a 75-nm tuning range," IEEE Photonics Technol. Lett., 9 (1997) 725–727.

15. Doug Anthon, J. D. Berger, Joe Drake, S. Dutta, Al Fennema, John D. Grade, Stephen Hrinya, Fedor Ilkov, Hal Jerman, David King, Howard Lee, Alex Tselikov, and K. Yasumura, "External cavity diode lasers tuned with silicon MEMS," Optical Fiber Communication Conference, paper TuO7, Anaheim, CA, March 17–22, 2002.
16. C. R. Giles and E. Desurvire, "Modeling erbium-doped fiber amplifiers," J. Lightwave Technol., 9 (1991) 271–283.
17. A. R. Chraplyvy and R. W. Tkach, "What is the actual capacity of single-mode fibers in amplified lightwave systems?," IEEE Photonics Technol. Lett., 5 (1993) 666–668.
18. J. E. Ford and J. A. Walker, "Dynamic spectral power equalization using micro-mechanics," IEEE Photonics Technol. Lett., 10 (1998) 1440–1442.
19. C. R. Giles, V. Aksyuk, B. Barber, R. Ruel, L. Stulz, and D. Bishop, "A silicon MEMS optical switch attenuator and its use in lightwave subsystems," IEEE J. Selected Topics in Quantum Electronics, 5 (1999) 18–25.
20. J. A. Walker, K. W. Goosen, and S. C. Arney, "Fabrication of mechanical antireflection switch for fiber-to-the-home systems," J. Micro-Electro-Mech. Syst., 5 (1996).
21. P. Yeh, Optical Waves in Layered Media (John Wiley, New York 1988).
22. C. K. Madsen, J. A. Walker, J. E. Ford, K. W. Goosen, T. N. Nielson, and G. Lenz, "A tunable dispersion compensating MEMS all-pass filter," IEEE Photonics Technol. Lett., 12 (2000) 651–653.
23. C. D. Poole and J. Nagel, "Polarization effects in lightwave systems," in Optical Fiber Communications, vol. IIIA (San Diego: Academic Press, 1997).
24. C. Pu, L. Y. Lin, E. L. Goldstein, N. J. Frigo, and R. W. Tkach, "Micromachined integrated optical polarization-state rotator," IEEE Photonics Technol. Lett., 12 (2000) 1358–1360.
25. J. Albert, F. Bildeau, D. C. Johnson, K. O. Hill, K. Hattori, T. Ktagawa, Y. Hibino, and M. Abe, "Low-loss planar lightwave circuit OADM with high isolation and no polarization dependence," IEEE Photonics Technol. Lett., 11 (1999) 346–348.
26. C. R. Doerr, L. W. Stulz, M. Cappuzzo, E. Laskowski, A. Paunescu, L. Gomez, J. V. Gates, S. Shunk, and A. E. White, "40-wavelength add-drop filter," IEEE Photonics Technol. Lett., 11 (1999) 1437–1439.
27. J. E. Ford, V. Aksyuk, D. J. Bishop, and J. A. Walker, "Wavelength add-drop switching using tilting micromirrors," J. Lightwave Technol., 17 (1999) 904–911.
28. C. R. Giles, B. Barber, V. Aksyuk, R. Ruel, L. Stulz, and D. Bishop, "Reconfigurable 16-Channel WDM DROP module using silicon MEMS optical switches," IEEE Photonics Technol. Lett., 11 (1999) 63–65.
29. A. R. Ranalli, B. A. Scott, and J. P. Kondis, "Liquid crystal-based wavelength selectable cross-connect," presented at 25th European Conference on Optical Communication, Nice, France (Sep. 26–30, 1999).

30. C. Pu, L. Y. Lin, E. L. Goldstein, and R. W. Tkach, "Client-configurable eight-channel optical add/drop multiplexer using micromachining technology," IEEE Photonics Technol. Lett., 12 (2000) 1665–1667.
31. L. Y. Lin, E. L. Goldstein, and R. W. Tkach, "Free-space micromachined optical switches with submillisecond switching time for large-scale optical crossconnects," IEEE Photonics Technol. Lett., 10 (1998) 525–527.
32. R. T. Chen, H. Nguyen, and M. C. Wu, "A low voltage micromachined optical switch by stress-induced bending," presented at 12th IEEE International Conference on Micro Electro Mechanical Systems, Orlando, FL (1999).
33. B. Behin, K. Y. Lau, and R. S. Muller, "Magnetically Actuated micromirrors for fiber-optic switching," presented at Solid-State Sensor and Actuator Workshop, Hilton Head Island, SC (1998).
34. H. Toshiyoshi and H. Fujita, "Electrostatic micro torsion mirrors for an optical switch matrix," J. Microelectromechanical Systems, 5 (1996) 231–237.
35. L. Y. Lin, E. L. Goldstein, and R. W. Tkach, "Free-space micromachined optical switches for optical networking," IEEE J. Selected Topics in Quantum Electronics: Special Issue on Microoptoelectromechanical Systems (MOEMS), 5 (1999) 4–9.
36. H. Laor, "MEMS mirrors application in optical cross-connects," presented at IEEE LEOS Summer Topical Meetings: Optical MEMS, Monterey, CA (1998).
37. J. T. Verdeyen, Laser Electronics, 2nd ed (Prentice-Hall, Englewood Cliffs, NJ, 1989).
38. L. Y. Lin, E. L. Goldstein, and R. W. Tkach, "On the expandability of free-space micromachined optical crossconnects," J. Lightwave Technol., 18 (2000) 482–489.
39. S. Pannu, C. Chang, R. S. Muller, and A. P. Pisano, "Closed-loop feedback-control system for improved tracking in magnetically actuated micromirrors," presented at International Conference on Optical MEMS, Kauai, Hawaii (Aug. 21–24, 2000).

Chapter 6 | PLC-type Thermooptic Switches for Photonic Networks

Yoshinori Hibino

NTT Photonics Laboratory 3-1, Monnosato Wakamiya Atsugi Kanagawa, 243-0198 Japan

6.1. Introduction

The rapid and global spread of the Internet is accelerating the growth of optical communication networks. Photonic networks based on wavelength division multiplexing (WDM) systems [1–3] have played a key role in increasing the capacity and flexibility of these networks. Typical types of photonic networks are shown schematically in Fig. 6.1. They began as point-to-point WDM transmission systems, and are now evolving into ring networks. Various optical devices have been developed for WDM-based photonic networks, and some have already been installed in commercial communication systems. Of these, filters and optical switches are two of the most important passive devices used in the networks. Filters are needed wherever signals with different wavelengths propagating in a fiber have to be multiplexed or demultiplexed in WDM networks. We also need various kinds of optical switches for optical add/drop multiplexing (OADM) systems [4, 5] and optical crossconnect (OXC) systems [6] to make networks more flexible. Moreover, the rapid progress made on networks has led to a demand for more channels at a lower cost. This means we must develop larger-scale optical devices that are less expensive.

Optical switches composed of silica-based planar lightwave circuits (PLCs) have been developed for a variety of applications in photonic networks, including OADM and OXC systems. The main feature

WDM TECHNOLOGIES: PASSIVE OPTICAL COMPONENTS
$35.00

ISBN: 0-12-225262-4

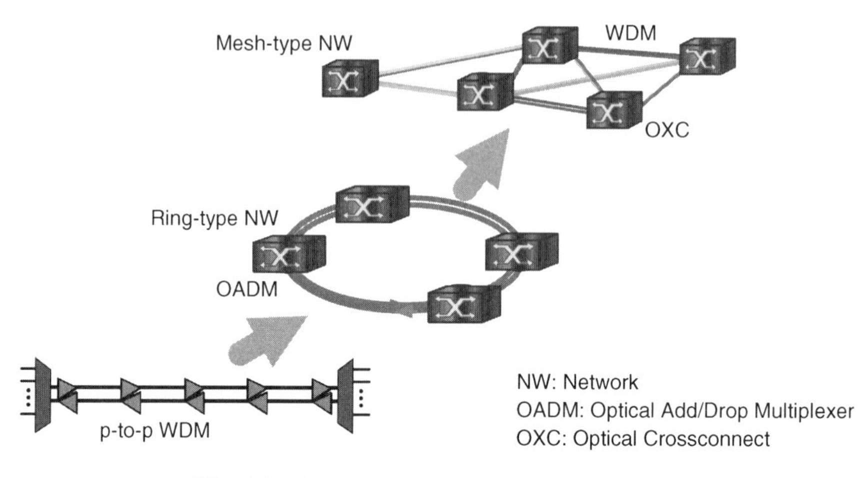

Fig. 6.1 Typical photonic network configurations.

distinguishing these applications is the required switching time, which varies from a few milliseconds to subnanosecond levels. PLC-type optical switches are mainly used to provide lightpaths and system protection because the switching speed of passive switches is usually only of millisecond order [7]. For protection applications, the switches are used to move the traffic stream from a primary fiber to another fiber should the former fail. Small 2×2 switches are usually sufficient for this purpose. For these applications, the switches are used inside optical crossconnects to reconfigure them to support new lightpaths. The challenge here is to realize larger switches.

In addition to switching time, other important parameters used to characterize the suitability of a switch for optical networking applications are insertion loss, the extinction ratio between on and off states, crosstalk, polarization-dependent loss (PDL) and wavelength dependence. These characteristics are reviewed here for PLC-type switches. The state of integration of optical switches is considerably less than that of electric switches, as illustrated by the fact that a 16×16 optical switch, is currently considered a large switch.

This section reviews the development of PLC-type optical switches for WDM-based photonic networks. While several bulk-type and fiber-type optical components have already been installed in practical systems, PLC-type devices have also been playing an important role recently because of their suitability for large-scale integration and mass production.

PLC characteristics are briefly described in Section 6.2, PLC-type switch characteristics are described in Section 6.3, and their applications are demonstrated in the subsequent sections.

6.2. Planar Waveguide Circuits

PLCs, in which fiber-matched silica-based waveguides are integrated, can provide various key devices for optical networks [8–10]. This is because they are suitable for large-scale integration, offer long-term stability, and can be mass produced. The PLC device family has been extended from the 1st to the 4th generation as shown in Fig. 6.2, and includes optical couplers, 1/N optical power splitters, thermooptic switches (TOSW), arrayed waveguide grating (AWG)-type multi/demultiplexers [11, 12], and lattice filters [13, 14] for high-speed transmission systems. AWGs are key components in high-capacity WDM networks, and we have recently fabricated a 400-channel 25-GHz spacing AWG using high-contrast waveguides [15]. Lattice-type interleave filters separate WDM signal channels with an equal spacing into two groups with twice the spacing, thus doubling the WDM channel number available using a conventional multi/demultiplexer. A flat-top and low-loss interleaver [16] has been constructed based on the lattice circuit theory. This interleaver is composed of directional couplers and waveguide delay lines. The PLC-type TOSW is one of the most important devices for photonic networks, and it is described in detail in subsequent sections.

The process used for fabricating an optical waveguide is shown in Fig. 6.3 [17]. First, glass soot consisting of an undercladding layer and a core layer

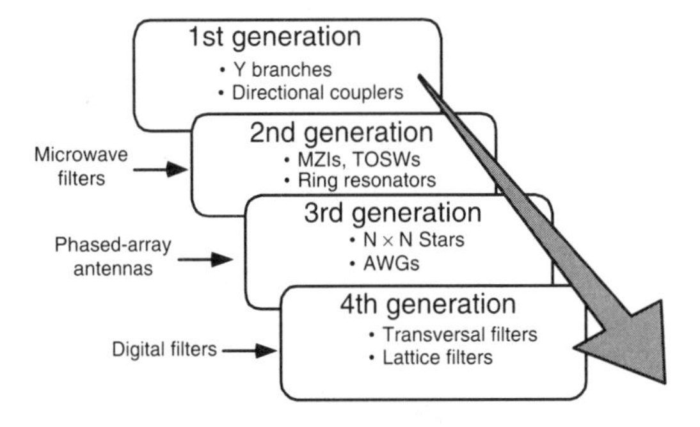

Fig. 6.2 Evolution of silica-based PLC devices.

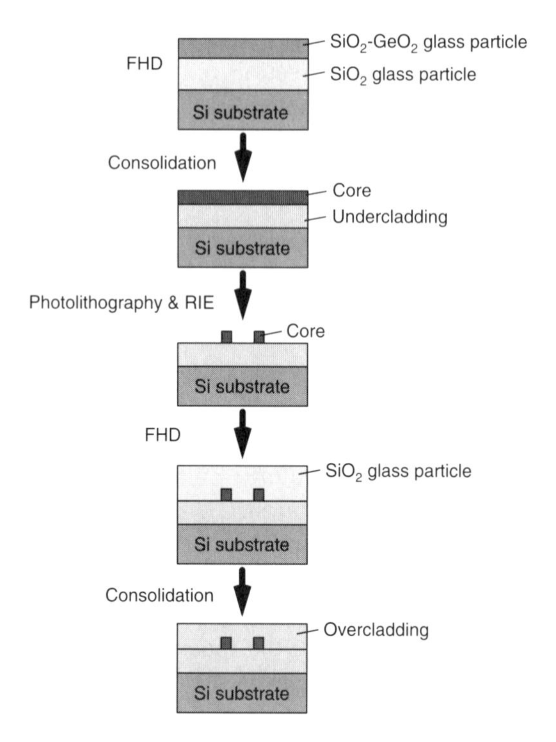

Fig. 6.3 PLC fabrication process.

is deposited on a silicon (Si) substrate by flame hydrolysis deposition (FHD). This method was developed by modifying the vapor-phase axial deposition (VAD) method used in the manufacture of optical fiber and planar waveguides. Specifically, the FHD method introduces gas materials such as $SiCl_4$ into an oxy-hydrogen burner, thus causing an oxidation reaction within the flame, and then deposits fine particles such as SiO_2 on the substrate. Here, the main component of the undercladding layer is SiO_2, but traces of B_2O_3 and P_2O_5 are included as dopant materials to lower the softening-point temperature below the melting temperature of the Si substrate while keeping the refractive index fixed. Both B_2O_3 and P_2O_5 work to lower the softening-point temperature of SiO_2, but the former lowers the refractive index of SiO_2 while the latter raises it. GeO_2 is included in the core layer to raise the refractive index slightly [18]. Next, the Si substrate with the deposited soot layers is heated in an electric furnace at temperatures ranging from 1200 to 1300°C to form transparent glass layers. In the next step, the top of the core layer is coated with photoresist, onto which a circuit pattern is transferred using the photolithographic technology employed in the manufacture of large-scale integrated (LSI) circuits. This resist is then used as an etching mask for reactive ion etching (RIE), which is employed to remove sections other than those that are to become the core. After this, the FHD method is used again to deposit soot and form an overcladding layer. The components of this layer are the same as those of the undercladding layer, but the amounts of B_2O_3 and P_2O_5 are increased slightly to reduce the consolidation temperature of the overcladding to below the temperature used to form the undercladding layer and prevent core deformation. Finally, the overcladding is also consolidated in an electric furnace to form transparent glass.

The advantages of the silica-on-Si waveguide are its low propagation loss [19] and low coupling loss with optical fibers. Specifically, we have realized a propagation loss of less than 0.02 dB/cm in a 10-m-long waveguide [20]. These waveguides with a well-defined structure and a low loss enable us to design various optical integrated circuits by using numerical simulation techniques. In terms of PLC device reliability, fiber-pigtailed 1×8 splitter modules have been the subject of a detailed investigation based on the Telcordia reliability requirements for passive optical devices, which include long-term damp heat and mechanical tests [21, 22]. The PLC chips themselves were chemically and physically stable, and their reliability mainly depended on fiber connections. The tests on the fiber-pigtailed splitter modules confirmed the high reliability of PLC-type devices. Moreover, the lifetime of the 1×8 splitter modules was estimated from tests in which temperature and humidity were varied. The results yielded a 30-year failure rate at 25°C-90% RH of as low as below 0.6 FIT.

6.3. Mach-Zehnder Interferometer Type TO Optical Switches

6.3.1. FUNDAMENTALS OF MZI

PLC-type optical switches consist of a Mach-Zehnder interferometer (MZI) with a phase shifter (thin-film heater). The MZI is an extremely effective circuit for achieving a variety of optical functions in an optical waveguide circuit. Figure 6.4 shows examples of optical circuits with MZIs. Each MZI consists of two couplers and two optical waveguide arms that connect the couplers. These couplers are generally 3-dB (50%) couplers and come in several types including the Y-junction [23], the directional coupler

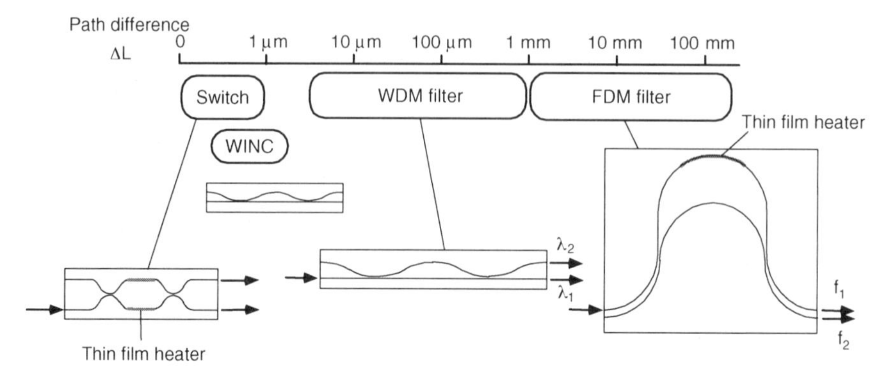

Fig. 6.4 Mach-Zehnder interferometer-type devices.

(DC) [24], and the multi-mode interferometer (MMI) coupler [25]. In this format, light incident on the MZI splits along two paths at the first-stage 3-dB coupler, then reunites and interferes at the second-stage coupler. Various functions can be achieved by controlling the interference conditions using couplers, switches, modulators, wavelength (frequency) filters, and other devices.

The MZIs are realized by using silica-based PLC technology, and their circuit functions are classified according to the difference in optical path length of the waveguide arms (Fig. 6.4). For a difference ΔL of zero, a circuit equipped with a phase shifter (thin-film heater) acts as an optical switch. An MZI with a ΔL of about 0.6 μm operates as a wavelength-insensitive coupler (WINC) [26]. This function is achieved by canceling out the wavelength dependence of the directional couplers with the wavelength dependence of the phase difference generated in the waveguide arms. For example, extremely flat characteristics at a coupling efficiency of $20 \pm 2\%$ can be obtained in this way in the 1.25 to 1.65 μm wavelength region. An MZI with a ΔL of several to several tens of micrometers functions as a wavelength division multiplexing (WDM) filter that multiplexes, for example, signals with wavelengths at 1.3 and 1.55 μm [27]. Finally, an MZI circuit with an even larger ΔL can function as a frequency division multiplexing (FDM) filter capable of handling signals at 10-GHz (0.08-nm) intervals [28].

Of the MZI-type devices fabricated using silica-based PLC technologies, this section focuses on the MZI-type optical switch with a phase shifter whose optical path-length difference is extremely small. Various kinds of optical switch have been demonstrated by integrating 2×2 MZI-type switch units. The following describes the 2 × 2 MZI switch, which is the most basic component of PLC-type optical circuits. First, we describe the fabrication of a PLC-type optical switch, including the process for fabricating an optical waveguide and a TO phase shifter. Then, we explain the basic principle and characteristics of the switch, which employs the TO effect.

6.3.2. PLC-TOSW FABRICATION

PLC-TOSWs are usually fabricated using optical waveguides with relative refractive-index difference Δ between the core and the cladding of either 0.25 or 0.75%. The core size of the 0.25%Δ waveguide is 8 μm × 8 μm, and the core of the 0.75%Δ waveguide is 6 μm × 6 μm. The cladding thickness is about 50 μm and the Si-substrate thickness is 1 mm. Furthermore, a waveguide with a 0.25%Δ has the same optical-intensity distribution as optical fiber, and as a result, the connection loss with optical fiber is no greater than 0.1 dB/point, thus enabling us to realize a low-loss circuit. However, as Δ is small in this case, light confinement is weak and the curvature radius of the curved waveguide must be as large as 30 mm or greater. This increases the circuit size, which is not conducive to a large-scale circuit. By contrast, a waveguide with a 0.75%Δ has strong light confinement, allowing us to realize a curvature radius as small as 5 mm. In this case, however, the connection loss with optical fiber is as large as 0.4 dB/point. So, we use a waveguide with a Δ of 0.25% for small-scale circuits such as 2 × 2 switches and a waveguide with a Δ of 0.75% for large-scale circuits such as 1 × N and N × N matrix switches.

In PLC-TOSWs, switching is the result of a TO effect achieved by heating optical waveguides. This requires a thin-film heater that functions as a TO phase shifter. Figure 6.5 shows the process used for fabricating the heaters and the gold wiring that provides electrical power to the heaters. Two types of film are used as thin-film heaters: Ta_2N sputter film and Cr vapor-deposited film. Figure 6.5(a) shows the process for fabricating the Ta_2N heater. This process begins with the formation of Ta_2N film on the waveguides by the sputtering method. The thickness of this film is controlled to obtain the resistance desired for the heating function. Next, the heater pattern is transferred to a resist by a photolithography technique and non-heater sections are removed by RIE using the resist as a mask. Figure 6.5(b) shows the liftoff fabrication process, which is used to pattern a vapor-deposited film for a Cr heater or gold wiring. The pattern for a Cr

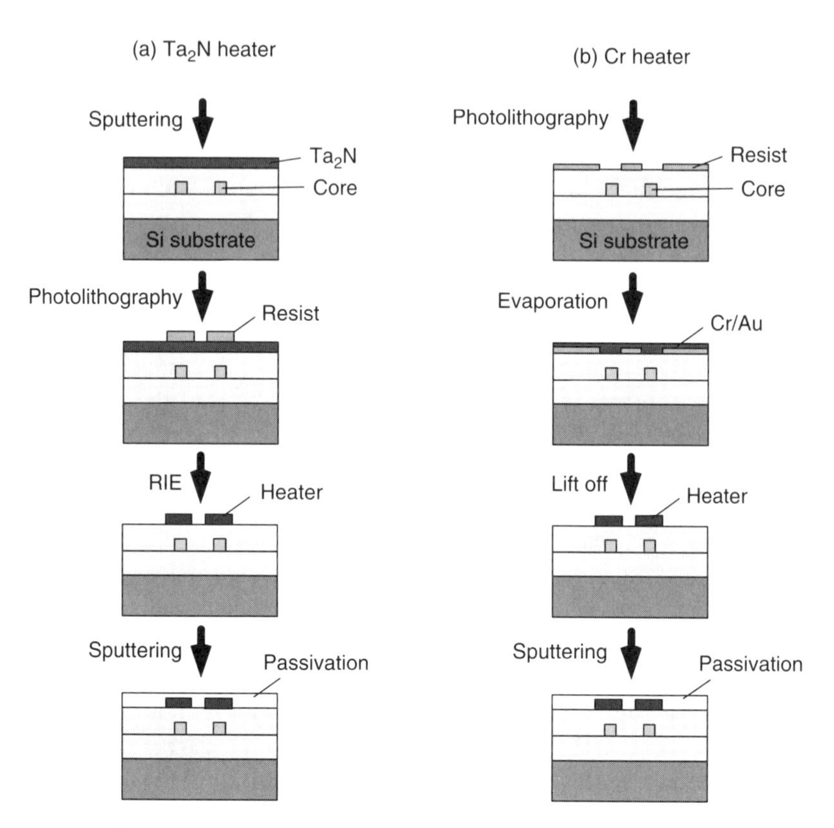

Fig. 6.5 Process for fabricating metal heaters and gold wiring.

heater or gold wiring is transferred to resist on top of a waveguide, again by a photolithographic process, and in this case, resist corresponding to the heater and wiring sections is removed. Next, a Cr or gold film is deposited over the entire wafer surface. Finally, the resist and the thin film above it are both removed by a resist-removal solution and the heater or wire is formed. An SiO_2 film may also be sputtered on top of the heater or wire to prevent heater oxidation or breaks in the wiring.

The Ta_2N sputter heater is superior to the Cr vapor-deposited heater in terms of power durability, and with little fluctuation in resistance due to oxidation, is also highly reliable. Consequently, while the easily fabricated Cr vapor-deposited heater was mainly used in the early stages of research, the highly reliable Ta_2N sputter heater is being used as the commercialization of 8×8 matrix switches progresses. A standard heater is about 5 mm long and 50 μm wide.

6.3.3. *OPTICAL CHARACTERISTICS OF TOSW*

The TO effect is a phenomenon by which the refractive index of a substance changes with temperature. In silica glass, this effect is characterized by an increase in the refractive index as the temperature rises. The silica waveguides targeted in this research also exhibit a TO effect in which the refractive index increases with increasing temperature. The value of this index is given by the following equation:

$$n(T) = 0.9 \times 10^{-8}T^2 + 1.02 \times 10^{-5}T + 1.4497 \tag{6.1}$$

This is an empirically derived equation with T being the temperature of the waveguide in degrees Celsius. Figure 6.6 shows the refractive index with respect to waveguide temperature as calculated by Eq. (6.1). As shown, the refractive index is about 1.45 for a waveguide temperature of 25°C. This equation tells us that the TO effect in silica waveguides is a nonlinear effect because temperature has a squared term. The coefficient of this term, however, is extremely small, which means that the term can be ignored near room temperature. The TO effect for silica glass may therefore be thought of as an almost linear effect. This plot in Fig. 6.6 confirms this—the change in the refractive index with temperature is basically linear.

Silica-based PLC-type optical switches can be made to perform a switching function by combining the interferometer configuration with the

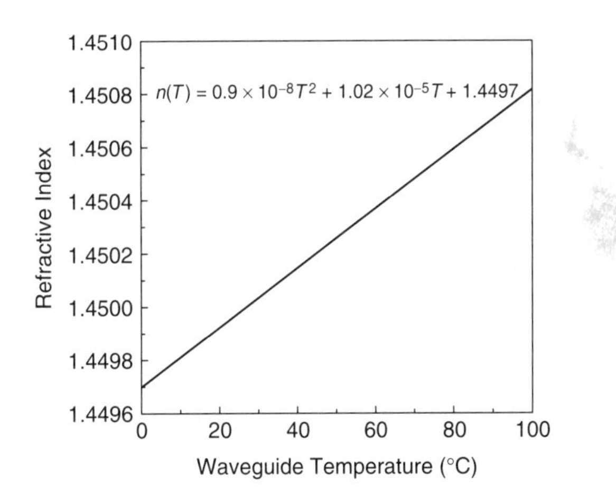

Fig. 6.6 Temperature dependence of refractive index in PLC.

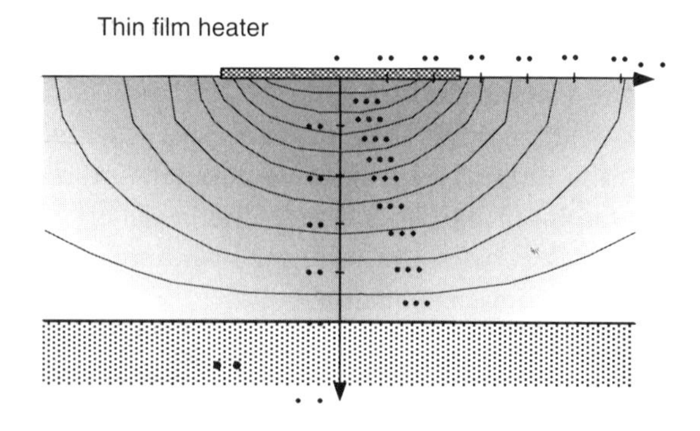

Fig. 6.7 Schematic view of temperature distribution in PLC.

TO effect. Specifically, switching is achieved by driving a thin-film metal heater deposited directly on a waveguide and changing the temperature in the vicinity of the core. Figure 6.7 shows the temperature distribution over the cross-section of a glass waveguide that has been heated by a thin-film heater. These results were obtained by finite element method (FEM) analysis and are normalized with the temperature of glass in contact with the center of the heater, which is taken to be 1. In the calculations, we set the thickness of the glass layer and the heater width both at 50 μm. The substrate was Si crystal, and the ambient environment of the thin-film heater and the overcladding was air. The thermal conductivity values of the Si substrate, silica glass, and air at room temperature (27°C) were 1.70 [W/(cm · deg)], 0.014 [W/(cm · deg)], and 2.61×10^{-4} [W/(cm · deg)], respectively. Because heat flows to the substance with the highest thermal conductivity value, most of the heat generated by the thin-film heater propagates through the glass and flows to the Si substrate. The Si substrate acts as a good heatsink because its thermal conductivity is higher than that of the glass, and because its volume is much larger than that of the glass layer. As a result, heat that reaches the Si substrate radiates after spreading out uniformly within the substrate, and the temperature distribution of the Si substrate can therefore be considered practically uniform and about the same as that of the outside air. As a consequence, the temperature distribution shown in Fig. 6.7 occurs in a glass waveguide sandwiched between a heater (heating element) and an Si substrate at outside air temperature. From the figure, we can see that the temperature near the core is about 40% that of the heater when the core is positioned at the center of the cladding, and that the core temperature rises or drops when the core is moved toward

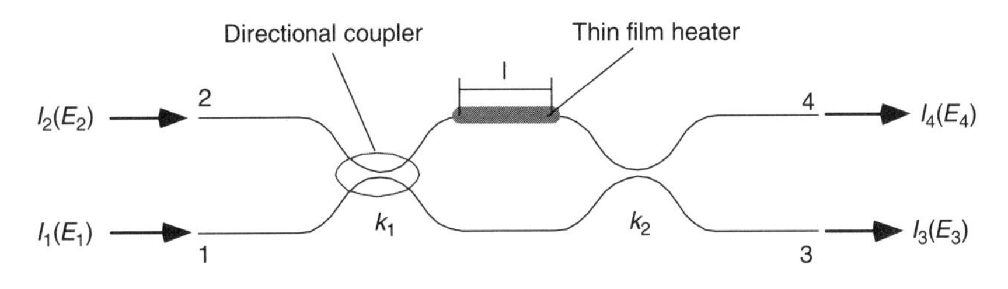

Fig. 6.8 Configuration of Mach-Zehnder interferometer.

or away from the heater, respectively. In short, if a certain core temperature is determined for switching, the heater temperature must be lowered if the core is situated near the heater and raised if it is far from the heater. In other words, the heater power consumption is small if the core is near the heater and large if it is far from it.

Next, to clarify the way in which an MZI switch operates, we first derive the optical output equations of an MZI. The MZI configuration is shown in Fig. 6.8. Here, light incident on the interferometer splits at the first-stage DC, and after passing through the waveguide arms, recombines and interferes at the second-stage DC before exiting at output ports. The state of the output light can be given by the following equation in terms of the directional couplers and waveguide arms [29]:

$$\begin{bmatrix} E_3 \\ E_4 \end{bmatrix} = M_{\mathrm{D1}} M_{\mathrm{AM}} M_{\mathrm{D2}} \begin{bmatrix} E_1 \\ E_2 \end{bmatrix} \tag{6.2}$$

In this equation, E_1 and E_2 denote light input from input ports 1 and 2; E_3 and E_4 the light output from output ports 3 and 4; M_{DC1} and M_{DC2} the transfer matrices of the first- and second-stage directional couplers; and M_{MZ} the transfer matrix of the MZ waveguide arms. The DC transfer matrix can be given by the following equation if the two waveguides making up the coupler have the same shape and if their propagation constants are the same:

$$M_{\mathrm{D1}} = \begin{bmatrix} \cos(\kappa d) & j\sin(\kappa d) \\ j\sin(\kappa d) & \cos(\kappa d) \end{bmatrix} \tag{6.3}$$

Here, κ is the coupling coefficient and d is the coupling length. Next, denoting the phase difference between the MZ waveguide arms as $\Delta\phi$, the transfer matrix for light passing through the MZ waveguide arms can be given as follows:

$$M_{\mathrm{AM}} = \begin{bmatrix} \exp(j\Delta\phi/2) & 0 \\ 0 & \exp(-j\Delta\phi/2) \end{bmatrix} \tag{6.4}$$

The phase difference $\Delta\phi$ is expressed as follows.

$$\Delta\phi = 2\pi n\Delta L/\lambda \tag{6.5}$$

Here, $n\Delta L$ is the optical path length difference generated between the waveguide arms, n is the waveguide's effective refractive index, ΔL is the difference between the optical path lengths of the waveguide arms, and λ is wavelength.

We now express the intensity of optical output in terms of electric power using Eqs. (6.2) to (6.5), where power is expressed as the product of the electric field and its complex conjugate, as shown following:

$$I = EE^* \tag{6.6}$$

Here, I_3 and I_4 indicate the optical output power from ports 3 and 4. The DC power transfer ratio (coupling efficiency) is given by the following equation:

$$k = \sin^2(\kappa d) \tag{6.7}$$

Accordingly, if we denote the DC coupling efficiency of stages 1 and 2 as k_1 and k_2, respectively, Eq. (6.2) takes on the following form.

$$\begin{aligned} I_3 = {} & \{(\sqrt{(1-k_1)(1-k_2)} - \sqrt{k_1 k_2}\,)^2 \\ & + 4\sin^2(\Delta\phi/2)\sqrt{k_1 k_2(1-k_1)(1-k_2)}\,\}I_1 \\ & + \{(\sqrt{k_1(1-k_2)} - \sqrt{(1-k_1)k_2})^2 \\ & + 4\cos^2(\Delta\phi/2)\sqrt{k_1 k_2(1-k_1)(1-k_2)}\,\}I_2 \end{aligned} \tag{6.8}$$

$$\begin{aligned} I_4 = {} & \{(\sqrt{k_1(1-k_2)} - \sqrt{(1-k_1)k_2})^2 \\ & + 4\cos^2(\Delta\phi/2)\sqrt{k_1 k_2(1-k_1)(1-k_2)}\,\}I_1 \\ & + \{(\sqrt{(1-k_1)(1-k_2)} - \sqrt{k_1 k_2})^2 \\ & + 4\sin^2(\Delta\phi/2)\sqrt{k_1 k_2(1-k_1)(1-k_2)}\,\}I_2 \end{aligned} \tag{6.9}$$

Here, I_1 and I_2 denote the input light power from ports 1 and 2.

In this section, we describe the switching operation of the TO 2×2 switch. Here, we assume a TO 2×2 switch with a symmetric MZ configuration in which the waveguide arms are equal in length. Also, for the sake of clarity, we limit the input light to port 1 and assume that the two directional couplers have the same coupling efficiency ($k = k_1 = k_2$). In this case, Eqs. (6.8) and (6.9) become as follows:

$$I_3/I_1 = (1-2k)^2 + 4k(1-k)\sin^2(\Delta\phi/2) \tag{6.10}$$

$$I_4/I_1 = 4k(1-k)\cos^2(\Delta\phi/2) \tag{6.11}$$

The phase difference $\Delta\phi$ in the preceding equations is expressed as follows:

$$\Delta\phi = 2\pi\Delta nl/\lambda \tag{6.12}$$

Here, Δnl is the difference in the optical path lengths generated between the waveguide arms, Δn is the change in the refractive index due to the TO effect, and l is the length of the thin-film heater. First, the difference in the optical path length between the waveguide arms when the thin-film heater is not operating is 0 ($\Delta\phi = 0$) and Eqs. (6.10) and (6.11) become as follows:

$$I_3/I_1 = (1-2k)^2 \tag{6.13}$$

$$I_4/I_1 = 4k(1-k) \tag{6.14}$$

Consequently, for a DC coupling efficiency of 50% ($k = 0.5$), we obtain the following equations:

$$I_3/I_1 = 0 \tag{6.15}$$

$$I_4/I_1 = 1 \tag{6.16}$$

As a result, all the light is output at cross-ports 1 to 4. By contrast, when the thin-film heater on one waveguide arm is operated to give an optical-path-length difference between the waveguide arms of half a wavelength (a phase difference of π), the output light becomes as follows, regardless of DC coupling efficiency:

$$I_3/I_1 = 1 \tag{6.17}$$

$$I_4/I_1 = 0 \tag{6.18}$$

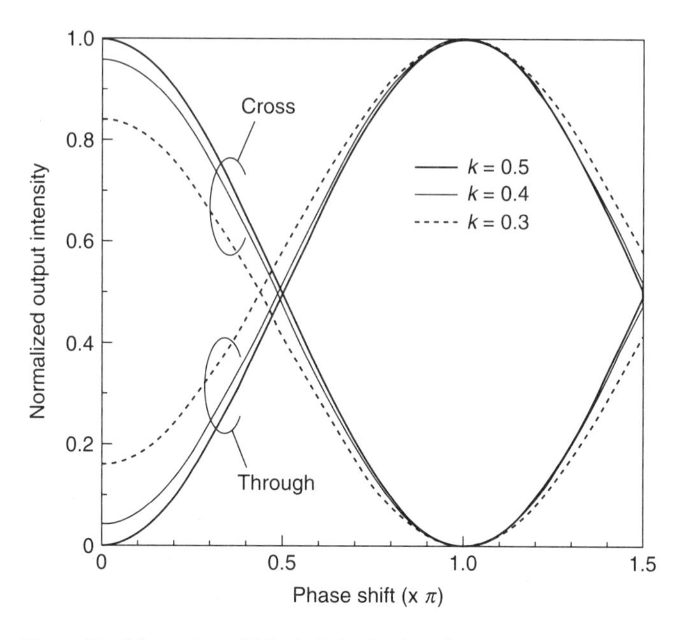

Fig. 6.9 Normalized intensity of Mach-Zehnder interferometer with different coupling ratios.

In other words, the output port switches from port 3 to 4. In the following discussion, the state during which the thin-film heater is not operated is called the "OFF-state," and that during which the heater is operated and the optical-path-length difference between the waveguide arms is half a wavelength is called the "ON-state."

Figure 6.9 shows the optical output characteristics versus the phase difference between the waveguide arms as calculated from Eqs. (6.10) and (6.11). Here, the phase difference on the horizontal axis may be read as a change in temperature near the core or a change in the effective refractive index of the waveguide as a result of the TO effect, or even the heater power consumption. These calculations were performed for three DC coupler efficiency values: $k = 0.5$, 0.4, and 0.3. As shown, some light leaks to the through-port (port 1 to 3 or port 2 to 4 in Fig. 6.2) as the coupling efficiency shifts from 0.5 at a phase difference of 0. Figure 6.10 shows the results of calculating the amount of light leaked to the through-port versus coupling efficiency k. This is the amount of leaked light with respect to input light, and we see, for example, that the DC coupling efficiency must be kept within $50\% \pm 5\%$ to hold the leaked light at below 20 dB. By contrast, for a phase difference of π, all light outputs pass to

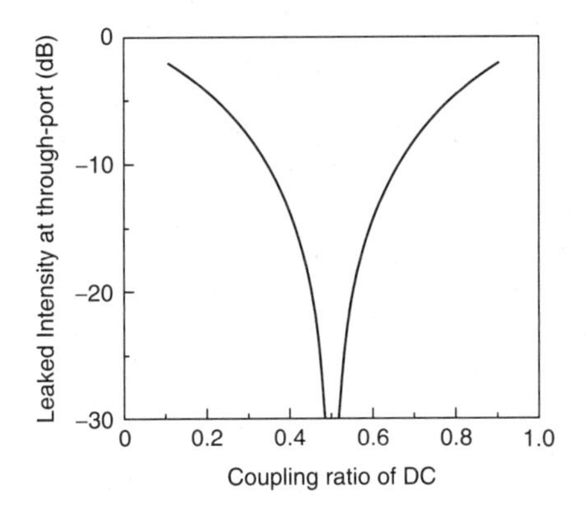

Fig. 6.10 Relationship between leaked intensity at through-port and coupling ratio in MZI.

the through-port regardless of the coupling efficiency. This relationship between phase difference and leaked light is a characteristic that must be considered when constructing a low crosstalk switch, and is particularly important when discussing and constructing a low-crosstalk matrix switch.

We consider the temperature change and power consumption required for switching in the 2×2 switch. The temperature of the waveguide arm directly under the heater required for switching is estimated to be about 40°C at 1.55 μm from Eqs. (6.1) and (6.2) if the heater length is 5 mm and the substrate temperature is 25°C. This means that switching will occur if the temperature near the core rises by about 15°C. Accordingly, the heater temperature change and power consumption at the time of switching for a cladding thickness of 50 μm with the core located at the center of the cladding center would be about 37.5°C and 0.5 W, respectively.

Another important parameter when evaluating a switch is the time needed for switching. This is called "switching time" or "switching speed," here defined as the time required to change the light intensity from 10 to 90%. In addition, the time involved in supplying power to the TO phase shifter and outputting the light (or quenching the light, depending on the port) is called "rise time," and the time involved in cutting off the power supplied to the TO phase shifter and quenching the light (or outputting the light, depending on the port) is called "fall time." Figure 6.11 shows the thermal response characteristics for a 2×2 optical switch. The response characteristics shown at the bottom of the figure depict the voltage applied to the TO phase shifter and those at the top of the figure the corresponding

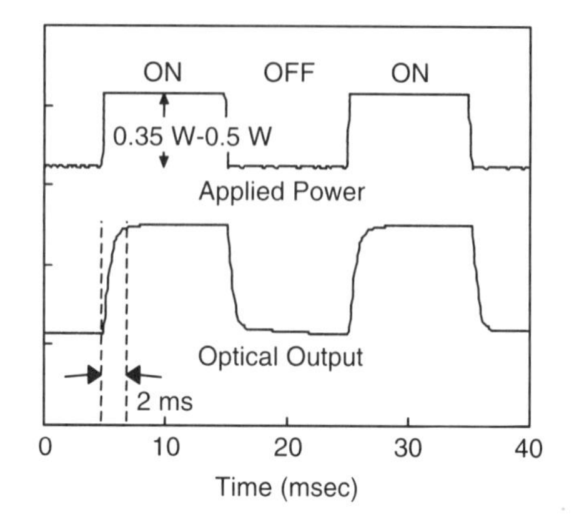

Fig. 6.11 Typical switching power and time of PLC-type MZI switch.

optical output. These are the characteristics for a through-port (port 1 to 3 in Fig. 6.7) in a symmetric MZ switch. The switching time in this example consists of a rise time of 1.1 ms and a fall time of 1 ms. The switching time is mainly determined by cladding thickness and core position. In general, though, the switching time increases with increases in cladding thickness. The switching times for the optical switches targeted in this research lie in the 1 to 3 ms range.

In addition to the switching time, there are other important parameters used to characterize the suitability of a switch for optical networking applications. They are listed following.

1. The extinction ratio of an ON-OFF switch is the ratio of the output power in the ON-state to the output power on the OFF-state. This ratio should be as large as possible.
2. The insertion loss of a switch is the power fraction (usually expressed in dB) that is lost because of the presence of the switch, and must be as small as possible.
3. For a given switching state or interconnection pattern, and output, the crosstalk is the ratio of the power at that output from the desired input to the power from all other inputs. Usually, the crosstalk of a switch is the worst-case crosstalk over all output and interconnection patterns.
4. As with other components, switches should have a low polarization-dependent loss (PDL).

6.3.4. PHASE-TRIMMING TECHNIQUE [30]

The MZI-type TO switch can be used as a gate switch, and its extinction ratio should be as high as possible. Equation (6.11) shows that the extinction ratio of an MZI-type switch is infinite for both ports with a DC coupling efficiency of 50% ($k = 0.5$). However, there is usually a slight deviation in the coupling efficiency ($k \neq 0.5$). If $k \neq 0.5$, a comparison of Eqs. (6.10) and (6.11) shows that the cross port (port 1 to port 4) has a higher extinction ratio. Because the gate switch is generally used as a normally closed state, that is, to shut down lights in the off state, the electric power should be zero in the off state. So, an MZI-type TO switch with a half-wavelength path difference has been proposed for use as a gate switch. This switch can shut down output lights with a high extinction ratio in the off state without power.

However, the transmittance of the MZI-type TO switch with a half-wavelength path difference is not always minimum when the heating power is zero. This is because slight fabrication errors cause a phase error in the waveguides and a slight deviation of the optical path length difference in the MZI from the designed value. Figure 6.12 shows the typical switching characteristics of a fabricated MZI with a half-wavelength path difference. It should be noted that the transmittance minimum shifts from an applied power of zero. Therefore, we apply a small bias power, even when the MZI

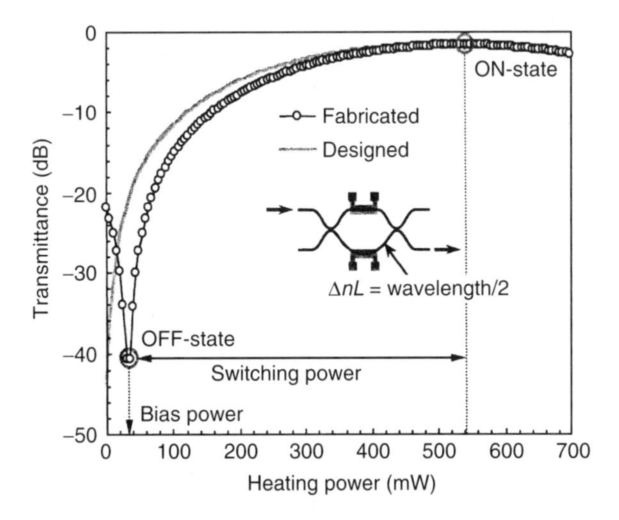

Fig. 6.12 Typical switching power offset in PLC-type MZI switch.

is in the OFF-state, to obtain a sufficient extinction ratio [31]. This electrical biasing, however, causes two serious problems in larger-scale matrix switches. One is the increased power consumption needed to maintain the switch state and the other is the complicated driving circuit.

A simple and practical phase-trimming technique, already developed [24], can permanently correct the phase-error in each MZI and enable us to avoid the need for electrical biasing. The phase-trimming technique employs a local heating method with a thin film heater [32]. The heating power for the phase-trimming is very high (4–7 W), about 10 times the usual heating power for switching (0.3–0.5 W). When a low power of around 0.5 W is applied to the heater, the refractive index change is a temporary change caused by the TO effect, which reverts completely once the power is switched off. However, the refractive index increases irreversibly when a high power of around 5 W is applied. Thus, some portion of the refractive index increase remains permanently after the power has been switched off. This permanent change does not occur in the low power region and is stable during thermal annealing for 90 min at 300°C [32].

While the physical mechanism of this permanent refractive index change is not clear, we suppose the index increase to be induced by the stress from the thin film heater or from overcladding glass in the vicinity of the heater. We estimated the temperature of a heater with an applied power of around 5 W to be roughly 450°C by means of a thermal analysis simulation. As a result, we believe that the structural change in the thin film heater or the overcladding glass is probably caused by the high temperature.

Figure 6.13 shows the permanent refractive index change dependence on heating time with heating power as a parameter. By carefully selecting the heating power and time, the phase-error can be eliminated and the bias power adjusted to zero, as described in the next subsection. The phase-trimming technique requires no additional optical equipment such as a high-power laser with focusing lenses. It needs only electrical equipment for measuring the switching characteristics.

The cross-path extinction ratio ER of an MZI without bias power is expressed as:

$$\mathrm{ER} = \sin^2(\phi_{\mathrm{err}}) \tag{6.19}$$

where, ϕ_{err} is the phase-error in the MZI. Calculations with this equation show that the phase-error must be reduced to under 3.6 or 2.0% of the half-wavelength phase change to obtain extinction ratios of over 25 or 30 dB, respectively, in an MZI without bias power. Therefore, high phase-shift

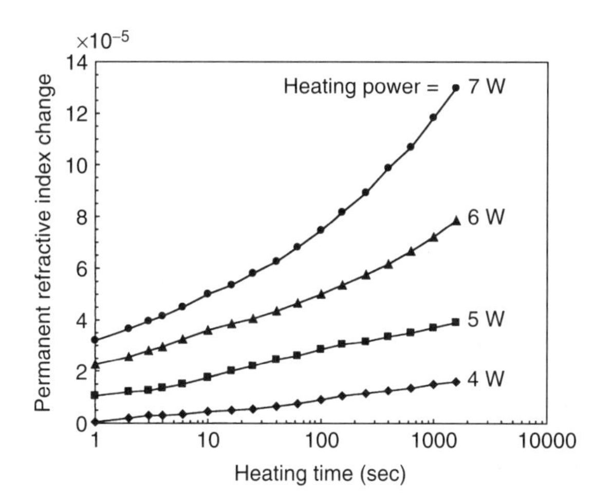

Fig. 6.13 Permanent index change in PLC by high-power heating.

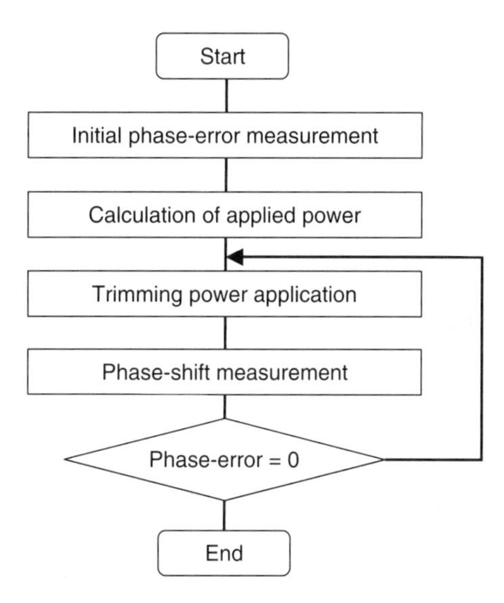

Fig. 6.14 Phase-trimming procedures.

accuracy is necessary for the phase-trimming. Figure 6.14 shows the phase-trimming procedure used in the automatically controlled trimming equipment developed to obtain high phase-shift accuracy. An appropriate heating power is selected, taking account of the required phase-shift value in each MZI in order to finish the phase-trimming in a practical time. Then the trimming power is applied while the phase-shift is monitored.

The preceding describes the basic characteristics of the silica-based PLC-type TOSW and also the fabrication technologies, optical output equations, and phase-trimming technique. In the following section, we describe various optical switches composed of MZI switch units in terms of their configuration and optical characteristics.

6.4. Application of 2 × 2 PLC-type TOSW

6.4.1. VARIABLE OPTICAL ATTENUATOR (VOA) [33]

Variable optical attenuators (VOAs) are now playing a more important role in the gain control of linear-repeaters for WDM networks and channel power equalization in WDM OADM nodes. Such applications require a relatively fast operating speed of ~1 ms and a compact size. While high-speed non-mechanical VOAs using a Faraday rotator have already been proposed, a silica-based PLC is another candidate for realizing high-speed compact VOAs. Because PLC technologies enable us to integrate arrayed VOAs, a PLC-type VOA is more advantageous with regard to size than a Faraday rotator-type VOA. In addition, a PLC is more thermally stable than a Faraday rotator. Here we describe the principle and characteristics of the PLC-type VOA.

The PLC-type VOA is composed of a simple symmetric MZI with TO phase shifters. To compensate for the polarization dependence of the VOA, a half-wavelength plate made of polyimide is inserted between the MZI arms. We can vary the attenuation by controlling the electric power applied to the phase shifter. We fabricated the VOA using conventional FHD and RIE techniques.

The measured transmittance of the fabricated VOA is shown in Fig. 6.15, in which the solid and broken lines represent the maximum and minimum output for various polarization states, respectively. The polarization sensitivity obtained from the two lines is also shown in Fig. 6.15 by a dot-dash line. We obtained a polarization sensitivity of less than 1 dB at an attenuation of 20 dB. The rise and fall times of the VOA were less than 2 ms. The VOA characteristics we obtained are summarized in Table 6.1. A wide wavelength range of over 20 nm with a wavelength dependence of 0.8 dB is sufficient for typical WDM systems. Thermal stability under 0.8 dB was obtained without any thermal compensation, while the ambient temperature was changed from 5 to 65°C. The attenuation reproducibility for the driving voltage we used was also good (<0.20 dB). Accordingly, the hysteresis, which is often observed in mechanical-type VOAs, was negligible.

Table 6.1 **Characteristics of Fabricated PLC-type 1 × N Switches**

Switch Scale	**1 × 8**	**1 × 16**	**1 × 32**
Switch Type	Tree	Tree + Tap (TTC)	Tree
Avg. Insertion Loss (dB)	1.2	1.9	1.6
Avg. Min. Extinction Ratio (dB)	52	59	62
Max. Power Consumption (W)	1.5	1.2	2.2
Waveguide Length (mm)	60	169	132
Circuit Number in Chip	2	2	1
Chip Size (mm^2)	18 × 60	35 × 74	68 × 70
Module Size (mm^3)	154 × 33 × 13	122 × 53 × 13	****

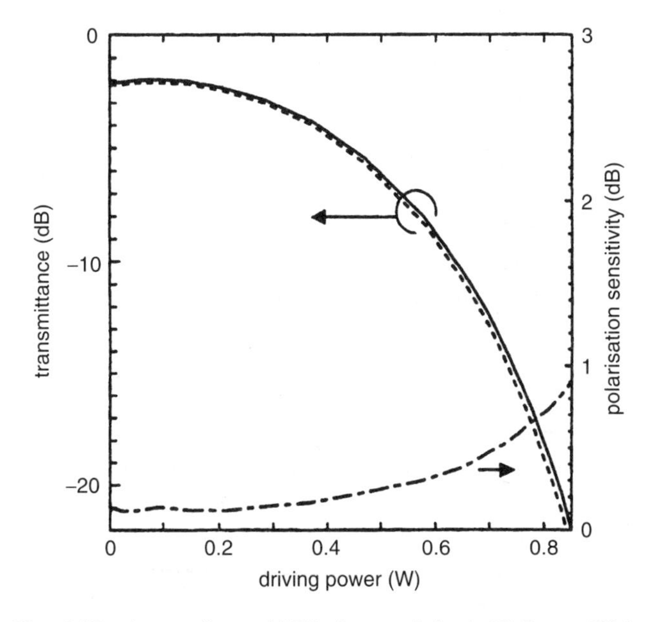

Fig. 6.15 Attenuation and PDL characteristics in PLC-type VOA.

6.4.2. EIGHT-ARRAYED 2 × 2 *SWITCH [34]*

An optical add/drop multiplexer (OADM), which adds and drops incoming and outgoing channels according to their wavelengths in WDM systems, is important in photonic networks. Various methods have been reported for OADM systems, including the combination of an AWG and a space

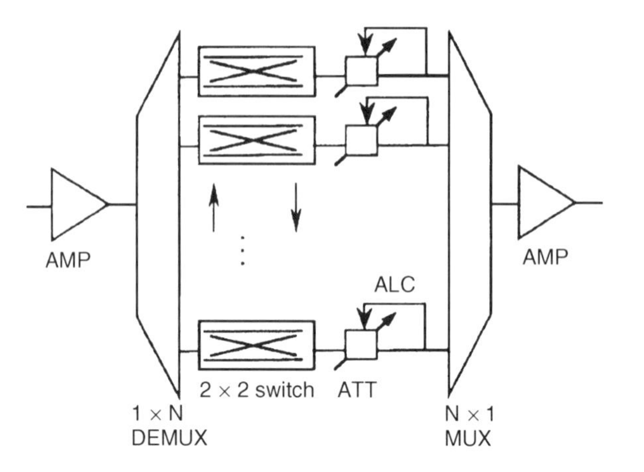

Fig. 6.16 Basic architecture of OADM system with MUX/DEMUX and 2 × 2 optical switches.

division optical switch, the combination of a fiber Bragg grating and a circulator, and a Bragg grating written in a Mach-Zehnder interferometer. Of these, the OADM with space switches between a pair of AWGs, which is shown in Fig. 6.16, has the advantage of allowing the switch state to be changed (add/drop or pass through) in service without affecting the other wavelength channels. A space division optical switch in the OADM node of an OADM system requires two functions. One is high-isolation switching for the add/drop operation. The switch must have a high extinction ratio in order to avoid any crosstalk-induced degradation in the bit error rate. The other requirement is channel-by-channel level equalization as shown in Fig. 6.16. Any power deviation among the WDM channels degrades the signal-to-noise ratio for weaker power channels while high-power channels suffer waveform distortion induced by fiber nonlinearities. Generally, the optical power of the input to the switch has a variation among the WDM signals, which arises from node loss or amplifier gain deviations. Therefore, level equalization is essential in order to keep the optical power of the WDM signals launched into the transmission fiber constant.

TOSWs are promising for an OADM system because of their compactness, low loss, high extinction ratio, and mass producibility. A novel TOSW configuration for an OADM system recently proposed and demonstrated is shown in Fig. 6.17(a), compared with a conventional switch circuit with a level equalizer. The new configuration consists of three TOSWs in a channel. One of the TOSWs is used as a component in a double-gate switch and as a level equalizer by utilizing the fact that the MZI has both optical

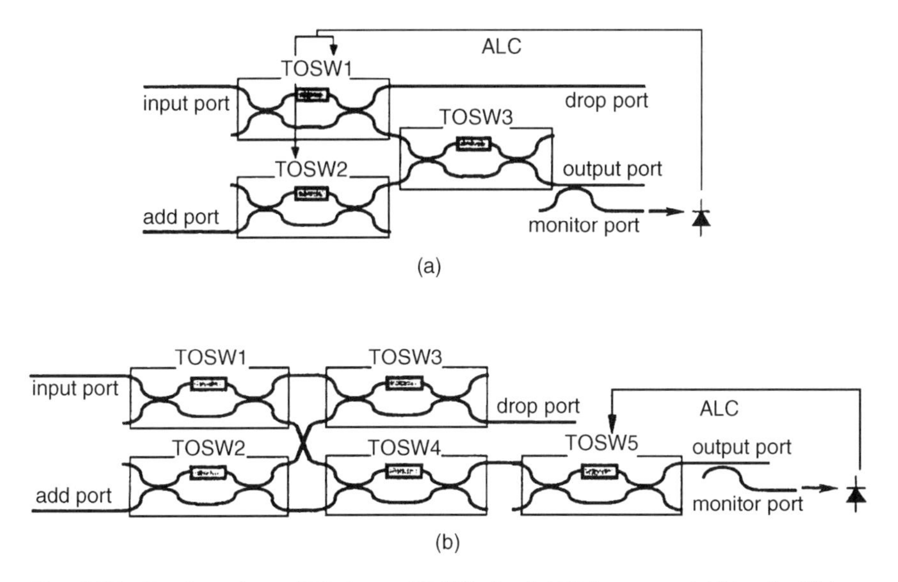

Fig. 6.17 Configurations of PLC-type TOSWs for OADM systems. (a) Novel add/drop switch. (b) Conventional add/drop switch.

switching and level equalization functions. By contrast, the conventional configuration is composed of five MZI-type TOSWs. The use of a small number of MZIs not only reduces the loss but also the scale and power consumption. Because the number of wavelength channels is large in an OADM system, reducing the switch size is beneficial in terms of system construction. The switch operation of the add/drop and through states of the new configuration is shown in Fig. 6.18. In the add/drop mode, the input signal and the added signal are led to the drop and output ports, respectively. TOSWs 1 and 3 operate as a double-gate switch for terminating the dropped signal. The added signal is terminated at the drop port because the paths are not connected. TOSW 2 performs level equalization for the added signal. In the through port the input signal is led to the output port, and TOSW 1 performs level equalization for the input signal. TOSWs 2 and 3 operate as a double-gate switch for terminating the added signal. As described in Fig. 6.18, the signal tapped at the output port is used for the auto level control (ALC) of both TOSWs 1 and 2. The ALC is switched between TOSW1 in the through mode and TOSW2 in the add/drop mode. This add/drop switch replaces the 2×2 switch and attenuator used in the OADM system.

The eight-arrayed add/drop switch with three symmetric MZIs per channel, shown in Fig. 6.17(a), was fabricated using silica-based PLCs with a

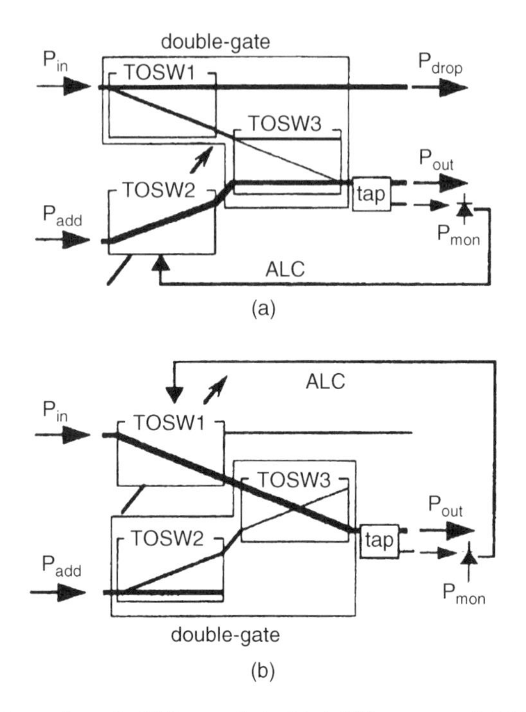

Fig. 6.18 Operation of add/drop switch. (a) Add/drop state. (b) Through state.

Δ value of 0.7%. The switch was provided with an electrical circuit for ALC.

Typical transmission characteristics of one TOSW are shown in Fig. 6.19. The maximum and minimum transmission curves were measured with a polarization-scrambled signal as a function of driving power. Two dips are clearly observed in the minimum transmission curve, corresponding to the maximum losses for the input signals with the TE and TM modes. The maximum extinction ratio of the TOSW was about 30 dB for any polarization state. To use the TOSW as a level equalizer, we need to set a controllable driving power range where the transmission decreases monotonically with the driving power. The maximum attenuation of this switch is 22 dB, which is large enough to equalize the signal level of the OADM.

The losses of the through, added, and dropped signals in the eight-channel arrayed switch are summarized in Fig. 6.20. The average losses of the input-to-output, input-to-drop, and add-to-output ports were 3.0, 3.1, and 3.2 dB, respectively. These loss values usually satisfy OADM

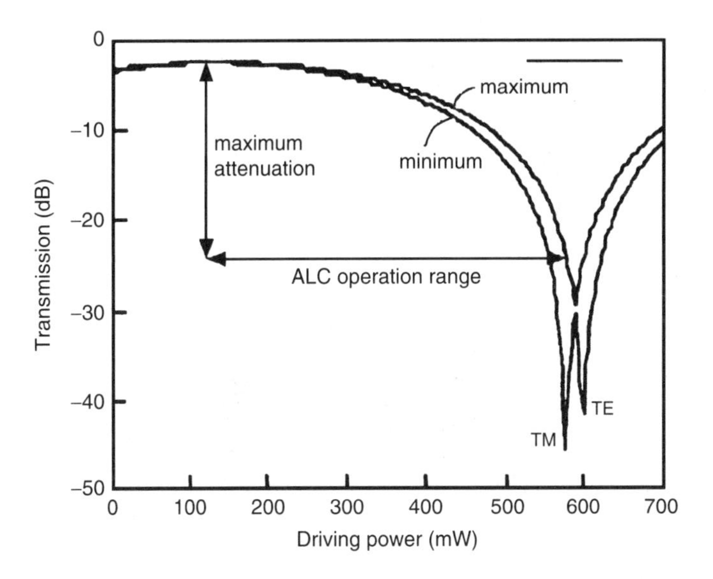

Fig. 6.19 Typical transmission characteristics of a PLC-TOSW.

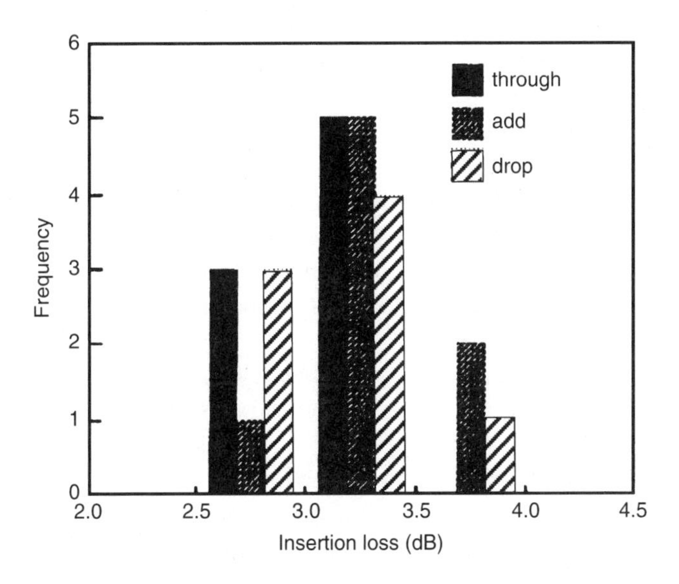

Fig. 6.20 Insertion loss of eight-channel arrayed add/drop switch.

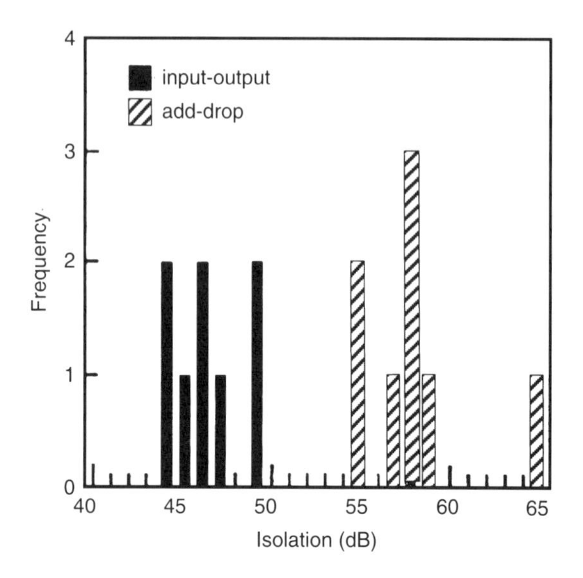

Fig. 6.21 Extinction ratio of 8-channel arrayed add/drop switch.

system requirements. The optical isolation of the eight-channel arrayed switch in the add/drop state is shown in Fig. 6.21. We measured the isolation defined by the loss ratio between the input-to-drop and input-to-output ports, or the loss ratio between add-to-output and add-to-drop ports, for eight channels, all in the add/drop state. The input signals terminated at more than 44 dB at the output port in any signal polarization state due to the double-gate switching. This is a typical value for a double-gate switch configuration. The added signal terminated at more than 54 dB at the drop port in any signal polarization state because the waveguide paths from the add port were not connected to the drop port. The extinction ratio of more than 44 dB resulted in the penalty due to in-band crosstalk being less than 0.03 dB.

The power crosstalk from the other channels was measured while signals were being launched into the seven other input or added signals in the through and add/drop states, respectively. A crosstalk of less than −42 dB was achieved under any launching conditions. To confirm that the dropped signal was unaffected by the neighboring channels, we estimated the power penalty caused by the power crosstalk. Seven crosstalk signals −42 dB lower than the dropped signal resulted in the power penalty being less than 0.002 dB. This value is much lower than the power penalty caused by in-band crosstalk because the signals from neighboring channels have different wavelengths. This low penalty value indicates that each add/drop switch integrated in the same chip was optically isolated.

In addition, we measured the ALC performance by varying the power level and the polarization angle of the linearly polarized input signal. The output power was measured under ALC operation at each input power. The variation in attenuation was below 0.3 dB, and polarization-independent ALC operation was successfully demonstrated. The slight deviation against polarization was due to the polarization-dependent loss of the monitor tap of 0.3 dB. As previously shown, we successfully developed a novel switch for OADM systems that provides high isolation between added and dropped signals and the polarization-independent automatic level control of output signals in through and add/drop states . The number of TOSWs required for the add/drop operation was reduced to only three. This resulted in higher integration density, lower loss, and reduced driving power consumption.

6.4.3. OPTICAL ADD/DROP MULTIPLEXER INTEGRATED WITH AWG

Recently we have developed an integrated OADM filter consisting of AWGs and TO switches [35]. This compact device provides simultaneous access to all wavelength channels in a WDM communication system. The configuration of a 16-channel optical OADM filter is shown in Fig. 6.22. Four AWGs and sixteen double-gate switches are integrated in a wafer. All these AWGs have the same grating parameters; a channel spacing of 100 GHz and a free spectral range of 3300 GHz in the 1.55 μm wavelength region. The double-gate TOSWs were used to improve the crosstalk characteristics of the OADM device. The configuration of the 2 $\times$ 2 double-gate TOSW is shown in Fig. 6.23. It consists of four MZIs with a TO phase shifter and an intersection. Any optical signal coupled into Port A_{in} or B_{in} passes through the cross-port of one of the four MZIs before reaching output port A_{out} or B_{out}. Therefore, the crosstalk of the switch is substantially better than that of a conventional single-stage TOSW. The OADM filter was fabricated with PLC technologies, and its size was 87 $\times$ 74 mm.

The function of the filter is as follows. WDM signals with 16 different wavelengths launched to the main input port are first demultiplexed by AWG1 and then 16 signals are introduced into the left-hand side arms of double-gate TOSWs. The cross angle of the intersecting waveguides was designed to be larger than 30 degrees to make the crosstalk and insertion loss negligible. When the double-gate switch is off, the light demultiplexed

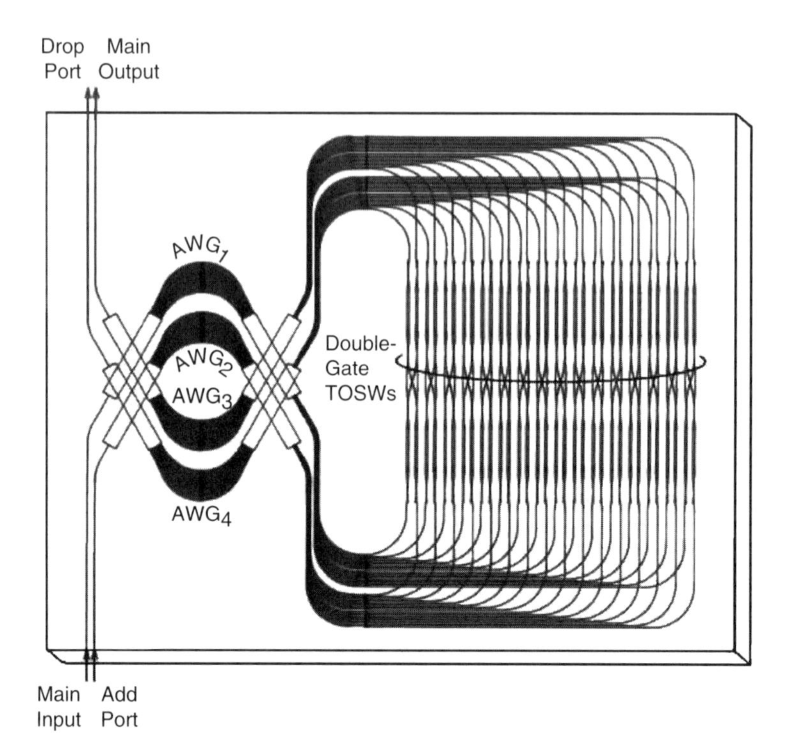

Fig. 6.22 Waveguide layout of 16-channel optical ADM with double-gate TO switches. (*See color plate*).

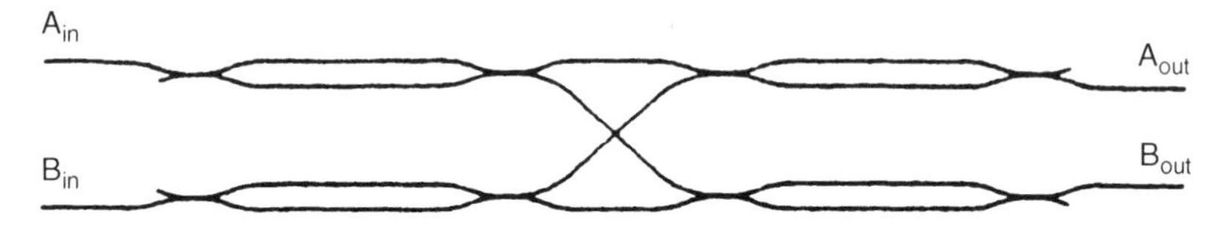

Fig. 6.23 Configuration of double-gate TO switch.

by AWG1 passes to the cross arm and is multiplexed again by AWG3. By contrast, if the double-gate TOSW is in the on state, the light demultiplexed by AWG1 passes to the through arm and is multiplexed by AWG4. Therefore, any specific wavelength signal can be extracted from the main output port and led to the drop port by changing the switch condition. A signal at the same wavelength as that of the dropped component can be added to the main output port when it is coupled into the add port in Fig. 6.22.

When switches SW_2, SW_4, SW_6, SW_7, SW_9, SW_{12}, SW_{13}, and SW_{15}, for example, are turned to the on state the selected signals λ_2, λ_4, λ_6, λ_7, λ_9, λ_{12}, λ_{13}, and λ_{15} are extracted from the main output port and led to the drop

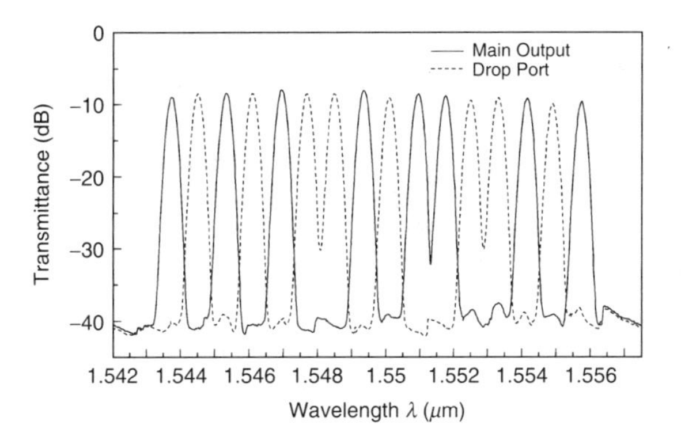

Fig. 6.24 Transmission spectra from main input port to main output port. (a) All switches are off. (b) TO switches SW_2, SW_4, SW_6, SW_7, SW_9, SW_{12}, SW_{13}, SW_{15} are on.

port as shown in Fig. 6.24. The on–off crosstalk is less than −28 dB with on-chip losses of 8–10 dB. This OADM filter is attractive for all optical WDM routing systems and allows the network to be transparent in terms of signal format and bit rate.

6.5. 1 × N PLC-type TOSW

This section reviews the characteristics of 1 × 8, 1 × 16, and 1 × 32 PLC-type TOSWs [36]. These TOSWs have advantages of compactness, good optical performance levels including low loss, low polarization dependent loss, and high extinction ratio, and fabrication repeatability. Moreover, these 1 × N switches have been incorporated into large-scale OXC systems and it has been confirmed that they are suitable for practical system uses.

6.5.1. CONFIGURATION OF 1 × *N TOSW*

Two types of logical arrangement have been proposed for PLC-type 1 × N TOSWs as shown in Fig. 6.25. Figure 6.25(a) shows a tree structure consisting of 1 × 2 switching units and gate-switching units. The gate-switching units are installed at every optical output port to reduce the optical crosstalk. The tree configuration was applied to 1 × 8 and 1 × 32 switches. The 1 × 8 switch consists of 3 stages comprising 7 1 × 2 switching units and 8 gate-switching units, and the 1 × 32 switch consists of 5 stages comprising 31 1 × 2 switching units and 32 gate-switching units. This configuration is advantageous in terms of realizing low loss and chip-size reduction because it has a small number of switch stages. Figure 6.25(b) shows another

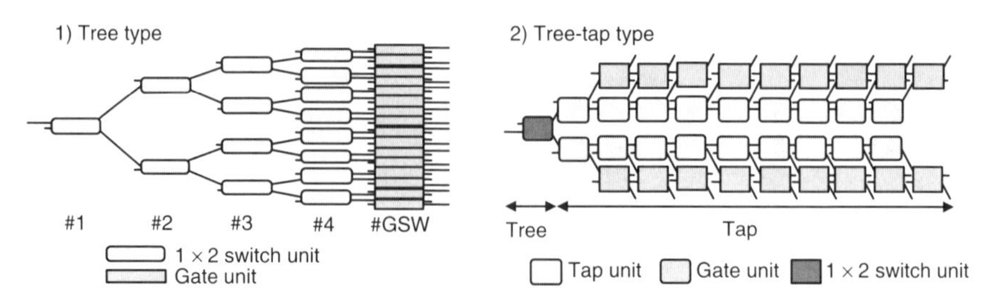

Fig. 6.25 Configuration of PLC-type 1 × N TOSWs. (a) Tree type. (b) Tree-tap type.

arrangement that consists of a tree structure part and a tap structure part. It is called the tree-tap combination (TTC) structure. A 1 × N switch can be realized by arranging N tapping units (switching units) on the main path. The tap-type switch offers the advantage of low power consumption because it needs only one active tapping unit regardless of switch scale. However, the tap-type switch chip is larger than the tree-type chip, because the former has a longer circuit. To overcome this problem, we adopted a structure that combined the tree and tap structures. This switch also has gate switching units at every optical output port to reduce the optical crosstalk. The tree-tap configuration was adopted for 1 × 16 TOSWs. These configurations are not unique, and should be determined based on system requirements, which include device size as well as optical performance. Thus, PLC technology is advantageous as regards flexibly designing the TOSW configuration.

The 1 × 2 switching unit used in the 1 × N switches has a symmetric MZI configuration whose waveguide arms are the same length. The gate-switching unit and tapping unit have an asymmetric MZI configuration with a half-wavelength path-length difference in two arms. The switching power of the unit is about 0.4 W and the switching time is about 3 ms. Because the switching characteristics are independent of environmental temperature fluctuations, it is unnecessary to control the chip temperature with thermo-control devices (e.g., a Peltier device).

6.5.2. *FABRICATION AND CHARACTERISTICS OF* 1 × *N TOSW*

We fabricated switch chips on a Si substrate using a combination of flame hydrolysis deposition (FHD) and reactive ion-beam etching (RIE). The relative refractive index difference was 0.75%, and the curvature radius was

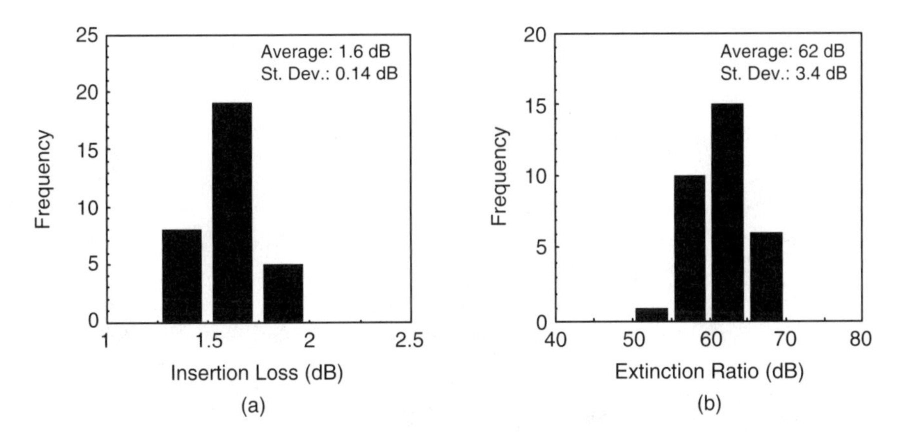

Fig. 6.26 Histograms of (a) insertion loss and (b) extinction ratio of 1×32 TOSW.

5 mm. The circuit parameters are listed in Table 6.1. Two 1 × 8 (1 × 16) circuits were integrated on a chip.

Table 6.1 also shows the optical characteristics and power consumption of 1 × N switches. These values were measured with a 1.55 μm-wavelength depolarized LD source. The average insertion losses and extinction ratios of the 1 × 8 switches were 1.2 and 52 dB, respectively. The corresponding values for the 1 × 16 switches were 1.9 and 59 dB, respectively. Figure 6.26 shows insertion loss and extinction ratio histograms of a 1 × 32 switch. We obtained an insertion loss of 1.6 dB and an extinction ratio of 62 dB for the switch. Moreover, the power consumption of 1.2 W for the 1 × 16 TTC-type switch was smaller than that for the 1 × 8 tree-type switch.

Figure 6.27 shows a photograph of a fiber-pigtailed 1 × 16 switch module [37]. The module was equipped with a switch chip that was fixed to fiber-ribbons at the input and output waveguides, a ceramic wiring board with a pin-grid array for the power supply, and a cooling fin. We realized very thin modules with a thickness of, for example, 13 mm. We used the switch modules in experimental OXC systems and confirmed that they had suitable characteristics for practical use as described in the following section.

6.5.3. SYSTEM APPLICATION [38, 39]

OXC systems are key network elements for realizing photonic networks. Recently, OXC systems have been developed based on the delivery and coupling (DC) switch in which PLC-TOSWs are used as key optical switches.

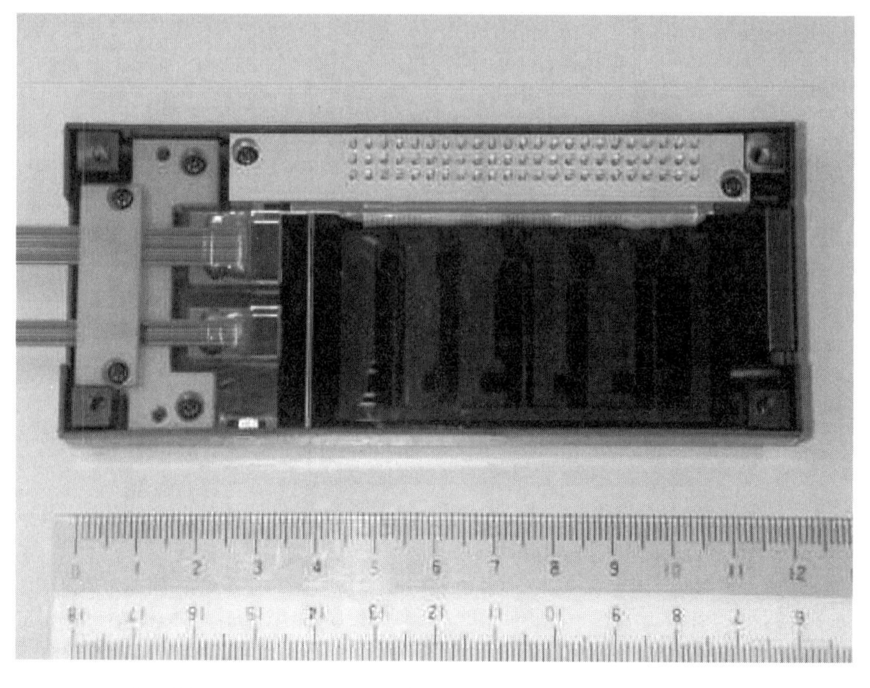

Fig. 6.27 Photograph of 1 × 16 TOSW module. (*See color plate*).

This system architecture offers modularity for incoming/outgoing link number (link modularity) and allows WP networks to be upgraded to VWP networks. Therefore, this architecture enables rapid introduction with minimal investment, while also supporting future incremental growth as traffic demand increases.

Figure 6.28 shows the DC-based OXC system architecture, in which N incoming/outgoing link pairs (N incoming and N outgoing links) are connected and M wavelengths are multiplexed into each link. Optical signals on an incoming link are demultiplexed into optical signals by a wavelength demultiplexer. Next, each optical signal is regenerated at the electrical level by optical receivers/senders (ORs/OSs). These optical signals are then delivered and coupled to the outgoing ports that correspond to the outgoing links by M × N DC-switches. Finally, optical signals delivered to the same outgoing port are coupled by an optical coupler.

The key component of this architecture is the DC-switch. This switch must allow any of the input optical signals to be optically connected to any of the output ports. Figure 6.29 shows the configuration of the M × N

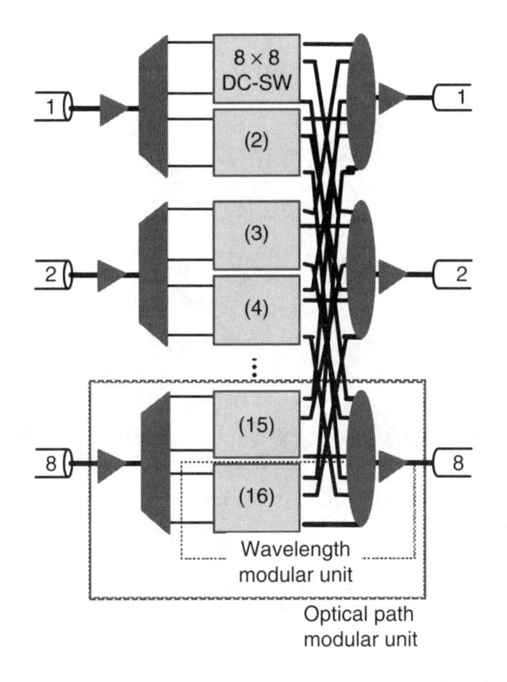

Fig. 6.28 Schematic structure of OXC system based on DC-SW.

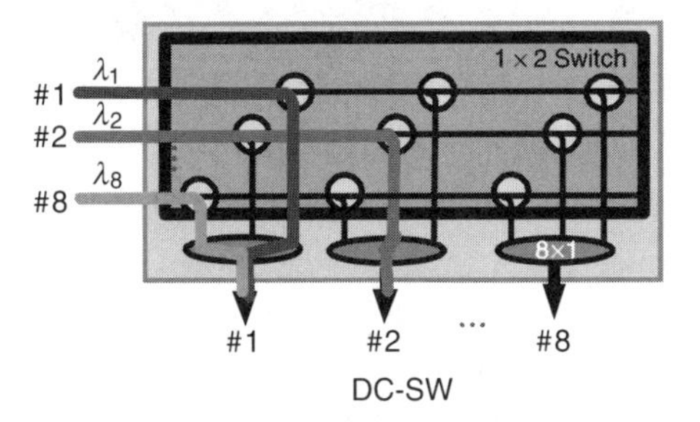

Fig. 6.29 Schematic configuration of DC-SW.

DC switch, which consists of M × N 1 × 2 switches and N 8 × 1 couplers. This DC switch is strictly non-blocking. By means of this characteristic, the restoration or establishment of new optical paths has no effect on previously established active optical paths.

We have recently developed an 8 × 16 DC switch board with which to construct an OXC system with a 320-Gb/s throughput. In the generic architecture depicted in Fig. 6.28, M and N correspond to 8 and 16, respectively. The actual configuration of the 8 × 16 DC switch consists of

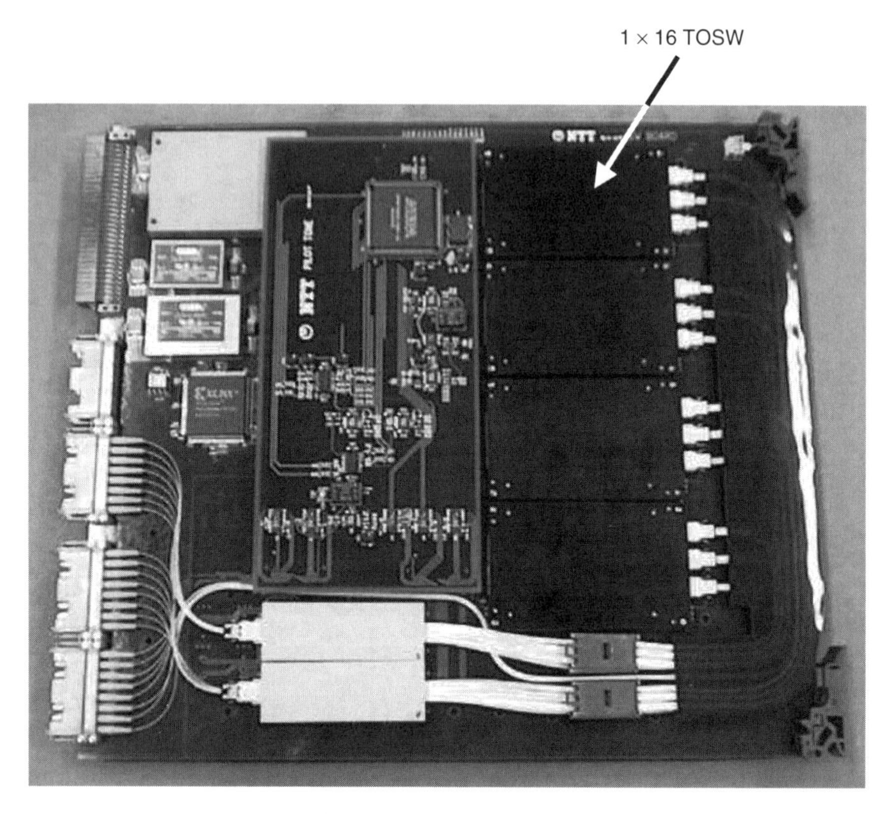

Fig. 6.30 Photograph of 8 × 16 DC-SW board. (*See color plate*).

8 1 × 16 TOSWs and 16 8 × 1 couplers as shown in Fig. 6.29. As described in the previous section, a gate switch is equipped at every output port of the 1 × 16 TOSW to realize a high extinction ratio. Each optical path has 2.5 Gb/s capacity, so we realized a total throughput of 320 Gb/s (=2.5 Gb/s × 8 × 16). Figure 6.30 shows a photograph of the 8 × 16 DC switch board we fabricated. The board size is 330 × 300 mm^2. Electrical driving circuits for the TO switches and on-board power supplies are mounted on the board.

Figure 6.31(a) shows typical insertion loss histograms for the 8 × 16 DC switch board. The average and worst insertion losses were 12.8 and 14.5 dB, respectively. Figure 6.31(b) shows typical extinction ratio histograms for the 8 × 16 DC switch board. The average and worst extinction ratios were 58.4 and 34.5 dB, respectively. A few ports just missed the 40 dB limit, but subsequent process refinement would ensure that all ports are satisfactory. Moreover, the PDL of the DC switch was less than 0.5 dB,

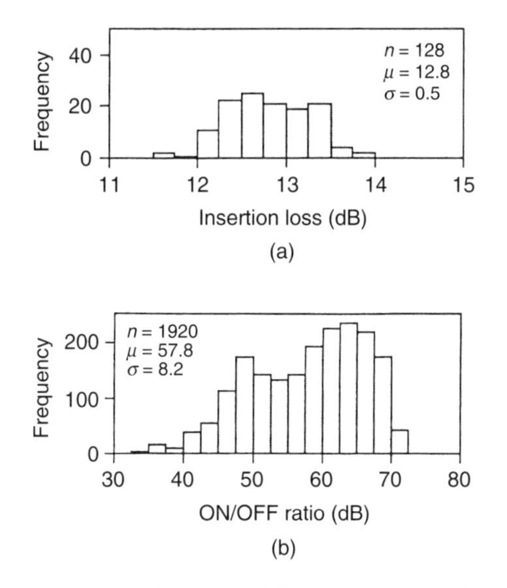

Fig. 6.31 Histograms of (a) insertion loss and (b) extinction ratio of 8 × 16 DC-SW board.

and the switching time was about 14 ms, including all necessary procedures such as communication between the control personal computer and the board. These values satisfy the system requirements well. Therefore, this successful switch board construction will ensure the development of photonic networks using the PLC-type TOSWs.

6.6. N × N Non-blocking Matrix Switch

We fabricated polarization insensitive TO 8 × 8 and 16 × 16 matrix switches using silica-based PLC technologies on a Si substrate. These switches have a strictly non-blocking structure and are useful for optical crossconnect systems.

6.6.1. 8 × 8 *MATRIX SWITCH*

6.6.1.1. Original (1st) Version [40, 41]

Configuration

The first version of the PLC-type 8 × 8 matrix TOSW, whose structure is shown in Fig. 6.32, was fabricated in 1993. The switch has a strictly non-blocking structure and consists of 64 active switch units that form a

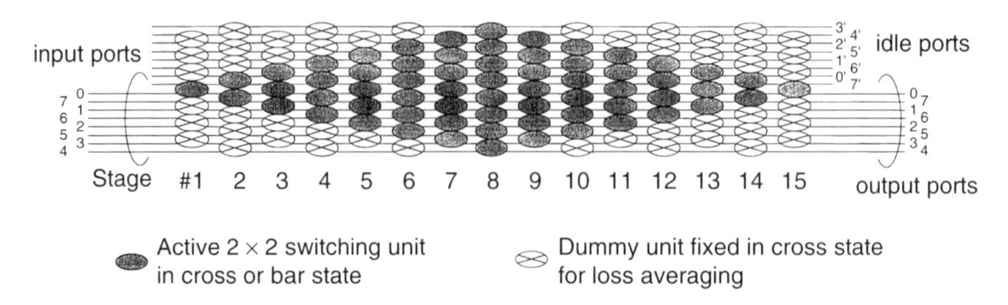

Fig. 6.32 Logical arrangement of 8 × 8 matrix switch.

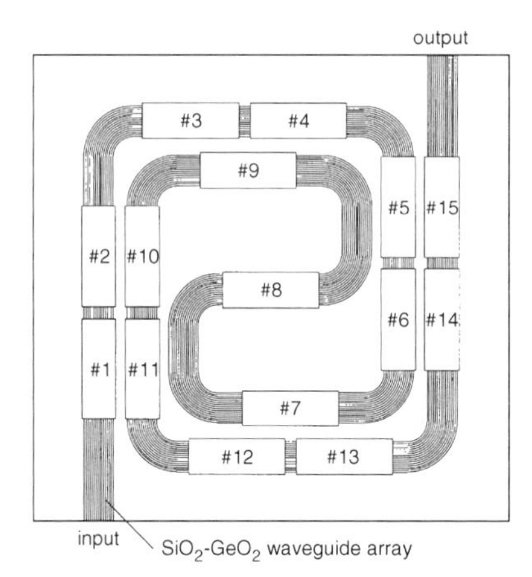

Fig. 6.33 Waveguide layout of 8 × 8 matrix switch.

diamond-shaped area and 48 dummy switch units at the four corners. The dummy switch units have almost the same loss as the active switch units and were arranged for loss averaging. When the matrix switch is composed only of active switch units, the number of units through which optical signal lights pass varies from 1 to 15 depending on the optical path. The switch unit has a slight excess loss that includes curvature loss. Therefore, the optical output powers differ depending on the optical path. However, the insertion losses were averaged by arranging the 48 dummy switch units at the four corners because then every signal light passes through 7 to 15 units.

The waveguide pattern of the fabricated 8 × 8 matrix TOSW is shown in Fig. 6.33. Because a single switch unit is about 12 mm long and the total length of the 15 switch unit stages becomes large, these stages are arranged along a 35-cm-long meander waveguide array. The curvature radius of the waveguide we used was 5 mm. The 8 × 8 matrix TOSW was fabricated on

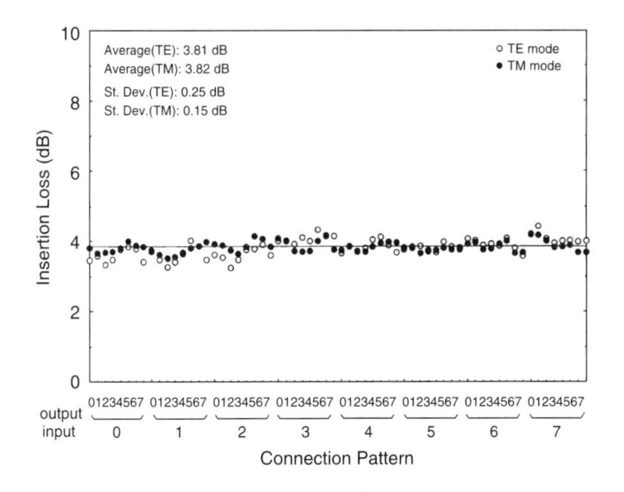

Fig. 6.34 Insertion loss of input/output connection pattern in 8 × 8 matrix switch.

a Si substrate using conventional FHD and RIE methods. The conventional MZI switching unit we used is shown in Fig. 6.33.

Characteristics

The switching characteristics were measured at 1.3 μm. The insertion loss of the matrix switch is shown against each input–output pair in Fig. 6.34. The matrix switch had a very low average loss of 3.8 dB for both the TE and TM modes, including the input and the output fiber coupling loss. This low insertion loss was obtained mainly because of the low propagation loss of SiO_2-GeO_2 waveguides and the lateral offset at waveguide inflection points such as the junction between a straight and a curved waveguide. Moreover, the extinction ratio of the matrix switch is shown against each input–output pair in Fig. 6.35. The average extinction ratio was over 22 dB. However, not every switching unit had an extinction ratio of over 20 dB.

Here we consider the origin of the crosstalk in the matrix switch. Figure 6.36 shows the crosstalk from the switching units to the output ports. Figure 6.36(a) shows the crosstalk from a cross-state (off-state) switching unit. The residual accumulated crosstalk, which leaks into a bar port, reaches the output ports. Figure 6.36(b) shows the crosstalk from a bar-state (on-state) switching unit. The unswitched crosstalk, which leaks to a cross-port, can pass safely to an idle output port. Therefore, the only switching unit in which there is no crosstalk is one that can pass 100% of the light to cross-ports in the off state. In a conventional switching unit, the signal

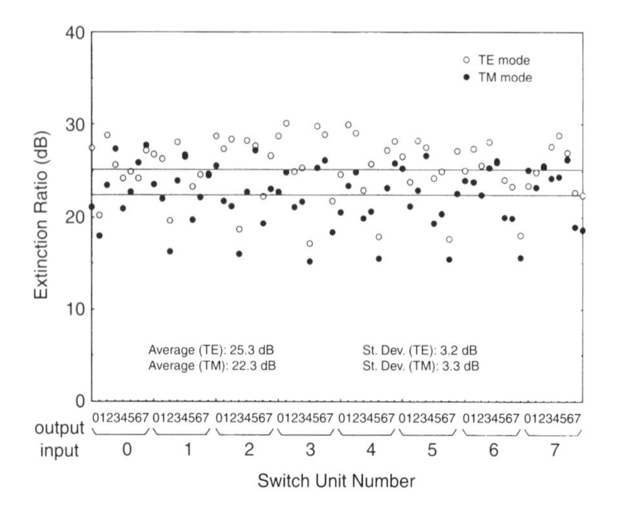

Fig. 6.35 Extinction ratio of input/output connection pattern in 8 × 8 matrix switch.

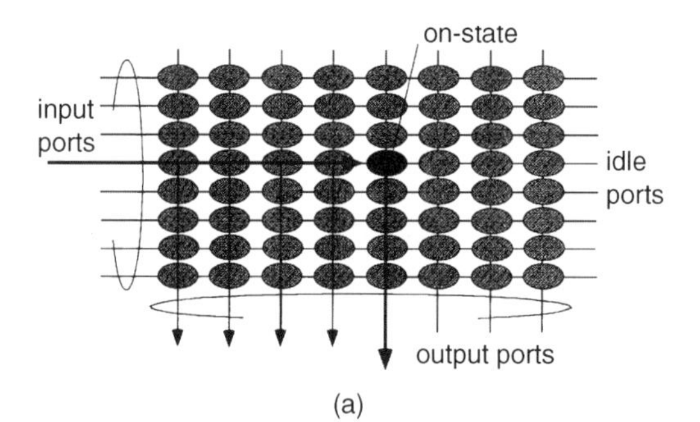

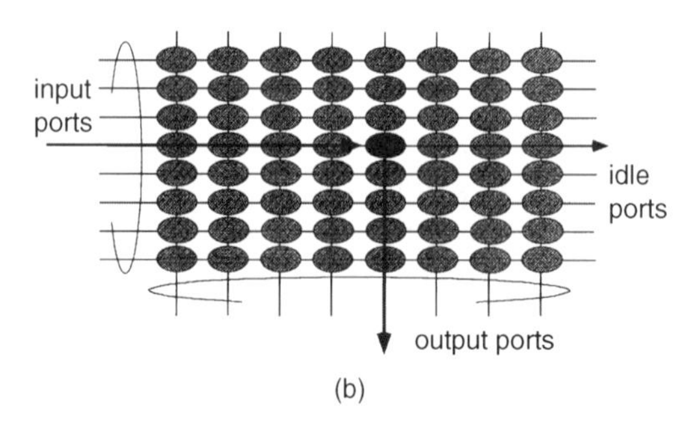

Fig. 6.36 Crosstalk from switching units to output ports. (a) Bar-state crosstalk. (b) Cross-state crosstalk.

light passes perfectly through the main-path (through-path) on condition that the switching unit is the on-state (waveguide arm length difference: half wavelength) and the two DCs have the same coupling ratio, even if the DC-coupling ratio is not 50% (k =/ 0.5). The signal light, however, leaks to the main-path in k =/ 0.5 and the off-state (wavelength arm length difference: 0). This is crosstalk and to reduce this, the DC-coupling ratio must be set precisely at 0.5. Silica-based PLCs offer better fabrication controllability and reliability than other waveguides, but not every switching unit had an extinction ratio of over 20 dB. Therefore, the conventional MZI switching unit is unsuitable for the matrix switch arrangement.

6.6.1.2. Improved (2nd) Version [42]

Configuration

In order to improve the crosstalk performance of the matrix switch, we proposed an asymmetric MZI switching unit with a half wavelength path difference between the waveguide arms. Figure 6.37 shows the configuration of our proposed switching unit. The unit consists of an MZI part with a half wavelength path difference between the waveguide arms and an intersecting waveguide part. The optical characteristics of the switching unit are opposite to those of a conventional switching unit because the half wavelength path difference is set in advance. Therefore, optical signals travel through the through-path in the off state. In the matrix switch arrangement in Fig. 6.32, optical signals have to travel through the cross-path in the off state. This is solved by locating an intersecting waveguide part behind the MZI switching part. Even if the intersecting waveguide part is positioned in front of the MZI switching part or the waveguide arms in MZI switching part intersect, the optical characteristics remain the same as those in Fig. 6.37.

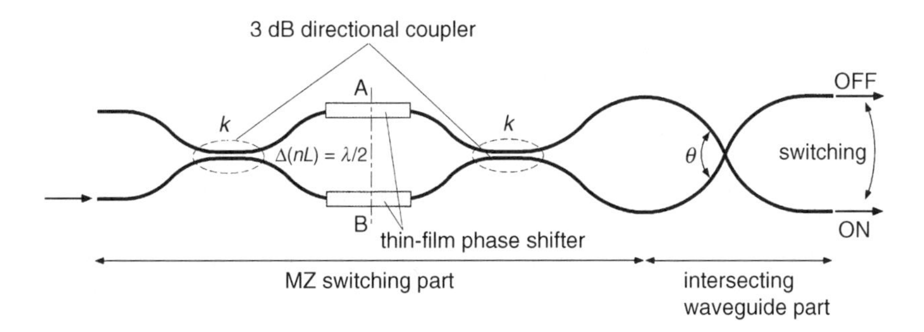

Fig. 6.37 Novel configuration of MZI switching unit with intersecting waveguides.

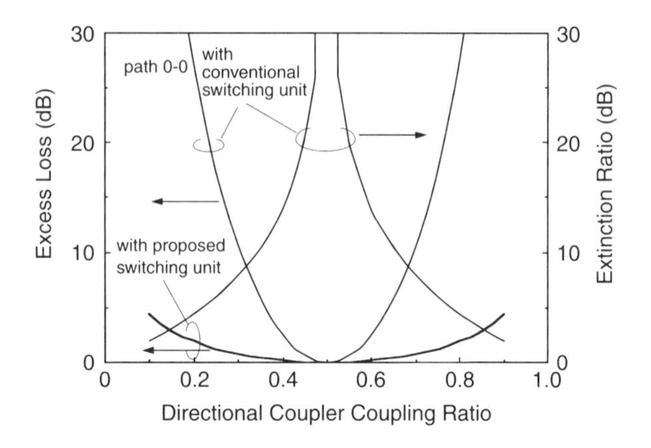

Fig. 6.38 Calculated excess loss and extinction ratio of 8×8 matrix switch with novel and conventional switching units as a function of DC-coupling ratio.

Figure 6.38 shows calculated excess losses and extinction ratios of the 0-0 path of the 8×8 matrix switches as a function of the DC-coupling ratio with conventional and proposed switching units. This figure does not show the extinction ratio of the 8×8 matrix switch with the proposed switching units because the value is infinite. As seen, the DC-coupling ratio dependence of the excess loss of the matrix switch is much smaller with the proposed switching units than with the conventional switching unit. The 0.5 dB excess loss range, for example, is the DC-coupling ratio $k = 0.5 +/- 0.04$ in the matrix switch with the conventional switching units and $k = 0.5 +/- 0.15$ in the matrix switch with the new switching unit. The excess loss in the matrix with the new units is only 2 dB even if the DC coupling ratio $k = 0.2$. Then the extinction ratio of the proposed matrix switch is theoretically infinite, whereas the range of the 20 dB extinction ratio of the conventional matrix switch is only $k = 0.5 +/- 0.05$. These results confirmed that the 8×8 matrix switch with the proposed switching units has much greater DC-coupling ratio tolerance than the conventional one.

The optical path length difference between the waveguide arms in the MZI switching part was designed to be 15% longer than the calculated half wavelength. While the optical path difference must be exactly a half wavelength, it is difficult to set it precisely at a half wavelength due to fabrication errors. So, the length in one of the arms is set shorter than a half wavelength, and a small amount of electric power, corresponding to the difference from the half wavelength path-length difference, is applied

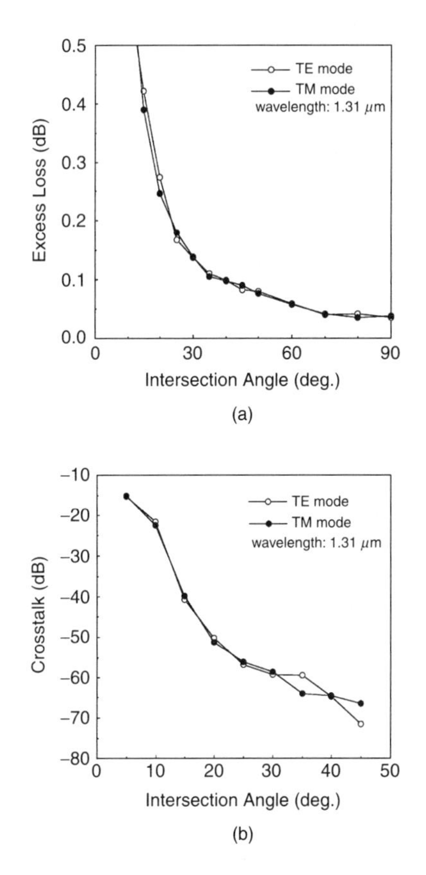

Fig. 6.39 Waveguide intersection angle dependence: (a) excess loss and (b) crosstalk.

to the heater in the shorter arm. The wavelength difference of the matrix switch is improved when the optical path-length difference between the waveguide arms is zero in the on state.

The intersecting waveguide, especially its angle, influences the loss and optical crosstalk of the switch. Therefore, it is necessary to select a suitable intersection angle. Figure 6.39(a) and (b) show the excess loss and crosstalk dependence on the waveguide intersection angle, respectively. These values were measured at 1.31 μm. As the angle increases, the excess loss first decreases rapidly and then, from 30 degrees, more gently. The crosstalk also decreases as the angle increases, and is under -60 dB at over 30 degrees. From these results, the intersection angle should be as large as possible in terms of reducing the excess loss and crosstalk. However, the length and width of the switching unit increases as the intersection angle increases. Therefore, the maximum intersection angle is limited by the wafer size and was 30 degrees when the circuit was arranged on a 4-inch wafer. With a waveguide intersection angle of 30 degrees, the excess loss was 0.14 dB and the crosstalk was under −60 dB (from Fig. 6.39).

The 8 × 8 matrix switches were fabricated on a Si substrate using a combination of FHD and RIE. A SiO_2-GeO_2 waveguide was used to reduce propagation loss. The core was 7.5 mm wide and 6 mm high. The cladding was 60 mm thick. The Δ was 0.75%, and the curvature radius was 5 mm. The DC coupling ratio of the switch was set at about 30% to confirm the feasibility of the asymmetric MZI configuration. Ta_2N thin-film heaters were fabricated by a sputtering method. The heater was 4 mm × 0.05 mm.

Characteristics

We measured the optical characteristics of the 8 × 8 matrix switches at 1.3 μm with a laser diode and a power meter. The insertion losses are shown against each input–output pair in Fig. 6.40. The average insertion loss was 7.3 dB for the TE mode and 7.5 dB for the TM mode. The insertion loss increased due to the excess loss of the intersecting waveguide structure. Histograms of the extinction ratios for the 8 × 8 matrix switch are plotted in Fig. 6.41. The average extinction ratios are 31.2 dB for the TE mode and 31.3 dB for the TM mode. As shown in Fig. 6.41, the extinction ratio was greatly improved for the switch with a 30% coupling ratio. While it

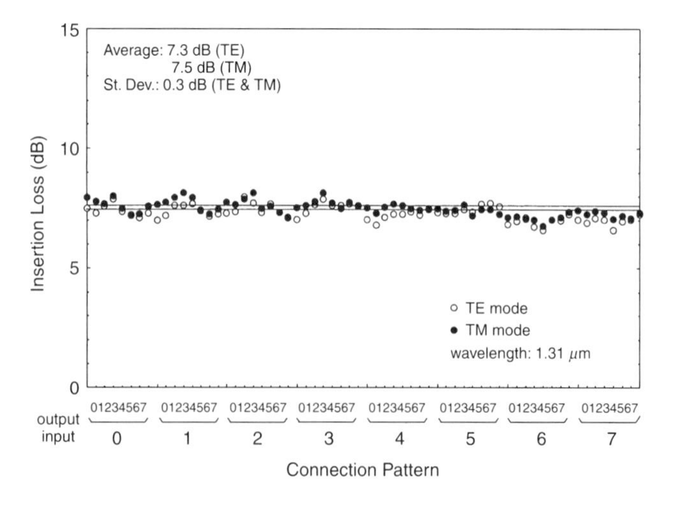

Fig. 6.40 Insertion loss of input/output connection pattern in 8 × 8 matrix switch with novel MZI switching units.

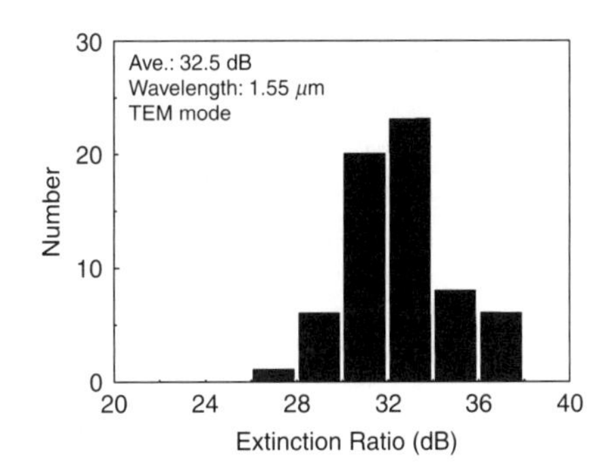

Fig. 6.41 Extinction loss of input/output connection pattern in 8 × 8 matrix switch with novel MZI switching units.

is theoretically infinite, the extinction ratio of the switch is limited to over 26 dB because the coupling ratios of the two DCs in the MZI were slightly different. These results confirmed that the asymmetric MZ switching unit with intersecting waveguides provides a high extinction ratio even if the DC coupling ratio deviates greatly from 50%. The switching response time was about 0.2 ms, and the average switching and offset powers of the asymmetric MZI unit were 0.41 and 0.037 W, respectively.

6.6.1.3. PI-loss (3rd) Version of 8 × 8 Matrix Switch [43–45]

Design of N × N Optical Matrix Switch

To improve the extinction ratio of the matrix TOSW further, we used a double-gate MZI switching unit for the 8 × 8 TOSW. The configuration of the double-gate MZI switching unit is shown in Fig. 6.42. This switching unit is composed of two asymmetrical MZIs with an optical path length difference of a half-wavelength and TO phase shifters, and an intersection. The switching unit is in the OFF-state when the two MZIs are simultaneously set in the bar-state. The unit can be switched to the ON-state when both the MZIs are simultaneously set in the cross-state by applying heating power to the phase shifters. In the OFF-state, lights launched into the two input ports are transmitted through the OFF-path, which is from input port 1 to output port 1 and from input port 2 to output port 2. In the ON-state, light launched into input port 1 is transmitted through the ON-path, which is from input port 1 to output port 2. Although light launched into input port 2 in the ON-state is not transmitted to output port 1 but to an idle port in the unit, this path is not required in the matrix switch. The extinction ratio

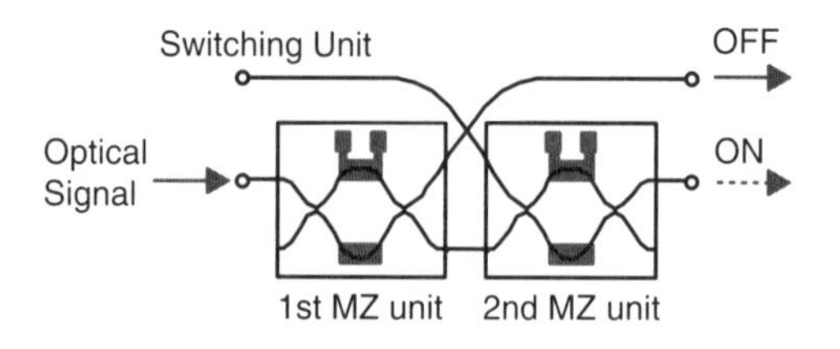

Fig. 6.42 Configuration of double-gate switching unit.

of the ON-path in the OFF-state is the most important factor as regards the crosstalk of the matrix switch because the light power leaked into the ON-path in the OFF-state is led to an output port of the matrix switch. In this proposed configuration, the leaked light power to the ON-path in the OFF-state is greatly reduced because the unwanted light power from the first MZI is blocked by the second MZI. Therefore, the expected extinction ratio of the matrix switch is twice that obtained with a conventional single-MZI switching unit.

In addition, the extinction ratio of a cross-path in an MZI is inherently large when the MZI is in the bar-state, even if the coupling ratios of the two 3-dB couplers deviate from the ideal value of 50%, as long as the two ratios are the same. In a conventional single-MZI switching unit, this path configuration has been realized by swapping the output ports with an intersection. Because the double-MZI switching unit also inherits this path configuration, as shown in Fig. 6.42, it also has large fabrication tolerances for the coupling ratio deviation.

Figure 6.43 shows the logical arrangement adopted for the strictly non-blocking matrix switch, which has a characteristic of path-independent insertion loss (PI-loss) [46, 47]. This is logically equivalent to the conventional crossbar matrix arrangement with a diamond shape shown in Fig. 6.32 because each input waveguide has one crosspoint, where the switching unit is in the ON-state, with each output waveguide. This arrangement requires only N switching unit stages in the $N \times N$ matrix switch, compared with the $2N - 1$ stages required with a conventional crossbar arrangement. This will be useful for reducing waveguide length. Moreover, to minimize the extra length with the double-MZI switching unit geometry, we locate the intersections between the stages in the vacant areas beside the switching units in the same stage. By using this proposed layout, the total waveguide length is effectively reduced to half that with the crossbar arrangement, and we can expect the insertion loss to be reduced but the high extinction ratio to be maintained.

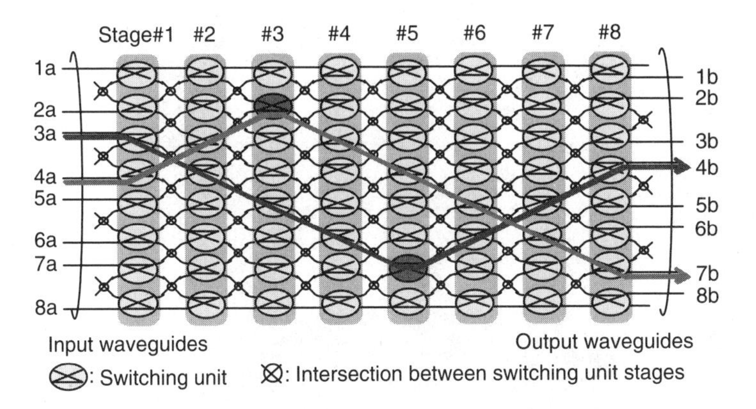

Fig. 6.43 Logical arrangement of 8 × 8 matrix switch with PI-loss structure. (*See color plate*).

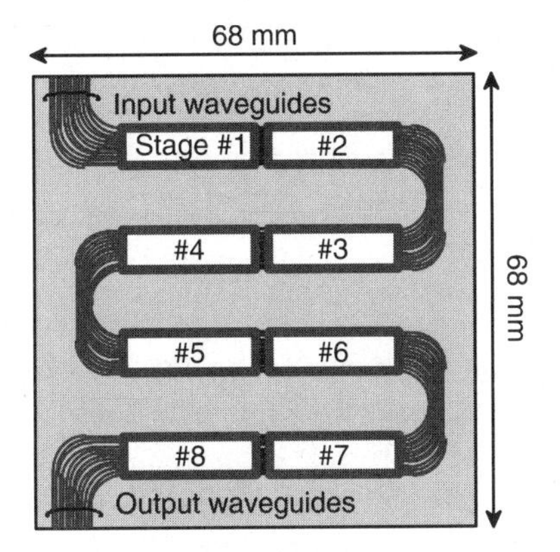

Fig. 6.44 Circuit layout of 8 × 8 matrix switch with PI-loss structure.

The circuit layout of our 8 × 8 matrix switch is shown in Fig. 6.44. Eight switching unit stages are located along the winding waveguides to minimize the total waveguide length on the 4-inch wafer. We used a minimum curvature bending radius of 5 mm and an intersection angle of about 30 degrees to avoid any loss increment and crosstalk. The total waveguide length was 290 mm and this is about 2/3 that of a conventional single-MZI crossbar matrix switch, despite the greater length of the switching unit. The chip size was 68 × 68 mm. This novel configuration matrix switch has almost the same number of intersection waveguides and MZIs on one optical path as a previously reported conventional single-MZI matrix switch. This means there is hardly any loss increment caused by the circuit configuration.

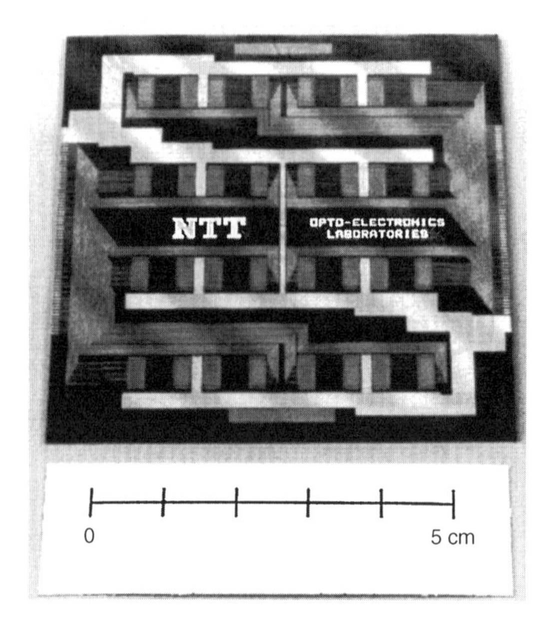

Fig. 6.45 Photograph of 8 × 8 matrix switch with PI-loss structure.

We fabricated the silica-based 8 × 8 TO matrix switch on a silicon wafer by using PLC technology. Ta_2N thin-film TO heaters and gold-wiring films were patterned on the overcladding. The core size was 7 × 7 μm and the relative refractive index difference between the core and cladding was 0.75%. The coupling ratio of the 3-dB directional coupler was intentionally moved to 33% at 1.55 μm to confirm the large fabrication tolerances. Figure 6.45 is a photograph of the fabricated 8 × 8 matrix switch chip.

Characteristics of 8 × 8 *Matrix Switch with PI-loss Structure*

We measured the characteristics of the fabricated 8 × 8 matrix switch chip with a depolarized light source at 1.55 μm. A polarization maintaining fiber and a conventional 1.3 μm zero dispersion single-mode fiber were butted to the input and output ports, respectively, of the matrix switch chip. Figure 6.46 shows typical transmittance against the heating power to the TO heater of the first MZI in which the heating power to the second MZI was changed as a parameter. The open circles indicate that the second MZI is in the cross-state (ON-state), and the filled circles indicate the bar-state (OFF-state). These data were recorded before phase trimming with the thin-film heater. The first and second MZI are in the cross-state (ON-state) when the

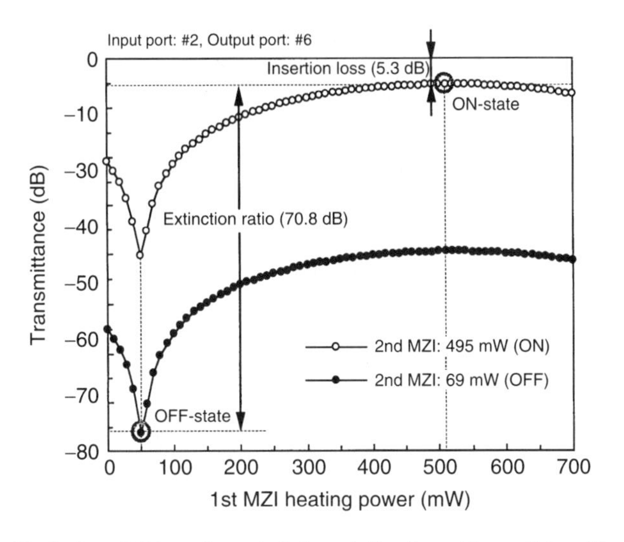

Fig. 6.46 Typical switching characteristics of 8×8 matrix switch with double gate-switching unit.

applied power is 506 and 495 mW, and in the bar-state (OFF-state) at 50 and 69 mW, respectively. The extinction ratio when the first and second MZIs are operated simultaneously is double that when only one MZI is operated. In this example, a 70.8 dB extinction ratio was successfully achieved. The insertion loss was 5.3 dB.

The transmittance of the MZI before the phase trimming is not at its minimum when the heating power is zero, as discussed in Section 6.3.4. So, a small bias heating power is necessary even in the OFF-state if the phase trimming is not carried out. In the 8×8 matrix switch, the optical path length difference of the MZI was designed to be 10% longer than a half-wavelength to ensure that the bias power was only applied to the shorter arm side heater in all the switching units. This is the same heater to which the switching power is applied, thus reducing the number of drive circuits.

The measured extinction ratios and insertion losses for all 64 possible optical paths from the input port to the output port are shown in Fig. 6.47(a) and (b), respectively. One switching unit at the crosspoint was operated (ON or OFF) in an optical path, the others were in the OFF state. The extinction ratio ranged from 49 to 76 dB, with an average value of 60.3 dB. The improvements in the average and worst extinction ratios when the leaked light obstructer was used were 2.5 and 0.4 dB, respectively. The insertion

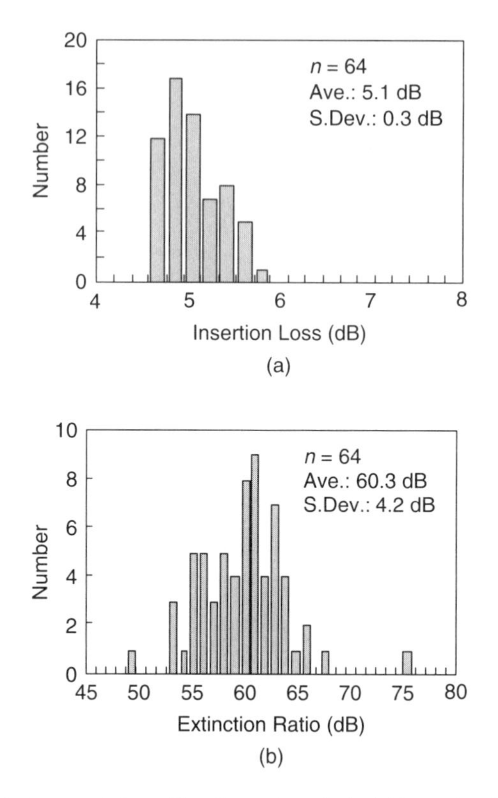

Fig. 6.47 Optical characteristics of 8 × 8 matrix switch. (a) Insertion loss. (b) Extinction ratio.

loss ranged from 4.7 to 5.9 dB, with an average value of 5.1 dB. The average and worst extinction ratios and insertion losses are summarized in Table 6.2 alongside the best measured characteristics of our previously reported single-MZI crossbar matrix switch. The extinction ratio is almost twice that of a single MZI crossbar matrix switch with a low insertion loss. Moreover, the average PDL of the switch was 0.3 dB.

The multiple crosstalk, which is a sum of leaked lights from the other N − 1 input signals, is important in the N × N matrix switch. While the multiple crosstalk depends on the optical path pattern with different switching configurations, a simple model is used to accumulate the crosstalk at all N − 1 switching units and all N − 1 intersections. Here the switching unit at Stage i is in the ON-state. The crosstalk values at the switching unit at Stage j before and after Stage i should be the same as the intersection crosstalk in the switching unit and the OFF-state transmittance, respectively.

Table 6.2 **Average and Worst Extinction Ratios and Insertion Losses in 8 × 8 Matrix Switches**

	PI-loss	*Conventional*
Number of switching unit	8	15
Switching unit length (mm)	23	17.5
Total waveguide length (mm)	290	430
Chip size (mm)	68 × 68	71 × 68
Number of intersections in a path	15	15
Extinction ratio (avg.) (dB)	<49.4 (60.3)	<24.4 (32.5)
Insertion loss (avg.) (dB)	<5.9 (5.1)	<8 (6.2)

The intersection crosstalk X is estimated to be XR/Lon_{av}. Here, XR is the intersection crosstalk ratio for the transmittance of straight light, and it was less than −65 dB in measured test waveguides on the same chip. Lon_{av} is the average insertion loss. The OFF-state transmittance Toff is defined as $Toff = 1/(Lon \cdot ER)$. Thus, multiple crosstalk ratios MXR_i for the input signal Ton_i are estimated as follows on the assumption that the input signal powers are uniform:

$$MXR_i = 2\left(\sum_{j=1,j>i}^{N} Toff_j + (N + i - 2)X\right) \Big/ Ton_i \qquad (6.20)$$

where $Ton = 1/Lon$. The doubling term is added to the equation because, in the worst case, noises leaking at two points from another optical path interfere with each other.

From Eq. (6.20), the MXR in each optical path in the 8 × 8 matrix switch was estimated to range from −43.3 to −48.4 dB with an average value of −46.0 dB. Despite this being a worst-case estimation, we obtained an excellent value of better than −43 dB. Because an MXR of less than −26 dB is required for a system penalty of 1 dB [11], this matrix switch has a nonuniformity margin of more than 17 dB in terms of input optical power.

The wavelength dependence characteristics were measured by employing 1.55 μm band wavelength light from a tunable laser diode with a high-speed polarization scrambler using a piezoelectric device. Figure 6.48 shows the wavelength dependence of the extinction ratios and insertion losses for all 64 optical paths. The filled squares indicate the average

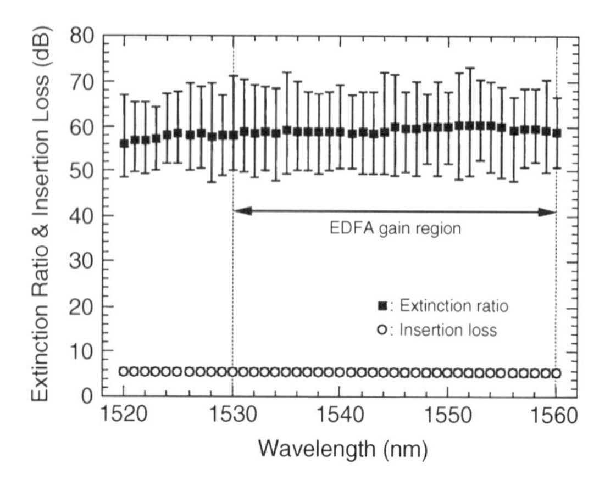

Fig. 6.48 Wavelength dependence of optical characteristics in 8 × 8 matrix switch with double gate-switching unit.

extinction ratio and the open circles show the average insertion loss. The error bars indicate the minimum and maximum extinction ratios and insertion losses. The worst extinction ratio and worst insertion loss were 47.6 and 6.4 dB, respectively, in the 1530 to 1560 nm range. This flat response covers the gain band of practical erbium-doped fiber amplifiers (EDFA). These results show that this device can be practically employed in a wavelength division multiplexing optical communication system.

The power consumption of the double-MZI switching unit is twice that of a single-MZI switching unit because the number of MZIs is doubled. The average ON-state and OFF-state power of the double-MZI switching units was 1.06 and 0.134 W, respectively. The total power consumption was 16.0 W under usual operating conditions in which eight switching units were in the ON-state with the others in the OFF-state. This value is not too large for practical use. The power consumption can be reduced by using a phase-trimming technique to eliminate OFF-state bias power.

The 8 × 8 matrix switch module was fabricated using the PLC-type TOSW chip and is shown in Fig. 6.49 [48]. The switch chip was fixed on a ceramic substrate integrated with 64 driving circuits. Input and output 8-fiber ribbons were pigtailed to the chip with UV-curable adhesive. A cooling fin was attached to the ceramic substrate. The module size was 145 × 156 × 22 mm. An insertion loss of less than 8 dB and an extinction ratio of more than 40 dB were achieved for the module, which provided stable operation from 0 to 65°C. This 8 × 8 matrix switch module is now commercially available.

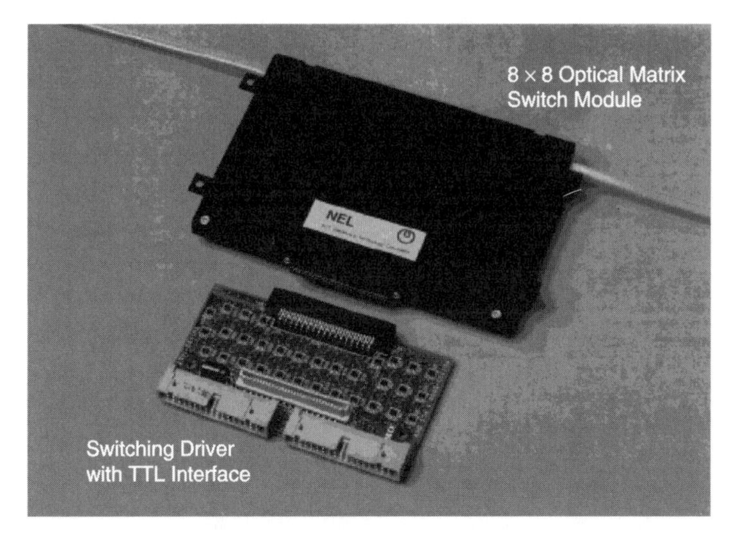

Fig. 6.49 Photograph of 8 × 8 matrix switch module. (*See color plate*).

6.6.2. 16 × 16 *MATRIX SWITCH*

6.6.2.1. Configuration of 16 × 16 Matrix Switch [49, 50]

To respond to the strong demand for a large-scale optical switch, we successfully fabricated a 16 × 16 matrix TOSW by developing the technologies we used for the 8 × 8 matrix TOSW. The configuration of the double-MZI switching unit used in the matrix switch is the same as that of the 8 × 8 matrix switch. The logical arrangement of the 16 × 16 matrix switch with the PI-loss configuration is shown in Fig. 6.50. The 16 × 16 TOSW is four times larger than the 8 × 8 TOSW. This arrangement reduces the total circuit length to half that with the conventional arrangement, making it effective for fabricating a low-loss large-scale matrix switch.

The circuit layout of a 16 × 16 matrix switch with 0.75% 7×7 μm waveguides on a 6-inch wafer is shown in Fig. 6.51. Sixteen switching unit stages were located along the twisting waveguides to minimize the total waveguide length, including the connective waveguides between the stages. One switching unit stage contains 16 switching units, and 256 switching units in total are integrated on the chip. The connective waveguide consists of 32 waveguides. We used a curvature bending radius of more than 5 mm and waveguide intersection angles of about 30 degrees to suppress the loss increment and crosstalk. A comparatively short total circuit length of

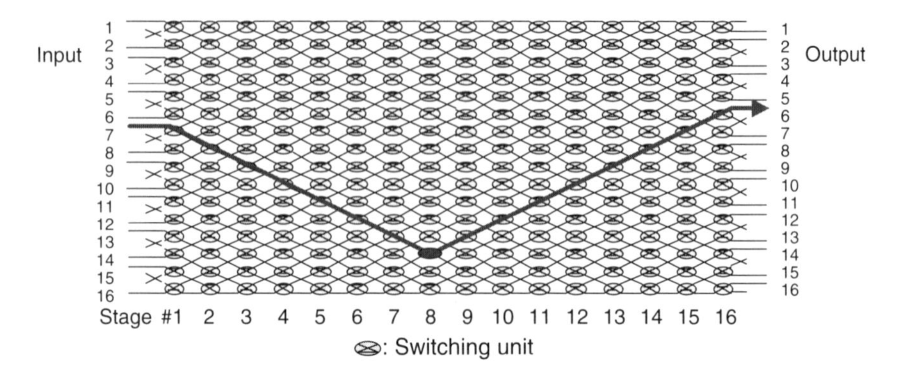

Fig. 6.50 Logical arrangement of a 16 × 16 strictly nonblocking matrix switch. (*See color plate*).

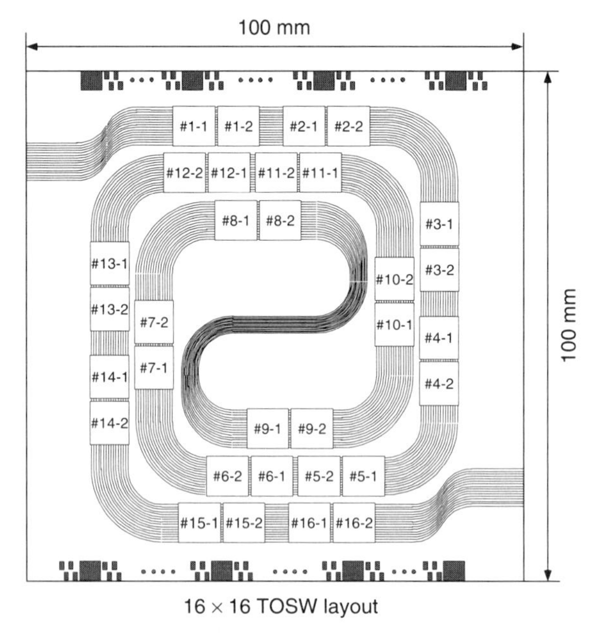

Fig. 6.51 Circuit layout of a 16 × 16 matrix switch with double-MZI switching units. (*See color plate*).

66 cm was realized, despite the greater curvature bending radius. The chip size was 100 × 107 mm. After fabricating the waveguides using the 6-inch wafer PLC process, we formed thin-film TO heaters and electrodes on the overcladding photolithographically. Then, we covered the heater with glass passivation film to enable it to endure the high power applied during the phase-trimming process described in the next section. Figure 6.52 shows a

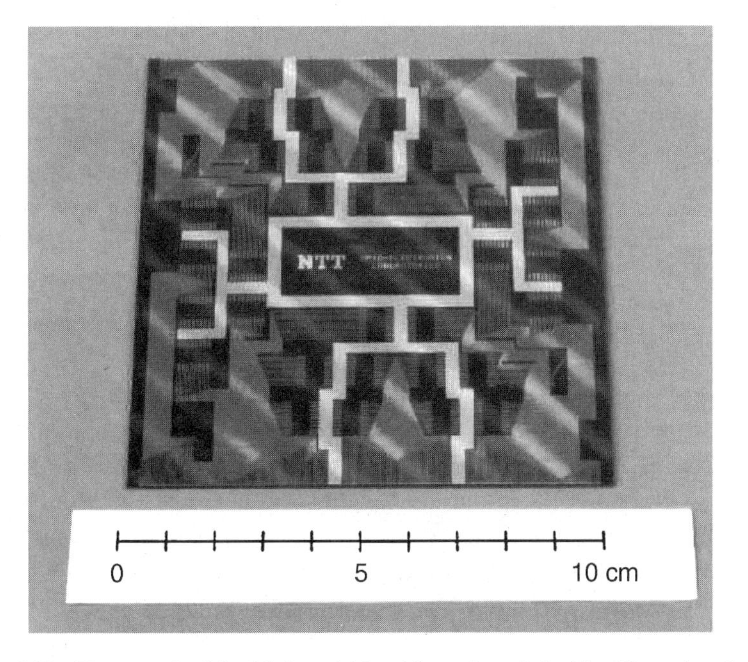

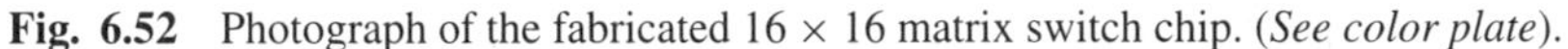

Fig. 6.52 Photograph of the fabricated 16 × 16 matrix switch chip. (*See color plate*).

photograph of the fabricated 16 × 16 matrix switch chip. The patterns on the chip are electrodes providing access to each MZI from electrode pads located on the chip edge.

As already mentioned in Section 6.3, if the phase trimming is not carried out in the matrix switch, the total bias power is approximately proportional to the square of the matrix size N, because the bias power is applied to N (N − 1) OFF-state switching units and is also applied to N ON-state switching units as a part of the ON-state power (= the bias power + the switching power). This is in contrast to the total switching power, which is proportional to N for N ON-state switching units. The average bias power is about 1/8 of the switching power in a usual matrix switch in which the optical path length difference of the MZI is designed to be 10–15% longer than a half-wavelength to ensure that the bias power is only applied to the shorter arm side heater in all switching units [14, 23]. Thus, the total bias power in an 8 × 8 matrix switch is nearly the same as the total switching power. Furthermore, the total bias power in a 16 × 16 matrix switch is about twice the total switching power. So, it is important to carry out phase-trimming for each MZI because the phase-error varies in each MZI in the matrix switch.

Phase-trimming was performed for all 512 MZIs with this automatic procedure. The phase-error was eliminated and the bias power was adjusted to less than 3 mW. This accuracy corresponds to 0.6% of the half-wavelength phase change, and is sufficient to obtain a high extinction ratio without bias power. The phase-trimming process caused no loss degradation. In the fabricated matrix switch chip, the required heating time was 25 and 1 sec using a heating power of 7 W to obtain the maximum and average phase shift, respectively. We estimated the total processing time for phase-trimming 512 MZIs to be around 30 min under this condition.

6.6.2.2. Characteristics of 16 × 16 TOSW Chip

We measured the characteristics of the fabricated 16 × 16 matrix switch chip using the same method that we used for the 8 × 8 matrix TOSW. Figure 6.53 shows example transmittance data in an optical path from the input port to the output port. A double-MZI switching unit was operated to form the optical path. The horizontal axis is the heating power applied to the TO heater of the first MZI in the switching unit. The parameter is the heating power to the second MZI. The open circles indicate that the second MZI is in the cross-state (ON-state), and the filled circles indicate the bar-state (OFF-state). The ON-state power at which the insertion loss

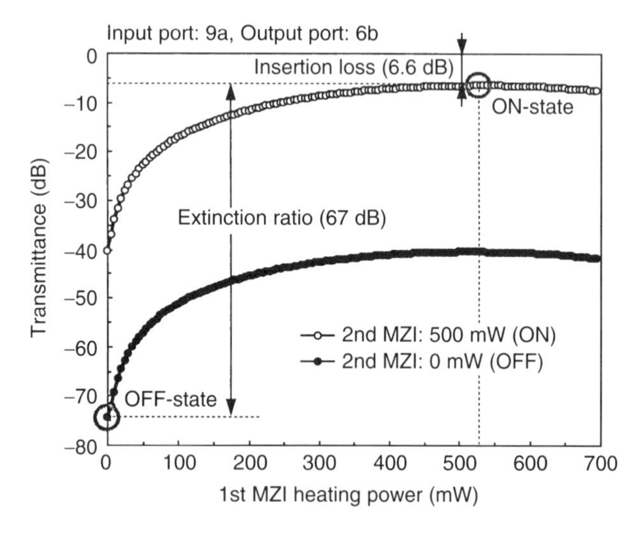

Fig. 6.53 Example switching characteristic of fabricated 16 × 16 matrix switch showing the transmittance versus the heating power to the thermooptic heater of the first MZI. The open circles indicate that the second MZI is in the cross-state (ON-state), and the filled circles indicate the bar-state (OFF-state).

was minimum was about 500 mW for an MZI. Because the transmittance is sufficiently small at 0 mW, it is clear that no electrical biasing is required in the OFF-state. In this example, we successfully achieved a 67 dB extinction ratio without bias power. The insertion loss of this optical path was 6.6 dB.

Figure 6.54(a) and (b) show the distributions of the measured insertion loss and extinction ratio, respectively, for all 256 possible optical paths from the input ports to the output ports. The insertion loss ranged from 6.0 to 8.0 dB, with an average value of 6.6 dB. The extinction ratio ranged from 40 to 63 dB, with an average value of 53 dB. Moreover, the PDL was less than 0.33 dB with an average value of 0.08 dB. These results show that we have successfully optimized the fabrication process for use with 6-inch wafers.

We also estimated the multiple crosstalk in the 16×16 matrix switch for all 256 optical paths in the worst polarization case based on Eq. (6.20), in which we assumed that the input signal powers were uniform. The multiple crosstalk ratio ranged from -50 to -33 dB, with an average value of -40 dB. As the number of points at which other optical paths cross in any actual optical path pattern is smaller than that in the model, the actual multiple crosstalk is better than the estimated value. Despite being the worst model estimation, we obtained an excellent value of better than -33 dB for the 16×16 matrix switch.

We also measured the wavelength dependence characteristics of the 16×16 TOSW. The extinction ratio and insertion loss at the worst optical path were better than 42 and 8.2 dB, respectively, in the 1530–1560 nm range. This flat response, which is similar to that of the 8×8 TOSW, covers the gain band of practical erbium-doped fiber amplifiers (EDFA). These results show that this device can be practically employed in WDM network systems.

The average ON-state power of the double-MZI switching units, which have two MZIs, was 1.04 W. Thus, the total power consumption on the chip needed to form any 16 optical paths was 16.6 W, where 16 double-MZI switching units are in the ON-state with the others in the ZERO-state. This consumption can be handled by simple fan-based air-cooling when the chip is packaged, and is not too large for practical use. Before phase-trimming, the total operating power was 61 W because bias power was required for all switching units to obtain a high extinction ratio. This clearly proves that the phase-trimming technique is very useful for reducing the power consumption of large-scale switches, such as $N \times N$ matrix switches and $1 \times N$ switches, in which there are many more OFF-state than ON-state switching units.

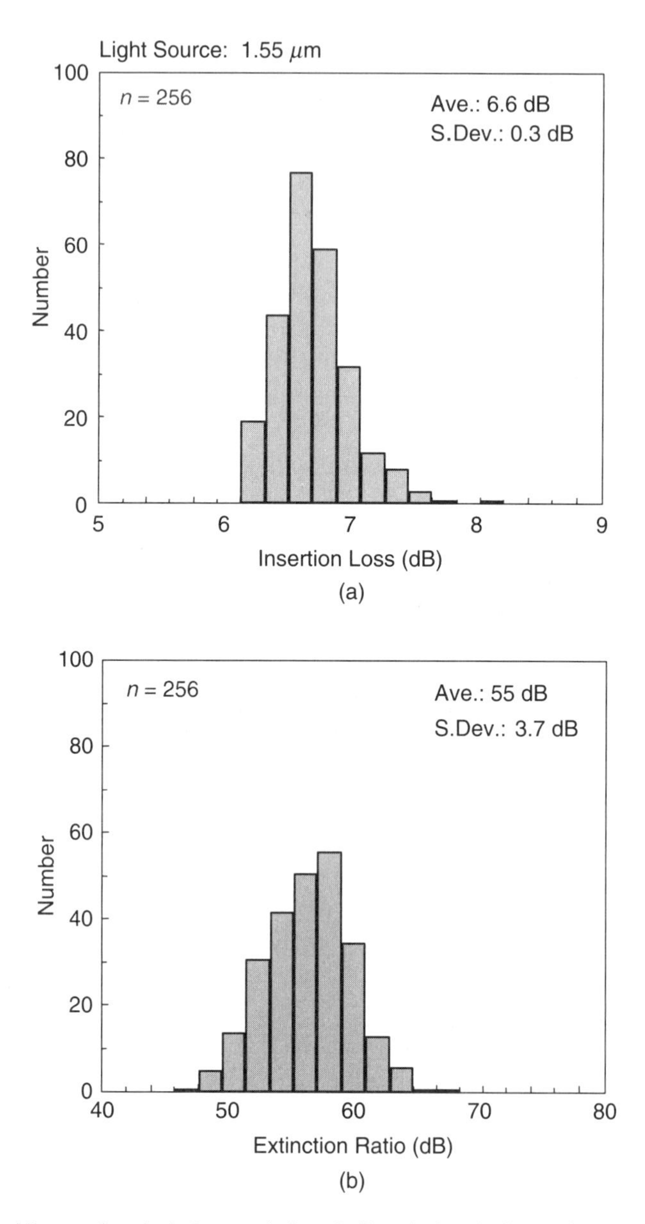

Fig. 6.54 Measured optical characteristics of all optical paths in the fabricated 16×16 matrix switch chip at 1.55 μm. (a) Insertion loss distribution. (b) Extinction ratio distribution.

6.6.2.3. 16 × 16 TOSW Module [51]

Driving circuits with TTL interfaces were used for the 8 × 8 TOSW module, which require the same number of control terminals as switching units. Therefore, if we adopt these conventional driving circuits in a large-integrated TOSW such as a 16 × 16 TOSW module, we need 256 or more terminals. Because such a large number of terminals complicates the module structure, a new driving circuit design must be developed to simplify the module structure. Here we describe a simple 16 × 16 TOSW module using newly developed driving circuits. This module is the largest integrated optical device ever reported among waveguide-type matrix switch modules. The new driving circuits are composed of 4 driving ICs with serial interfaces, and they enable us to reduce significantly the required number of control terminals, from 256 to only 3.

A schematic diagram of the driving circuits for the 16 × 16 TOSW module is shown in Fig. 6.55(a). In the module with the new driving circuits composed of 4 driving ICs, only 3 control terminals are necessary because of the serial interfaces. The serial control signals are transformed into parallel signals with the driving ICs. In addition, because an arbitrary and independent switching power can be supplied to each switching unit, it is possible to flatten the optical output power level even if the optical input power level fluctuates. This new circuit design can be easily applied to a larger-integrated optical module.

A time chart of the control signals is shown in Fig. 6.55(b). The control signals consist of three signals: CLOCK, DATA, and LATCH. The CLOCK signal is a series of 1-MHz pulses. The DATA signal is a series of 256-bit pulses synchronized with the CLOCK signal, and determines the switches that should turn on. The LATCH signal is a trigger pulse to begin/end the transmission of CLOCK and DATA signals to the ICs. For example, if the DATA signal is 00100000... (only the 3rd pulse is 1, and the remaining 255 pulses are 0), the switching power is supplied to switching unit #3 when the LATCH pulse is input. It takes about 0.26 ms to register a DATA signal at the ICs. Although all the switches are idling during registration, this does not affect the switching performance because 0.26 ms is much shorter than the switching time of this switch.

A photograph of the fabricated module is shown in Fig. 6.56. The PLC-TOSW chip was mounted on a multi-layered ceramic substrate on which the driving circuits were integrated. The PLC chip and the driving circuit were connected with gold wires. Both the terminals of the serial interfaces

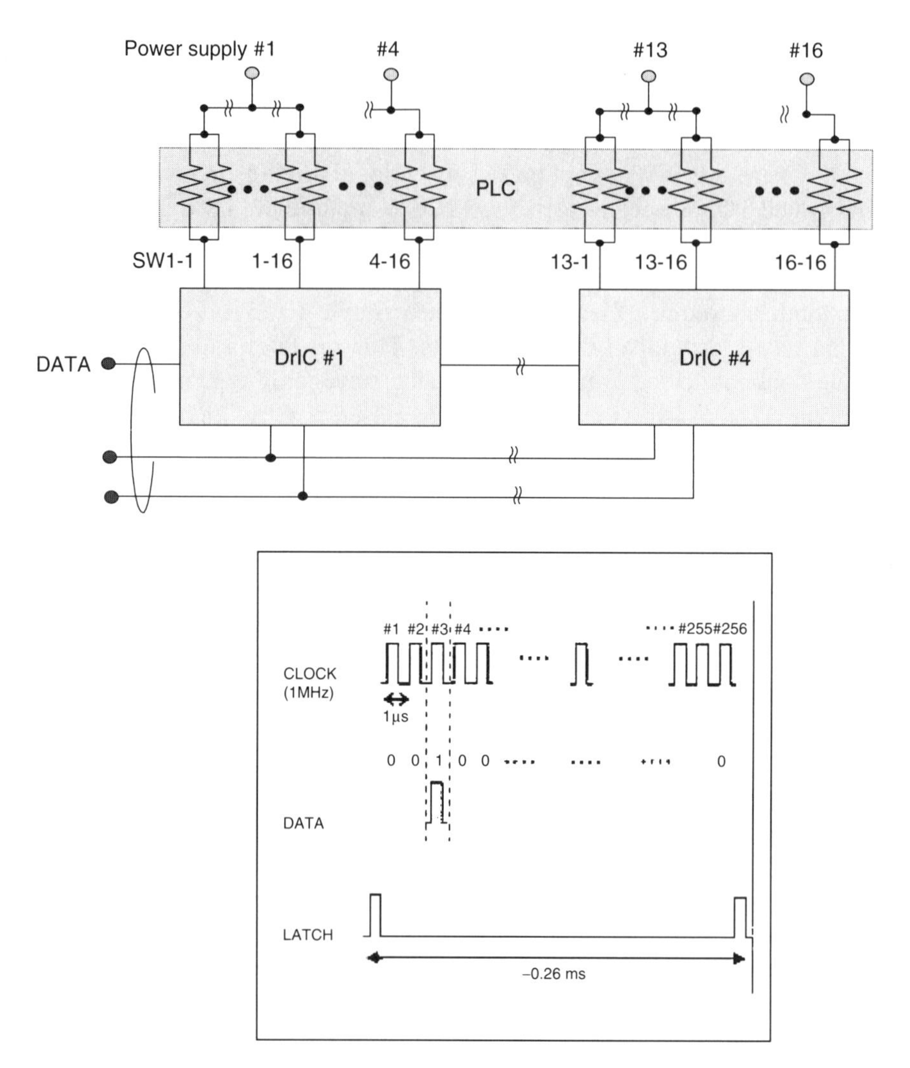

Fig. 6.55 Schematic diagram of driving circuits and time chart of the control signals.

and those for the switching power supply were connected to an electrical connector. Two fiber ribbons were butted and fixed to the facets of the input and the output ports of the chip. The reverse surface of the ceramic substrate was equipped with a cooling fin. The module was $165 \times 160 \times 23$ mm including the cooling fin.

We measured the performance of the fabricated TOSW module at 1.55 μm. We obtained a low insertion loss of 7.3 dB, a high extinction

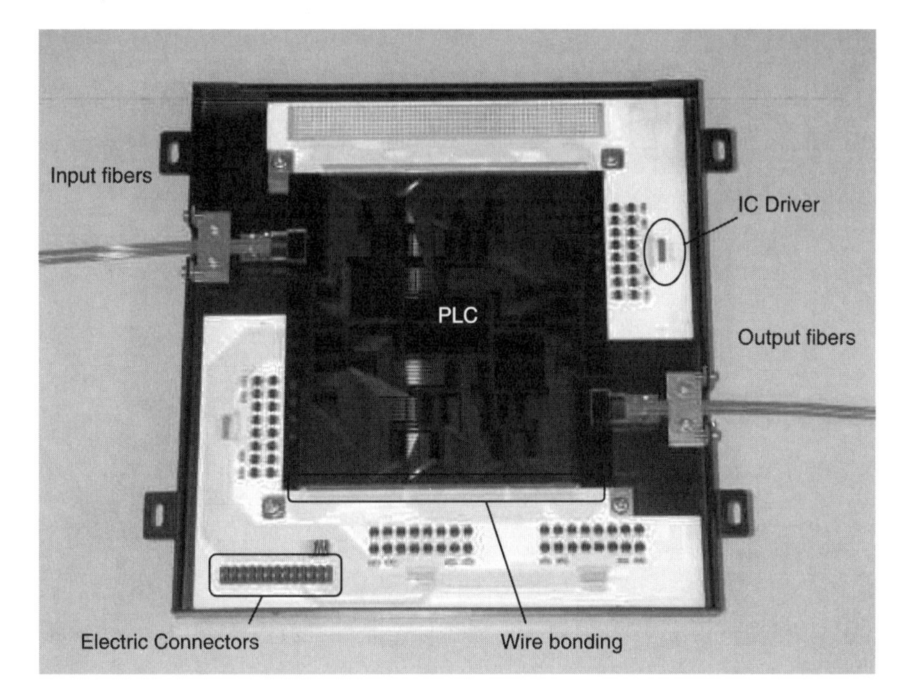

Fig. 6.56 Photograph of 16 × 16 matrix TOSW module. (*See color plate*).

ratio of 60.7 dB and a low PDL of 0.11 dB. The switching power per switch unit was 0.85 W, corresponding to a total switching power of 13.6 W (= 0.85 × 16). A switching time of less than 4.1 ms was realized, which was measured from a LATCH pulse input. This module is a very promising component for realizing large photonic network systems.

6.7. TOSW with Low Power Consumption [52]

Recently, a novel configuration has been demonstrated for reducing the power consumption in a silica-based PL-type TOSW without any insertion loss increase. The schematic structure of the proposed TOSW is shown in Fig. 6.57. This structure consists of a conventional MZI, heat insulating grooves, and trenches on the surface of the substrate. The trenches are filled with silica glass and are formed only under the heaters to thicken the under-cladding without increasing substrate warp. These trenches and the grooves, are formed symmetrically with respect to the MZI arms to avoid degrading

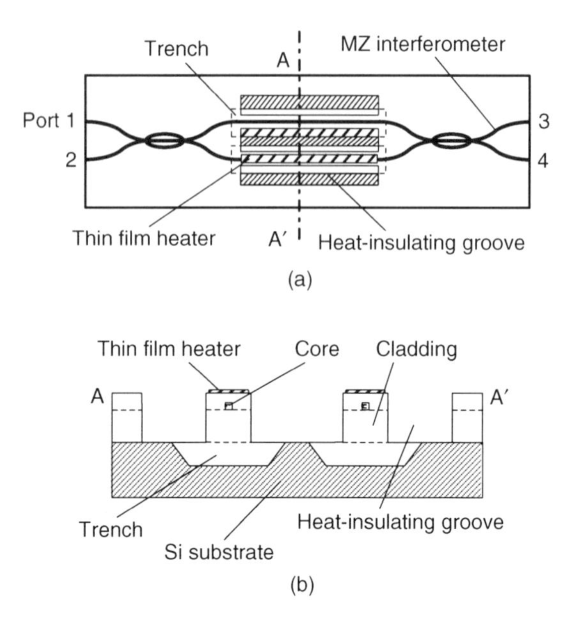

Fig. 6.57 Schematic structure of MZI switching unit with trench and heat-insulating groove.

the switching characteristics. The distance between the core center and the heater is set at 15 μm. The undercladding thickness of the proposed switch is set at 40 μm, which is the maximum value for the process because of wafer warp. The trench depth was also set at 40 μm to make it twice the total undercladding thickness. The trenches filled with silica glass are laid down under the heaters to suppress heat diffusion into the Si substrate. Heat-insulating grooves are also formed to restrict lateral heat diffusion.

We fabricated the proposed switch using PLC technologies and a mechanical polishing technique. The heat-insulating grooves were formed with RIE. The measured transmission characteristics of both the proposed and conventional TOSW at 1.55 μm are shown in Fig. 6.58 against applied powers. The fabricated switch needs only 90 mW switching power, 80% less than that of a conventional TOSW. The insertion losses were as low as 1 dB for both TOSWs, and we obtained an extinction ratio of more than 30 dB for the proposed TOSW. Moreover, the switching time of the proposed TOSW was 5 ms. While the switching time was increased because of the trade-off with the lower power consumption, the switching time is still sufficient for photonic network applications. This switch will therefore

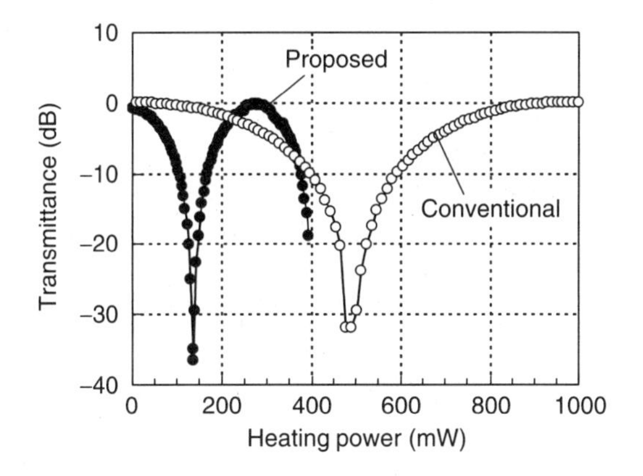

Fig. 6.58 Switching characteristics of conventional and proposed switches.

enable the realization of large-scale and high-density optical switches with low power consumption.

6.8. Conclusion

This chapter has reviewed the development of PLC-type TO switches for WDM-based photonic networks. Various kinds of optical devices have been developed to make it possible to construct flexible and large-capacity networks, and the optical switch is one of the key components for photonic networks. PLC-type TO switches have been developed to provide lightpaths and protection for such systems as OADM and OXC systems because the switching speed of passive switches is usually only of millisecond order. In this chapter, we began by describing the fabrication technologies and basic characteristics of PLC-type TO switches with simple calculations. Then, we reported the optical performance and applications of various types of switches including 2×2 switches, a $1 \times N$ switch, and $N \times N$ matrix switches. In terms of optical performance, insertion loss, extinction ratio, crosstalk, PDL, and wavelength dependence are important. These characteristics are reviewed here for PLC-type switches. Each device has been or will be installed in a suitable application in WDM systems. The next step is to meet demands that devices be made less expensive, on a larger scale, and with expanded functions. Further advances in optical components and related technologies will contribute greatly to the construction of photonic networks.

References

1. H. Toba, K. Oda, K. Nakanishi, N. Shibata, K. Nosu, N. Takato, and M. Fukuda, "A 100-ch optical WDM transmission/distribution at 622 Mbits/s over 50 km," IEEE J. Lightwave Technol., 8 (1990) 1396–1401.
2. I. P. Kaminow, "A wideband all-optical WDM network," J. Select. Areas in Commun., 14 (1996) 780–799.
3. T. Miki, "Photonic transport networks," Proc. IEEE, 81 (1993) 1594–1601.
4. N. Nagatsu, S. Okamoto, K. Koga, and K. Sato, "Flexible OADM architecture and its impact on WDM ring evolution for robust and large-scale optical transport networks," IEICE Trans. Commun., E82-B (1999) 1105–1114.
5. H. Toba, K. Oda, K. Inoue, K. Nosu, and T. Kitoh, "An optical FDM-based self-healing ring network employing arrayed-waveguide grating filters and EDFA's with level equalizers," J. Select. Areas in Commun., 14 (1996) 800–813.
6. M. Koga, Y. Hamazumi, A. Watanabe, S. Okamoto, H. Obara, K. Sato, M. Okuno, and S. Suzuki, "Design and performance of an optical path cross-connect system based on [a?] wavelength path connect," IEEE J. Lightwave Technol., 14 (1996) 1106–1119.
7. K. Sato, "Photonic transport network OAM technologies," IEEE Communications Magazine, Special issue on "Operation and management of broadband networks," 34 (1996) 86–94.
8. A. Himeno, K. Kato, and T. Miya, "Silica-based planar lightwave circuits," J. Selected Topics in Quantum Electron., 4 (1998) 913–924.
9. T. Miya, "Silica-based planar lightwave circuits: Passive and thermally active devices," J. Selected Topics in Quantum Electron., 6 (2000) 38–45.
10. Y. Hibino, "Passive optical devices for photonic networks," IEICE Trans. Electron., E83-B (2000) 2178–2190.
11. H. Takahashi, S. Suzuki, K. Kato, and I. Nishi, "Arrayed waveguide grating for wavelength division multi/demultiplexer with nanometer resolution," Electron. Lett., 26 (1990) 87–88.
12. M. K. Smit, "New focusing and dispersive planar component based on an optical phased array," Electron. Lett., 24 (1988) 385–386.
13. K. Jinguji and M. Kawachi, "Synthesis of coherent two-port lattice-form optical delay-line circuit," IEEE J. Lightwave Technol., 13 (1995) 73–82.
14. K. Takiguchi, K. Jinguji, and Y. Ohmori, "Variable group-delay dispersion equaliser based on a lattice-form programmable optical filter," Electron. Lett., 31 (1995) 1240–1241.
15. Y. Hida, Y. Hibino, T. Kitoh, Y Inoue, M. Itoh, T. Shibata, A. Sugita, and A. Himeno, "400-channel arrayed waveguide grating with 25 GHz spacing using 1.5%-Δ waveguides on 6-inch Si wafer," Electron. Lett., 37 (2001).

16. M. Oguma, K. Jinguji, T. Kitoh, T. Shibata, and A. Himeno, " Flat-passband interleave filter with 200 GHz channel spacing based on planar lightwave circuit-type lattice structure," Electron. Lett., 36 (2000) 1299–1300.
17. M. Kawachi, "Silica waveguides on silicon and their application to integrated components," Opt. and Quantum Electron., 22 (1990) 391–416.
18. T. Kominato, Y. Ohmori, H. Okazaki, and M. Yasu, "Very low loss GeO_2-doped silica waveguides fabricated by flame hydrolysis deposition method," Electron. Lett., 26 (1990) 327–328.
19. Y. Hibino, H. Okazaki, Y. Hida, and Y. Ohmori, "Propagation loss characteristics of long silica-based optical waveguides on 5 inch Si wafers," Electron. Lett., 29 (1993) 1847–1848.
20. Y. Hida, Y. Hibino, H. Okazaki, and Y. Ohmori, "10 m long silica-based waveguide with a loss of 1.7 dB/m," in Proc. IPR, Dana Point, paper IthC6 (1995).
21. Y. Hibino, F. Hanawa, H. Nakagome, M. Ishii, and N. Takato, "High reliability optical splitters composed of silica-based planar lightwave circuits," J. Lightwave Technol., 13 (1995) 1728–1735.
22. H. Hanafusa, F. Hanawa, Y. Hibino, and T. Nozawa, "Reliability estimation for PLC-type optical splitters," Electron. Lett., 33 (1997) 238–239.
23. N. Takato, K. Jinguji, M. Yasu, H. Toba, and M. Kawachi, "Silica-based single-mode waveguides on silicon and their application to guided-wave optical interferometer," J. Lightwave Technol., 6 (1988) 1003–1010.
24. N. Takato, T. Kominato, A. Sugita, K. Jinguji, H. Toba, and M. Kawachi, "Silica-based integrated optical Mach-Zehnder multi/demultiplexer family with channel spacing of 0.01–250 nm," J. Selected Areas in Commun., 8 (1990) 1120–1127.
25. L. B. Soldano and E. C. M. Pannings, "Optical multi-mode interference devices based on self-imaging: principles and applications," J. Lightwave Technol., 13 (1995) 615–627.
26. K. Jinguji, N. Takato, A. Sugita, and M. Kawachi, "Mach-Zehnder interferometer type optical waveguide coupler with wavelength-flattened coupling ratio," Electron. Lett., 26 (1990) 1326–1327.
27. N. Takato, T. Kominato, A. Sugita, K. Jinguji, H. Toba, and M. Kawachi, "Silica-based integrated optical Mach-Zehnder multi/demultiplexer family with channel spacing of 0.01–250 nm," J. Selected Areas in Commun., 8 (1990) 1120–1127.
28. H. Toba, K. Oda, K. Inoue, K. Nosu, and T. Kitoh, "Demonstration of optical FDM based self-healing ring network employing arrayed waveguide-grating ADM filters and EDFA," Proc. ECOC'94, (1994) 263–265.
29. K. Okamoto, Fundamentals of Optical waveguide. Academic Press, San Diego, 2000.

30. T. Goh, M. Yasu, K. Hattori, A. Himeno, and Y. Ohmori, "Low loss and high extinction ratio silica-based strictly nonblocking 16 × 16 thermo-optic matrix switch," IEEE Photon. Technol. Lett., 10 (1998) 810–812.
31. K. Moriwaki, M. Abe, Y. Inoue, M. Okuno, and Y. Ohmori, "New silica-based 8 × 8 thermo-optic matrix switch on Si that requires no bias power," in Tech. Digest OFC'95, San Diego, USA, paper WS1 (Mar. 1995), 211–212.
32. M. Abe, Y. Inoue, K. Moriwaki, M. Okuno, and Y. Ohmori, "Optical path length trimming technique using thin film heaters for silica-based waveguides on Si," Electron. Lett., 32:19 (Sept. 1996) 1818–1819.
33. T. Kawai, M. Koga, M. Okuno, and T. Kitoh, "PLC type compact variable optical attenuator for photonic transport network," Electron. Lett., 34 (1998) 264–265.
34. K. Hattori, M. Fukui, M. Jinno, M. Oguma, and K. Oguchi, "PLC-based optical ad/drop switch with automatic level control," J. Lightwave Technol., 17 (1999) 2562–2571.
35. K. Okamoto, M. Okuno, A. Himeno, and Y. Ohmori, "16-ch optical add/drop multiplexer consisting of arrayed waveguide gratings and double-gate switches," Electron. Lett., 32 (1996) 1471–1472.
36. M. Okuno, T. Watanabe, T. Goh, T. Kominato, T. Shibata, T. Kawai, M. Koga, and Y. Hibino, "Low-loss and high extinction ratio silica-based 1 × N thermo-optic switches," Tech. Dig. OECC/IOOC2001, Sydney Australia (July 2001) paper 13C2-1, 60–61.
37. T. Kominato, A. Himeno, M. Ishii, T. Goh, K. Hattori, F. Hanawa, and K. Kato, "Silica-based optical switch with a single-facet array fiber connection," in Proc. CLEO Pacific Rim, Makuhari Japan, paper FH2 (1997).
38. A. Watanabe, S. Okamoto, M. Koga, K. Sato, and M. Okuno, "Design and performance of delivery and coupling switch board for large scale optical path cross-connect system," IEICE Trans. Commun., E81-B (1998) 1203–1212.
39. K. Koga, A. Watanabe, T. Kawai, K. Sato, and Y. Ohmori, "Large-capacity optical path cross-connect system for WDM Photonics transport network," J. Selected Areas in Commun., 16 (1998) 1260–1269.
40. M. Okuno, A. Sugita, T. Matsunaga, M. Kawachi, Y. Ohmori, and K. Kato, "8 × 8 optical matrix switch using silica-based planar lightwave circuits," IEICE Trans. Electron., E76-C (1994) 1215–1223.
41. M. Okuno, K. Kato, Y. Ohmori, M. Kawachi, and T. Matsunaga, "Improved 8 × 8 integrated optical matrix switch using silica-based planar lightwave circuits," J. Lightwave Technol., 12 (1994) 1597–1606.
42. M. Okuno, K. Kato, R. Nagase, A. Himeno, Y. Ohmori, and M. Kawachi, "Silica-based 8 × 8 optical matrix switch integrating new switching unit with large fabrication tolerance," J. Lightwave Technol., 17:5 (May 1999) 771–781.

43. T. Goh, A. Himeno, M. Okuno, H. Takahashi, and K. Hattori, "High extinction ratio and low loss silica-based 8 × 8 thermo-optic matrix switch," IEEE Photon. Technol. Lett., 10 (1998) 358–360.
44. A. Himeno, T. Goh, M. Okuno, H. Takahashi, and K. Hattori, "Silica-based low loss and high extinction ratio 8 × 8 thermooptic matrix switch with path-independent loss arrangement using double Mach-Zehnder interferometer switching units," in Proc. ECOC'96, Oslo, Norway, 4, paper ThD.2.2 (Sep. 1996) 149–152.
45. T. Goh, A. Himeno, M. Okuno, H. Takahashi, and K. Hattori, "High-extinction ratio and low-loss silica-based 8 × 8 thermooptic matrix switch," J. Lightwave Technol., 17 (1998) 1192–1199.
46. T. Shimoe, K. Hajikano, and K. Murakami, "A path-independent-insertion-loss optical space switching network," in Tech. Dig. ISS'87, 4, paper C12.2 (1987), 999–1003.
47. T. Nishi, T. Yamamoto, and S. Kuroyanagi, "A polarization-controlled free-space photonic switch based on a PI-loss switch," IEEE Photon. Technol. Lett., 5:9 (Sep. 1993) 1104–1106.
48. R. Nagase, A. Himeno, M. Okuno, K. Kato, K. Yukimatsu, and M. Kawachi, "Silica-based 8 × 8 optical matrix switch module with hybrid integrated driving circuits and its system application," J. Lightwave Technol., 12:9 (Sept. 1994) 1631–1639.
49. T. Goh, M. Yasu, K. Hattori, A. Himeno, and Y. Ohmori, "Low loss and high extinction ratio silica-based strictly nonblocking 16 × 16 thermo-optic matrix switch," IEEE Photon. Technol. Lett., 10 (1998) 810–812.
50. T. Goh, M. Yasu, K. Hattori, A. Himeno, M. Okuno, and Y. Ohmori, "Low loss and high extinction ratio strictly nonblocking 16 × 16 thermo-optic matrix switch on 6-in wafer using silica-based planar lightwave circuit technology," J. Lightwave Technol., 19 (2001) 371–379.
51. T. Shibata, M. Okuno, T. Goh, M. Yasu, M. Ishii, Y. Hibino, A. Sugita, and A. Himeno, "Silica-based 16 × 16 optical matrix switch module with integrated driving circuits," Tech. Digest OFC'01, Anaheim, USA, paper WS1 (Mar. 2001), 211–212.
52. R. Kasahara, M. Yanagisawa, A. Sugita, T. Goh, M. Yasu, A. Himeno, and S. Matsui, "Low-power consumption silica-based 2 × 2 thermooptic switch using trenched silicon substrate," IEEE Photon. Technol. Lett., 11:9 (Sept. 1999) 1132–1134.

Chapter 7 | Lithium Niobate and Electrooptic Guided-wave Optical Switch

Hideaki Okayama

Oki Electric Industry Co. Ltd., 550-5, Higashiasakawa-cho, Hachioji-shi, Tokyo 193-8550, Japan

ABSTRACT

This chapter describes an optical switch using electrooptic material, especially lithium niobate crystal. The start of the chapter describes a guided-wave optical switch using a single routing stage, and then devices implemented by connecting multiple ranks of switching elements. The single routing stage switch can be constructed using deflection or diffraction. The number of modes used in operation and the device structure can classify switching elements. Representative switching network architectures constructed by connecting switching elements are shown. Some device demonstrations using lithium niobate and other materials such as compound semiconductors are introduced.

7.1. Introduction

Optical switch refers to a device that controls the direction of the light signal or changes states between transmitting and cutting off the light signal. Optical switches can be classified into devices with or without optoelectronic signal conversion. In this chapter, only devices without optoelectronic signal conversion are explained. In these devices, the light signal is transmitted through the device as a light wave. By doing so, the device becomes inherently insensitive to the format of the light signal. Signals with any format and speed can be transmitted through the device so long as the dispersion effect doesn't distort the signals.

The switching function can be accomplished in time, space, or frequency domain. This chapter mainly describes the spatial switch.

Like other types of optical devices, the optical switch can be implemented using bulk components or optical waveguide technology. Conventional components such as prisms, lenses, mirrors, and so on are used as bulk components. The light signal is transmitted in a free space. In the device using optical waveguide technology, the light is confined into a region called the core, where the refractive index is made higher than the

WDM TECHNOLOGIES: PASSIVE OPTICAL COMPONENTS
$35.00

ISBN: 0-12-225262-4

surrounding region. Waveguide technology avoids the, problematic optical axis alignment associated with the bulk type components. Optical waveguide devices are classified into either optical fiber or planar substrate types. The planar substrate type device is fabricated using a process similar to that for integrated circuits, which enables mass production of relatively complex optical circuits.

$LiNbO_3$ is a ferroelectric crystal, which exhibits a large electrooptic (EO) effect, acoustooptic (AO) effect, nonlinear effect, and thermooptic (TO) effect. Crystal large enough to obtain a wafer of more than four inches in diameter is grown regularly. With this material, a low loss optical waveguide can be fabricated relatively easily by Ti diffusion or proton exchange methods. These features have made this material the most used as a substrate for fabricating optical switches from the early days of optical guided-wave device development.

This chapter reviews development of guided-wave optical switch technology using electrooptic material, especially $LiNbO_3$ substrate. Through its long history of development, almost all of the guided-wave optical switch operation principles have been verified using $LiNbO_3$ as described briefly here.

7.2. Overview of Guided-wave Optical Switching Technology

Optical switches can be grouped into four types by their schemes for routing the optical signal (Fig. 7.1). The figures depict devices based on optical waveguide technology, but the concept is also applicable to devices using bulk components or optical fiber.

The first method (Fig. 7.1(a)) uses an optical deflector or scanner. The direction of the light beam is controlled by an analog signal. The deflection angle is proportional to the control signal. The second method (Fig. 7.1(b)) uses diffraction of the light by electrically induced gratings. In the first two methods, the deflected light travels through free space or a planar optical waveguide to the destination.

In the last two methods (Fig. 7.1(c, d)), the destination of the light signal is selected by alternating between two possible switching states of the switching elements. In the method using 1×2 (One input two output), 2×1 (Two input one output), or 2×2 (Two input two output) switching elements (Fig. 7.1(c)), a device with many port numbers is constructed by

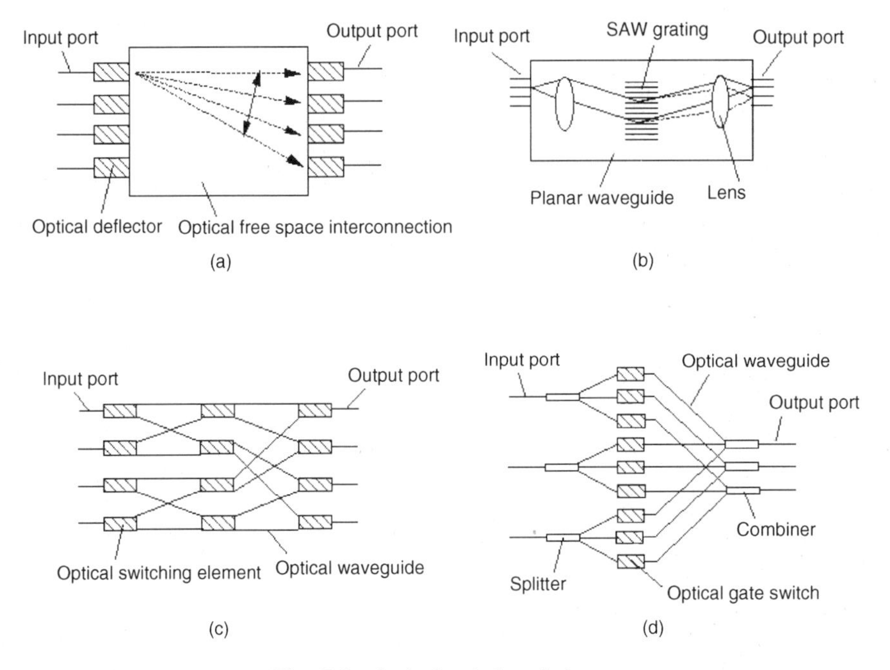

Fig. 7.1 Optical switch variations.

connecting switching elements in multiple ranks with optical waveguide interconnections to form a switching network. Because a switching element has two output ports, two switching states exist. (In some applications the intermediate state is used to split optical signals to two directions.) Two levels of control signal are sufficient to control a switching element. The control scheme can be regarded digital, in contrast to the analog type of the deflector. In Fig. 7.1(d), an optical gate switch is used to set up a path between input and output ports. An optical gate switch is a one-input one-output device that is capable of changing states between transparent and cut off by a control signal. The device has two switching states corresponding to two control signal states, so is regarded as a digital type control scheme. A light signal at the input port is split into the number of output ports by an optical coupler. The light signals are recombined at the output port by an optical coupler. Input and output couplers are connected so as to set up every possible connection between input and output ports. Optical gate switches placed at each connection establish the desired path by setting an optical gate switch on the desired path to the transparent state and leaving other optical gate switches in the cut-off state.

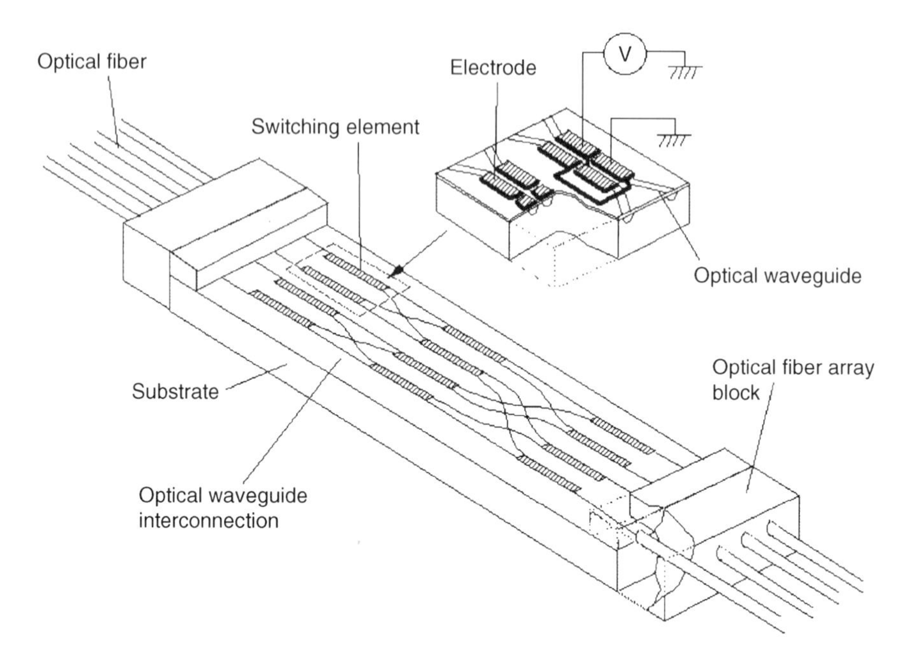

Fig. 7.2 Optical switch chip with multistage switching elements.

The direction or intensity of the light signal can be controlled by several schemes, including a mechanical one such as moving mirrors, and schemes without mechanics. Nonmechanical schemes include the electrooptic (EO) effect, acoustooptic (AO) effect, nonlinear effect, and thermooptic (TO) effect, all with no moving parts involved.

Many types of optical switches have been implemented by myriad combinations of routing, lightwave direction control, and confinement schemes. The most studied among them are devices constructed on a substrate by connecting switching elements with a EO or TO effect. A schematic of such a device structure is shown in Fig. 7.2. The use of EO effect with a solid-state structure improves the reliability and switching speed by avoiding moving parts. The drive power is reduced, optical axis alignment between elements is avoided, many switching elements are integrated, and mass production becomes possible by adopting waveguide structure fabricated on a substrate. Elements other than switches, such as wavelength filters, wavelength converters, and such, can also be integrated into a chip. Switch network architecture using multiple rank switching elements is selected mostly for its crosstalk requirement and its analogy to electronic devices.

7.3. Single-stage Routing Multi-port Optical Switch

The device uses a single-stage element for routing the optical signal to the output port. Deflection by phase control of the lightwave, or diffraction, can be used to change the direction of a light signal.

7.3.1. *DEFLECTOR TYPE*

There are two configurations to attain deflection of the lightwave in a guided-wave device. The first configuration uses collimated light beam and an electrooptically refractive-index-controlled prism. The second configuration uses waveguide arrays and electrodes placed on waveguides to control the phase.

The typical structure of the device using the collimated light beam and EO prism is shown in Fig. 7.3. An element such as a waveguide lens is used to generate a collimated light beam. The light beam is propagated in a planar waverguide that is fabricated using electrooptic material. A prism-shaped electrode is placed on top of the planar waveguide. By applying voltage to the electrode, the refractive index of the prism-shaped region changes. When the light beam passes through the slanted interface of the prism, the light beam is refracted due to Snell's law [1]. The angle Θ of the refraction is given by

$$\Theta = \Delta n L/(nW) = rn^3(V/g)L/(nW)/2 \qquad (7.1)$$

with Δn being the refractive index change, L the prism length, W the prism width, r the electrooptic coefficient, n the planar waveguide refractive index, V the applied voltage, and g the distance between electrodes.

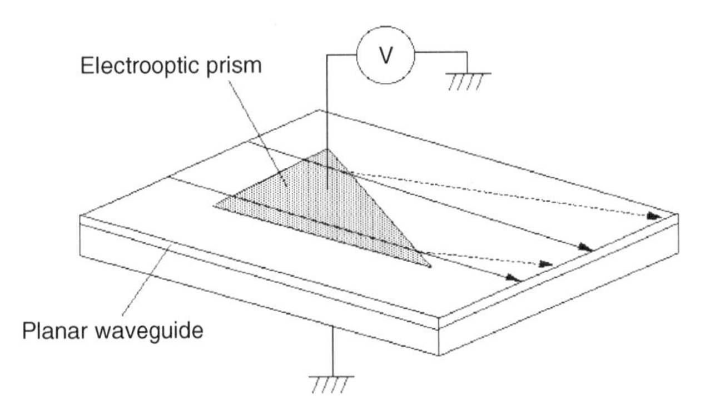

Fig. 7.3 Deflection-type optical switch using electrooptic prism.

In the ideal case, the electric field is generated between the electrode on the surface and the ground electrode under the planar waveguide. The distance between electrodes must be short enough to produce an electric field with sufficient strength to attain the desired deflection angle at a moderate applied voltage. A device using conducting substrate has been fabricated to implement such a configuration. For an LT substrate of $g = 500\ \mu m$ and $L/W = 13$, $\Theta = 4$ mrad is attained with 600 V applied voltage [2]. Using PZT 3.6-μm-thick waveguide and semiconducting Nb:$SrTiO_3$ substrate, a deflection angle of 40 mrad is attained with 40 V applied voltage for $L/W = 10$ [3].

The refractive index change, due to the Pockels effect, reverses its polarity when the direction of the electric field against the crystal axis is reversed. Several methods, originally developed for a second harmonic generation device, can be used to fabricate a region with reversed polarization. Push–pull configuration to enlarge the refraction angle is attained by cascaded or parallel prisms with alternating polarizations. Deflection angle of plus and minus 44 mrad is attained by plus and minus 70 V-applied voltage in 200 μm thick LN [4].

The waveguide array device (Fig. 7.4) consists of an electrode placed on each waveguide (sometimes called a phaser). The input light is divided

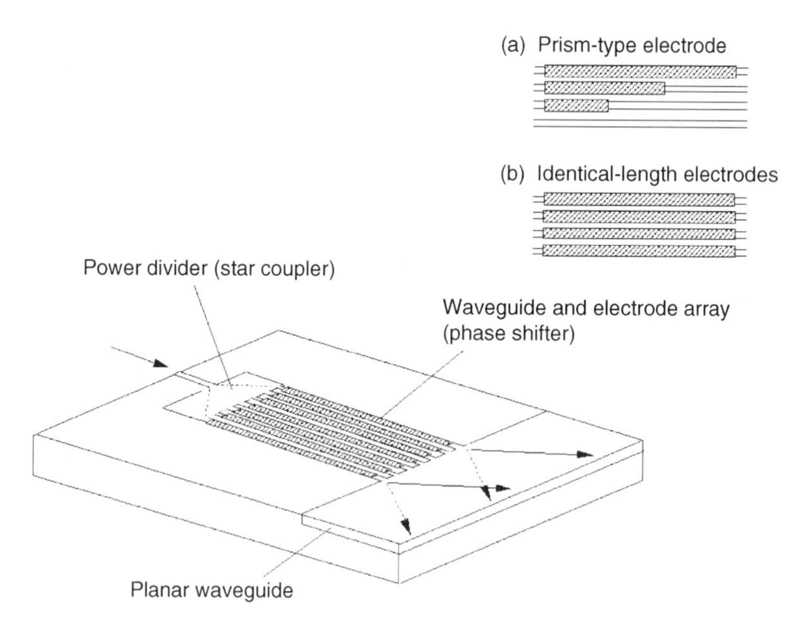

Fig. 7.4 Waveguide array deflection type optical switch.

among waveguides at the optical coupler placed at the input of the waveguide array. The input coupler can be a star coupler or branching waveguides. By applying voltage to electrodes, the refractive index changes and the phase shift accumulates along the electrode length as the light travels the waveguide. The amount of phase shift is set to increase from the first to the last row of the waveguide array. A planar waveguide is connected to the end of the waveguide array. Light output from the waveguide array forms a wavefront according to the phase. A plane wave is generated if the phase difference between adjacent waveguides is constant.

The deflection angle Θ is given by $\Theta = \delta(\Delta nL)/(nd)$, as will be shown, where d is the distance among adjacent waveguides at the interface between waveguide array output and the planar waveguide, and $\delta(\Delta nL)$ denotes the ΔnL difference between adjacent waveguides. Δn is the refractive index change through the electrooptic effect from applied voltage and L is the electrode length. If the electrode length is the same for all the waveguides, $\Theta = \delta\Delta nL/(nd)$. A different amount of voltage is applied to electrodes to attain a refractive index change difference $\delta\Delta n$ between waveguides. When allocating different lengths for electrodes $\Theta = \Delta n\delta L/(nd)$. Here, δL denotes the electrode length difference between adjacent electrodes. All the electrodes are driven by the same voltage to generate constant refractive index change Δn. The structure is similar to a device having a prism-shaped electrode to deflect the collimated light beam.

With N_A being the number of waveguides, $W = N_A d$ and the total electrode length $L = N_A \delta L$. The deflection angle $\Theta = \Delta nL/(nW)$. However, the lightwave at the output waveguide is not identical to the collimated light beam of the prism electrode device, because the wavefront at the output planar waveguide is constructed from an ensemble of lightwaves from waveguides.

The lightwave generated in the planar waveguide is approximated at large propagation distance z as

$$I(\Theta) = |\Sigma_i A_i(\Theta - id/z)\exp(-jk_0 n \sin\Theta id + j\phi_i)|^2 \qquad (7.2)$$

When the light angle distribution from each waveguide $A_i(\Theta - id/z)$ is set to be constant $A(\Theta)$ in the waveguide array for simplicity, and $\phi_i = k_0 i\delta(\Delta nL)$, then Eq. (7.2) becomes

$$I(\Theta) = A(\Theta)\sin^2\{k_0 N[n\sin\Theta d - \delta(\Delta nL)]/2\}/ \sin^2\{k_0[n\sin\Theta d - \delta(\Delta nL)]/2\} \qquad (7.3)$$

The light beam is deflected to the direction determined by $k_0[n \sin \Theta d - \delta(\Delta nL)]/2 = m\pi$ (m: integer), so that if $\Theta \ll 1$

$$\Theta = m\lambda/(nd) + \delta(\Delta nL)/(nd). \tag{7.4}$$

The light beam is deflected into multiple directions for nonzero m (high-order deflected beam), which is the result of discrete waveguide output positions. However, if the light diffraction angle from each waveguide is narrow enough so that the angle of the high-order deflected beam is above the diffraction angle, most of the light power is in the $m = 0$ main light beam. The light is also deflected into multiple directions in the parallel prism-shaped electrode deflector device due to cyclic light output structure [5].

Three types of waveguide array devices have been demonstrated so far (Fig. 7.5). The first one is a 1 × N type, the second a nonblocking N × N type, and the third a N × N type, non-blocking only for one input signal. The last type of structure is similar to an arrayed waveguide grating wavelength (AWG) multiplexer and has been implemented as a TO switch. EO devices have been demonstrated as the first (1 × N) and the second (N × N) types.

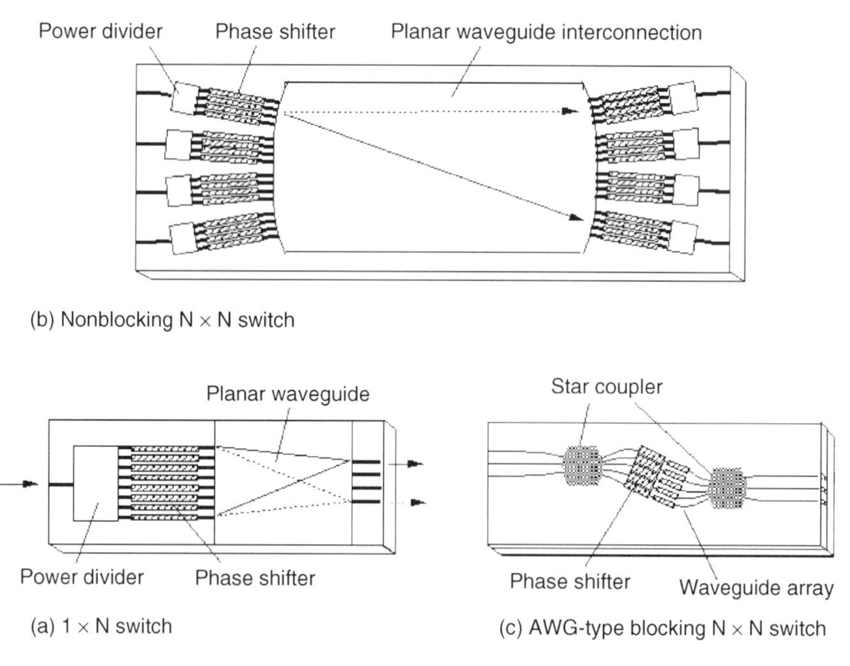

Fig. 7.5 Waveguide array optical switches.

The 1 × N device consists of an input coupler to divide the input light into the waveguide array, the waveguide array with electrodes, an output planar waveguide that propagates the deflected light beam, and output waveguides connected at the far end of the planar waveguide. An AWG-type N × N device is implemented by connecting additional input ports to the 1 × N device.

A strictly nonblocking N × N device is constructed by placing an N deflector at the input and an N deflector (selector) at the output [6]. The input and output deflectors are connected by planar waveguide interconnections. Each deflector is composed of a coupler to divide light between the waveguide array and a phaser (waveguide array with electrodes). The input deflector drives the light beam to a desired output deflector (selector). The output deflector (selector) selects the light beam from the desired input port and sends it to the output port. Focusing of the light into the output waveguide at the output deflector (selector) is the reverse process of the light deflection at the input deflector. The tilted lightwave front input to the selector is aligned into a lightwave front that focuses the light into the output waveguide by the phaser.

Several methods can be adopted to increase the available channel number N over the basic structure. Cascades of deflectors [7, 8] or electrodes [8] are simple examples. Two deflectors with different beam steering capability are connected in series for the cascaded deflector configuration. The first and second deflectors perform the coarse wide and fine small angle deflection, respectively. In the cascaded electrode structure, electrodes are grouped into different combinations at each rank. Coarse and fine-tuning are done at first- and second-rank electrodes, respectively.

In a GaAs-AlGaAs device, a 1 × 9 device with 30 waveguide array was demonstrated [9]. A 1 μm wide waveguide on 3 μm pitch is capable of scanning the light beam through 0.33 rad. The input light is fed into the waveguide array through a planar waveguide and the deflected light is focused into 9 output waveguides at the end of a 1-mm-long planar waveguide. The crosstalk was –15 dB.

A 4 × 4 device was demonstrated in $LiNbO_3$ [6]. The device consists of 4 deflectors at the input and output ports, respectively. Each deflector is composed of a 4-channel waveguide array, electrodes with different lengths and push–pull configuration [10], and a coupler to feed the input light into the waveguide array. The interconnection planar waveguide is 14 mm long and the electrode is 7.5 mm long, with a total device length of 45 mm. The required drive voltage was less than 26 V. The crosstalk was –14 to –21 dB.

The AWG-type device fabricated so far uses SiON-SiO_2 waveguide technology and TO effect to control the phase required for switching, although the device can also be implemented using the EO effect. The deflector-type switch is an analog control device in which the deflection angle is proportional to the drive voltage. The deflection angle should be set in high precision for a large channel number device. The dc drift should be suppressed in this type of device.

7.3.2. DIFFRACTION TYPE

Acoustooptic diffraction has been used to implement single-stage deflection of the light signal. Diffractions in planar waveguide [11] and two-dimensional free-space beam steering [12] have been demonstrated using $LiNbO_3$ waveguide technology. The grating period generated by the surface acoustic wave (SAW) can be tuned by changing the RF drive frequency. The light beam can be scanned by changing the frequency. The Bragg condition is given by

$$\Theta_B = \sin^{-1}(\lambda/2\Lambda) \tag{7.5}$$

with Λ and λ being the wavelengths of the grating and the light beam, respectively. The diffraction angle change is obtained by differentiating the equation as

$$\Delta\Theta = \lambda\Delta f/(2V_{SAW}\cos\Theta_B), \tag{7.6}$$

where Δf is the frequency change and V_{SAW} the SAW velocity.

The schematic structure using diffraction in a planar waveguide is shown in Fig. 7.6. The input light is fed into the planar waveguide through the channel waveguide. A collimating waveguide lens converts the light into parallel beams. Because the input channel waveguide position against the collimating lens is different, the propagation angle of the light beam is different for each input port. SAW transducers generate acoustooptic gratings in the planar waveguide. The acoustooptic grating diffracts the light beam. The diffracted light beam is focused into an output channel waveguide by a focusing waveguide lens. The light is diffracted when the light beam incident angle is in the Bragg condition. The Bragg condition is given by Eq. (7.5) with $\Theta_B = \Theta_{LB} + \Theta_{SAW}$, Λ being the SAW wavelength. Θ_{LB} and Θ_{SAW} are the angles of the light beam and SAW against the axis parallel to the device end face, respectively. The angle Θ_{LB} is different for light beams from different input ports. Multiple tilted SAW

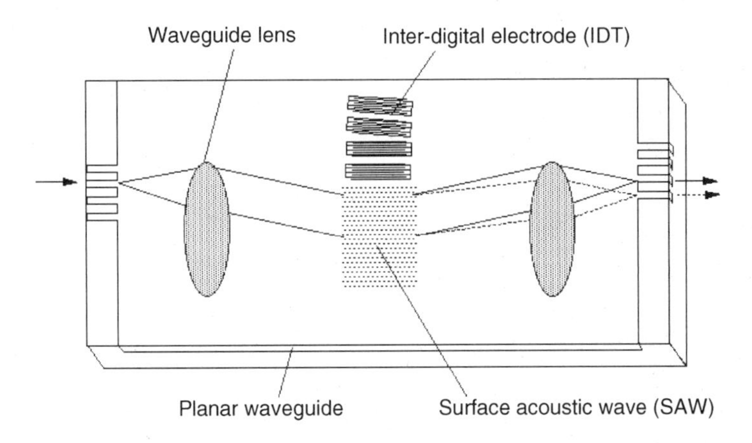

Fig. 7.6 Multiport acoustooptic diffraction optical switch (Tsai 1998). Appl. Phys. Lett., 60 (4), 27 Jan. 1992. © 1992 American Institute of Physics.

transducers with different Θ_{SAW} were used for multiple input ports so that one transducer was devoted to one input port. Different SAW wavelengths Λ (Frequency) are generated to drive light from an input port to many output ports.

A 4×4 and 8×8 switch were demonstrated using Y-cut $LiNbO_3$ at 0.6328 μm wavelength. The size of the substrates were 1×3 and 1×3.7 cm^2, respectively. The 13.5 MHz RF drive power of 125 mW yielded 25% efficiency. The crosstalk was –12.2 dB and switching time 0.4 μs in an 8×8 device.

The structure of the 2-dimensional (2D) beam steering device is shown in Fig. 7.7. The device uses lateral diffraction by the acoustooptic grating, similar to the planar waveguide device, and vertical diffraction by acousto-optic grating traveling along the light beam propagation axis in a proton-exchanged (PE) waveguide. The lateral diffraction is done using Bragg diffraction. The vertical diffraction is done using the guided TM mode to substrate radiation TE mode conversion. To phase match the two modes, the parallel vector component of the substrate radiation mode should be equal to the propagation constant of the TE mode at the PE waveguide. The condition gives $\cos \Theta = (|k_{TM}| - |K_{SAW}|)/|k_s|$, where K_{SAW} represents the SAW wave vector, k_{TM} the guided TM mode, and k_s the substrate radiation mode wave vectors, respectively. Tuning of the SAW wavelength ($\Lambda = 2\pi/K_{SAW}$) can change the substrate radiation mode propagation angle Θ. Two-dimensional scanning of 18 horizontal and 40 vertical beam spots for 0.6328 μm light was demonstrated. An electrooptic Bragg grating diffraction switch is shown in Fig. 7.8.

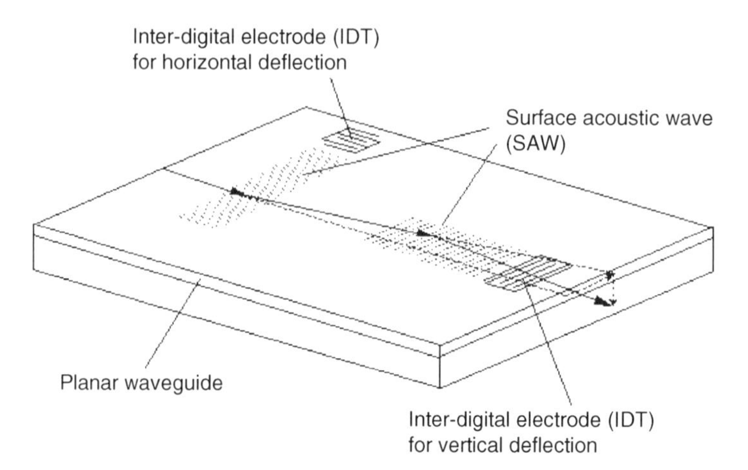

Fig. 7.7 3D beam steering acoustooptic switch.

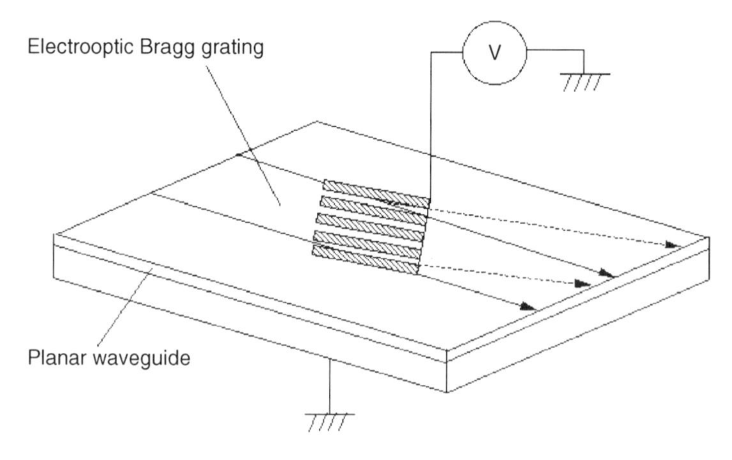

Fig. 7.8 Electrooptic Bragg grating diffraction switch.

The EO effect can also be used for diffracting the light. The EO grating is generated by an interdigital or comb-shaped electrode. The period of the grating is defined by the period of the electrodes. The grating period cannot be changed, and tuning of the diffraction angle is usually difficult. The grating was used as a 1×2 or 2×2 crosspoint switching element for a multi-stage switching network [13]. Domain reversal technology is used to generate an EO grating with a strip-shaped single electrode [14].

In bulk devices, several studies have been done using AO, LCD, and EO devices [15–17]. A deflector-type TO switch has also been studied [18].

7.4. Optical Switching Elements for Multistage Switching Network

Many types of optical switching elements based on guided-wave phenomena have been studied for constructing the switching network. Most of the devices studied are 2×2, 1×2, or 2×1 switching elements. For the 2×2 optical switching elements two switching states exist, corresponding to the combinations of input and output ports (Fig. 7.9). In the "bar" state, parallel paths connect input and output ports. In the "cross" state, crossed paths connect input and output ports. Optical switching elements can be classified according to what type of normal mode phenomena they use. Normal mode phenomena used for switching are (a) propagation constant change, (b) field distribution change, and (c) mode conversion. Normal modes can be distinguished by the field distribution type at the waveguide cross-section. Normal modes are excited in the optical guided-wave system, similar to the microwave guide. A certain optical field in an optical waveguide system is represented as a super position of normal modes. An example of normal modes is shown in Fig. 7.10.

Devices are classed first by the number of modes used for switching. Representative examples of switching elements are shown in Figs. 7.11 and 7.12. Devices using single or double normal modes are shown in Fig. 7.11. Devices using more than three normal modes are shown in Fig. 7.12.

The most well-known device using single normal mode is a Y-branch optical switch. The device operates using the deformation of the optical

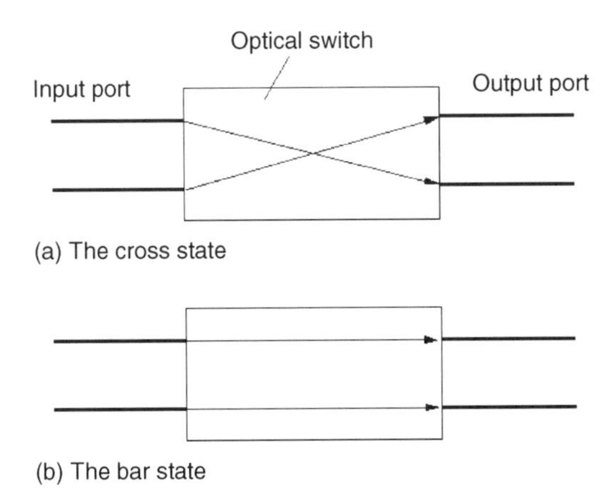

Fig. 7.9 Two switching states in a 2×2 optical switching element.

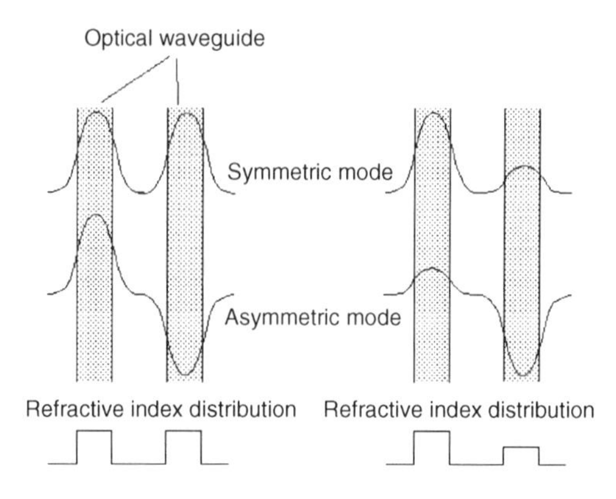

Fig. 7.10 Normal modes in directional coupler.

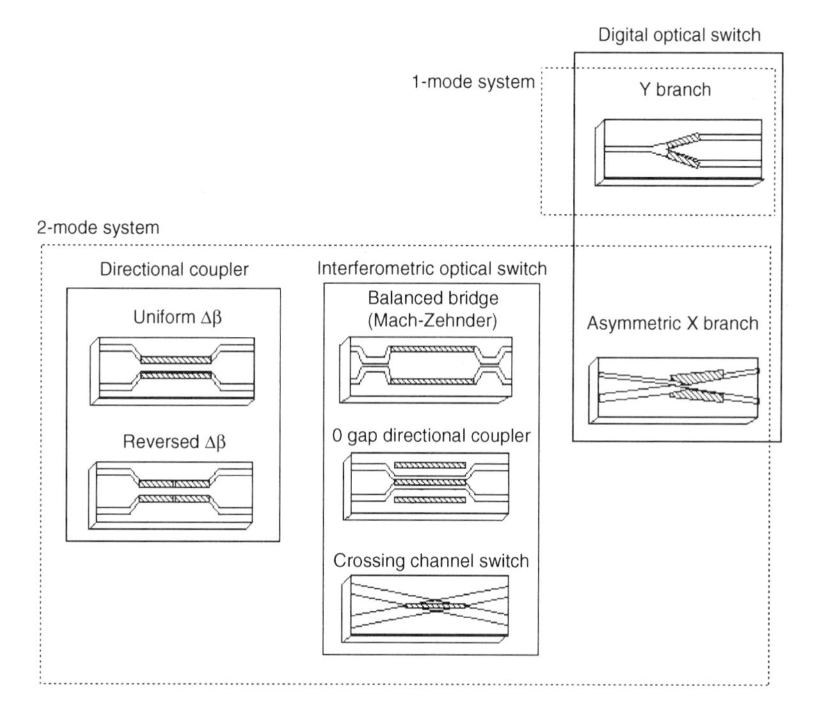

Fig. 7.11 Optical switching elements using one and two normal modes.

light field by changing the refractive index difference between waveguides forming the branch. The devices using double normal modes are represented by the directional coupler, balanced bridge (Mach-Zehnder) switch, and asymmetrical X-branch (digital) switch. The directional coupler uses

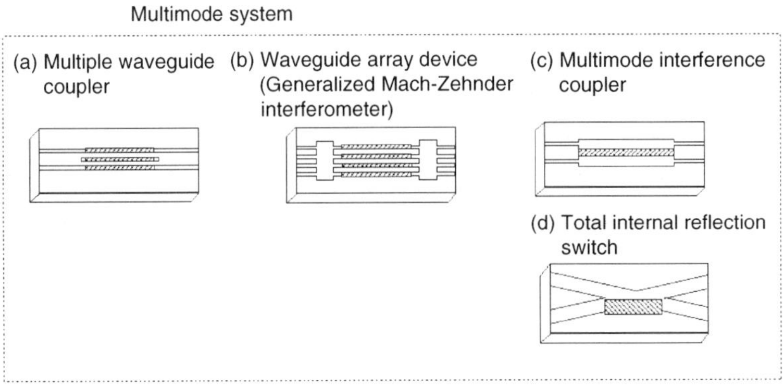

Fig. 7.12 Optical switching elements using multimode effect.

propagation constant change due to induced refractive index difference between waveguides. The balanced bridge uses normal mode conversion for switching by changing the refractive index difference between two light paths. The asymmetrical X-branch uses deformation of the optical light field with changing of the refractive index difference between waveguides forming the branch, similar to the Y-branch switch.

Well-known devices among those using more than three normal modes are the multiple guide directional coupler, Multi-Mode Interference (MMI) coupler, total internal reflection (TIR), and multi-leg Mach-Zehnder devices.

Taking drive voltage (or current) for the horizontal axis and optical output power observed at an output for the vertical axis results in plotting a switching curve. Devices can be classified into groups by their switching curve (Fig. 7.13). Many switching voltages exist in the interferometer-type switching curve. Switching states are attained at phase conditions separated by 2π. In the digital-type switching curve, a switching state changes to and stays in another switching state above a certain threshold voltage. A device using a digital type switching curve is called a digital optical switch, and has a switching voltage range that is wide for both the cross and bar states. The switching curve exhibiting a wide switching voltage range for only one switching state is a wide-sense digital type switching curve, also called a digital switching curve in some cases. The digital switching curve is valuable in applications due to its immunity to drive voltage variation and drift.

The most basic structure for the digital optical switch is the 1×2 Y-branch optical switch. A 2×2 switching element is constructed

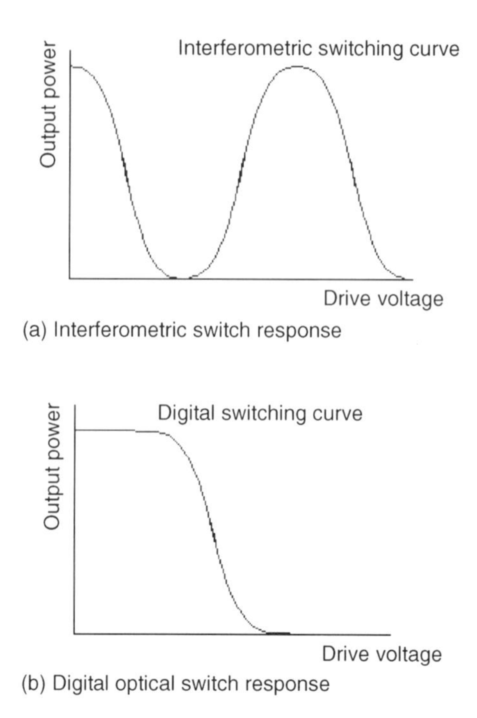

Fig. 7.13 Switch response.

connecting Y-branch switches. The Y-branch optical switch uses optical field deformation caused by a refractive index between two waveguides. The power of the fundamental mode concentrates in a waveguide with higher refractive index. A digital optical switch can also be constructed using design procedures proposed for wavelength filters. Because the optical switching elements are connected in many ranks to implement the multiport switching network, complex switching element structure should be avoided to suppress insertion loss increase. A digital optical switch tends to require large refractive index change (voltage) for switching. Design methods to reduce drive voltage have been proposed for a Y-branch switch.

In the early stages of research, the directional coupler type switch was studied for its simple structure and low loss characteristics. In recent years, emphasis is on studies for implementing digital optical switches that are useful in applications. Studies have been done on devices with one or two modes in the past. Recent studies aim at devices using more than three modes. Digital optical switching, fabrication ease, or multi-port devices are realized using multi-mode effects.

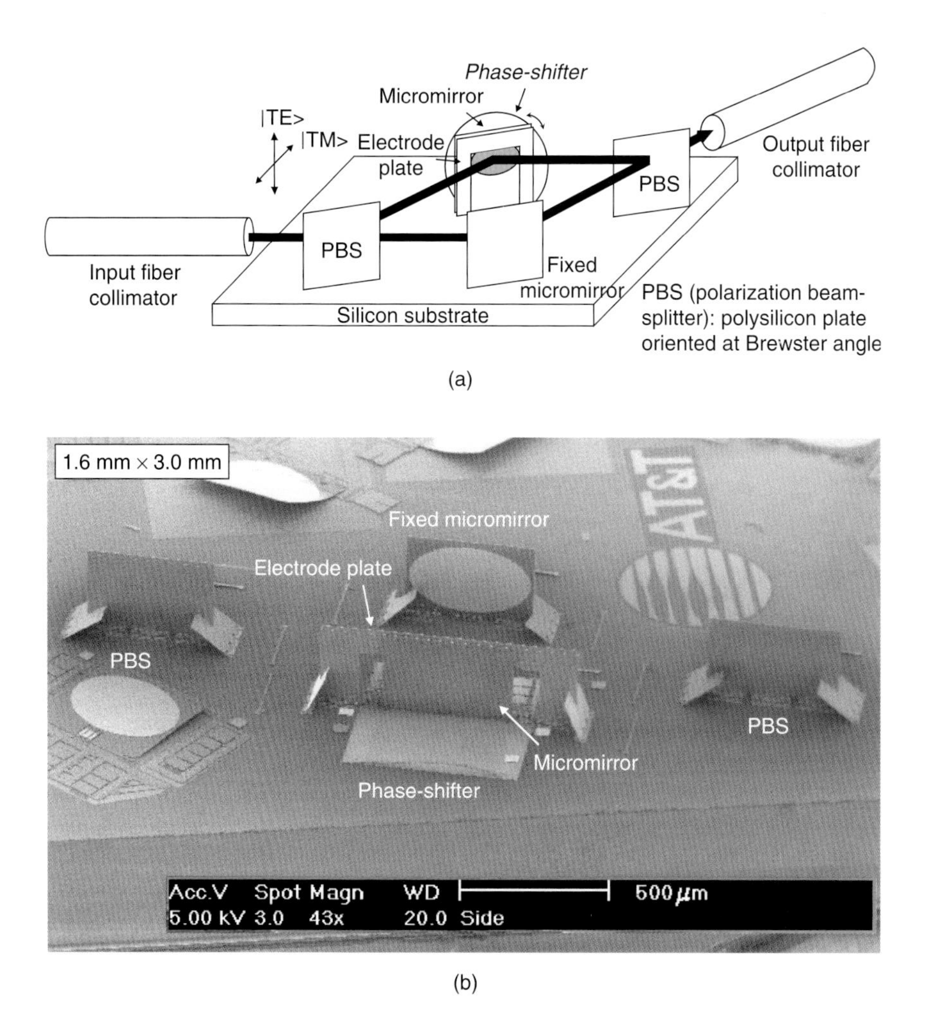

Fig. 5.22 (a) Schematic drawing, and (b) SEM photograph of a MEMS polarization rotator.

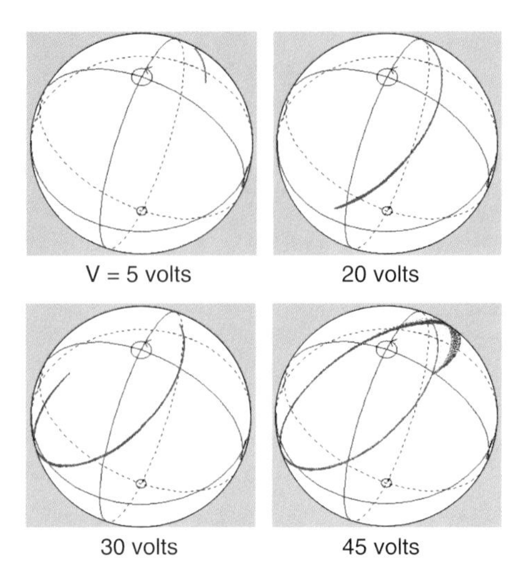

Fig. 5.24 Polarization rotation on the Poincaré sphere by applying bias to the phase shifter in the MEMS polarization rotator.

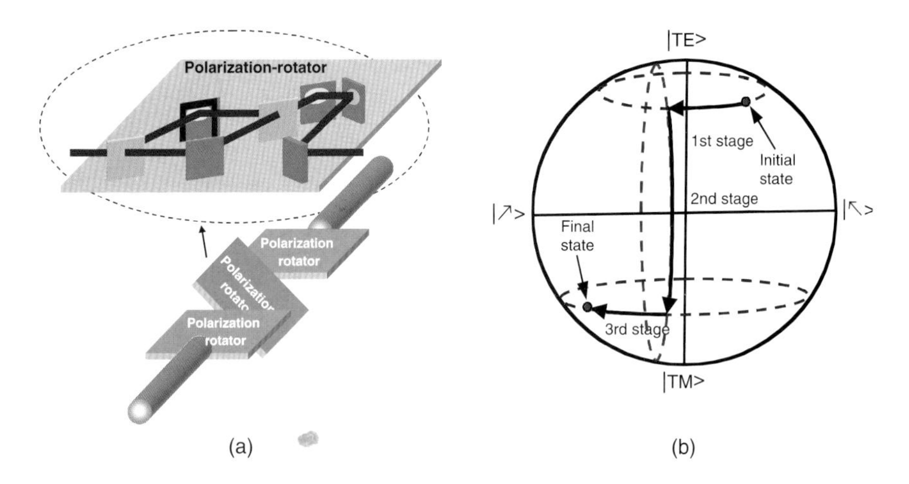

Fig. 5.25 (a) A full polarization controller obtained by cascading three stages of MEMS polarization rotators. (b) Action on the Poincaré sphere by the proposed MEMS polarization controller.

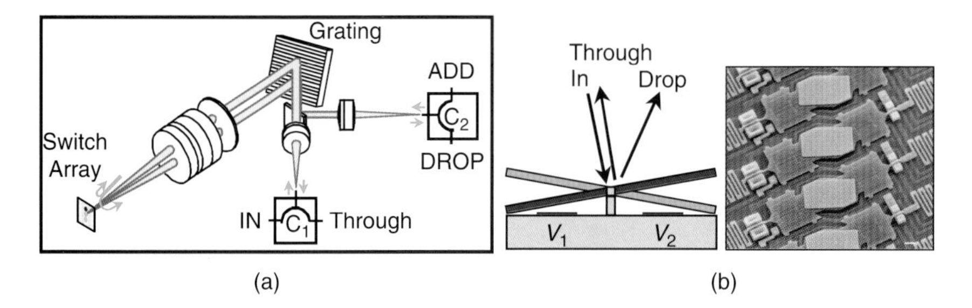

Fig. 5.26 (a) A WADM utilizing an array of tilting MEMS mirrors. (b) SEM and working principle of the MEMS mirror array [27].

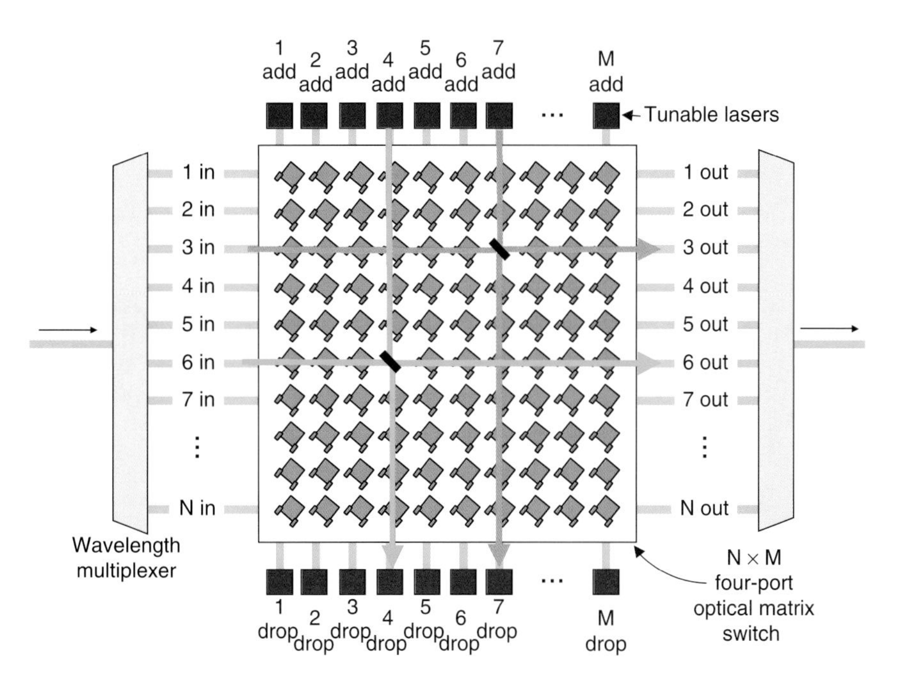

Fig. 5.27 2-D MEMS crossbar matrix switch for WADM with client configurability.

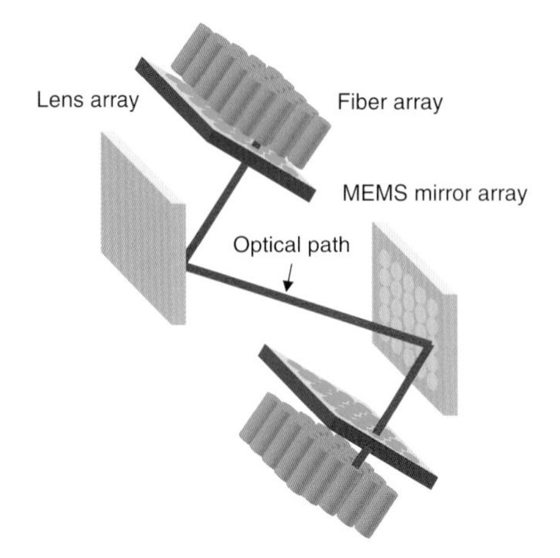

Fig. 5.34 Schematic drawing of a 3-D analog beam-steering switch.

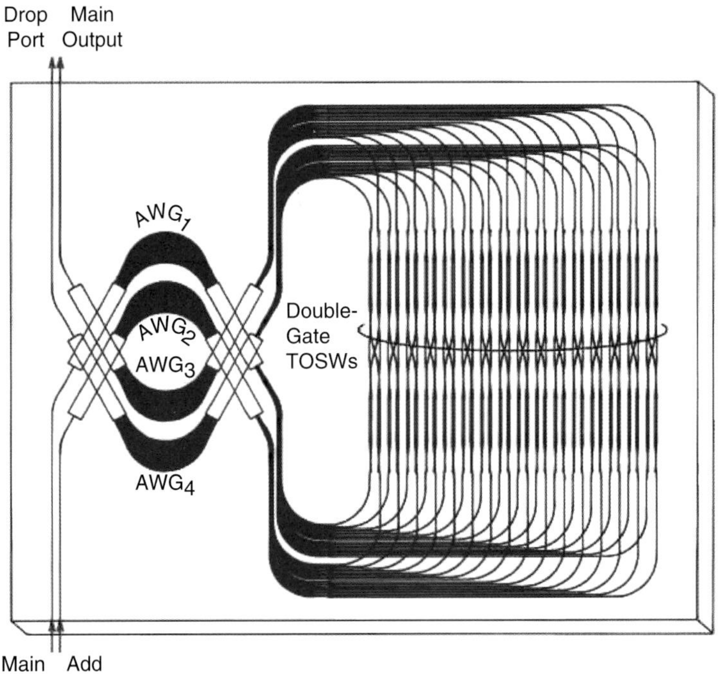

Fig. 6.22 Waveguide layout of 16-channel optical ADM with double-gate TO switches.

Fig. 6.27 Photograph of 1 × 16 TOSW module.

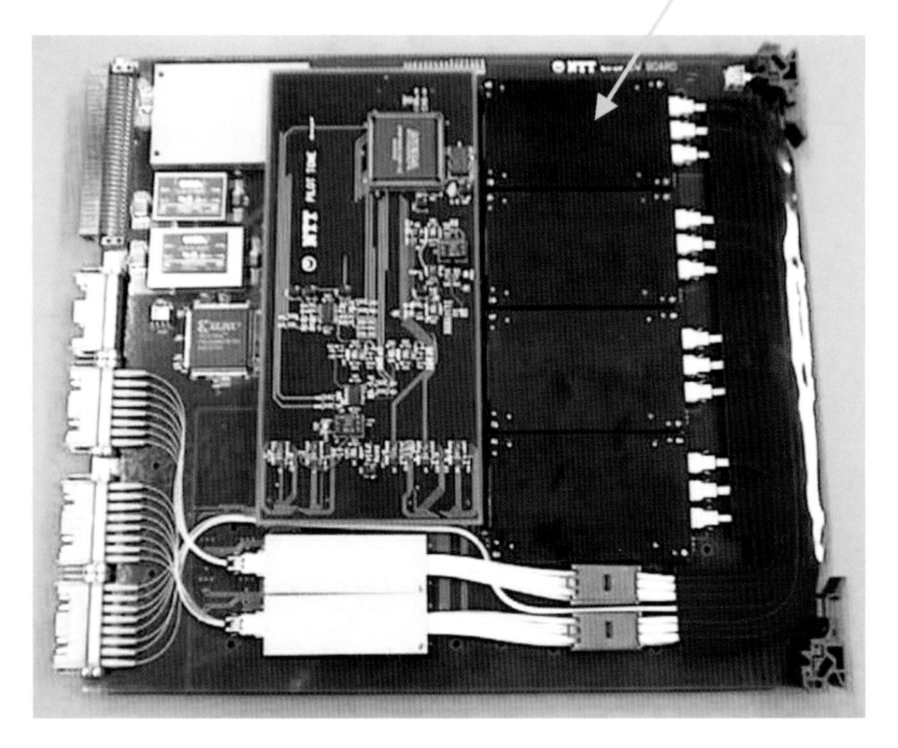

Fig. 6.30 Photograph of 8 × 16 DC-SW board.

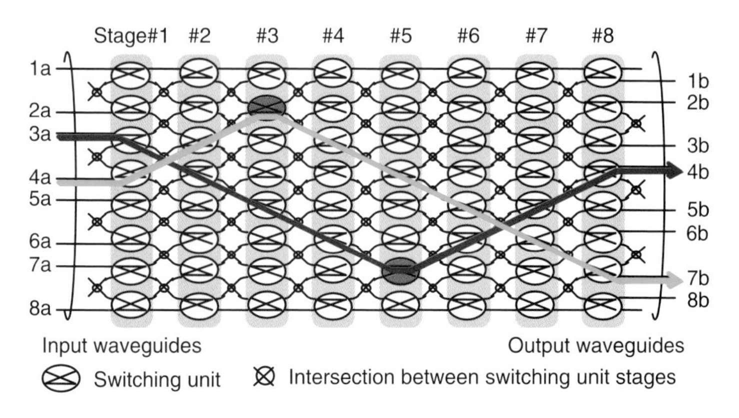

Fig. 6.43 Logical arrangement of 8 × 8 matrix switch with PI-loss structure.

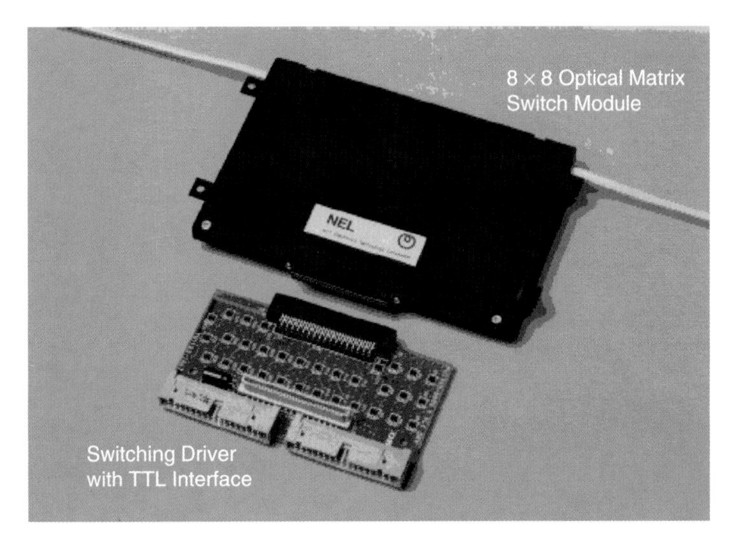

Fig. 6.49 Photograph of 8 × 8 matrix switch module.

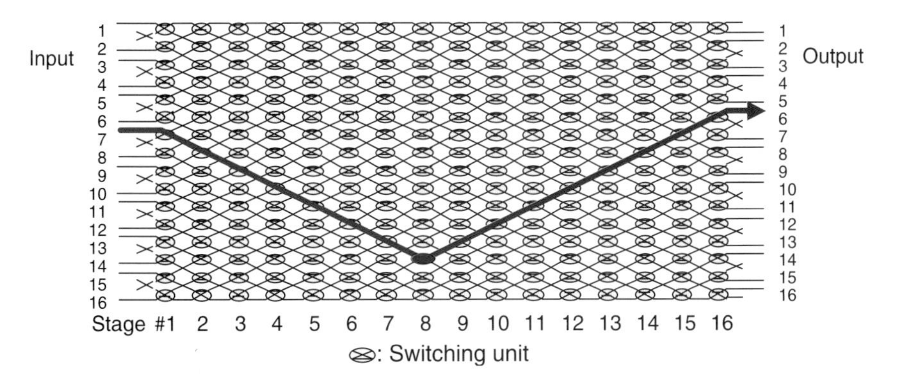

Fig. 6.50 Logical arrangement of a 16 × 16 strictly nonblocking matrix switch.

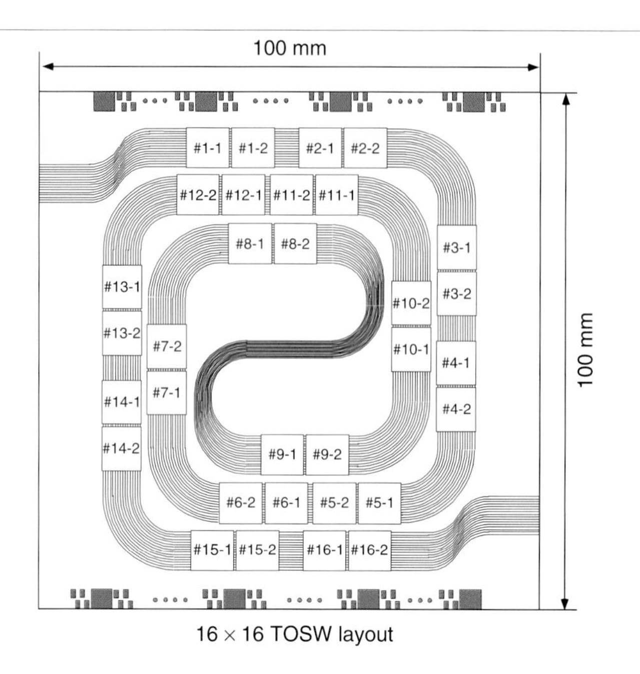

Fig. 6.51 Circuit layout of a 16 × 16 matrix switch with double-MZI switching units.

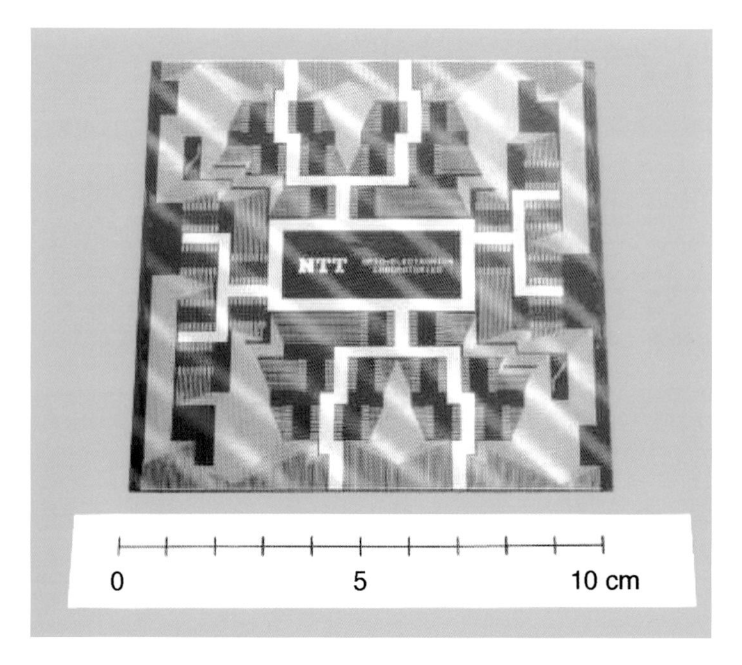

Fig. 6.52 Photograph of the fabricated 16 × 16 matrix switch chip.

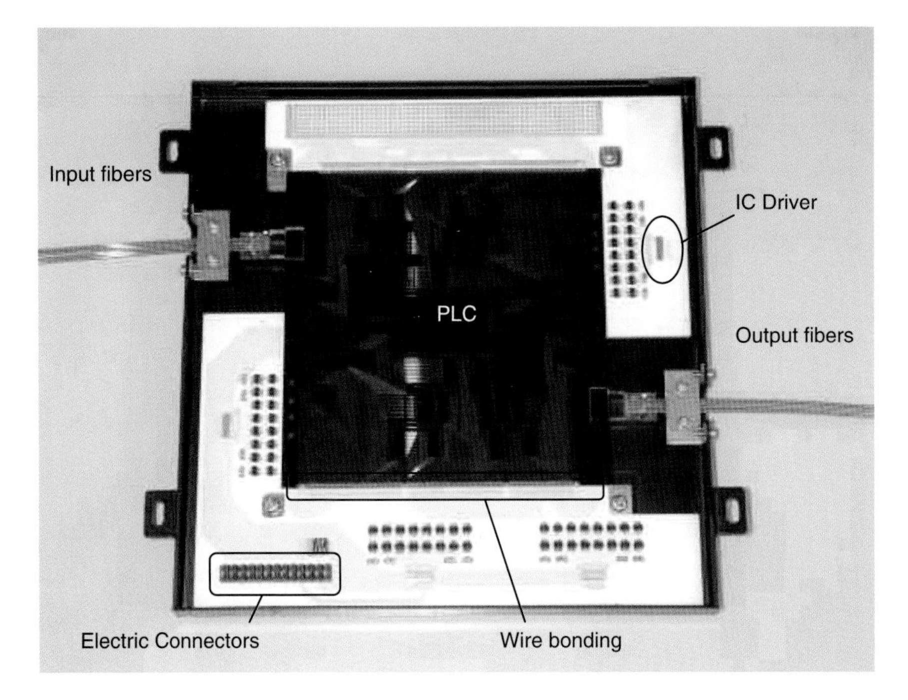

Fig. 6.56 Photograph of 16 × 16 matrix TOSW module.

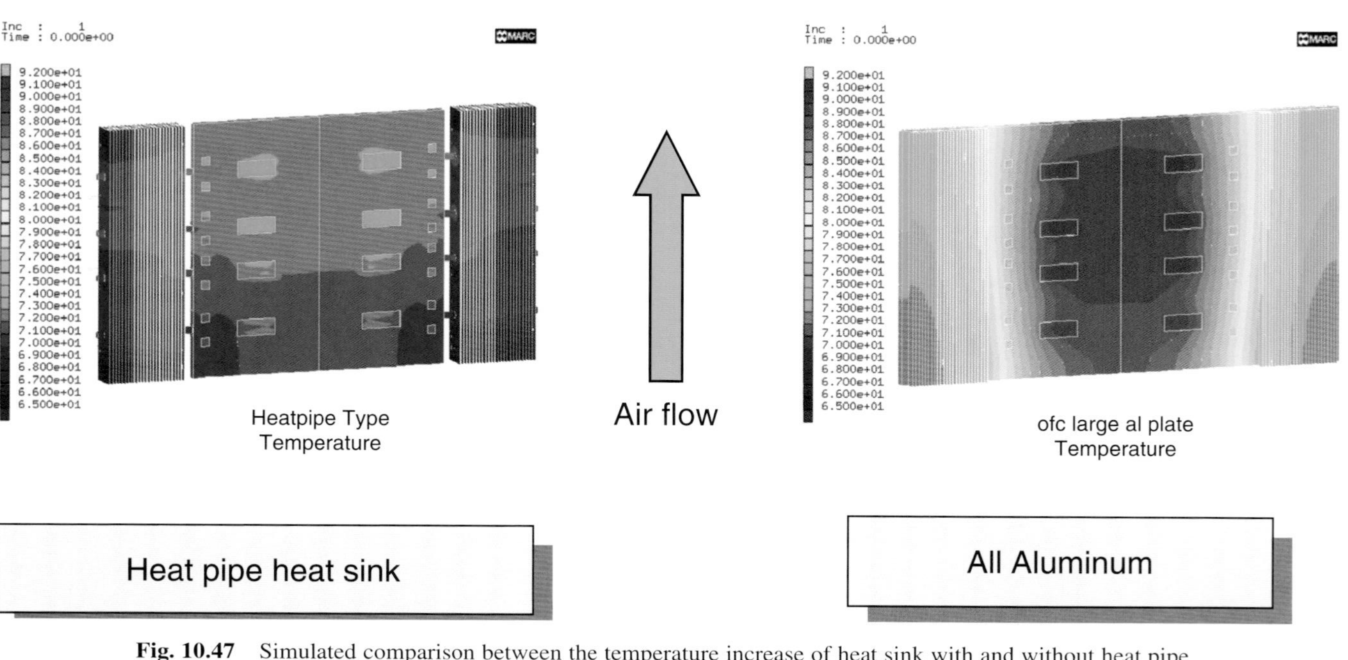

Fig. 10.47 Simulated comparison between the temperature increase of heat sink with and without heat pipe.

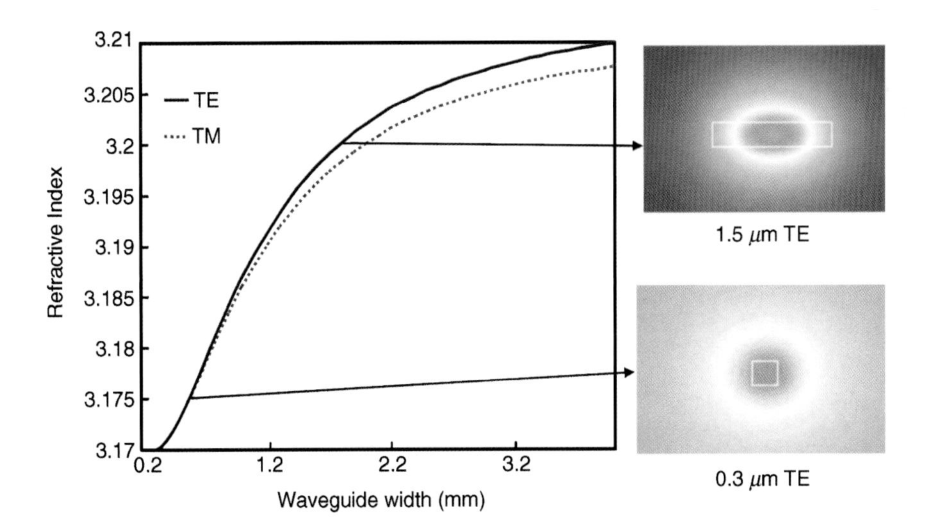

Fig. 11.4 Optical modes in buried heterostructure semiconductor waveguides calculated by the finite difference method. The left figure shows the equivalent refractive index as a function of the waveguide width. Photographs on the right show optical field profiles in the waveguide.

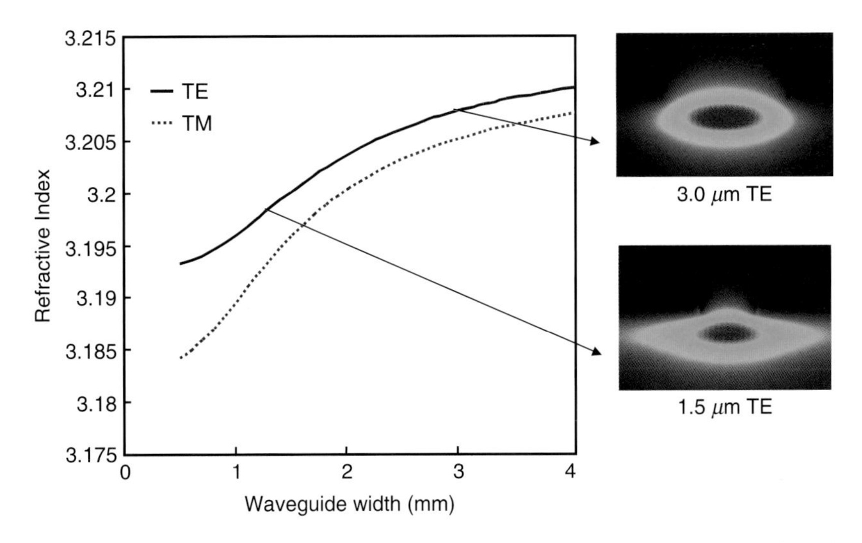

Fig. 11.5 Optical modes in a ridge-structure semiconductor waveguide calculated by the finite difference method. The left figure shows the equivalent refractive index as a function of waveguide width. Photographs on the right show optical field profiles in the waveguide.

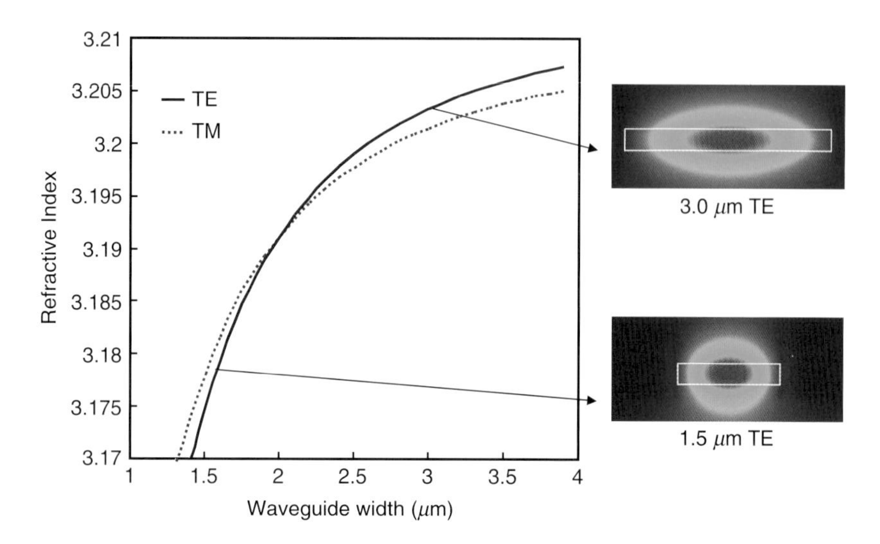

Fig. 11.6 Optical modes in a deep ridge semiconductor waveguide calculated by the finite difference method. The left figure shows the equivalent refractive index as a function of waveguide width. Photographs on the right show optical field profiles in the waveguide.

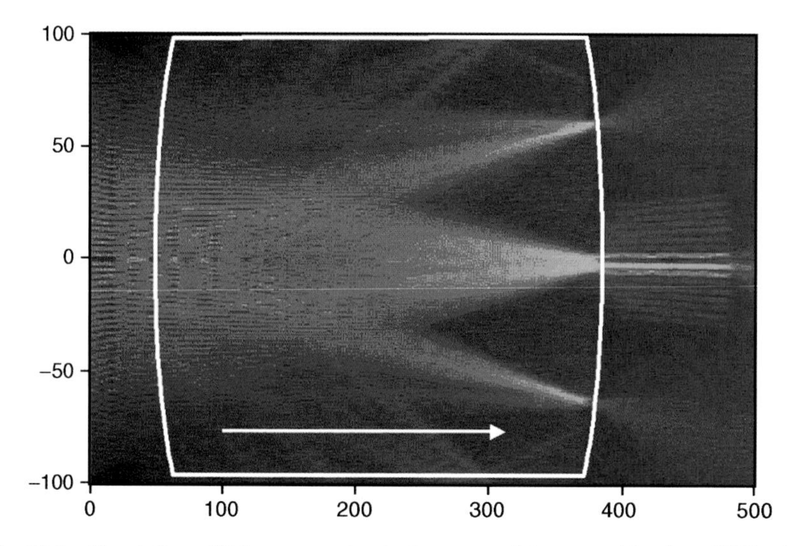

Fig. 11.8 Simulation of light propagation in the output slab waveguide of an AWG calculated by the beam propagation method (BPM). Light wavelength λ is the center in the FSR, i.e., $\lambda = \lambda_0^m$.

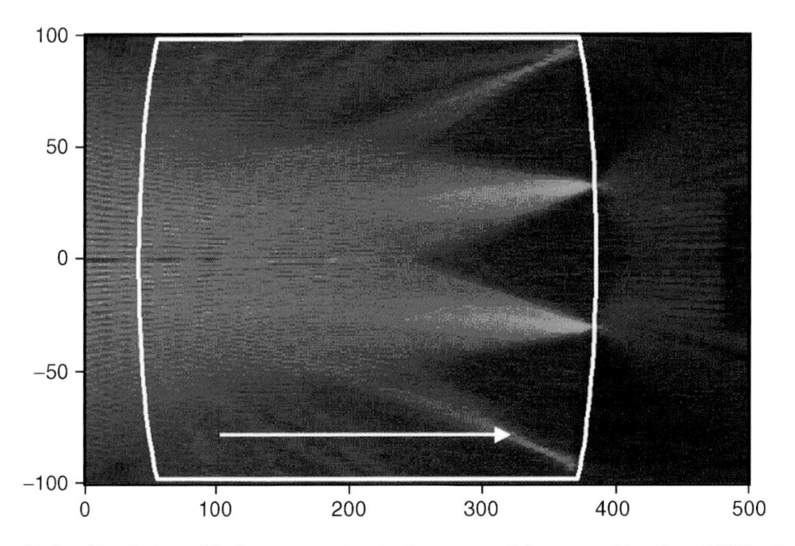

Fig. 11.9 Simulation of light propagation in the output slab waveguide of an AWG calculated by the beam propagation method (BPM). Light wavelength λ is the edge in the FSR, i.e., $\lambda = (\lambda_0^m + \lambda_0^{m+1})/2$.

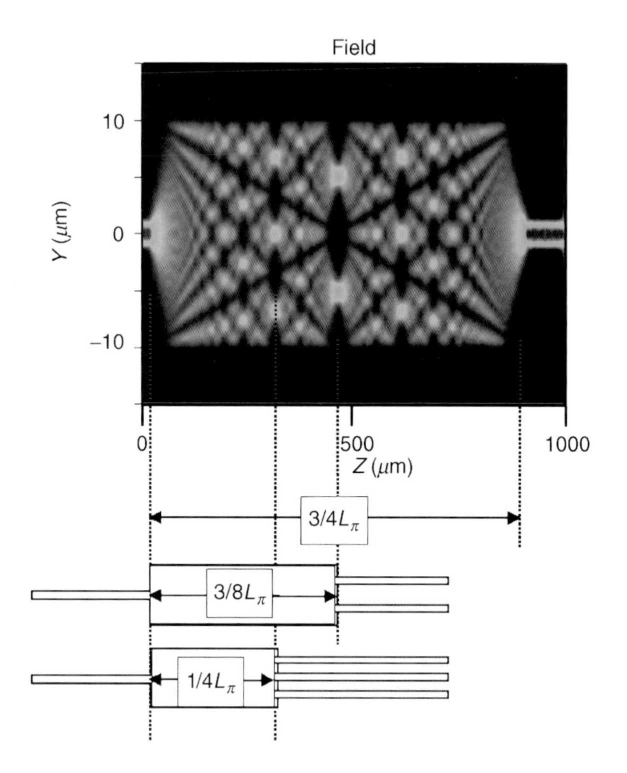

Fig. 11.25 Simulated light propagation in an multi-mode interference (MMI) coupler calculated by the beam propagation method (BPM). The device contains one input port at the center of the MMI waveguide. The symmetry of the device reduces device length by a factor of four. The simulated device (top) has a length of $1.5L_{pN}$ or forming a focus point on the output side. The light in the MMI waveguide forms several focusing points at the intermediate positions. The device with the intermediate length therefore works as a $1 \times N$ coupler (lower figures).

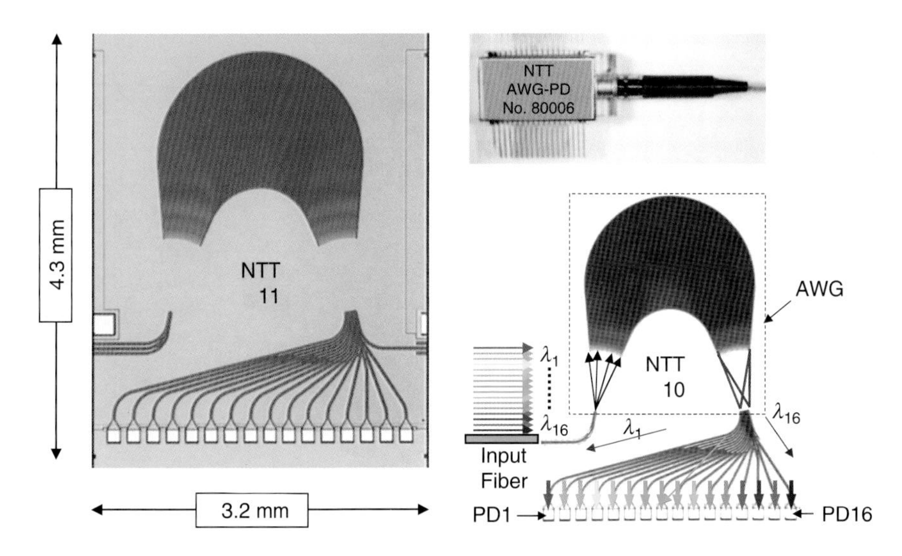

Fig. 11.26 An AWG integrated with photodiodes. On the left is a photograph of the chip, and on the top right a photograph of the packaged device. The drawing explains the operation of the device.

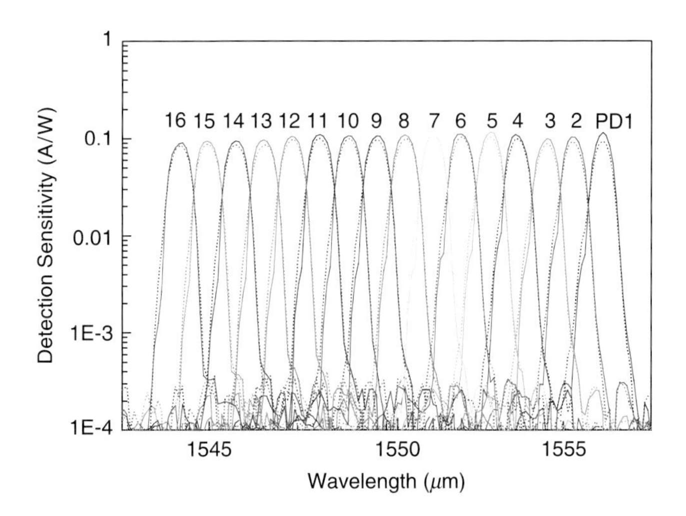

Fig. 11.28 Spectrum response of the efficiencies in an AWG integrated with photodetectors. The device is designed for equally spaced WDM systems with channel separation of 100 GHz. In the figure, the 16 response curves for different detectors overlap.

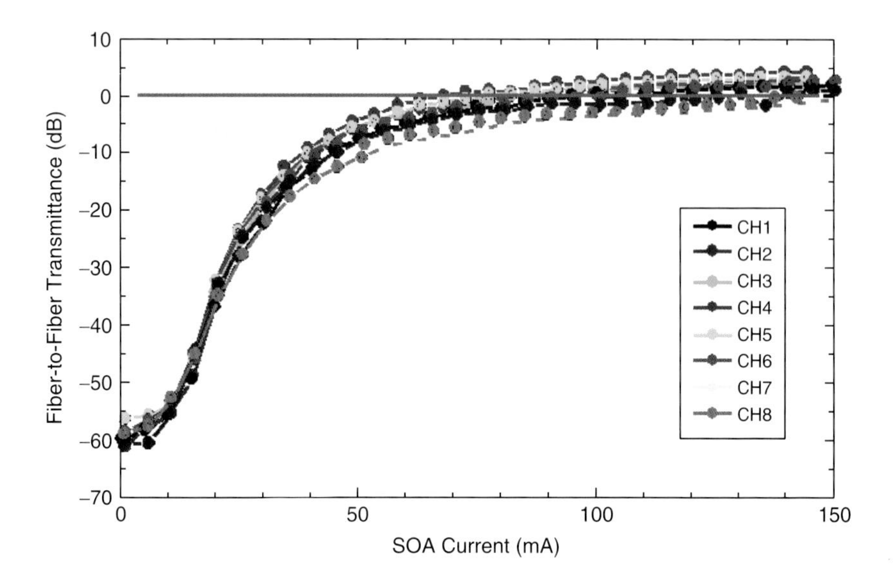

Fig. 11.34 Fiber-to-fiber gain of the AWG–SOA integrated device as a function of the injection current to an optical amplifier. Lensed fibers are coupled to the input and the output waveguides of the device. The fiber-to-fiber gain was measured for eight different wavelength channels with the corresponding amplifier excited.

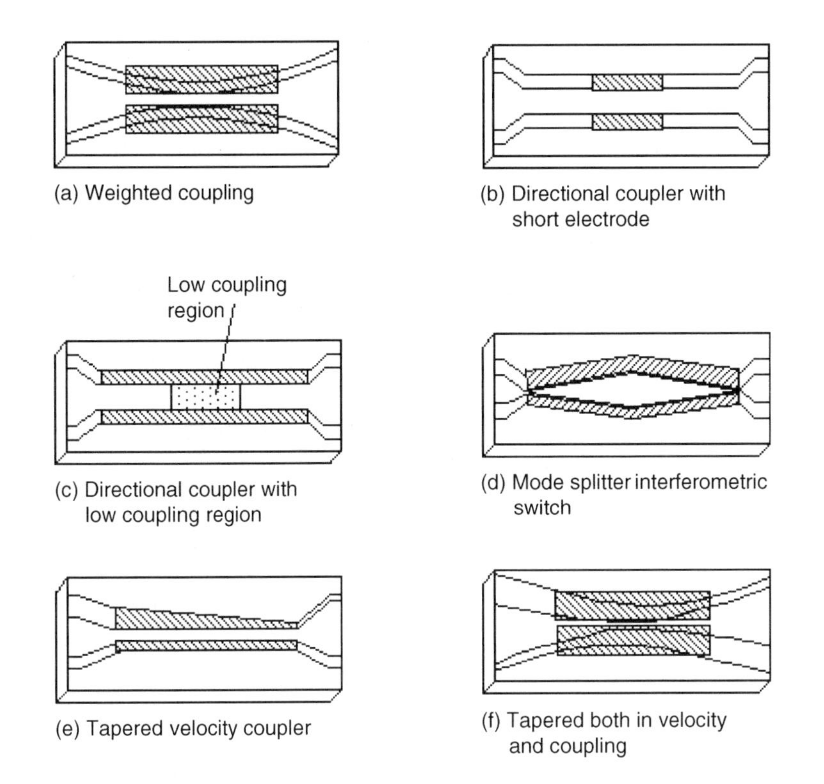

Fig. 7.14 Optical switching element variations.

Each type of device is described in greater detail in the following sections. Many variations of switches (Fig. 7.14) have been proposed to improve some characteristics.

7.4.1. SINGLE AND TWO-MODE EFFECT

7.4.1.1. Directional Coupler

The device is composed of two waveguides placed close to each other. A light propagated in one of the waveguides excites propagated light in another waveguide when the distance between the waveguides is sufficiently small and the two waveguides are identical. In this case, all the light power is transferred to another waveguide after some propagation distance called coupling length L_c. After the light power is transferred completely to another waveguide at the coupling length, the light power is regained in the original waveguide at twice the coupling length by the same process when the light is propagated along the waveguide system. The light power is transferred periodically from one waveguide to another and back again. If the length of the coupling region is set to one coupling length, the light fed into one waveguide is ejected from the output port of another waveguide. The cross state is achieved. By adding electrodes to the waveguide, the refractive index difference is generated through the electrooptic effect by applied voltage. If the refractive indexes of the two waveguides are different, the power transfer is reduced. The light fed into a waveguide remains in the same waveguide and emerges from the output port connected to the waveguide. The bar state is achieved.

The device principle can be described by the normal mode representation. Two normal modes, symmetrical and asymmetrical, are excited in the two-waveguide system. The propagation constant is different for the two modes. The interference between the two modes as they propagate along the waveguide system describes the periodical power transfer between two waveguides. The optical field in one waveguide is diminished when the optical field polarity in the waveguide is the opposite for the two modes, and enhanced when the polarity is the same. When a light is fed into a waveguide, two modes are excited at the input. After the light is propagated a distance equal to the coupling length, the phase between the two modes becomes π and the light field in the opposite waveguide is extinguished (cross state). Applying voltage to the electrode changes the propagation constants as well as the light field distributions of the normal modes. The coupling length changes (reduces) with the applied voltage. The bar state is achieved when twice the coupling length is equal to the coupling region length.

The device operation is analyzed by the coupling equation.

$$\begin{aligned} dR/dz - j\delta R &= -j\kappa S \\ dS/dz + j\delta S &= -j\kappa R \end{aligned} \tag{7.7}$$

where R and S are light fields in the first and the second waveguides respectively, κ is the coupling coefficient, $2\delta = \Delta\beta$ the propagation constant difference between waveguides, and z the propagation distance. The solution to Eq. (7.7) is given by the matrix form shown in Eq. (7.8.)

$$\begin{gathered} V = [\mathrm{M}]V_0 \\ V_0 = {}^{\mathrm{t}}(R_0 S_0), \quad V = {}^{\mathrm{t}}(RS) \end{gathered} \tag{7.8a}$$

where R_0 and S_0 are the initial value of R and S respectively. [M] is the transfer matrix with its components

$$\begin{aligned} &\mathrm{m}_{11} = A, \quad \mathrm{m}_{12} = \mathrm{m}_{21} = -jB \quad \text{and} \quad \mathrm{m}_{22} = A^* \\ &A = \cos[(\kappa^2+\delta^2)^{1/2}z] + j\delta \sin[(\kappa^2+\delta^2)^{1/2}z]/(\kappa^2+\delta^2)^{1/2} \\ &B = \kappa \sin[(\kappa^2+\delta^2)^{1/2}z]/(\kappa^2+\delta^2)^{1/2} \end{aligned} \tag{7.8b}$$

where * denotes a complex conjugate. The cross state is attained when $A = A^* = 0$, leading to $(\kappa^2+\delta^2)^{1/2}z = -\pi/2+\nu\pi$ and $\delta/(\kappa^2+\delta^2)^{1/2} = 0$ where $\nu =$ integer so that $\kappa z = -\pi/2 + \nu\pi$. The propagation length z at $\nu = 1$ is equal to the coupling length $L_c = \pi/(2\kappa)$. The bar state is attained when $B = 0$ leading to $(\kappa^2+\delta^2)^{1/2}z = \nu\pi$. Putting $\kappa = \pi/(2L_c)$, $[\pi z/(2L_c)]^2 + (\delta z)^2 = (\nu\pi)^2$. Drawing a switching diagram (Fig. 7.16) with z/L_c being the vertical axis and $2\delta z/\pi$ the horizontal axis, the cross states become isolated points $z/L_c = -1 + 2\nu$ on the vertical axis and the bar states become arcs with their diameters equal to 2ν. When the light is fed into one waveguide at the input ($R_0 = 1$, $S_0 = 0$), the output becomes $S = \kappa \sin[(\kappa^2+\delta^2)^{1/2}z]/(\kappa^2+\delta^2)^{1/2}$.

The coupling coefficient is a function of waveguide separation g, approximated as

$$\kappa = \kappa_0 \exp(-\gamma g) \tag{7.9}$$

where κ_0 and γ are constants [19]. The coupling coefficient κ in Eq. (7.7) can be a function of the propagation distance $\kappa(z)$. κ can be varied along the propagtion distance z by changing g. Equation (7.7) can be transformed to Eq. (7.10) using $\rho = S/R$.

$$d\rho/dz = -j2\delta\rho + j\kappa(z)(\rho^2 - 1) \tag{7.10}$$

when $\rho \ll 1$, so the solution for Eq. (7.10) is approximated as

$$\rho = \rho_0\left[-j\int \kappa(z)\exp(j2\delta z)dz\right]\exp(-j2\delta z) \tag{7.11}$$

Because $|S|^2 = |\rho|^2/(1+|\rho|^2) \fallingdotseq |\rho|^2$, the solution $|S|$ as a function of δ is approximated by the Fourier transform of $\kappa(z)$. Using coupling coefficient changing along the propagation distance (weighted coupling) can modify the switching curve ($|S|$ versus δ) of the directional coupler [20]. The sidelobe of the switching curve, conspicuous in a parallel waveguide directional coupler, can be decreased by introducing waveguide structure in which the waveguides are closed together gradually from input and output toward the center of the coupling region (Fig. 7.14(a)). The δ (drive voltage) range attaining low crosstalk for the bar state becomes wider. The switching curve is a wide-sense digital type.

The δ change required for switching parallel waveguide device with $z/L_c = 1$ is given by $2\delta L = 3^{1/2}\pi$. For the structure in which the electrode length L_e is the same as the coupling region length, $2\delta L_e = 3^{1/2}\pi$. Using an electrode shorter than the coupling region can lower the δL_e value required for switching (Fig. 7.14(b)) [21]. When $L_e/L \ll 1$, $2\delta L_e$ becomes almost equal to π. Placing a low coupling strength section in the middle of the coupling region can also lower the δL_e value required for switching (Fig. 7.14(c)) [22]. The structure does not require an additional coupling region. Increasing the coupling strength at this section minimizes the length of the additional coupling region required to produce sufficient coupling. Calculated $\Delta\beta$ required for switching decreases as the coupling coefficient at the low coupling coefficient region (line in Fig. 7.15) is reduced. In an experiment, the low coupling region was implemented using separation of the waveguide gap. Lowering of the switching voltage (dot in Fig. 7.15) was observed in the measurement.

The periodic power transfer between waveguides can also be described using an interference of the normal modes. The normal modes are obtained by diagonalizing the transfer matrix [M] in Eq. (7.8). Laplace transform of Eq. (7.7) can also be used to obtain normal mode propagation constants as eigenvalues. There are two modes with relative propagation constants $\beta_{e,o}$ equal to plus and minus $(\kappa^2 + \delta^2)^{1/2}$. For $\delta = 0$, $\beta_e - \beta_o = 2\kappa$ so that

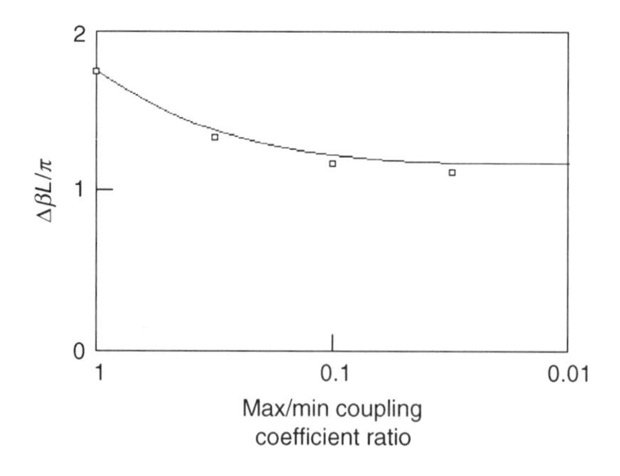

Fig. 7.15 Drive voltage reduction in directional coupler with low coupling region (Fig. 7.14(c)) [22]. J. Lightwave Tech., vol. 9, No. 11. © 1991 IEEE.

$L_c = \pi/(\beta_e - \beta_o)$. The coupling length corresponds to a length in which the π shift in phase difference between two normal modes is generated.

7.4.1.2. Reversed $\Delta\beta$ Coupler

In the directional coupler switch, the coupling region length should be exactly equal to the coupling length to attain low crosstalk in the cross state. A split electrode structure was proposed to attain a low-crosstalk cross state [23]. The adjacent electrode pairs are driven by opposite polarity voltages. The output is given by

$$V = [\mathrm{M}^-][\mathrm{M}^+]V_0$$
$$V_0 = {}^{\mathrm{t}}(R_0 S_0), \quad V = {}^{\mathrm{t}}(RS) \tag{7.12}$$

where the signs in front of δ differ for $[\mathrm{M}^-]$ and $[\mathrm{M}^+]$. The cross state is achieved when $2B^2 - 1 = 0$. This indicates that the optical light power is shared equally between two waveguides at the middle of the device in the cross state. Each electrode section is a 3 dB coupler. To achieve the bar state, it is required that $A = 0$ or $B = 0$. The first condition implies that each electrode section is in the cross state. The second condition implies that each electrode section is in the bar state. With L being the coupling region length, the cross state is achieved when $\sin^2\{\pi[(L/L_c)^2 + (2L\delta/\pi)^2]^{1/2}/4\} = [1 + (2L\delta/\pi)^2]\sin^2(\pi/4)$. The bar state is achieved at $L/L_c = 4\nu - 2$ or $(L/L_c)^2 + (2L\delta/\pi)^2 = (4\nu)^2$. The cross state becomes lines rather than isolated points in the switching diagram and is attained by suitable δ value (drive voltage) (Fig. 7.16).

The δL_e value required for switching (drive voltage) can be lowered by the same scheme adapted to a single electrode pair directional coupler switch. By composing the device with cascading directional coupler with low coupling strength at the middle of the coupling region, δL_e value to attain conditions required for the cross state or the bar state at each electrode pair can be lowered [24].

A device with digital switching curve can be attained by several methods [24, 25]. The range of δ attaining the cross state is wide around $L/L_c = 1 + 2\nu$ where the cross state is achieved at $\delta = 0$. In one method, the range of δ attaining the bar state can be widened by introducing the weighted coupling for each electrode section. In another method, directional couplers with low coupling strength in the middle of the coupling region are connected in series (Fig. 7.17).

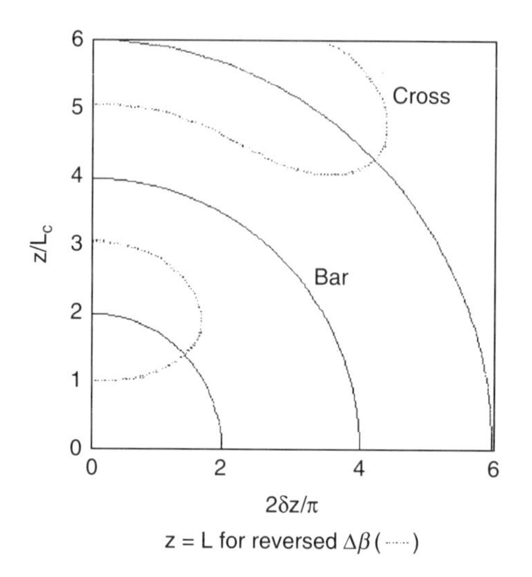

Fig. 7.16 Switching diagram for directional coupler and reversed $\Delta\beta$ switch.

7.4.1.3. Zero-gap Directional Coupler and Crossing Channel Device

As shown in the previous section, the coupling length can be obtained from the difference in the normal mode propagation constant $\beta_e - \beta_o$ as $L_c = \pi/(\beta_e - \beta_o)$. The symmetric normal mode field between waveguides increases, while the asymmetric normal mode field remains the same, when decreasing the gap between waveguides in a directional coupler. The propagation constant difference change between normal modes due to refractive index change between waveguide centers increases as the gap is decreased. Placing the electrode between the waveguide centers, the coupling length can be changed by applied voltage through the electrooptic effect. If the coupling length is an odd or even multiple of the coupling length, the device is in the cross state or bar state.

The zero gap directional coupler [26] consists of a waveguide sustaining two normal modes. Two branching waveguides are connected to the end of the two-mode waveguide separating the light field into two ports. The device was originally named Birfucation Optique Active (BOA) [27].

The same switching principle is also attained using crossing channel waveguides [28]. In the x-switch [29], the refractive index at the crossing wavguide is enhanced to strengthen the confinement of the higher order mode.

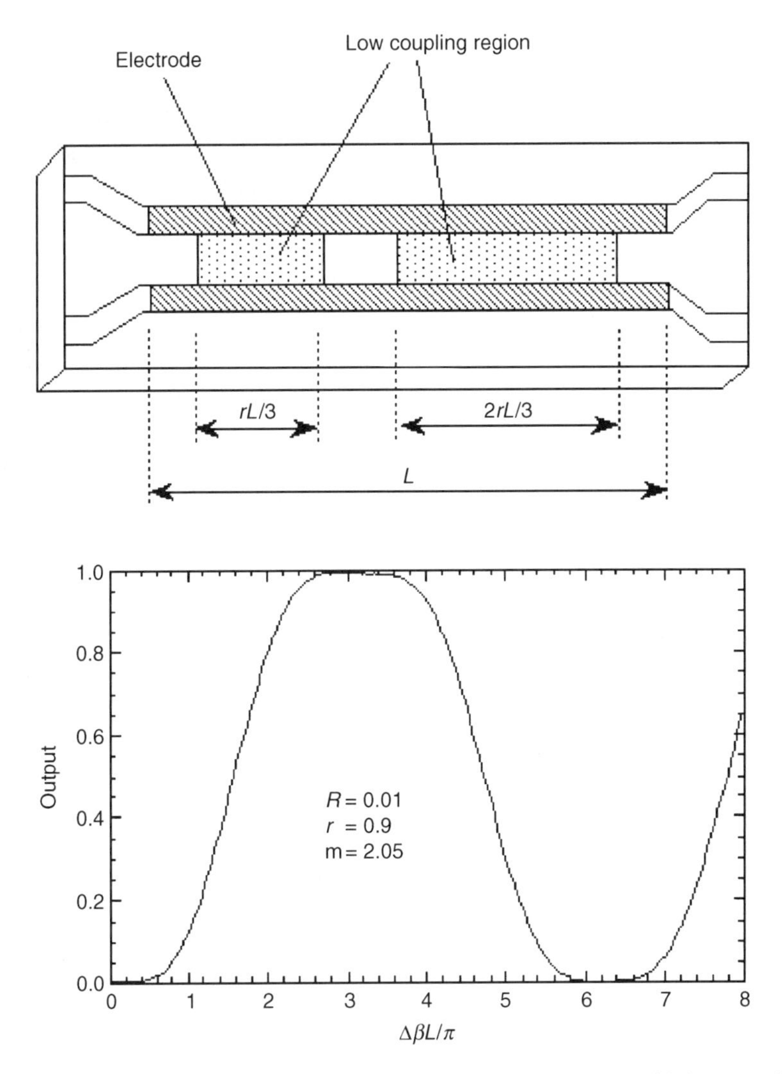

Fig. 7.17 Digital optical switch with cascading directional coupler with low coupling region.

7.4.1.4. Mach-Zehnder (Balanced Bridge)

The output of a directional coupler can be controlled by the phase difference of the lights fed into both the input ports. With input S_0 retarded for phase Φ against R_0 under the condition $\delta = 0$, the power $|R|^2 = (1 + \sin 2\kappa z \sin \Phi)/2$ and $|S|^2 = (1 - \sin 2\kappa z \sin \Phi)/2$ when $S_0 = R_0 = (1/2)^{1/2}$

as obtained from Eq. (7.8). For $\Phi = m\pi$ with m being an integer, no power transfer occurs. For $\Phi = (1/2 + 2m)\pi$ or $(-1/2 + 2m)\pi$, the power is transferred completely to one of the waveguides according to the phase at $z = (1/2 + \nu)\pi/(2\kappa) = (1/2 + \nu)L_c$. The output waveguide is switched according to the sign in front of $\pi/2$ in Φ. The output lights from the two output ports of the directional coupler with length $L_c/2$ excited from one input have equal power and phase difference $\Phi = \pi/2$, as can be seen in Eq. (7.8). By connecting directional couplers with length $L_c/2$ via phase modulator, an optical switch is implemented [30, 31]. The phase change for switching required at the phase shifter is π. The switching principle can be viewed as a mode exchange between symmetric and asymmetric modes at the phase shifter.

The coupling between the waveguide pair at the phase shifter (interferometer arm with electrodes) should be sufficiently small to attain pure interferometeric response. The coupling between waveguides can be reduced with a wide waveguide gap. In the basic structure, the waveguide curve structure is used to connect the narrow gap coupler and the wide gap interferometer arm sections. An etched groove between waveguides [32] or waveguides with different propagation constants [33, 34] can be used to avoid coupling between waveguides and eliminate waveguide curve sections to shorten the device. The waveguide curve section can also be avoided using a multimode interference (MMI) coupler. The 3dB coupler with widely separated input and output waveguide is realized using MMI [35].

A device similar to the Mach-Zehnder can be constructed using the normal mode-splitting function of the branching waveguide composed of waveguides with different widths (Fig. 7.14(d)) [36, 37].

7.4.1.5. Y-branch Optical Switch

The device consists of branching waveguides and electrodes to control the refractive index difference between waveguides. The symmetric mode light field dominates in a higher refractive index waveguide at large waveguide separation. At small waveguide separation the symmetric mode light field is almost equal in both waveguides. The light field changes its shape adiabatically as the light is propagated through the waveguide branch if the waveguide separation is increased gradually from the start to the end of the waveguide branch. When a symmetric mode is excited at the start of the branch, the light power is concentrated to the output where the refractive index is higher, enabling 1×2 switch operation. The light power of the asymmetric mode dominates in a waveguide with lower refractive index. The waveguide branch angle should be sufficiently small to ensure adiabatic conditions to suppress mode conversion between symmetric and asymmetric modes. If the mode conversion occurs, the extinction ratio is deteriorated. The condition is given as [38]

$$\Delta\beta/(\Theta\gamma) > 0.43 \tag{7.13}$$

where $\Delta\beta$ is the propagation constant difference between waveguides, Θ the branching angle, and γ the coefficient (Eq. (7.9)) showing the waveguide gap dependence of the coupling coefficient. Small branching angle Θ (typically smaller than 1 mrad) and large $\Delta\beta$ is required. The condition shows that above some threshold of $\Delta\beta$ the switch remains in the same switching state. A digital switching curve is attained.

When the light is fed into a waveguide branch from the wide waveguide gap side, the inverse process of the just-described switching principle occurs. Symmetric normal mode is excited at the beginning of the branch when the light is fed into the waveguide with higher refractive index. Asymmetric normal mode is excited at the beginning of the branch when the light is fed into the waveguide with lower refractive index. The width difference has a similar effect as a refractive index difference. A 2×2 switch is implemented, by connecting the branching waveguide with different widths or the refractive index to branching waveguides with electrodes [39, 40].

The $\Delta\beta$ value required for switching tends to be high in the Y-branch optical switch. There are two methods to decrease $\Delta\beta$ required for switching (Fig. 7.18).

In the first method [41] the small angle branching waveguide begins with a non-zero waveguide gap and is connected to branching or tapered waveguide with higher branching angle to excite symmetric mode at the start of the branch. The schematic structure is shown in Fig. 7.18(a). The length of the device becomes shorter than the conventional structure by removing the narrow gap section. The effect of the narrow gap section on switching performance is relatively small and can be replaced by a branch with a larger divergence angle. The structure is a Y-fed directional coupler [42] with a weighted coupling. $\Delta\beta$ change required for switching versus waveguide curve function is shown in Fig. 7.19.

In the second method [43], a curved waveguide structure is used to reduce the $\Delta\beta$ required for switching (Fig. 7.18(b)). The mode conversion generating crosstalk is similar to Eq. (7.8), with ρ being the amount of

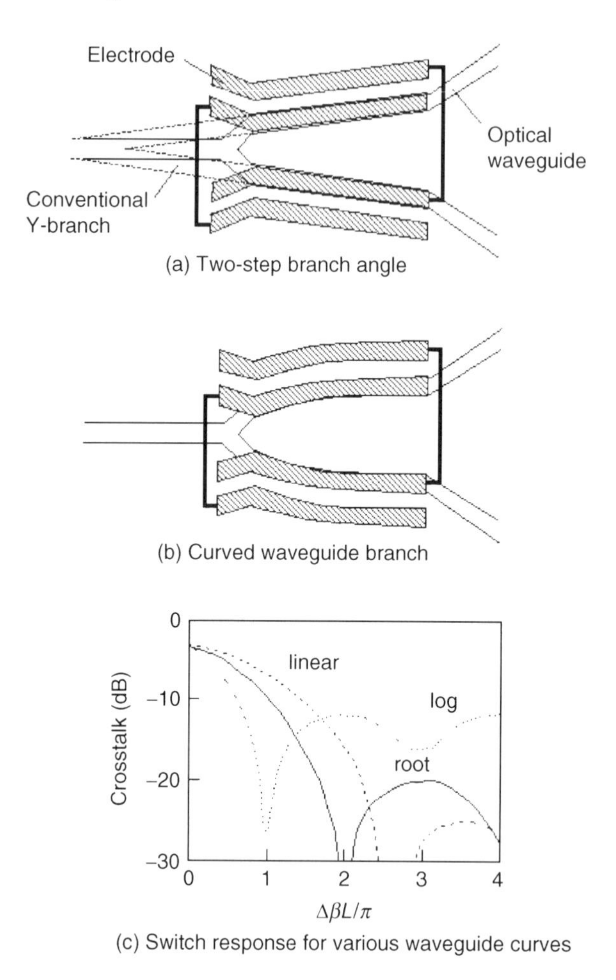

Fig. 7.18 Y-branch optical switch with reduced drive voltage [43].

mode conversion. The maximum mode conversion rate can be estimated with a ratio of δ over K, the propagation constant difference and coupling coefficient between normal modes, respectively. The mode conversion is large when K is large and δ is small. The coupling coefficient is proportional to branching angle Θ and can be denoted as $K = C\Theta$. The coupling coefficient K is a function of the gap between the waveguides and peaks at a certain gap value. In a conventional constant branch structure, the mode conversion takes place at a wide gap portion of the branch at low δ (low applied voltage) and the narrow gap portion at large δ (high applied voltage). The ratio of δ over K increases as δ is increased,

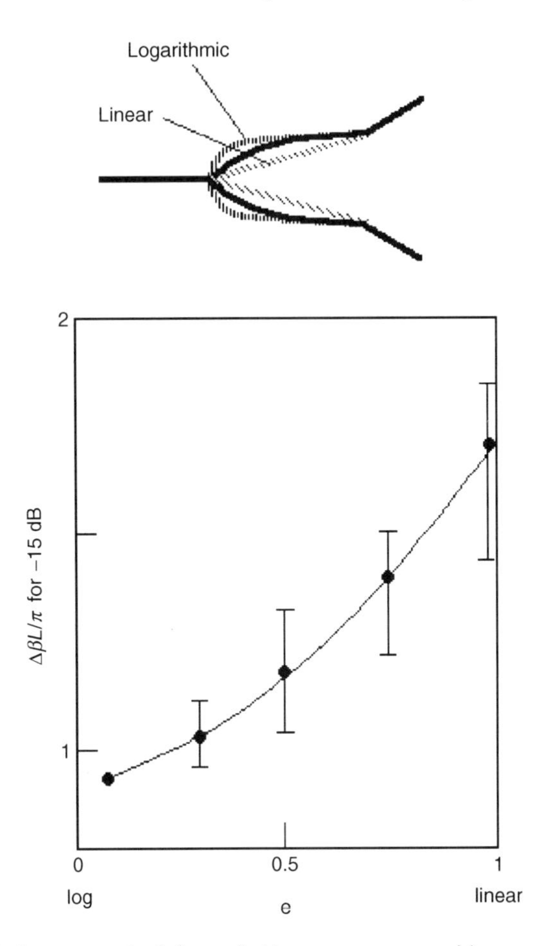

Fig. 7.19 $\Delta\beta$ change required for switching versus waveguide curve function [43]. J. Lightwave Tech., vol. 11, No. 2. © 1993 IEEE.

indicating that mode conversion becomes low at large δ. This is why the crosstalk is decreased at large δ or high applied voltage. Because, K is proportional to the local branching angle Θ, the crosstalk is expected to reduce at low δ when Θ is small at the wide gap region of the branch. This consideration leads to a curved waveguide branch structure as shown in Fig. 7.18(b). Calculation shows that the δ or drive voltage (for −15 dB) is reduced by bending the waveguide (Figs. 7.18 and 7.19). The drive voltage and crosstalk are in a trade-off relation so that an optimum waveguide curve may exist. About 30% reduction of the drive voltage in the experiment was reported.

This construction has also been adopted for compound semiconductor devices [44, 45]. The dual taper and double-etching processes were used to control the coupling coefficient between waveguides.

A waveguide curve can be generated for certain $\Delta\beta$ yielding the shortest length for a given conversion magnitude. Intermediate structure between the waveguide curve and straight waveguide shows reduced $\Delta\beta$ required for switching [46].

Some other computing procedures or codes to generate Y-branch structure, aiming at the realization of specific device performance, have been investigated. Design considerations have lead to devices with lower crosstalk under –40 dB [47].

7.4.1.6. Tapered Velocity Coupler

When the waveguide width or refractive index difference is gradually changed along the propagation distance, the normal mode light field changes its shape as the light is propagated. Consider a structure in which the waveguide width difference is decreased from the input and output toward the center, where the width difference is zero (Fig. 7.14(e)). At the input, symmetrical or asymmetrical normal mode is excited when the light is fed into wide or narrow waveguides. At the output, the light power is concentrated into wide or narrow waveguides for symmetric or asymmetric normal mode. The cross state is attained with symmetrical waveguide width structure. A weighted coupling version of the device that exhibits lower crosstalk at the cross state (Fig. 7.14(f)) has been demonstrated in InP [49–51]. By placing electrodes to the coupler and applying voltage, the refractive index of one waveguide can be made to be always larger or smaller than the other throughout the coupler, so that the light fed into one of the waveguides stays in the same waveguide, attaining the bar state. A demonstration was reported using $LiNbO_3$, attaining wide-sense digital switching [48].

7.4.2. *MULTIMODE EFFECT*

Although two normal modes are sufficient to implement a 1×2 or 2×2 switch, some improvements are anticipated in a device using many waveguide modes. Examples are described in some detail in the following.

7.4.2.1. Directional Coupler and Reversed $\Delta\beta$

A directional coupler switch with a single electrode pair or reversed $\Delta\beta$ operation is theoretically possible, similar to the device using two waveguides. However, due to increased parameters' degree of freedom, the switch operation becomes more complex than a device with two waveguides [52]. The number of normal modes is equal to the number of waveguides in a system composed of well-confined waveguides.

The simplest form in this category is a device using three waveguides. There are three normal modes in the system. The fundamental mode exhibits all the light fields with the same polarity in all waveguides, the first-order mode with its field only in the outermost waveguides having opposite polarity, and the second-order mode with its field in all the waveguides but having opposite polarity in the middle waveguide. When a light is fed into an outer waveguide, those three modes are excited. The three modes have different propagation constants. As the light is propagated through the coupler, the phase relation between normal modes changes due to the propagation constant difference. If the propagation constant difference between the fundamental and first-order mode is the same as between first-order mode and second-order mode, the field distribution becomes the mirror symmetry of the input light at some propagation distance. The cross state is attained at this coupler length. By placing electrodes on the outer waveguide and changing the refractive index in the opposite direction, the propagation constant difference is changed to attain the bar state [53].

Several reports on more complex functions other than 2×2 switching have been published [54, 55].

For the system with more than three waveguides, the coupling coefficient between waveguides should be set in a special relation to attain the full power transfer between outermost waveguides at the shortest coupling length [56]. The propagation constant differences between the normal modes are uniform in this condition [57]. The refractive index change for the bar state should be that the change among adjacent waveguides is the same for all the waveguides, equal to zero at the center waveguide, and asymmetric at the center, reversing the polarity (Fig. 7.20) [56]. When the uniform coupling coefficient is used, the light propagation in coupled waveguides resembles that in multimode waveguide with uniform refractive index. Multi-port beam steering is shown to be possible using this coupled waveguides system [58].

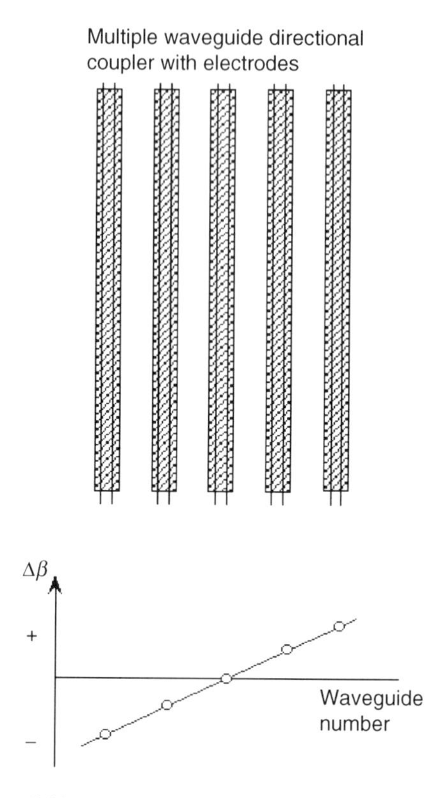

Fig. 7.20 Multiple waveguide directional coupler.

A reversed $\Delta\beta$ switch can be implemented using multiple waveguides. The three-waveguide device operation has been analyzed by transfer matrix as similar to the two-waveguide coupler. In the two-electrode-pair device, the bar state is attained when each section is in the cross state or the bar state. The cross state is attained when half of the light power is in the center waveguide and the remaining power is shared among outer waveguides at the middle of the coupling region. It is anticipated that a reversed $\Delta\beta$ switch can be implemented for a general number of waveguides [59].

A polarization-independent $LiNbO_3$ switch was demonstrated using improved tolerance of the three-waveguide coupler compared to the two-waveguide version [60].

Tapering of the waveguide width and gap can also be adopted to multi-waveguide directional couplers.

7.4.2.2. Multichannel Waveguide Branch

A branching waveguide device can be constructed using multiple waveguides [61, 62]. The field distribution of the normal mode in a waveguide array can be changed by the refractive index allocation of the waveguides. The normal mode excited at the start of the branch, where the waveguide gap is small, changes its field distribution adibatically as it is propagated along the waveguide branch. For an example, the power of the fundamental symmetric mode field is concentrated at the waveguide with the highest refractive index at the end of the branch, where the waveguide gap is the widest. By properly setting the refractive index of the waveguides by electrooptic effect, an output waveguide is selected to perform switching. A tapered waveguide or multimode interference coupler is connected to the start of the branch to excite the fundamental mode.

7.4.2.3. Multimode Interference (MMI) Coupler

The device uses interference of normal modes in a multimode waveguide. An interferometric optical switch similar to the zero-gap directional coupler can be implemented by placing a narrow electrode in the middle of the multimode waveguide [63]. We consider multimode waveguide with width W, thickness b, and refractive index n. The propagation constants of the low-order normal modes are approximated as $\beta_{\mathrm{pq}} \doteqdot [(2\pi n/\lambda)^2 - (\pi p/W)^2 - (\pi q/b)^2]^{1/2}$, with p, q being the mode numbers. In the simplest case, where the waveguide is single mode in depth ($q = 1$), the phase shift difference between modes at distance z is given as $\Delta\beta_{\mathrm{ij}} z = [\pi\lambda z/(4W^2 n)](p_{\mathrm{i}}^2 - p_{\mathrm{j}}^2)$ for modes far from cutoff. Between adjacent modes ($p_{\mathrm{i}} = p + 1, p_{\mathrm{j}} = p$), $\Delta\beta_{\mathrm{m,m+1}} z = [\pi\lambda z/(4W^2 n)](2p + 1)$ so that at $z = 4nW^2/\lambda$, the phase between symmetric and asymmetric mode is $(2p + 1)\pi$. At this propagation distance the field distribution of the asymmetric mode changes its sign and becomes opposed to the symmetric mode, attaining the cross state. The symmetric mode propagation constant is changed through the electrooptic effect by applying voltage. An induced phase shift of π can reverse the phase relation between the symmetric and asymmetric modes to attain the bar state. As can be seen in the formula of propagation constant β_{pq}, the switching function is also attainable for a device with input and output waveguides multimode both in width and depth.

A device with a wide higher refractive index region placed at the middle of the MMI is proposed to attain short device length for the cross state [64–66]. The cross state is attained in the condition where the differential mode propagation constants are almost uniform, similar to the coupled waveguides with appropriate coupling coefficients for full transfer between outer waveguides.

The Mach-Zehnder type device is implemented using MMI. Two MMI couplers with multiple inputs and outputs are connected by multiple waveguides with electrodes for the phase control. With mode number $m(= p-1)$, it is shown that the phase shift of π between fundamental ($m = 0$) and first-order ($m = 1$) modes is attained at $z = 4nW^2/(3\lambda)$, which is called the beat length L_π between these modes. The propagation constants of normal modes are given by $\beta_m - \beta_0 = \pi m(m+2)/(3L_\pi)$ in the paraxial conditions. An N-fold image of the input is obtained at the coupler length of $3L_\pi/N$ and an N $\times$ N coupler is attained at this length [67]. Relative phase of the light from input i to an output j is given as $\Psi_{\mathrm{ij}} = -(\pi/2)(-1)^{\mathrm{i+j}+N} + [\pi/(4N)][\mathrm{i}+\mathrm{j}-\mathrm{i}^2-\mathrm{j}^2+(-1)^{\mathrm{i+j}+N}(2\mathrm{ij}-\mathrm{i}-\mathrm{j}+1/2)]$ [68]. When identical MMI couplers are placed at the input and output, the phase required at the p_{th} waveguide to route the light from input I to output J is derived using the transfer matrix as

$$\phi_{\mathrm{p}} + \Psi_{\mathrm{pJ}} + \Psi_{\mathrm{Ip}} = 2\nu\pi \tag{7.14}$$

where ϕ_{p} is the phase shift required at the multiple waveguide Mach-Zehnder arm and ν is the integer. The switch operation is analogous to the AWG-type switching device in Section 7.3.1.

7.4.2.4. Total Internal Reflection (TIR)

The device structure is composed of crossing waveguides similar to the single-mode crossing waveguide device. However, in this device, multimode waveguides are used instead. Electrodes are placed at the intersection to induce a low refractive index region. When the induced refractive index difference is sufficiently large, the light becomes totally reflected at the low refractive index region's interface.

The switching curve characteristics change with the design conditions [69]. When the crossing angle is low—smaller than around a condition of obtaining phase shift of π between fundamental (m $=$ 0) and first-order (m $=$ 1) modes—in a waveguide supporting a few modes, the switching curve shows sinusoidal-interference-type characteristics. When the crossing angle is large in a highly multimode waveguide, the switching curve shows characteristics having a threshold in accordance with the description of total reflection phenomenon. Digital switching is attained in certain design conditions, but the refractive index change required to achieve switching is large for a TIR device. In Fig. 7.21, cross-state crosstalk and the conditions for digital switching are shown in a

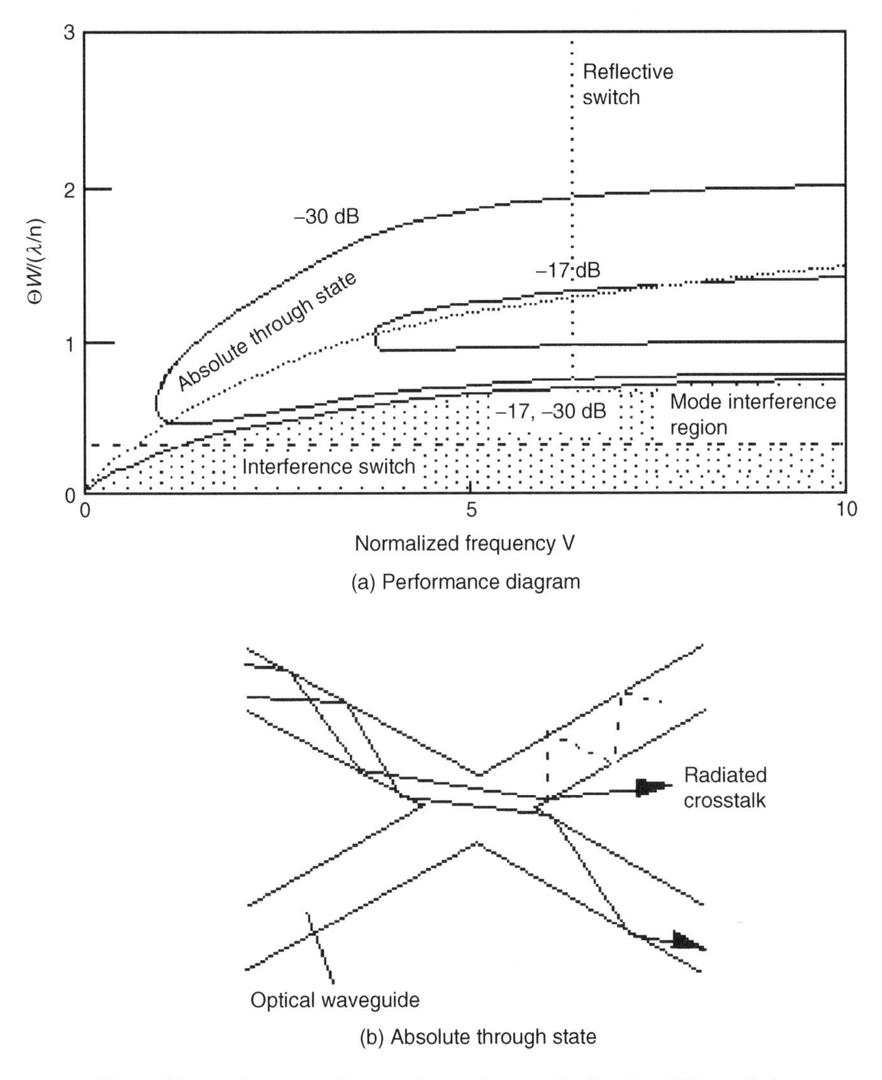

Fig. 7.21 Performance diagram for total internal reflection (TIR) switch.

crossing-angle Θ versus normalized-frequency V diagram. The crosstalk is calculated using multiple scattering interaction analysis. From both the multiple scattering interaction analysis and ray tracing method, it is shown that the characteristics can be given by normalized parameter $\Theta W/(\lambda/n)$ and $V = 2\pi W(2ndn)^{1/2}/\lambda$, with W being the waveguide width, n the refractive index, and dn the refractive index difference between waveguide and substrate. The sinusoidal switch response region is depicted as the shaded region in Fig. 7.21. The absolute through state shown in the figure is a state in which the crosstalk light leaked into the crossed waveguide is in a cut-off condition and radiated into the substrate. The low crosstalk cross state is attained above a certain waveguide crossing angle.

The device has been demonstrated using $LiNbO_3$ at 0.6328 μm wavelength [70]. At the longer wavelengths devoted to optical communication, semiconductor materials achieving larger refractive index change are used.

7.5. Optical Switch Arrays

Optical switch arrays are multi-port devices constructed using the switching elements described in Section 7.4. The optical switching network is fabricated into a chip by integrating switching elements and interconnection waveguides.

Optical switching networks are classified by their ability to establish paths between input and output ports (Fig. 7.22) as strictly nonblocking, wide-sense nonblocking, rearrangeably nonblocking, and blocking types. The crossbar architecture in Fig. 7.22 is often classified as the wide-sense nonblocking type. The strictly nonblocking network can establish any paths in any order. In the wide-sense nonblocking network, paths can be established at a demand only by a proper switching algorithm. In the rearrangeably nonblocking network, all the new paths can be established if one is allowed to change the state of switching elements not corresponding to the new path. The switching states should be reset to establish new connections. Finally, some connections between input and output ports cannot be established simultaneously in the blocking network.

A structure with a higher ability to establish paths between input and output ports tends to have more complex structure, making it difficult to integrate into one chip. In applications, the strictly nonblocking type is preferred in most cases. Architecture with fewer switching element ranks and waveguide crossings is advantageous for low insertion loss. The total waveguide and device length is also small for architectures with fewer

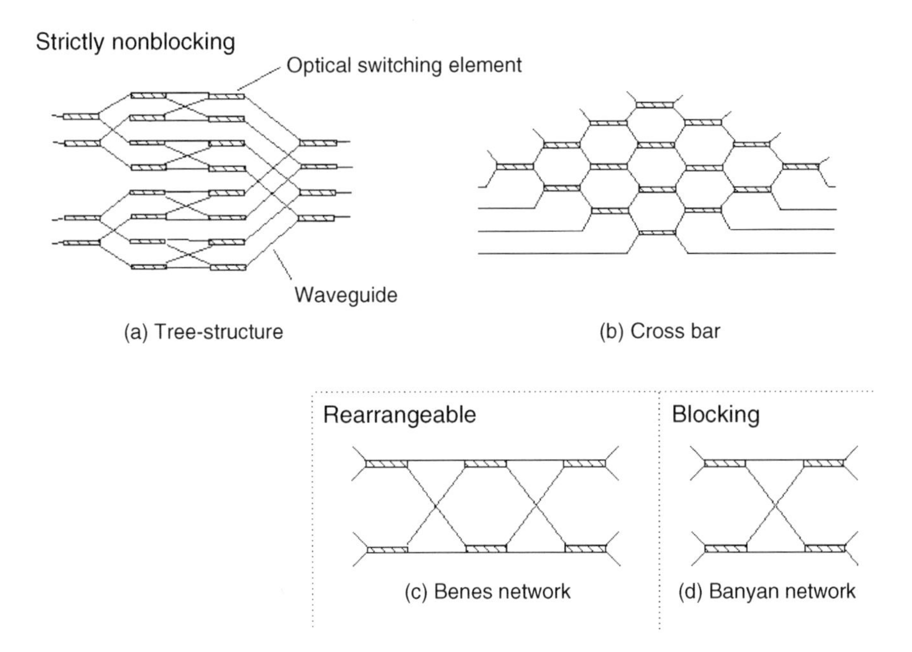

Fig. 7.22 Various optical switch arrays.

switching element ranks. But crosstalk below −40 dB is required in many cases. Optical switching elements are often used in pairs to attain a crosstalk level impossible with a single element.

These optical switch array chips are connected to implement larger switching networks with more input and output ports, to attain lower crosstalk or higher ability to establish paths between input and output ports.

Interconnection is needed to connect switch array chips. Interconnection schemes can be classified into several groups (Fig. 7.23). The free space interconnection scheme can attain a large interconnection number because it uses three-dimensional space. However, the optical axis alignment is problematic, especially for interconnecting single-mode waveguide devices. In the scheme using waveguide interconnections fabricated on a substrate, the optical axis alignment is done only at the interface between the switching device and waveguide interconnection substrate. Due to a fabrication procedure similar to an electronic integrated circuit, a mass production of complex interconnection is possible. However, for a large number of interconnections, the waveguide pattern becomes complex, involving many waveguide crossings. Because the waveguide crossing generates

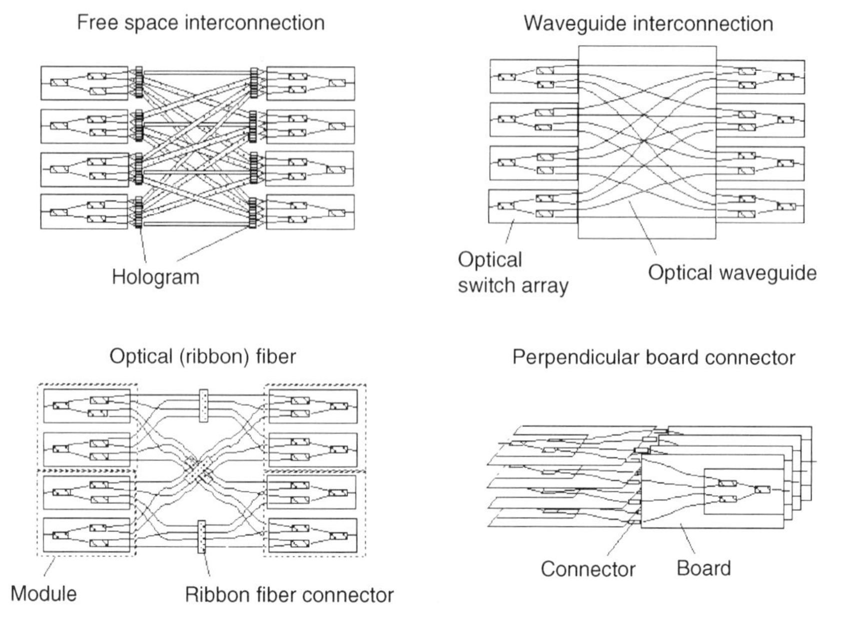

Fig. 7.23 Interconnection schemes.

some loss, the total insertion loss becomes high. A three-dimensional waveguide structure may overcome the problem, but at the expense of cumbersome fabrication procedures.

The interconnection using optical fiber is low loss and optical axis alignment is done only at the interface between device and optical fiber. The number of optical axis alignments at the interface is reduced by an optical fiber array structure with fibers aligned on a vehicle chip. V grooves are fabricated on the vehicle chip to align optical fibers. For a large number of interconnections, the handling of the optical fibers becomes troublesome. A scheme using tape fibers has been proposed to overcome the handling problem.

Details of each technology are described in the following sections. Considerations on switching network architecture dedicated for wavelength routing in a WDM network node will be given.

7.5.1. SINGLE CHIP ARCHITECTURE

The switching elements can be integrated into a chip with waveguide interconnections to implement a device capable of connecting input and output ports on demand. Due to the limited size of the substrate available, the network size that can be integrated into a chip is also limited. Network architectures with simpler structure and fewer element ranks are desirable to implement a device with a channel number as large as possible for the given element size.

7.5.1.1. Blocking

This type of network is classified as either partially connected or fully connected. In the partially connected network, the signal can be routed from an input port to only a limited fraction of the output ports. In the fully connected network, the signal can be routed from an input port to all of the output ports. However, some simultaneous routings of the signals from multiple input ports to output ports are impossible in the blocking-type network because some routings are blocked by an already-existing connection.

Construction of a $N \times N$ fully connected network by 2×2 switching elements requires $\log_2 N$ ranks of switching elements and $\log_2 N-1$ ranks of intermediate interconnections between switching elements. The switching element number totals $(N/2) \log_2 N$. The representative network structure of this kind is a banyan network (named for a South Asian tree with many roots hanging from its branches). Several topologies similar to the banyan network that differ in their ways of interconnecting switching elements are known, such as SW-banyan Omega, Baseline, and n-cube networks.

The banyan network has been used in the electronic packet switch, incorporating route finding, sorting, and buffering techniques to avoid congestion. In the optical switch technology, the banyan network has been used as a building block for constructing larger nonblocking switching networks.

The interconnection waveguide topology must be designed to minimize device length. In the simplest case, the switching elements are uniformly spaced apart for every rank, and switching elements are connected by curved waveguides with constant curvature of their radius (more optimized curve structure has been proposed to minimize light power loss associated with waveguide bends). With S as the switching stage spacing, and R the waveguide curvature of radius, the total interconnection length is given as the j_{th} rank interconnection length $L_{ic}(j) = [4RS2^{j-1} - S^2 2^{2(j-1)}]^{1/2}$ summed over all interconnection ranks $R_n = \log_2 N$. The total interconnection length is approximated as $2(RS)^{1/2}(1 - 2^{R_n/2})/(1 - 2^{1/2})$ when $S \ll R$. The signal light traverses $N - \log_2 N - 1$ interconnection waveguide crossings. The minimum waveguide crossing angle must be large enough to attain low loss and low crosstalk. In most cases, the insertion loss is of primary concern for selecting the crossing angle, because it decreases less steeply when increasing the crossing angle than does the crosstalk [71]. The minimum crossing angle in this structural example is given as $\sin(\Theta_x/2) = [S/R - (S/R)^2/4]^{1/2}$.

The crosstalk is low due to a small number of switching element ranks. The crosstalk is given as $X_s \log_2 N$ or $X_{slog} - 10 \log_{10}(\log_2 N)$ in dB, with X_s being the crosstalk generated at each switching element ($X_{slog} = -10 \log_{10} X_s$). The insertion loss is also relatively small due to the small switching element ranks and waveguide crossing numbers in this architecture.

7.5.1.2. Rearrangeable

This class of architecture permits all possible combinations between input and output ports. (However, some already existing connections should be rerouted when establishing a new connection between input and output ports.) Several architectures have been proposed for this class of network, including Benes [72], dilated-Benes [73], and n-stage planar permutation [74] networks.

The Benes network is composed of N/2 rows of switching elements arranged in $2 \log_2 N - 1$ ranks. The structure is obtained by a $J \times J$ unit composed of two $J/2 \times J/2$ switches connected to J output and input ports via a 2×2 switch, repeating the procedure recursively from $J = N$ to $n = 4$, replacing the $J/2 \times J/2$ switch with the $J/2 \times J/2$ unit. The total interconnection length is approximated as $4(RS)^{1/2}(1 - 2^{R_n/2})/(1 - 2^{1/2})$ when $S \ll R$. The signal light traverses $2N - 2 \log_2 N - 2$ interconnection waveguide crossings. 16×16 switch has been reported using $LiNbO_3$ [75] and 4×4 in InP [76].

Dilated-Benes is a modification of the Benes network, proposed to reduce the crosstalk via a method that will be shown in Section 7.5.2. 16×16 and 8×8 switches have been reported using $LiNbO_3$ [77, 78].

An n-stage planar permutation network is composed of N/2 ranks of unit, each with two ranks of switching elements. In a unit, first-stage N/2 switches in a row and second-stage $N/2 - 1$ switches in a row are connected to each other except for the outer interconnections. There is no waveguide crossing in the architecture, but the switching element rank number equaling N is large compared to the Benes network. The total interconnection length is approximated as $(2RS)^{1/2}(N - 1)$ when $S \ll R$. 4×4 switch has been reported using $LiNbO_3$ [79] and InP [80].

Before moving on to strictly nonblocking, we note that there is a class of architectures called wide-sense nonblocking, which are intermediate between rearrangeable and strictly nonblocking. All of the possible combinations between input and output ports can be established without affecting already-existing connections if the switching element is driven by a special crossbar-type algorithm using minimum elements. Typical architecture of this class is a 4×4 network composed of 8 switching elements arranged in 4 ranks [81] (see Fig. 7.28(a)). All four of the switching elements at the two middle stages should not be in the same switching state to drive the device as a wide-sense nonblocking network. The crossbar architecture can also be regarded as a network belonging to this class, when the algorithm of switching only the element connecting the desired input and output ports is taken into account.

7.5.1.3. Strictly Nonblocking

In this class of architecture, all possible combinations between input and output ports can be established in any order without affecting already-existing connections. Representative architectures are the crossbar, tree, and simplified tree (Fig. 7.24).

In the binary tree type the input light signal is routed by 1×2 switching elements connected in $\log_2$ N ranks, whereas in the crossbar type switching elements are connected in N ranks. The crossbar type has been used in the past but, due to its long device length, it is seldom used in $LiNbO_3$. A hybrid between these two types may be possible but in the following we concentrate on the basic architectures.

Crossbar

The basic $N \times M$ crossbar architecture is constructed by crossing N parallel input waveguides and M parallel output waveguides and placing NM switching elements at waveguide cross points. The input signal light is

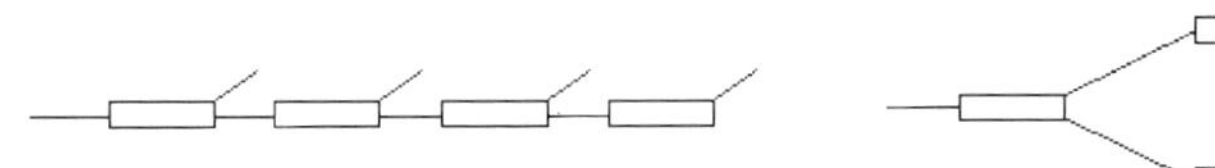

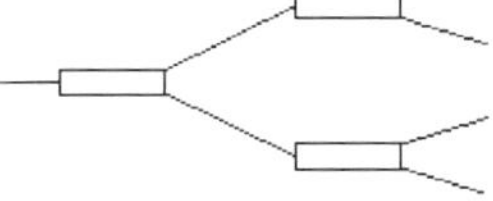

(a) Basis of crossbar type architecture

(b) Basis for binary tree type architecture

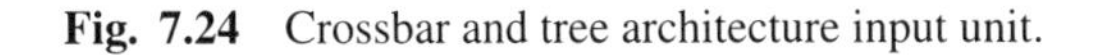

Fig. 7.24 Crossbar and tree architecture input unit.

routed to a desired output waveguide by driving the switching element connected to the desired output. The structure is simple, involving no waveguide crossings other than at the switching elements. However, because the light signal travels through N + M − 1 switching elements from the input to the output, the insertion loss and crosstalk accumulation is large. The number of switching elements on the course differs with the combination of the input and output ports (or path). Dummy switching elements and waveguide lengths are used to compensate for the insertion loss difference depending on the path the light signal travels. The crosstalk is given as NX_s or $X_{slog} - 10\ \log_{10}$ N in dB, with X_s being the crosstalk at each switching element ($X_{slog} = -10\ \log_{10} X_s$) for the structure with those dummy elements. The crosstalk characteristic deteriorates when there is an insertion loss difference affecting the attenuated signal to the crosstalk ratio. Due to its large switching element rank number N + M − 1, the crossbar architecture is disadvantageous for fabricating the device into a chip of limited area. A reflection structure was proposed to halve the device length [82]. 4 × 4 [83, 84] and 8 × 8 [85] switches were demonstrated using $LiNbO_3$.

An architecture called PILOSS (Path-Independent LOSS) (Fig. 7.25) or square crossbar has been proposed to overcome the problem of crossbar architecture [86]. A total of N^2 switching elements arranged in N ranks is used to construct the N × N network. The interconnection waveguides connect adjacent row switching elements in the adjacent ranks. There are N input waveguide paths and N output waveguide paths connected by switching elements placed at cross points of these paths. The light signal travels through N switching elements and N − 1 waveguide crossings, the number of which is independent of the path. The insertion loss becomes independent of the paths without using dummy elements. The crosstalk characteristics

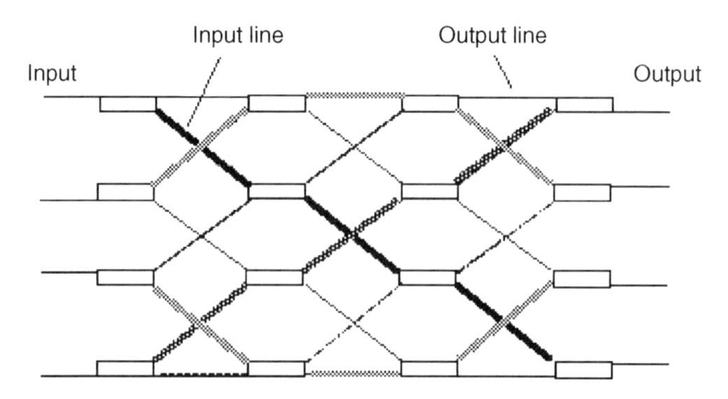

Fig. 7.25 Path-independent loss (PILOSS) switching network.

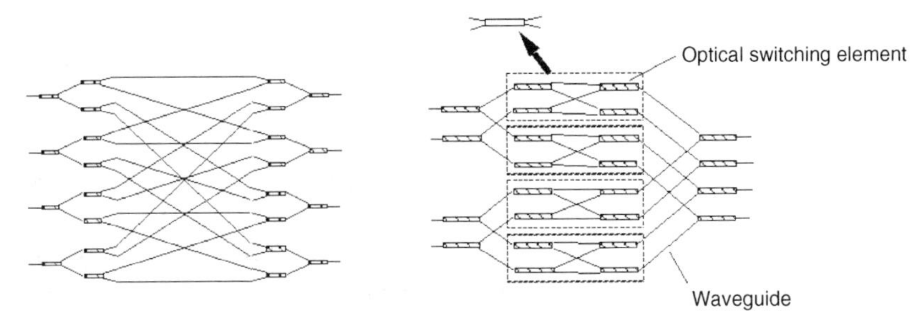

(a) Tree structure (router-selector architecture) (b) Tree structure with minimum crossings

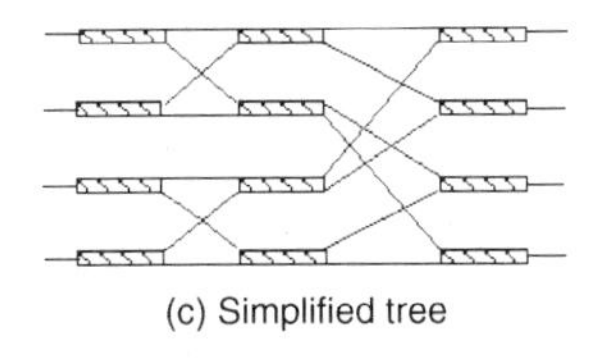

(c) Simplified tree

Fig. 7.26 Tree-type architectures.

are the same as with the crossbar with loss compensation scheme. The total interconnection length is approximated as $2(\mathrm{N}-1)(RS)^{1/2}$ when $S \ll R$. A 4×4 switch was demonstrated using $\mathrm{LiNbO_3}$ [86].

Tree Architecture

The $\mathrm{N} \times \mathrm{M}$ tree architecture or router/selector [87] architecture (Fig. 7.26) is constructed by connecting $1 \times \mathrm{M}$ binary tree switches totaling N and $\mathrm{N} \times 1$ binary tree switches totaling M. A passive power divider to implement broadcasting can replace the input binary tree switch. The switching element ranks total $\log_2 \mathrm{N} + \log_2 \mathrm{M}$, which is smaller than for the crossbar. The total switching element number is $2\mathrm{NM} - \mathrm{N} - \mathrm{M}$. The light signal is routed to the desired output by the input binary $1 \times \mathrm{M}$ tree switch. The output binary $\mathrm{N} \times 1$ tree switch selects the light signal to be routed to the output port. The crosstalk light is rejected at the output binary tree switch, gaining excellent crosstalk characteristics. The crosstalk is given as $\log_2 \mathrm{M}X_\mathrm{s}^2$ or $2X_\mathrm{slog} - 10\ \log_{10}(\log_2 \mathrm{M})$ in dB, with X_s being the crosstalk at each switching element ($X_\mathrm{slog} = -10\ \log_{10} X_\mathrm{s}$).

There are two arrangements for switching elements and waveguide interconnections. In the first arrangement [88], binary tree switches are placed at input and output without overlapping (Fig. 7.26(a)). The structure is simple but the light signal traverses a huge number of waveguide crossings, i.e., $MN - N - M + 1$ at maximum. The total interconnection length is approximated as $2(RS)^{1/2}(N-1)$ when $S \ll R$ for an $N \times N$ switch. In the second arrangement (Fig. 7.26(b)), the structure is optimized to reduce the number of maximum waveguide crossings on a path to $2N + M - \log_2 N - \log_2 M - 3$. The total interconnection length is approximated as $2(RS)^{1/2}[(1+2^{1/2})N/2 - 2^{1/2}]$ when $S \ll R$ for an $N \times N$ reduced waveguide crossing structure. The length is longer than the first arrangement. To reduce the insertion loss, a structure with a lower waveguide crossing number is used in a $LiNbO_3$ device. 16 × 16 [89], 8 × 8 [90], and 4 × 4 [91] switches in $LiNbO_3$, and a 4 × 4 switch in InP have been reported [92]. The polymeric 8 × 8 switch has been demonstrated [93].

Simplified Tree Architecture

In the middle rank of the tree structure with reduced waveguide crossings, a unit with 4 switches and 4 interconnection waveguides exhibits a routing function similar to a 2 × 2 switch. The unit can be replaced by a single 2 × 2 switching element. The structure is called a simplified tree structure (Fig. 7.26(c)). The number of switching elements is reduced to $N(5N/4-2)$ for an $N \times N$ switch. The switching element rank becomes $2\log_2 N - 1$. The maximum waveguide crossing number on a path is $3N - 2\log_2 N - 4$. The overall crosstalk characteristics are determined mainly by the crosstalk of the middle stage 2 × 2 switch. The crosstalk is given as $-10 \log_{10} X_s - 10 \log_{10}[X_{s2\times2}/X_s + X_s(\log_2 N - 1)]$ in dB, with $X_{s2\times2}$ being the crosstalk of 2 × 2 and X_s the other switching elements. The total interconnection length is approximated as $2(RS)^{1/2}[(1+2^{1/2}/2)N/2 - 1]$ when $S \ll R$ for $N \times N$ switch, which is almost 1/1.4 of that for the tree architecture. The 8×8 switch has been demonstrated using $LiNbO_3$ [94, 95] and GaAs/AlGaAs [96].

7.5.2. *CROSSTALK REDUCTION IN A SINGLE CHIP DEVICE*

In most EO waveguide optical switching elements, it is hard to achieve a crosstalk level under −40 dB. A crosstalk reduction scheme using additional switching elements other than those for routing is used in many cases. The crosstalk light generated at a switching element is rejected by the additional switching element (Fig. 7.27). The crosstalk reduction scheme is inherently incorporated into the tree structure.

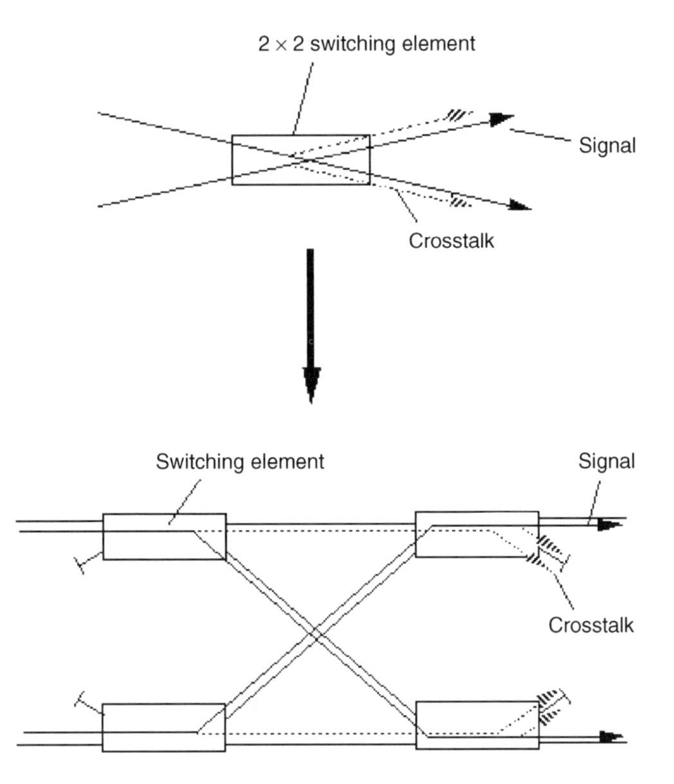

Fig. 7.27 Crosstalk reduction scheme.

The crosstalk of a 2 × 2 switching element can be reduced by using a 2 × 2 routing element composed of two 1 × 2 switching elements at the input and two 2 × 1 switching elements at the output (2 × 2 tree structure) as a switching element in a network. In a network composed of multi-stage 2 × 2 switching elements, such as banyan, Benes, 4 × 4 wide-sense nonblocking networks, or square crossbar architectures, 2 × 1 and 1 × 2 switching elements of the adjacent low-crosstalk 2 × 2 routing unit can be connected to form a 2 × 2 switching element in which only one light signal is routed. The switching element ranks are reduced by this modification (Fig. 7.28) [97]. The dilated Benes network can be obtained from a Benes network with this procedure.

In the crossbar type architectures, a 1 × 2 switching element on the input waveguide and a 2 × 1 switching element on the output waveguide are connected at crossing points [98]. In $LiNbO_3$ a parallel input and output waveguide arrangement [98] is required because only a large curvature of

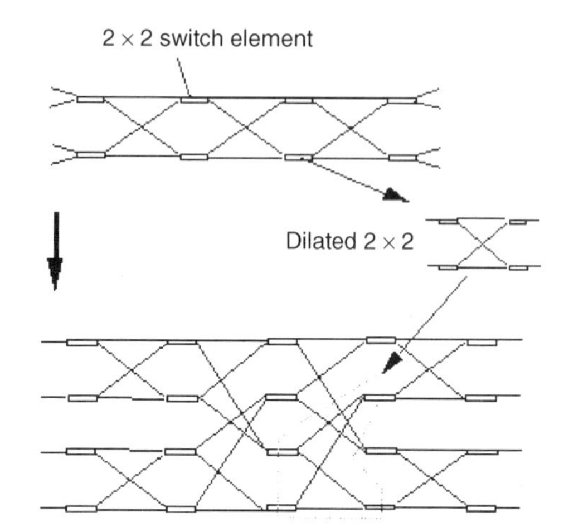

(a) Dilated 4 × 4 wide-sense nonblocking network

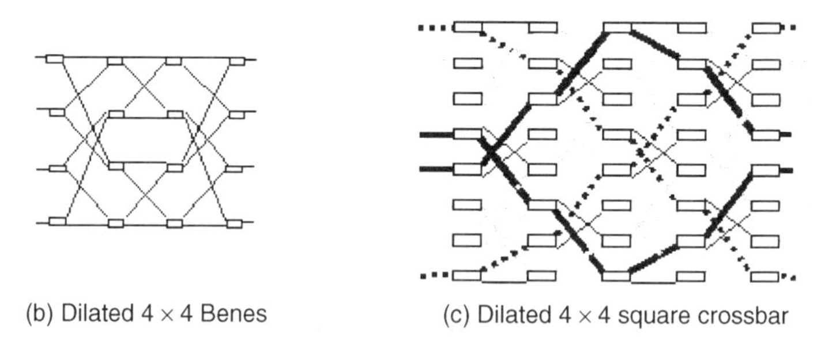

(b) Dilated 4 × 4 Benes

(c) Dilated 4 × 4 square crossbar

Fig. 7.28 Low crosstalk switch arrays [97]. J. Lightwave Tech., vol. 18, No. 4. © 2000 IEEE.

radius is possible for Ti-diffused $LiNbO_3$ waveguide interconnection. More compact devices using crossing input and output waveguide arrangements have been realized in compound semiconductor materials [99–102]. As the port number is increased, the crosstalk degrades more rapidly in the crossbar than in the tree structure, even if the crosstalk reduction scheme is used in the crossbar-type architectures.

7.5.3. MULTIPLE CHIP ARCHITECTURE

An increment of the port number over that available with a single N × N device, or improvement of the switching functionality, is attained by interconnecting several N × N device chips with optical fibers or waveguide interconnection boards. Many networks described in Sections 7.5.1 and 7.5.2 can be divided into several chips whenever the network sections are

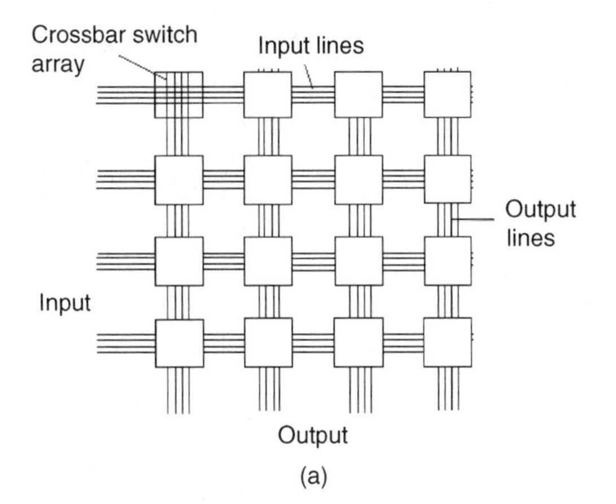

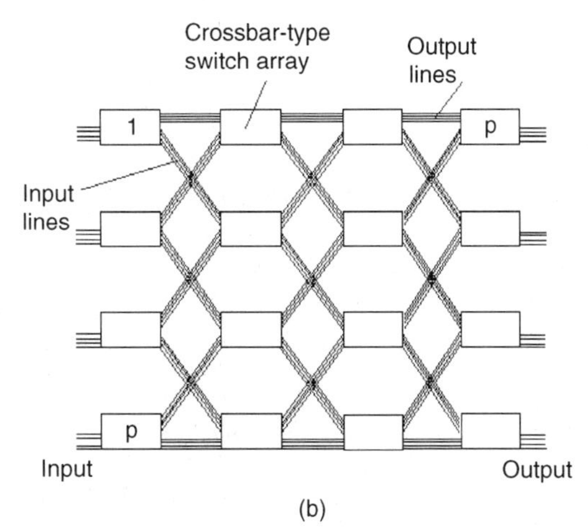

Fig. 7.29 Crossbar-type extension [97]. J. Lightwave Tech., vol. 18, No. 4. © 2000 IEEE.

separable. In the following, the most popular architectures designed for multiple chips are described.

7.5.3.1. Crossbar-type Extension

For the crossbar architecture, the simplest extension method is to arrange p^2 chips in p rows and p ranks and connect them together (Fig. 7.29(a)). The output of the input waveguide is connected to the input of the next crossbar and the output of the output waveguide is connected to the input of the output waveguide in the next crossbar. An $N = pn$ network is obtained using $n \times n$ chip. However, in this configuration, light traveling the shortest path traverses only one chip, while light traveling the longest path traverses

$2p - 1$ chips. The insertion loss and crosstalk differences among paths are large in this structure.

To overcome the problem, a square crossbar-type architecture has been proposed (Fig. 7.29(b)). Chips are arranged in p rows and p ranks. The output of the input waveguide of the ith row chip is connected to the input of the $(i + 1)$th and $(i - 1)$th row chip input waveguide. The output of the output waveguide of the ith row chip is connected to the input of the $(i + 1)$th and $(i - 1)$th row chip output waveguide. Every path from the input to the output traverses p chips, avoiding the insertion loss and crosstalk deviations.

To construct $N \times N$ networks from $n \times n$ chips using these schemes, the light must travel through many chips $p = N/n$ so that the insertion loss and chip number p^2 become large. For a larger extension, the Clos network described in the following is advantageous.

7.5.3.2. Clos Network

This network architecture (Fig. 7.30(a)) was discovered more than 40 years ago [103] and has been used in electronic telephone exchanges. The architecture is still a basic procedure to expand the port number using smaller nonblocking switching networks. A three-stage $N \times N$ Clos network is constructed exclusively connecting $r = N/n$ of $n \times m$ switches at the input stage, m of $r \times r$ switches in the middle stage, and $r = N/n$ $m \times n$ switches at the output stage. When m is larger than or equal to n the network is rearrangeable, and when m is larger than or equal to $2n - 1$ the network is strictly nonblocking. When the largest switch size is limited to $m \times m$ the available network size is $N = m^2$ for the rearrangeable or $N = m(m + 1)/2$ for the strictly nonblocking. Networks with more than 5 stages are obtained by recursive procedure to further increase the port number at the expense of an increment of the loss. In three-stage architecture there are mN/n interconnections at each stage. A portion of the 128-port Clos network was demonstrated using $LiNbO_3$ [104].

7.5.3.3. Extended Generalized Shuffle (EGS) Network

The network (Fig. 7.31) was originally proposed for switching using a self-electrooptic effect device (SEED) [105]. The EGS class network involves a broad spectrum of structures using $n \times m$-type switching elements. An EGS network used in constructing a switching network, with 1×2, 2×2, and 2×1 switching elements, is composed of $1 \times F$ fan-out section,

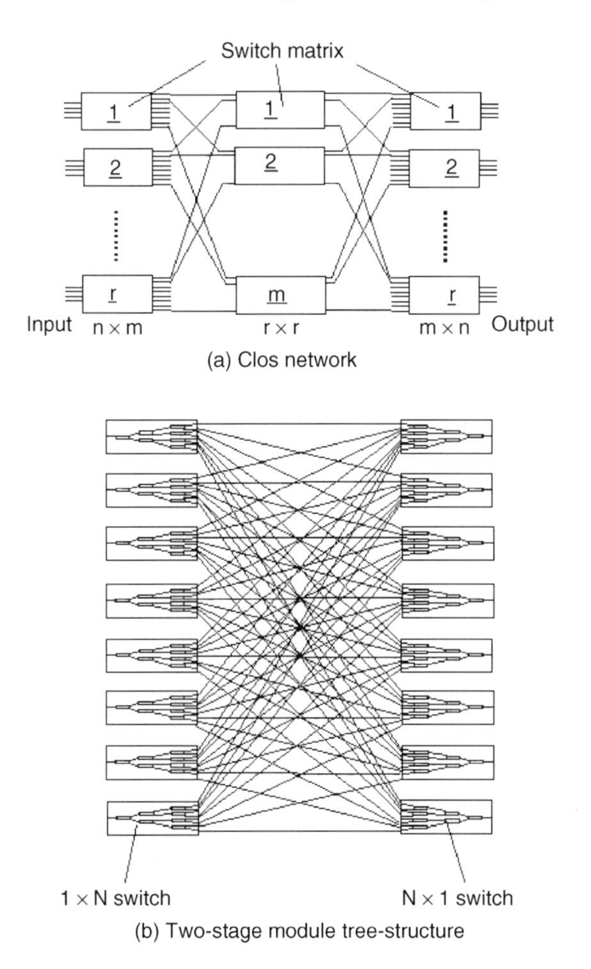

Fig. 7.30 Multi-module architectures with less module ranks [97].

S-rank NF/2-row switching section, and F × 1 fan-in section. The blocking probability of the EGS is given as $\{1 - [1 - (F2^{\alpha-1})^{-1}]^{S-\alpha+1}\}^{\wedge}(F2^{S}/N)$, with α being a parameter denoting the signal number handled in a switching element [105]. A strictly nonblocking network is obtained with proper combinations of F and S.

A low-crosstalk 16 × 16 strictly nonblocking optical switching system was demonstrated [106] using 8 dual binary tree 1 × 8 (used as 1 × 7: F = 7) switch arrays, 16 × 16 banyan, and 8 dual binary tree 8 × 1 (used as 7 × 1: F = 7) switch arrays. Inputting only one light signal to a switching element attains the low crosstalk. The 4-rank 56-row switching section is

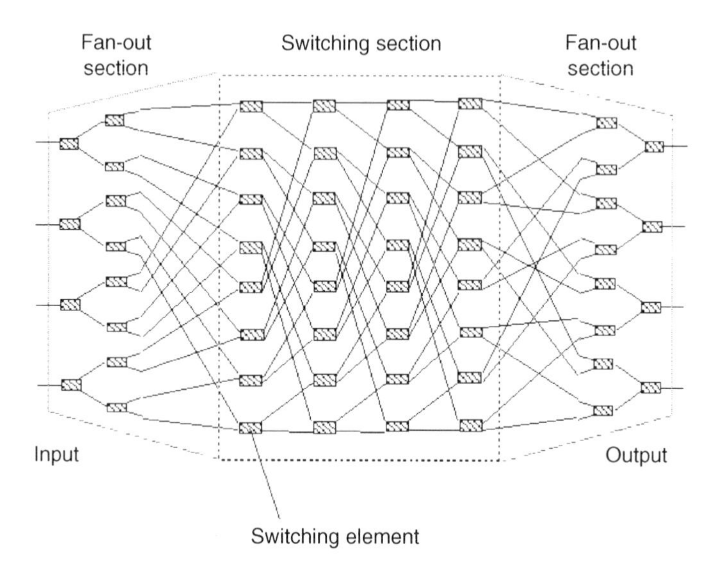

Fig. 7.31 Extended generalized shuffle (EGS) switch network [105].

implemented with the equivalent 7 16×16 banyan networks. A total of NF interconnections are required between switch modules at each stage.

Similar networks are obtained by K layers of $N \times N$ Benes or banyan network connected to input and output ports via $1 \times K$ and $K \times 1$ switches. A strictly nonblocking (Cantor network) and rearrangeable network (vertical replicated banyan network) is obtained using K larger than or equal to $\log_2 N$ layers of Benes and K larger than or equal to $N^{1/2}$ layers of banyan networks respectively [107].

7.5.3.4. Two-module Stage Network

The most well-known two-module stage network is a tree or router/selector architecture (Fig. 7.30(b)) (Section 7.5.1) with $1 \times N$ switch array module in the input stage, $N \times 1$ switch array module in the output stage, and N^2 interconnections between them. Various network architectures can be divided in the middle and rearranged in input and output sections to obtain a two-module stage network.

A class of two-module stage networks based on the Clos network has been reported (Fig. 7.32) [97]. The middle-stage $r \times r$ switch in the Clos network is divided and combined with an input stage $n \times m$ switch or output stage $m \times n$ switch. When a 4×4 wide-sense nonblocking network is used as a $r \times r$ switch, an 8×8 wide-sense nonblocking switch using only 48 elements is obtained. To attain a low-crosstalk network, the dilation method

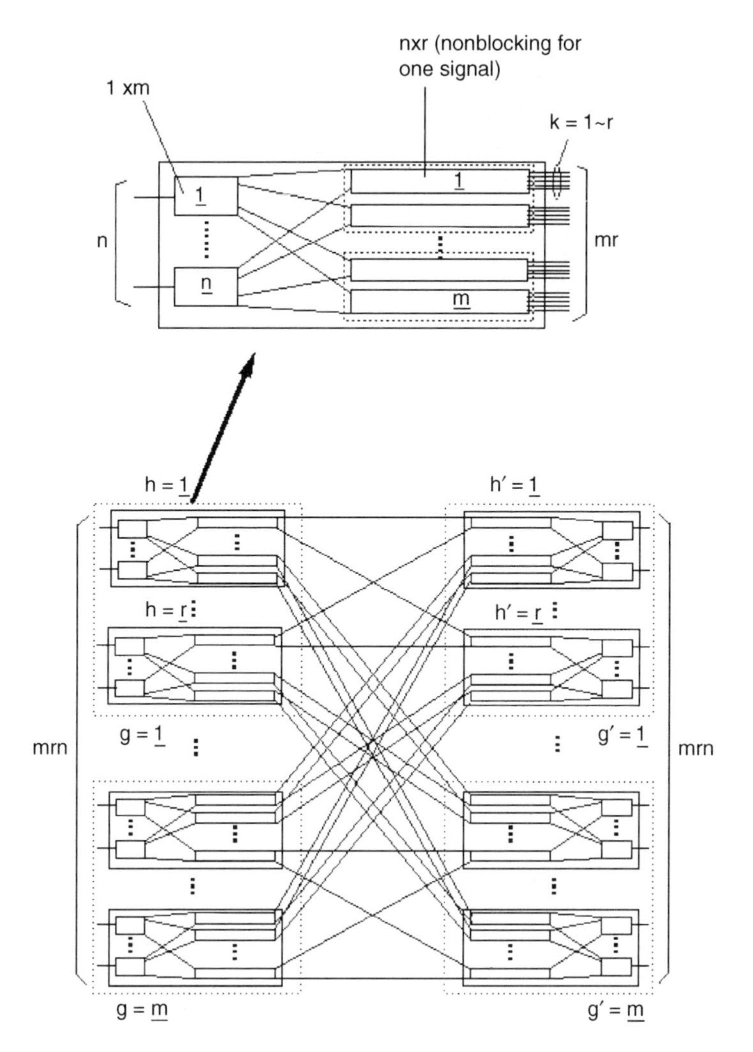

Fig. 7.32 Two-module stage network based on Clos network [97].

described in Section 7.5.2 can be used. A tree structure is used to obtain a useful network [108].

The input-stage n × m switch can be constructed using n input 1 × m switches and m output n × 1 switches connected together. The output-stage m × n switch can be constructed using m input 1 × n switches and n output m × 1 switches connected together. The middle-stage r × r switch can be constructed using r input 1 × r switches and r output r × 1 switches connected together. The n × 1 switch in the input-stage switch can be combined with $1 \times r$ switch in the middle-stage switch to construct a $n \times r$ switch in which only one light signal is routed. The $1 \times n$ switch in the output-stage switch can be combined with $r \times 1$ switch in the middle-stage switch to construct an $r \times n$ switch in which only one light signal is routed. The thus-obtained two-module network consists of an input module composed of n of $1 \times m$ switches connected to m of $n \times r$ switches and an output module composed of m $r \times n$ switches connected to n $m \times 1$ switches.

There are mrN/n interconnections between modules. For a strictly non-blocking network, the interconnection number is approximated as $2N^2/n$, which can be made smaller than for the tree architecture.

Many types of switches can be used as $n \times r$ and $r \times n$ switches routing only one signal, such as banyan network, AWG-type single-stage routing switch, or tilting mirror deflector (Fig. 7.33). AWG-type input routing module is equivalent to a tilting mirror device as shown in Fig. 7.33(b).

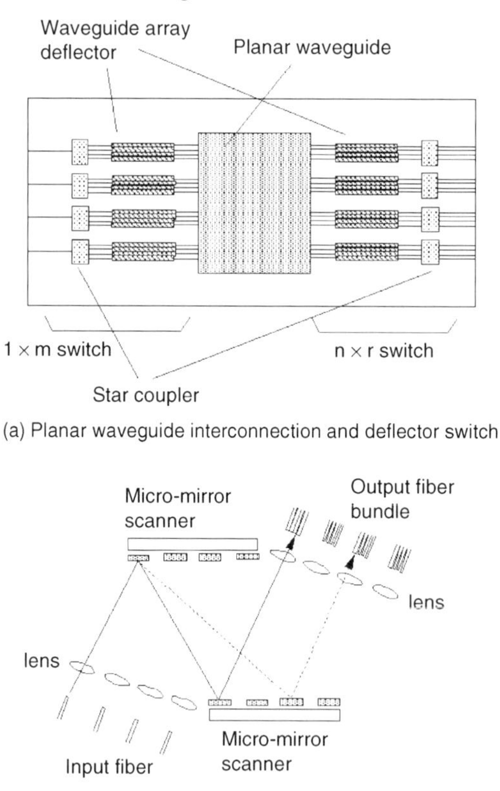

Fig. 7.33 Module structure using beam steering elements for architecture in Fig. 7.32 [97]. J. Lightwave Tech., vol. 18, No. 4. © 2000 IEEE.

7.5.4. ARCHITECTURE FOR WAVELENGTH ROUTING

An optical switch array designed especially for routing the light signal has been proposed (Fig. 7.34). Binary tree switches routing individual wavelength signals are integrated into a chip, and fiber bundle (ribbon fiber) interconnections are used to connect input- and output-stage chips. The connection between wavelength demultiplexer (or multiplexer) and the optical switch array is done by the fiber bundle to reduce the number of connectors. Units, each composed of an input optical switch array, output

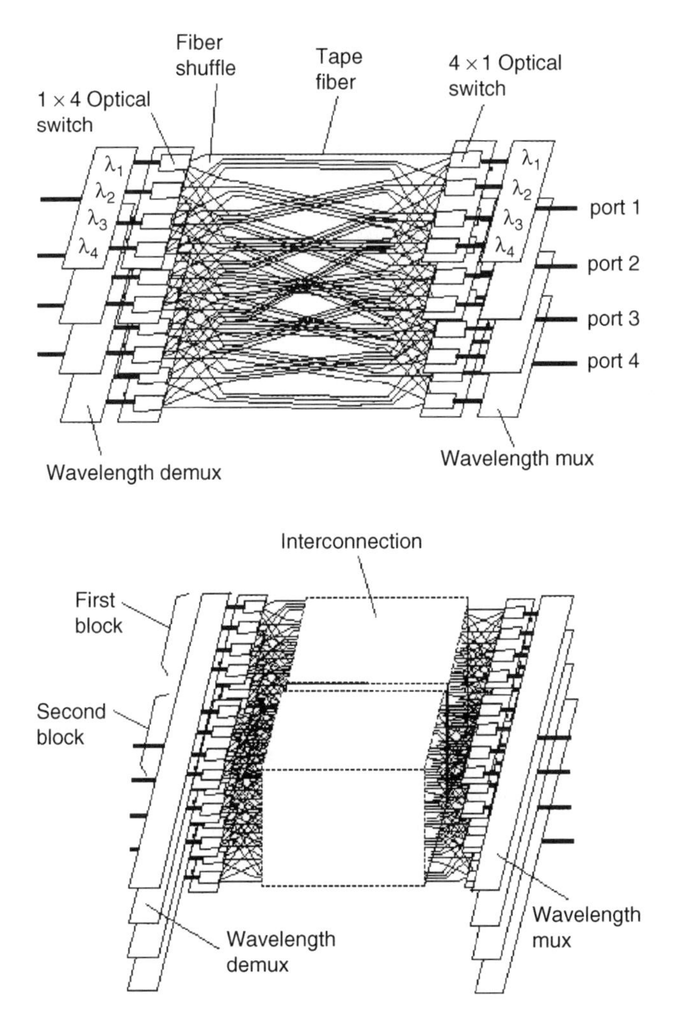

Fig. 7.34 Switch array structure for WDM routing.

optical switch array, and interconnections, can be installed successively to upgrade the wavelength channel number with no limits. The optical switches can be replaced by passive branch for multicasting or wavelength conversion similar to a cross-connect node structure proposed in Ref. 109. In this case, optimized allocation of routing switches, redundant crosstalk-reduction switches, and passive branch into two (input and output) chips can maximize the electrode length at each optical switch element to reduce the drive voltage.

7.6. Device Based on $LiNbO_3$

Materials mainly used in fabricating EO optical switches are semiconductor, ferroelectric crystal, and organic materials. Features of these materials are shown in Table 7.1.

Among several ferroelectric materials, $LiNbO_3$ has been used extensively from the beginning of the electrooptic guided-wave optical switch history. As for the optical switch network, simple blocking network was used first, followed by crossbar architecture. Recently, switch network architecture with fewer switching element ranks, such as tree or banyan architecture, is used mainly. The largest nonblocking network fabricated so far is the 16×16 tree structure [89], and the largest rearrangeable network is 16×16 dilated Benes [78]. The largest blocking banyan network is 32×32 [110]. 16×16 device using waveguide interconnection has also been reported [111]. Examples of the $LiNbO_3$ switch arrays reported so far are shown in Table 7.2. The typical value of a voltage-length product is 5–10 Vcm at 1.3–1.55 μm wavelength for the polarization with largest electrooptic coefficient. The insertion loss ranges from 5 to 15 dB depending on the switch array size. Crosstalk lower than -40 dB is attained in the low-crosstalk switching network architecture.

Several ferroelectric PLZT optical switches have been reported. The drive voltage is reduced compared to $LiNbO_3$ due to larger electrooptic coefficient. A TIR switch was demonstrated [112], but with somewhat high propagation loss. An improved process yielding lower propagation loss was reported more recently [113]. A Y-branch switch fabricated with the process shows polarization-independent switching with voltage-length product about 1/10 of that attained in $LiNbO_3$. Further reduction of the propagation loss in the channel waveguide is required to implement a large channel number optical switch array.

Table 7.1 **Various Materials Used for Optical Switch**

	Compound Semiconductor (InP, GaAs etc.)	***$LiNbO_3$***	***Organic Material (Polymer etc.)***	***Silica***
Advantage	*Integration with PD and LD *Improvement by artificial material design such as quantum well *Exhibits electrooptic and nonlinear optical effect *Small structure	*Large electrooptic, acoustooptic, nonlinear optic, and thermooptic effects *Easy fabrication of low loss waveguide	*Can be fabricated on various substrates *Cost reduction possibility *Electrooptic effect can be realized by material design *Possibility of size reduction	*Low loss *Good mode and refractive index matching with optical fiber *Size reduction possible
Disadvantage	*Difficulty in fabricating low loss waveguide for functional device *Large refractive index and mode mismatch with optical fiber	*Difficulty in reducing the structure size *Optical damage and DC-drift	*Higher loss than Silica *Material degradation	*Basically passive (Light wave control is done by TO effect)

Table 7.2 **Optical Switch Array Using $LiNbO_3$**

LN Switch Arrays

Blocking	*Rearrangeable*	*Strictly Nonblocking*
banyan 16×16 [Δβ] '93 [142] banyan 32×32 [MZ] '94 [110]	Benes 16×16 [Δβ] '91 [75] n-stage 4×4 [Δβ] '88 [79] dilated Benes 8×8 [Δβ] '90 [77] dilated Benes 16×16 [Δβ] '89 [78] **Customized tree 4×4 [Y] '94 [143]	crossbar 4×4 '79 [13] [Br], '86 [83] [Δβ], '85 [98], '88 [86], '90 [84] [Δβ], '87 [29] crossbar 8×8 [Δβ] '86 [85] **tree 4×4 [Y] '90 [91] **tree 8×8 [Y] '94 [90] **tree 16×16 [Y] '97 [111] **tree 16×16 [Y] 2000 [89] *simplified tree 4×4 '88 [144] [DC] **simplified tree 8×8 [Y, X] '93 [94] **simplified tree 8×8 '90 [145] [DC], '96 [95] [MZ]

1×N switches
1×4 '85 [141] (Δβ) 1×16 '91 [147] (MZ) *1×16 '86 [148] (DC) **1×16 '91 [146] (Y) **1×32 '91 [146] (Y)

MZ : Mach-Zehnder
Δβ : Reversed Δβ
DC : Directional coupler
Y : Y-branch, X : Asymmetric X
Br : Bragg

*Polarization independent
**Digital response, Polarization independent

7.7. Polarization-independent Structures

In the standard single-mode fiber used in optical communication networks, the polarization state changes randomly due to many types of perturbations. A devicc inserted into the optical signal transmission line should be insensitive to the input state of polarization. The EO coefficient of the electrooptic material differs by the polarization angle against the crystal axis. In $LiNbO_3$ crystal the largest EO coefficient for the light polarized parallel to the ferroelectric dipole axis (c-axis, z-axis) is several times larger than for the light polarized perpendicular to it. The crystal is birefringent, meaning that the refractive index also differs with the light polarization.

Many methods have been proposed to overcome polarization dependence (Fig. 7.35). Methods are classified by the utilized EO coefficient and the operation principle (Figs. 7.35, 7.36, and 7.37). Two orthogonal polarizations can be divided at the input, controlled separately, and combined at the output [114, 115]. Cascaded devices, each controlling separate polarizations, can be used [116]. Polarization conversion can be utilized to achieve polarization independence [117]. In the polarization converter, the hybrid modes of two polarizations are propagated as normal modes in a phase-matched condition. The propagation constants of hybrid modes can be controlled by the coupling strength between polarizations and the principle can be applied to directional coupler, reversed $\Delta\beta$, and Mach-Zehnder type switches.

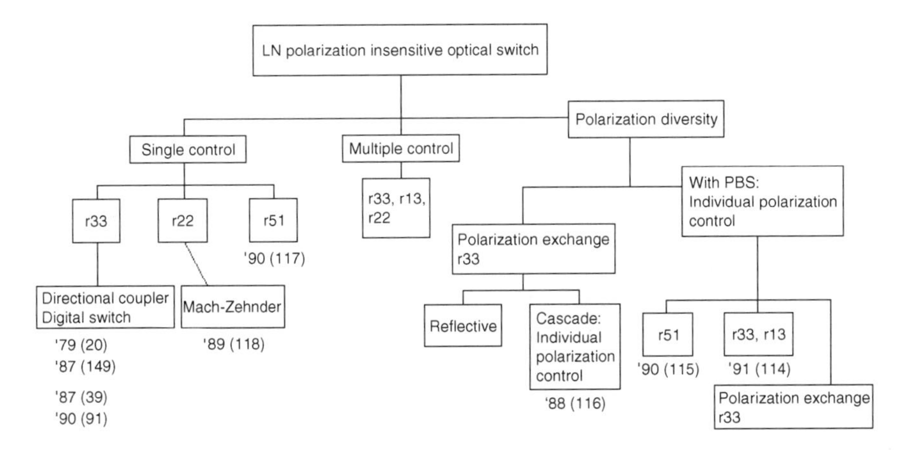

Fig. 7.35 Schemes for attaining polarization-independent optical switch in $LiNbO_3$.

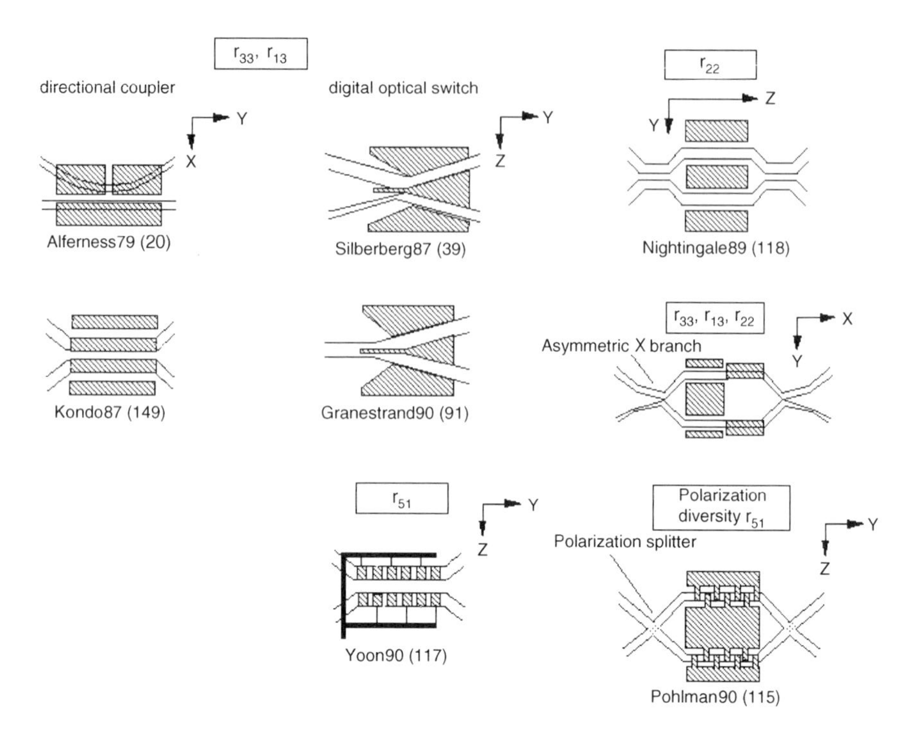

Fig. 7.36 Examples of polarization-independent optical switch using $LiNbO_3$.

Among the many methods, two methods with the simplest structure and control schemes are mainly used. The first one is a Mach-Zehnder device using an x-cut substrate and light propagated along the z-axis [118, 119]. The device uses r_{22} EO coefficient that is more than three times smaller than the largest r_{33}. Although with the opposite polarity, the magnitude of the EO coefficient becomes equal for both orthogonal polarizations in this configuration. The refractive indexes are also almost equal for both polarizations. A truly polarization-independent device may be obtained.

The second method is a digital switching device using a z-cut substrate in which waveguide with good light confinement is easily fabricated [39, 90]. The largest r_{33} and second largest r_{13} are used. Because the magnitude of r_{13} is about three times smaller than r_{33}, the digital switching curve is required to switch both polarizations. For the voltage with which the light associated with a smaller EO coefficient is switched, the light associated with the larger EO coefficient is already switched. The wide switching voltage range of the digital optical switch permits the device to switch light polarizations with different EO coefficients simultaneously.

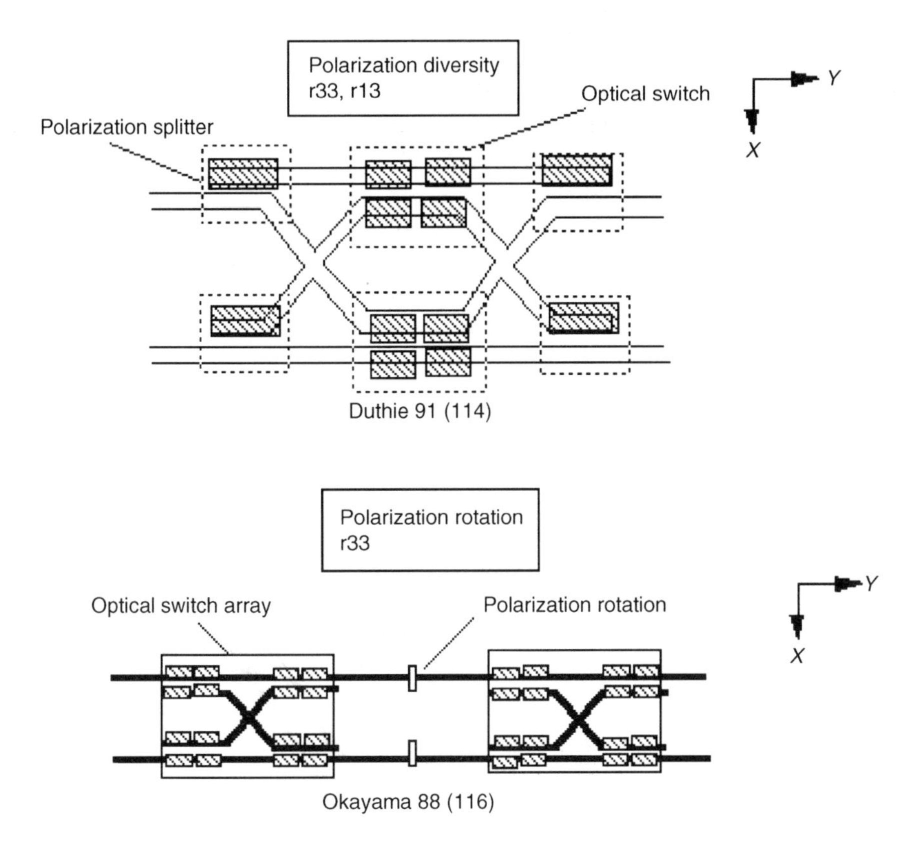

Fig. 7.37 Polarization-independent $LiNbO_3$ optical switch using polarization diversity.

In an electrooptic semiconductor device, suitable waveguide and crystal axis orientation or plasma effect due to carrier injection are used to attain polarization independence.

7.8. Other Materials with Electrooptic Effect

Compound III-V semiconductor is the most widely used among semiconductor materials. The refractive index is changed by the EO effect induced by applied voltage or plasma effect induced by carrier injection. The refractive index change is larger for carrier injection and reaches 10^{-2}, which is two orders of magnitude larger than that attainable with $LiNbO_3$. The EO effect with comparable refractive index change can be attained using multiple quantum well structure. Even in the device using EO effect, the switching voltage is reduced compared to $LiNbO_3$ due to small form factor and small gap between electrodes. However, the propagation and insertion loss tends to be high in the device structure designed for optical switches. The largest switch array fabricated so far is an 8×8 switch using GaAs [96]. The insertion loss is reduced using GaAs by large separation between band-gap and signal wavelength. A device integrated with amplifiers has been demonstrated using InP material [102]. In spite of less attractive noise characteristics compared to the erbium-doped fiber amplifier, integration of the semiconductor amplifier enables the fabrication of a loss-less optical device. Recently, the optical switch using a semiconductor amplifier gate switch has been studied most extensively due to its excellent crosstalk characteristics [120–122]. The study on 2×2 switching element technology is continuing, but the emphasis is now on devices using multimode interference [35, 123, 124]. The multimode interference coupler relaxes the fabrication tolerance compared to the directional coupler type. A spot-size converter can reduce the coupling loss between a semiconductor waveguide and an optical fiber. The small spot size of a semiconductor waveguide is enlarged to a large spot size by several types of waveguide width taper structures to eliminate spot-size mismatch. A switch array integrated with spot-size converter was demonstrated [76]. Some of the device demonstrations are listed in Table 7.3.

Studies have also been done on devices using silicon [125], though for fewer than those on III-V material.

Organic materials can be used to implement EO devices [126, 127]. However, devices using the TO effect are dominant in recent studies and have already been commercialized. The switching speed of a TO device is several hundred μs to several ms, which is slower than an EO device, but polarization independence is easily attained by the TO effect. The drive voltage drift phenomena caused by anisotropy of conductance and permittivity coefficients in the EO device is not a problem in a TO device. The largest switch array fabricated into a chip is an 8×8 switch [93]. The commercialized 8×8 switch box is constructed using 1×8 and 8×1 switches connected by optical fibers to avoid insertion loss and crosstalk problems associated with the one-chip device.

Devices using liquid crystal are also studied [128], but the number of studies for waveguide-type devices is much less than those for the bulk-type device.

Table 7.3 **Optical Switch Array Using Compound Semiconductor**

Semiconductor Electrooptic Switch Arrays

Blocking	*Rearrangeable*	*Strictly Nonblocking*
4 × 4 InP banyan '96 [150] (MMI) 1 × 4 '92 [153] '95 [154] '99 [155]	4 × 4 InP n-stage '91 [80], '92 [151] 4 × 4 InGaAsP/InAlAs Benes '95 [76]	8 × 8 GaAs/AlGaAs simplified tree '92 [96] 4 × 4 InP tree '92 [92] (Y) 4 × 4 InP crossbar '91 [152] 4 × 4 InP double crossbar '99 [101], '99 [100] 4 × 4 double crossbar with SOA '94 [102] (TIR)

A 16 × 16 TO switch array has been achieved with SiO_2 [129]. The crossbar-type switching network architecture is used to suppress the drive power. Only the switching element that connects the desired input and output is required to be driven. The number of switching elements that consume drive power is the same as the number of input ports. The device length in crossbar architecture tends to be long, but the low loss nature of the SiO_2 material allows for a long device. More details about SiO_2 and TO devices are described in other chapters.

7.9. Nonlinear Optical Switch

Devices described so far use electric control to switch the device. All optical switching [130] is attained using a nonlinear optical effect (Fig. 7.38) [131, 132]. The device uses cascaded sum-frequency-generation (SFG) and difference-frequency-generation (DFG). To attain a large wavelength conversion efficiency, periodically poled $LiNbO_3$ (PFLN) is used as a nonlinear material [133–136]. The gate light is introduced into a PPLN with its polarization parallel to the z-axis to utilize the largest nonlinear coefficient d_{33}. The optical signal light is fed into PPLN with its polarization inclined 45 degrees against the z-axis. The signal propagates as two separate lights in the PPLN crystal, i.e., one parallel and the other perpendicular to the z-axis. The signal light parallel to the z-axis is converted to another wavelength through SFG and then reconverted to its original wavelength

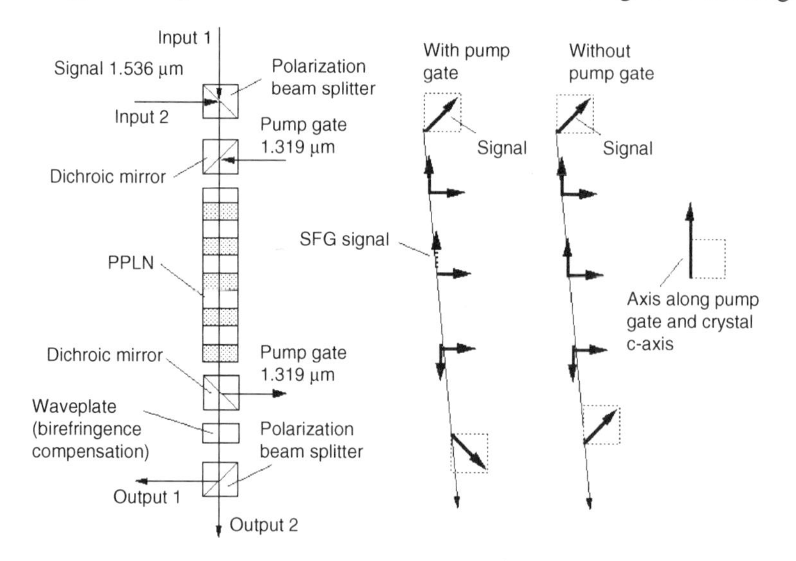

Fig. 7.38 Nonlinear optical switch using periodically poled $LiNbO_3$ (PPLN).

through DFG by an interaction with the gate light. The process generates the phase shift of π. The waveplate compensates for a birefringent retardation to combine the light parallel and perpendicular to the c-axis. The polarization direction of the light becomes 90 degrees against the original wavelength light. By switching the gate light on and off, the polarization of the signal light is rotated 90 degrees. A 1×2 optical switch is implemented by placing a polarization beam splitter at the output. In the experiment, a bulk-type PPLN was used to switch the 1.536 μm wavelength signal light by a 3 kW 1.319 μm wavelength gate light.

A switching system can be constructed using a wavelength conversion device as a switching element [137, 138]. Selecting wavelengths to be converted enables the switching function.

Other types of nonlinear optical switch are implemented using mostly nonlinearity in the semiconductor optical amplifier (SOA) [139, 140]. In a interferometric scheme, which is gaining much attention, SOA is placed in the arms of a Mach-Zehnder interferometer. The control signal changes the phase of the light to switch the outputs. To overcome the limit posed by the carrier recovery time, push-pull or delayed pulse schemes have been proposed.

7.10. Conclusion

Features of guided-wave optical switch arrays based on different technologies are depicted in Fig. 7.39. The devices demonstrated as one-chip devices are shown. For the EO device the drive voltage increases along with the

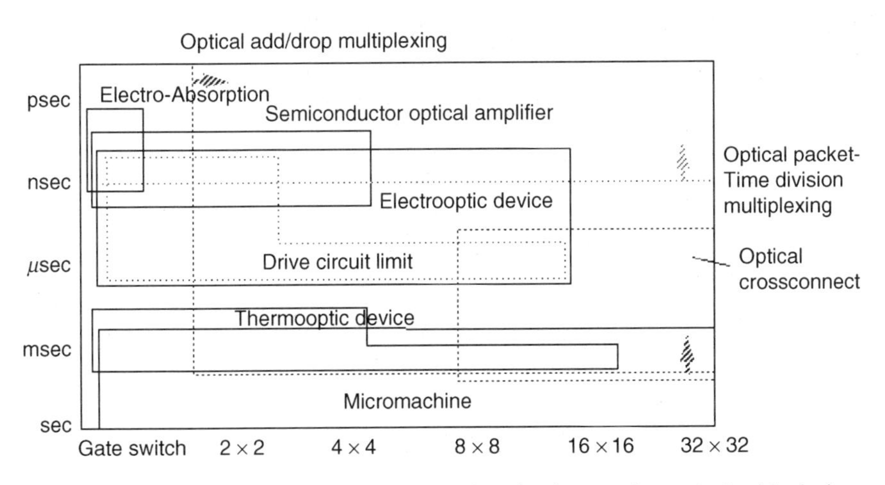

Fig. 7.39 Classified optical switch demonstrations implemented as a single chip device.

number of ports. The high-speed switching for large EO switch arrays is difficult at present. There is a trade-off between switching speed and the switching network size. Technologies to enable both are required [156]. The application area for an optical switch array can be from small-scale add drop multiplexer to large-scale slow-switching crossconnect. In some applications high-speed switching of the packet signal is required. The low power consumption nature of the EO device and the potential of attaining high-speed switching will find some applications in the ever-progressing field of optical communications.

References

1. M. Gottlieb, C. L. M. Ireland, and J. M. Ley, Chapters 3 and 4 in Electro-optic and acousto-optic Scanning and Deflection, Optical Eng. Series Vol 3 (Marcel Dekker Inc., New York, 1983).
2. Q. Chen, Y. Chiu, D. N. Lambeth, T. E. Schlesinger, and D. D. Stancil, "Guided-wave electro-optic beam deflector using domain reversal in $LiTaO_3$," J. Lightwave Tech., 12 (1994) 1401–1403.
3. K. Nashimoto, S. Nakamura, T. Morikawa, H. Moriyama, M. Watanabe, and E. Osakabe, "Fabrication of electro-optic $Pb(Zr, Ti)O_3$ heterostructure waveuides on Nb-doped $SrTiO_3$ by solid-phase epitaxy," Appl. Phys. Lett., 74 (1999) 2761–2763.
4. M. Yamada, M. Saitoh, and H. Ooki, "Electric-field induced cylindrical lens, switching and deflection devices composed of the inverted domains in $LiNbO_3$ crystals," Appl. Phys. Lett., 69 (1996) 3659–3661.
5. C. H. Bulmer, W. K. Burns, and T. G. Giallorenzi, "Performance criteria and limitations of electrooptic waveguide array deflectors," Appl. Opt., 18 (1979) 3282–3295.
6. H. Okayama and M. Kawahara, "Experiment on deflector-selector optical switch matrix," Electron. Lett., 28 (1992) 638–639.
7. C. R. Doerr and C. Dragone, "Proposed optical cross connect using a planar arrangement of beam steerers," IEEE Photon. Tech. Lett., 11 (1999) 197–199.
8. H. Okayama, Y. Okabe, T. Arai, and T. Tsuruoka, "Waveguide array optical switching element," Tech. Report IEICE PS99-5 (1999) 25–30 (in Japanese).
9. J. M. Heaton, D. R. Wight, J. T. Parker, B. T. Hughes, J. C. H. Birbeck, and K. P. Hilton, "A phased array optical scaninig (PHAROS) device used as a 1-to-9 way switch," IEEE J. Quantum Electron., 28 (1992) 678–685.
10. H. Sasaki and R. M. De La Rue, "Electro-optic multichannel waveguide deflector," Electron. Lett., 13 (1977) 295–296.

11. A. Kar-Roy and C. S. Tsai, "8 × 8 symmetric nonblocking integrated acousto-optic space switch module on $LiNbO_3$," IEEE Photon. Tech. Lett., 4 (1992) 731–734.
12. C. S. Tsai, Q. Li, and C. L. Chang, "Guided-wave two-dimensional acousto-optic scanner using proton-exchanged lithium niobate waveguide," Fiber and Integrated. Opt., 17 (1998) 157–166.
13. R. A. Becker and W. S. C. Chang, "Electrooptical switching in thin film waveguides for a computer communication bus," Appl. Opt., 18 (1979) 3296–3300.
14. H. Gnewuch, C. N. Pannell, G. W. Ross, P. G. R. Smith, and H. Geiger, "Nanosecond response of Bragg deflectors in periodically poled $LiNbO_3$," IEEE Photon. Tech. Lett., 10 (1998) 1730–1732.
15. S. Gosselin and J. Sapriel, "Versatile acousto-optic vector-matrix architecture for fast optical space switches," Tech. Digest ECOC'98 (1998) 253–254.
16. H. Yamazaki, T. Matsunaga, S. Fukushima, and T. Kurokawa, "4 × 1204 holographic switching with an optically addressed spatial light modulator," Appl. Opt., 36 (1997) 3063–3069.
17. B. Pesach, G. Bartal, E. Refaeli, A. J. Agranat, J. Krupnik, and D. Sadot, "Free-space optical cross-connect switch by use of electroholography," Appl. Opt., 39 (2000) 746–758.
18. Y. Fujii, K. Kajimura, S. Ishihara, and H. Yajima, "Thermal gradient induced optical deflection in TiO_2 crystal," Appl. Phys. Lett., 47 (1982) 217–219.
19. C. Bulmer and W. K. Burns, "Polarization characteristics of $LiNbO_3$ channel waveguide directional couplers," J. Lightwave Technol., LT-1 (1983) 227–236.
20. R. Alferness, "Polarization-independent optical directional coupler switch using weighted coupling," Appl. Phys. Lett., 35 (1979) 748–750.
21. L. McCaughan and S. K. Korotky, "Three-electrode Ti:$LiNbO_3$ optical switch," J. Lightwave Tech., LT-4 (1986) 1324–1327.
22. H. Okayama, T. Ushikubo, and T. Ishida, "Directional coupler with reduced voltage-length product," J. Lightwave Tech., 9 (1991) 1561–1566.
23. H. Kogelnik and R. V. Schmidt, "Switched directional coupler with alternating $\Delta\beta$," IEEE J. Quantum Electron., QE-12 (1976) 396–401.
24. H. Okayama, T. Kamijoh, and T. Tsuruoka, "Reversed and uniform $\Delta\beta$ directional coupler with periodically changing coupling strength," Jpn. J. Appl. Phys., 39 (2000) 1512–1515.
25. S. Thaniyavarn, "A synthesized digital switch using a 1 × 2 directional coupler with asymmetric $\Delta\beta$ phase reversal electrode," Tech. Digest IGWO'88, Paper TuC6 (1988).
26. R. A. Forber and E. Marom, "Symmetric directional coupler switches," IEEE J. Quantum Electron., QE-22 (1986) 911.

27. M. Papuchon, A. M. Roy, and D. B. Ostrowsky, "Electrically active optical bifurcation," Appl. Phys. Lett., 31 (1977) 266.
28. H. Nakajima, I. Sawaki, M. Seino, and K. Asama, "Bipolar-voltage-controlled optical switch using Ti:$LiNbO_3$ intersecting waveguides," Tech. Digest IOOC, Paper 29C4-5 (1983).
29. E. Voges and A. Neyer, "Integrated-optic devices on $LiNbO_3$ for optical communications," J. Lightwave Tech., LT-5 (1987) 1229–1238.
30. V. Ramaswamy and R. D. Standley, "A phased, optical coupler-pair switch," Bell Sys. Tech. J., 55 (1975) 767–775.
31. V. Ramaswamy, M. D. Divino, and R. D. Standley, "Balanced bridge modulator switch using Ti-diffused $LiNbO_3$ strip waveguides," Appl. Phys. Lett., 32 (1978) 644–646.
32. M. Minakata, "Efficient $LiNbO_3$ balanced bridge modulator/switch with an ion-etched slot," Appl. Phys. Lett., 35 (1979) 40–42.
33. O. Mikami and S. Zembutsu, "Modified balanced-bridge switch with two straight waveguides," Appl. Phys. Lett., 35 (1979) 145–146.
34. J. L. Jackel and J. J. Johnson, "Nonsymmetric Mach-Zehnder interferometers used as low-drive-voltage modulators," J. Lightwave Technol., 6 (1988) 1348–1351.
35. J. E. Zucker, K. L. Jones, T. H. Chiu, B. Tell, and K. B-Goebeler, "Strained quantum wells for polarization-independent electrooptic waveguide switches," J. Lightwave Technol., 10 (1992) 1926–1929.
36. H. Okayama, H. Yaegashi, I. Asabayashi, and M. Kawahara, "Balanced bridge optical switch composed of mode splitters," Tech. Digest MOC/GRIN'93, paper G19 (1993).
37. M.-C. Oh, H.-J. Lee, M.-H. Lee, J.-H. Ahn, and S. G. Han, "Asymmetric X-junction thermooptic switches based on fluorinated polymer waveguides," IEEE Photon. Tech. Lett., 10 (1998) 813–815.
38. W. K. Burns and A. F. Milton, "Waveguide transitions and junctions," In Guded-wave Optoelectronics, Chapter 3, Springer series in electronics and photonics 26. (Springer-Verlag, Berlin, 1988).
39. Y. Silberberg, P. Perlmutter, and J. E. Baran, "Digital optical switch," Appl. Phys. Lett., 51 (1987) 1230–1232.
40. A. N. M. M. Choudhury, W. H. Nelson, M. Abdalla, M. Rothman, R. Bryant, W. Niland, and W. Powazinik, "1.3 μm InP/InGaAsP digital optical switches with extinction ratio of 30 dB," Tech. Digest LEOS'93, paper OS2.1 (1993).
41. H. Okayama and M. Kawahara, "Y-fed directional coupler with weighted coupling," Electron. Lett., 27 (1991) 1947–1948.
42. S. Thaniyavarn, "Modified 1×2 directional coupler waveguide modulator," Electron. Lett., 22 (1986) 941–942.
43. H. Okayama and M. Kawahara, "Reduction of voltage-length product for Y-branch digial optial switch," J. Lightwave Tech., 11 (1993) 379–387.

44. M. N. Khan, J. E. Zucker, T. Y. Chang, N. J. Sauer, M. D. Divino, T. L. Koch, C. A. Burrus, and H. M. Presby, "Design and demonstration of weighted coupling digital Y-branch optical switches in InGaAs/InGaAsAlAs electron transfer waveguides," J. Lightwave Tech., 12 (1994) 2032–2039.
45. M. N. Khan, B. I. Miller, E. C. Burrows, and C. A. Burrus, "Crosstalk-, loss-, and length-reduced digital optical Y-branch switches using a double-etch waveguide structure," IEEE Photon. Tech. Lett., 11 (1999) 1250–1252.
46. W. K. Burns, "Shaping the digital switch," IEEE Photon. Tech. Lett., 4 (1992) 861–863.
47. R. Krahenbuhl and W. K. Burns, "Enhanced crosstalk suppression for Ti:$LiNbO_3$ digital optical switches," Tech. Digest Photonics in Switching, Paper PFA1 (1999).
48. S. K. Kim and V. Ramaswamy, "Tapered both in dimension and in index, velocity coupler: theory and experiment," IEEE J. Quantum Electron., 29 (1993) 1158–1167.
49. R. R. Syms, "The digital optical switch: analogus directional coupler devices," Opt. Commun., 69 (1989) 235–238.
50. A. Syahari, V. M. Schneider, and S. Al-Bader, "The design of mode evolution couplers," J. Lightwave Tech., 16 (1998) 1907–1914.
51. S. Xie, H. Heidrich, D. Hoffmann, H.-P. Nolting, and F. Reier, "Carrier-injected GaInAsP/InP directional coupler optical switch with both tapered velocity and tapered coupling," IEEE Photon. Tech. Lett., 4 (1992) 166–169.
52. R. G. Peall and R. R. A. Syms, "Further evidence of strong coupling effects in three-arm Ti:$LiNbO_3$ directional couplers," IEEE J. Quantum Electron., 25 (1989) 729–735.
53. H. A. Haus and C. G. Fonstad, "Three-waveguide couplers for improved sampling and filtering," IEEE J. Quantum Electron., QE-17 (1981) 2321–2325.
54. H. Ogiwara, "Optical waveguide 3×3 switch: theory of tuning and control," Appl. Opt., 18 (1979) 510–515.
55. S. Rushin and D. Meshulach, "Voltage-controlled $N \times N$ coupled waveguide switch and power splitter," IEEE Photon. Tech. Lett., 5 (1993) 203–206.
56. H. Haus and L. Molter-Orr, "Coupled multiple waveguide sysytems," IEEE J. Quantum Electron., QE-19 (1983) 840–844.
57. M. Kuznetsov, "Coupled waveguide analysis of multiple waveguide sysytems: the discrete harmonic oscillator," IEEE J. Quantum Electron., QE-21 (1985) 1893–1898.
58. T. Pertsch, T. Zengtraf, U. Peschel, A. Brauer, and F. Lederer, "Beam steering arrays," Appl. Phys. Lett., 80 (2002) 3247–3249.
59. L. A. Molter-Orr and H. Haus, "Multiple coupled waveguide switches using alternating $\Delta\beta$ phase mismatch," Appl. Opt., 24 (1985) 1260–1264.

60. H. Okayama, T. Ushikubo, and T. Ishida, "Three guide directional coupler as polarization independent optical switch," Electron. Lett., 27 (1991) 810–811.
61. E. Kapon and N. R. Thurston, "Multichannel waveguide junctions for guided-wave optics," Appl. Phys. Lett., 50 (1987) 1710–1712.
62. T. Ramadan, R. Scarmozzino, and R. M. Osgood, Jr., "A novel 1 × 4 coupler-multiplexer permutation switch for WDM applications," J. Lightwave Tech., 18 (2000) 579–588.
63. J. C. Campbell and T. Li, "Electro-optic multimode waveguide modulator or switch," J. Appl. Phys., 50 (1979) 6149–6154.
64. H. Okayama, H. Yaegashi, and M. Kawahara, "Digital optical switch using multimode interference coupler," Tech. Digest OEC'94, Paper 15B3-6 (1994).
65. H. H. El-Refaeli and D. A. M. Khalil, "Design of strip-loaded weak-guiding multimode interference structure for an optical router," IEEE J. Quantum Electron., 34 (1998) 2286–2290.
66. S. Nagai, O. Morishima, M. Yagi, and K. Utaka, "InGaAsP/InP multi-mode interference photonic switches for monolithic photonic integrated circuits," Jpn. J. Appl. Phys., 38, Pt. 1 (1999) 212–215.
67. E. C. M. Pennings, R. van Roijen, B. H. Verbeek, R. J. Deri, and L. B. Soldano, "Ultracompact multimode interference waveguide devices," Tech. Digest LEOS'93, Paper IO2.1 (1993).
68. N. S. Lagali, M. R. Paiam, R. I. MacDonald, K. Worhoff, and A. Driessen, "Analysis of generalized Mach-Zehnder interferometers for variable-ratio power splitting and optimized switching," J. Lightwave Tech., 17 (1999) 2542–2550.
69. G. E. Betts and W. S. C. Chang, "Crossing-channel waveguide electrooptic modulators," IEEE J. Quantum Electron., QE-22 (1986) 1027–1038.
70. C. S. Tsai, B. Kim, and F. R. El-Akkari, "Optical channel waveguide switch and coupler using total internal reflection," IEEE J. Quantum Electron., QE-14 (1978) 513–517.
71. G. A. Bogert, "Ti: $LiNbO_3$ intersecting waveguides," Electron. Lett., 23 (1987) 72–73.
72. V. E. Benes, Mathematical Theory of Connecting Networks and Telephone Traffic (Academic Press, New York, 1965).
73. K. Padmanabhan and A. N. Netravali, "Dilated networks for photonic switching," IEEE Trans. Commun., COM-35 (1987) 1357–1365.
74. R. A. Spanke and V. E. Benes, "N-stage planar optical permutation network," Appl. Opt., 26 (1987) 1226–1229.
75. P. J. Duthie and M. J. Wale, "16 × 16 single chip optical switch array in Lithium Niobate," Electron. Lett., 27 (1991) 1265–1266.
76. K. Kawano, S. Sekine, H. Takeuchi, M. Wada, M. Kohtoku, N. Yoshimoto, T. Ito, M. Yanagibashi, S. Kondo, and Y. Noguchi, "4 × 4 InGaAlAs/InAlAs

multiple-quantum well (MQW) directional coupler waveguide switch modules with spotsize converters and their 10 Gbit/s operation," Electron. Lett., 31 (1995) 96–97.

77. J. E. Watson, M. A. Milbrodt, K. Bahadori, M. F. Dautartas, C. T. Kemmerer, D. T. Moser, A. W. Schelling, T. O. Murphy, J. J. Veselka, and D. A. Herr, "A low-voltage 8 × 8 Ti: $LiNbO_3$ switch with a dilated-Benes architecture," J. Lightwave Tech., 8 (1990) 794–801.
78. T. O. Murphy, C. T. Kemmerer, and D. T. Moser, "A 16 × 16 Ti:$LiNbO_3$ dilated Benes photonic switch module," Tech. Digest Topical meeting on photonic switching. Paper PD3 (1989).
79. D. Hoffman, H. Heidrich, H. Ahlers, and M. K. Fluge, "Performance of rearrangeable nonblocking 4×4 switch matrices on $LiNbO_3$," IEEE J. Select. Areas Commun., 6 (1988) 1232–1240.
80. E. Lallier, A. Enard, D. Rondi, G. Glastre, R. Blondeau, M. Papuchon, and N. Vodjdani, "InGaAsP/InP 4 × 4 optical switch matrix with current injection tuned directional couplers," Tech. Digest ECOC/IOOC'91 (1991) 44–47.
81. V. E. Benes and R. P. Kurshan, "Wide-sense nonblocking network made of square switches," Electron. Lett., 17 (1981) 697.
82. P. J. Duthie, M. J. Wale, and I. Bennion, "A new architecture for large integrated optical switch-arrays," Tech. Digest Topical meeting on photonic switching, Paper ThD4 (1987).
83. G. A. Bogert, E. J. Murphy, and R. T. Ku, "Low crosstalk 4×4 Ti:$LiNbO_3$ optical switch with permanently attached polarization maintaining fiber array," J. Lightwave Tech., LT-4 (1986) 1542–1545.
84. P. P. Pedersen, J. L. Nightingale, B. E. Kincaid, J. S. Vrhel, and R. A. Becker, "A high-speed 4 × 4 Ti:$LiNbO_3$ integrated optic switch at 1.5 μm," J. Lightwave Tech., 8 (1990) 618–621.
85. P. Granestrand, B. Stoltz, L. Thylen, K. Bergvall, W. Doldissen, H. Heinrich, and D. Hoffmann, "Strictly nonblocking 8 × 8 integrated optical switch matrix," Electron. Lett., 22 (1986) 816–818.
86. I. Sawaki, T. Shimoe, H. Nakamoto, T. Iwama, T. Yamane, and H. Nakajima, "Rectangularly configured 4 × 4 Ti:$LiNbO_3$ matrix switch with low drive voltage," IEEE J. Select. Areas Commun., 6 (1988) 1267–1272.
87. R. A. Spanke, "Architectures for large nonblocking optical space switches," IEEE J. Quantum Electron., QE-22 (1986) 964–967.
88. K. Habara and K. Kikuchi, "Geometrical design considerations for a tree-structured optical switch matrix," Electron. Lett., 23 (1987) 376–377.
89. T. O. Murphy, S.-Y. Suh, B. Comissiong, A. Chen, R. Irvin, R. Grencavich, and G. Richards, "A strictly non-blocking 16 × 16 electrooptic photonic switch module," Tech. Digest ECOC'2000 (2000) 93–94.

90. P. Granestrand, B. Lagerstrom, P. Svensson, H. Olofsson, J.-E. Falk, and B. Stolz, "Pigtaled tree-structured 8 × 8 $LiNbO_3$ switch matrix with 112 digital optical switches," IEEE Photon. Tech. Lett., 6 (1994) 71–73.
91. P. Granestrand, B. Lagerstrom, P. Svensson, L. Thylen, B. Stolz, K. Bergvall, J.-E. Falk, and H. Olofsson, "Integrated optics 4×4 switch matrix with digital optical switches," Electron. Lett., 26 (1990) 4–5.
92. J.-F. Vinchant, M. Renaud, M. Erman, J. L. Peyre, P. Jarry, and P. Pagnod-Rossiaux, "InP digital optical switch: key element for guided-wave photonic switching," IEE Proc., 140 (1993) 301–307.
93. A. Borreman, T. Hoekstra, and M. Diemeer, "Polymeric 8 × 8 digital optical switch matrix," Tech. Digest ECOC'96, Paper ThD3.2 (1996).
94. H. Okayama and M. Kawahara, "Ti:$LiNbO_3$ digital optical switch matrices," Electron. Lett., 29 (1993) 765–766.
95. Y. Nakabayashi, M. Kitamura, and T. Sawano, "DC-drift free-polarization independent Ti:$LiNbO_3$ 8×8 optical matrix switch," Tech. Digest ECOC'96, Paper ThD2.4 (1996).
96. K. Hamamoto, T. Anan, K. Komatsu, M. Sugimoto, and I. Mito, "First 8 × 8 semiconductor optical matrix switches using GaAs/AlGaAs electro-optic guided-wave directional couplers," Electron. Lett., 28 (1992) 441–443.
97. H. Okayama, Y. Okabe, T. Arai, T. Kamijoh, and T. Tsuruoka, "Two-module stage optical switch network," J. Lightwave Tech., 18 (2000) 469–476.
98. M. Kondo, N. Takado, K. Komatsu, and Y. Ohta, "32 switch-elements integrated low-crosstalk $LiNbO_3$ 4 × 4 optical matrix switch," Tech. Digest IOOC-ECOC'85 (1985) 361–364.
99. S. Yu, M. Owen, R. Varrazza, R. V. Penty, and I. H. White, "Demonstration of high-speed optical packet routing using vertical coupler crosspoint space switch array," Electron. Lett., 36 (2000) 556–558.
100. I. Betty, R. Rouisiana-Webb, and C. Wu, "A robust, low-crosstalk, InGaAsP/InP total-internal-reflection switch for optical cross-connect applications," Tech. Digest Photonics in switching, Paper JWA2 (1999).
101. G. A. Fish, B. Mason, L. A. Coldren, and S. P. DenBaars, "Monolithic InP optical crossconnects: 4 × 4 and beyond," Tech. Digest Photonics in switching, Paper JWB2 (1999).
102. T. Kirihara, M. Ogawa, H. Inoue, H. Kodera, and K. Ishida, "Lossless and low-crosstalk characteristics in an InP-based 4 × 4 optical switch with integrated single-stage optical amplifiers," IEEE Photon. Tech. Lett., 6 (1994) 218–221.
103. C. Clos, "A study of non-blocking switching networks," Bell Sys. Tech. J., 32 (1953) 406–424.
104. C. Burke, M. Fujiwara, M. Yamaguchi, H. Nishinoto, and H. Honmou, "128 line photonic switching system using $LiNbO_3$ switch matrices and semiconductor traveling wave amplifiers," J. Lightwave Tech., 10 (1992) 610–615.

105. H. S. Hinton, An Introduction to Photonic Switching Fabrics, Chapter 6.4 (Plenum Press, New York, 1993).
106. E. J. Murphy, T. O. Murphy, A. F. Ambrose, R. W. Irvin, B. H. Lee, P. Peng, G. W. Richards, and A. Yorinks, "16 × 16 strictly nonblocking guided-wave optical switching system," J. Lightwave Tech., 14 (1996) 352–358.
107. A. Pattavina, Switching Theory, Chapters 2–4 (John Wiley & Sons, New York, 1998).
108. R. I. MacDonald, "Large modular expandable optical switching matrices," IEEE Photon. Tech. Lett., 11 (1999) 668–670.
109. S. Okamoto, A. Watanabe, and K. Sato, "Otpical path cross-connect node architectures for photonic transport network," J. Lightwave Tech., 14 (1996) 1410–1422.
110. H. Okayama and M. Kawahara, "Prototype 32 × 32 optical switch matrix," Electron. Lett., 30 (1994) 1128–1129.
111. S. Thaniyavarn, J. Lin, W. Dougherty, T. Traynor, K. Chiu, G. Abbas, M. LaGasse, W. Chaczenko, and M. Hamilton, "Compact, low insertion loss 16 × 16 optical switch array modules," Tech. Digest OFC'97, Paper TuC1 (1997).
112. H. Higashino, T. Kawaguchi, H. Adachi, T. Makino, and O. Yamazaki, "High speed optical TIR switches using PLZT thin-film waveguide on sapphire," Supplements to Jpn. J. Appl. Phys., 24, paper P3-403 (1985) 284–286.
113. K. Nashimoto, H. Moriyama, S. Nakamura, M. Watanabe, T. Morikawa, E. Osakabe, and K. Haga, "PLZT electro-optic waveguide and switches," Tech. Digest OFC2001, paper PD10 (2000).
114. P. J. Duthie and C. Edge, "A polarization independent guided-wave $LiNbO_3$ electrooptic switch employing polarization diversity," IEEE Photon. Tech. Lett., 3 (1991) 136–137.
115. T. Pohlman, A. Neyer, and E. Voges, "Polarization-independent switches on $LiNbO_3$," Tech. Digest Integrated Photonics Research (IPR'90), Paper MH7 (1990).
116. H. Okayama, A. Matoba, R. Shibuya, and T. Ishida, "Polarization independent optical switch with cascaded optical switch matrices," Electron. Lett., 24 (1988) 959–960.
117. D. W. Yoon, O. Eknoyan, and H. F. Taylor, "Polarization-independent $LiTaO_3$ guided-wave electrooptic switches," J. Lightwave Tech., 8 (1990) 160–163.
118. J. L. Nightingale, J. S. Vrhel, and T. E. Salac, "Low-voltage, polarization-independent optical switch in Ti-indiffused Lithium Niobate," Tech. Digest IGWO'89, Paper MAA3 (1989).
119. Y. Nakabayashi, "Polarization independent-DC drift free Ti: $LiNbO_3$ matrix optical switches," Tech. Digest OECC'98, Paper 13C2-1 (1998).
120. G. Soulage, A. Jourdan, P. Doussiere, M. Bachman, J. K. Emery, J. Da Loura, and M. Sotom, "4 × 4 space-switch based on clamped-gain semiconductor

optical amplifiers in a 16 × 10 Gbit/s WDM experiment," Tech. Digest ECOC'96, Paper ThD2.1 (1996).
121. K. Hamamoto, T. Sasaki, T. Matsumoto, and K. Komatsu, "Insertion-loss-free 1 × 4 optical switch fabricated using bandgap-energy-controlled selective MOVPE," Electron. Lett., 32 (1996) 2265–2266.
122. T. Kato, J. Sasaki, T. Shimoda, H. Hatakeyama, T. Tamanuki, S. Kitamura, M. Yamaguchi, T. Sasaki, K. Komatsu, M. Kitamura, and M. Itoh, "Hybrid-integrated 4 × 4 optical matrix switch module on silica based planar waveguide plat from," IEICE Trans. Electtron., 82-C (1999) 305–312.
123. N. Yoshimoto, Y. Shibata, S. Oku, S. Kondo, and Y. Noguchi, "Design and demonstration of polarization-independent Mach-Zehnder switch using a lattice-matched InGaAlAs/InAlAs MQW and deep-etched high-mesa waveguide structure," J. Lightwave Technol., 17 (1999) 1662–1669.
124. D. H. P. Maat, Y. C. Zhu, F. H. Gruen, H. van Brag, H. J. Frankena, and X. J. Leijtens, "Polarization-independent dilated InP based space switch with low crosstalk," IEEE Photon. Tech. Lett., 12 (2000) 284–286.
125. R. A. Soref, "Silicon-based optoelectronics," Proc. IEEE, 81 (1993) 1687–1706.
126. E. van Tomme, P. van Daele, R. Baets, G. R. Mohlmann, and M. B. J. Diemeer, "Guided wave modulators and switches fabricated in electro-optic polymers," J. Appl. Phys., 69 (1991) 6273–6276.
127. S.-Y. Shin, S.-S. Lee, M.-C. Oh, W.-Y. Hwang, and J.-J. Kim, "Polymer waveguide devices fabricated by electric poling," Tech. Digest CLEO/Pacific Rim, paper FB1 (1997).
128. H. Terui, M. Kobayashi, and T. Edahiro, "4 × 4 guided wave liquid crystal optical switch," Natl. Conv. IECE, paper 1047 (1983) (in Japanese).
129. T. Goh, M. Yasu, K. Hattori, A. Himeno, M. Okuno, and Y. Ohmori, "Low-loss and high-extinction-ratio silica-based strictly nonblocking 16 × 16 thermooptic matrix switch," IEEE Photon. Tech. Lett., 10 (1998) 810–812.
130. Y. Baek, R. Schiek, G. I. Stegeman, G. Krijnen, I. Baumann, and W. Sohler, "All-optical integrated Mach-Zehnder switching due to cascaded nonlinearities," Appl. Phys. Lett., 68 (1996) 2055–2057.
131. I. Yokohama, M. Asobe, A. Yokoo, H. Itoh, and T. Kaino, "All-optical switching by use of cascading of phase-matched sum-frequency generation and difference-frequency generation processes," J. Opt. Soc. Am. B, 14 (1997) 3368–3377.
132. H. Kanbara, H. Itoh, M. Asobe, K. Noguchi, H. Miyazawa, T. Yanagawa, and I. Yokohama, "All-optical switching based on cascading of second-order nonlinearities in a periodically poled Titanium-diffused Lithium Niobate waveguide," IEEE Photon. Tech. Lett., 11 (1999) 328–330.
133. C. Q. Xu, H. Okayama, and M. Kawahara, "1.5 μm band efficient broadband wavelength conversion by difference frequency generation in a periodically

domain-inverted $LiNbO_3$ channel waveguide," Appl. Phys. Lett., 63 (1993) 3559–3561.

134. S. J. B. Yoo, "Wavelength conversion technologies for WDM network applications," J. Lightwave Tech., 14 (1996) 955–966.
135. K. R. Parameswaran, M. Fujimura, M. H. Chou, and M. M. Fejer, "Low-power all-optical gate based on sum frequency mixing in APE waveguides in PPLN," IEEE Photon. Tech. Lett., 12 (2000) 654–656.
136. M. H. Chou, K. R. Parameswaran, I. Brener, and M. M. Fejer, "Optical signal processing and switching with second-order nonlinearities in waveguides," Tech. Digest OECC2000, Paper 13C1-1 (2000).
137. H. Okayama, C. Q. Xu, and T. Kamijoh, "Architecture for nonblocking wavelength switch using multi-wavelength conversion device," Tech. Digest OEC'96, Paper 17P-10 (1996).
138. N. Antoniades, S. J. B. Yoo, K. Bala, G. Ellinas, and T. E. Stern, "An architecture for a wavelength-interchanging cross-connect utilizing parametric wavelength converters," J. Lightwave Technol., 17 (1999) 1113–1125.
139. K. Tajima, "Ultrafast demultiplexing and bit-wise digital operation with symmetric Mach-Zehnder all-optical switch," Tech. Digest OECC2000, Paper 14D1–2 (2000).
140. H. N. Poulsen, A. T. Clausen, A. Buxens, L. Oxenloewe, C. Peucheret, C. Rasmussen, F. Liu, J. Yu, A. Kloch, T. Fjelde, D. Wolfson, and P. Jeppesen, "Ultra fast all-optical signal processing in semiconductor and fiber based devices," Tech. Digest ECOC2000, 3 (2000) 59–62.
141. K. Habara and K. Kikuchi, "Optical time-division space switches using tree-structured directional couplers," Electron. Lett., 21 (1985) 631–632.
142. E. J. Murphy, G. W. Richards, T. O. Murphy, M. T. Fatehi, and S. S. Bergstein, "Ti: $LiNbO_3$ photonic switch modules for large, strictly non-blocking architectures," Tech. Digest Photonic in Switching, Paper PMD3 (1993).
143. T. O. Murphy, E. J. Murphy, and R. W. Irvin, "An 8×8 Ti: $LiNbO_3$ polarization-independent photonic switch," Tech. Digest ECOC'94, 2 (1994) 549–552.
144. H. Nishimoto, S. Suzuki, and M. Kondo, "Polarisation-independent $LiNbO_3$ 4×4 matrix switch," Electron. Lett., 24 (1988) 1122–1123.
145. H. Nishimoto, M. Iwasaki, S. Suzuki, and M. Kondo, "Polarisation independent $LiNbO_3$ 8×8 matrix switch," IEEE Photon. Tech. Lett., 2 (1990) 634–636.
146. A. C. O'Donnell, "Polarisation independent 1×16 and 1×32 Lithium Niobate optical switch matrices," Electron. Lett., 27 (1991) 2349–2350.
147. A. C. O'Donnell and N. J. Parsons, "1×16 Lithium Niobate optical switch matrix with integral TTL compatible drive electronics," Electron. Lett., 27 (1991) 2367–2368.

148. J. E. Watson, M. A. Milbrodt, and T. C. Rice, "A polarization-independent 1 × 16 guided-wave optical switch integrated on Lithium Niobate," J. Lightwave Technol., LT-4 (1986) 1717–1721.
149. M. Kondo, Y. Ohta, Y. Tanisawa, T. Aoyama, and R. Ishikawa, "Low drive voltage and low loss polarization independent $LiNbO_3$ optical waveguide switches," Electron. Lett., 23 (1987) 1167–1169.
150. R. Krahenbuhl, R. Kyburz, W. Vogt, M. Bachmann, T. Brenner, E. Gini, and H. Melchior, "Low-loss polarisation-insensitive InP-InGaAsP optical space switches for fiber optical communication," IEEE Photon. Tech. Lett., 8 (1996) 632–634.
151. A. Greil, H. Haltenorth, and F. Taumberger, "Optical 4 × 4 InP switch module with fiber-lens-arrays for coupling," Tech. Digest ECOC'92, Paper WeP2.16 (1992).
152. P. J. Duthie, H. Shaw, M. J. Wale, and I. Bennion, "Guided wave switch array using electro-optic and carrier depletion effects in Indium Phosphide," Electron. Lett., 27 (1991) 1747–1748.
153. D. A. O. Davies, P. S. Mudhar, M. A. Fisher, D. A. H. Mace, and M. J. Adams, "Integrated loss less InP/InGaAsP 1 to 4 optical switch," Electron. Lett., 28 (1992) 1521–1522.
154. J.-F. Vinchant, A. Jourdan, J. Le Bris, G. Soulage, T. Fillion, and E. Grard, "InP 4 × 1 digital-optical-switch module for multiwavelength cross-connect applications," Tech. Digest OFC'95, paper ThK2 (1995).
155. M. N. Khan, B. I. Miller, E. C. Burrows, and C. A. Burrus, "Integrated high-speed 1 × 4 digital Y-branch switch array with passive taper fanouts for fibre interconnection," Electron. Lett., 35 (1999) 484–485.
156. H. Okayama, "Ring light beam deflector," Tech. Digest ODF2002, Paper TP11, (2002).

Part 3 | Optical Amplifier Technologies

Chapter 8 | Semiconductor Optical Amplifiers (SOA)

Niloy Dutta

Department of Physics and Photonics Research Center,
University of Connecticut, Storrs, CT 06269-3046, USA

8.1. Introduction

Significant advances in research and, development and application of semiconductor lasers, amplifiers, and modulators have occurred over the last decade. The fiber optic revolution in telecommunication, which provided several orders of magnitude improvement in transmission capacity at low cost, would not have been possible without the development of reliable semiconductor lasers.

Optical amplifier, as the name implies, is a device that amplifies an input optical signal. The amplification factor or gain can be higher than 1000 (>30 dB) in some devices. There are two principal types of optical amplifiers. They are the semiconductor optical amplifier and the fiber optical amplifier. In a semiconductor optical amplifier, amplification of light takes place when it propagates through a semiconductor medium fabricated in a waveguide form. In a fiber amplifier, amplification of light occurs when it travels through a fiber doped with rare earth ions (such as Nd^+, Er^+ etc.). Semiconductor optical amplifiers (SOA) are typically less than 1 mm in length whereas fiber amplifiers are typically 1 to 100 m in length. The operating principals, design, fabrication, and performance characteristics of InP-based semiconductor amplifiers are described in this chapter.

WDM TECHNOLOGIES: PASSIVE
OPTICAL COMPONENTS
$35.00

ISBN: 0-12-225262-4

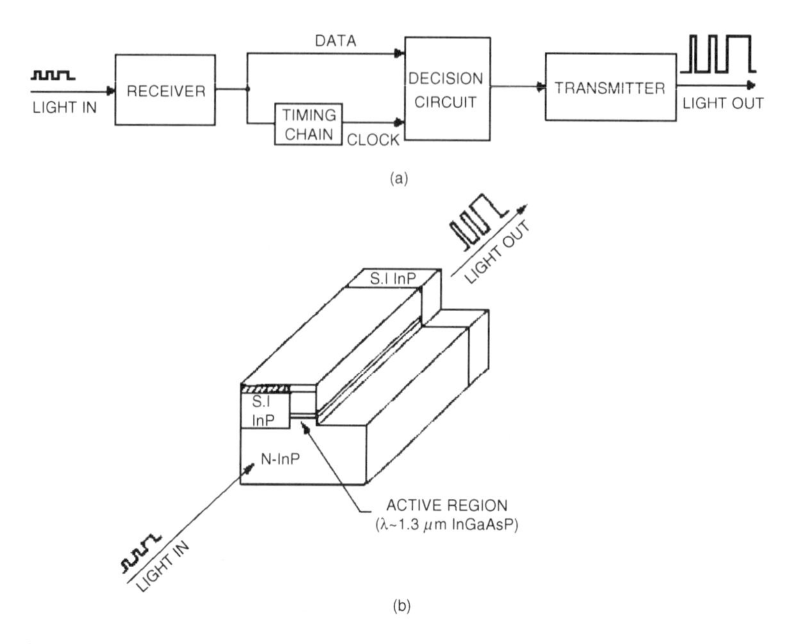

Fig. 8.1 (a) Block diagram of a lightwave regenerator, (b) schematic of a semiconductor optical amplifier.

In a lightwave transmission system, as the optical signal travels through the fiber, it weakens and gets distorted. Regenerators are used to restore the optical pulses to their original form. Figure 8.1(a) shows the block diagram of a typical lightwave regenerator. Its main components are an optical receiver, an optical transmitter, and electronic timing and decision circuits. Optical amplifiers can nearly restore the original optical pulses and thereby increase the transmission distance without using conventional regenerators. An example of a semiconductor amplifier that functions as a regenerator is shown schematically in Fig. 8.1(b). The semiconductor amplifiers need external current to produce gain, and fiber amplifiers need pump lasers for the same purpose. Because of its simplicity, an optical amplifier is an attractive alternative for new lightwave systems.

Semiconductor optical amplifier [1–7] is a device very similar to a semiconductor laser. Hence their operating principals, fabrication, and design are also similar. When the injection current is below threshold, the laser acts as an optical amplifier for incident light waves; above threshold it undergoes oscillation.

8.2. Semiconductor Optical Amplifier Designs

Semiconductor optical amplifiers can be classified into two categories; the Fabry-Perot (FP) amplifier and the traveling wave (TW) amplifier. A FP amplifier has considerable reflectivity at the input and output ends, resulting in resonant amplification between the end mirrors. Thus, a FP amplifier exhibits very large gain at wavelengths corresponding to the longitudinal modes of the cavity. The TW amplifier, by contrast, has negligible reflectivity at each end, resulting in signal amplification during a single pass. The optical gain spectrum of a TW amplifier is quite broad and corresponds to that of the semiconductor gain medium. Most practical TW amplifiers exhibit some small ripple in the gain spectrum, arising from residual facet reflectivities. TW amplifiers are more suitable for system applications. Therefore much effort has been devoted over the last few years to fabricating amplifiers with very low effective facet reflectivities. Such amplifier structures either utilize special low effective reflectivity dielectric coatings, or have tilted or buried facets. Fabrication and performance of these devices are described later.

Extensive work on optical amplifiers was carried out in the 1980s using the AlGaAs material system. Much of the recent experimental work on semiconductor optical amplifiers has been carried out using the InGaAsP material system with the optical gain centered around 1.3 μm or 1.55 μm. The amplifiers used in lightwave system applications, either as preamplifiers in front of a receiver or as in-line amplifiers as a replacement for regenerators, must also exhibit equal optical gain for all polarizations of the input light. In general, the optical gain in a waveguide is polarization dependent although the material gain is independent of polarization for bulk semiconductors. This arises from unequal mode confinement factors for the light polarized parallel to the junction plane (TE mode) and that for light polarized perpendicular to the junction plane (TM mode). For thick active regions, the confinement factors of the TE and TM mode are nearly equal. Hence the gain difference between the TE and TM modes is smaller for amplifiers with a thick active region (Fig. 8.2) [4].

The residual facet reflectivity produces a variation in gain as a function of wavelength of a semiconductor amplifier. This variation in gain appears in the form of a ripple (periodic variation in gain at the cavity mode spacing). The calculated gain ripple as a function of residual reflectivity (R) for two values of internal gain is shown in Fig. 8.3.

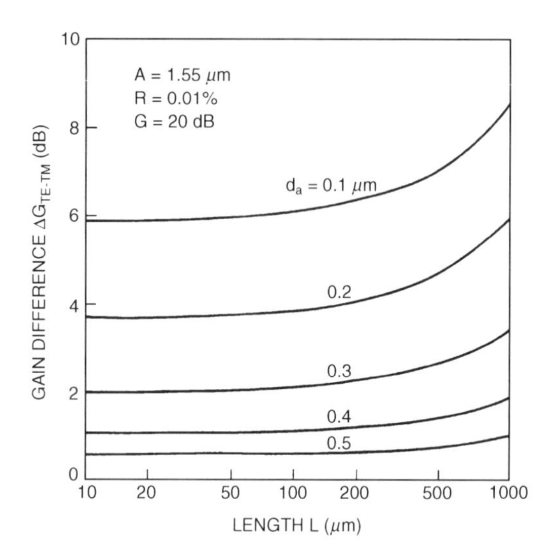

Fig. 8.2 The optical gain difference between the TE and TM mode of a semiconductor amplifier is plotted as a function of device length for different active layer thicknesses (O'Mahoney [7]).

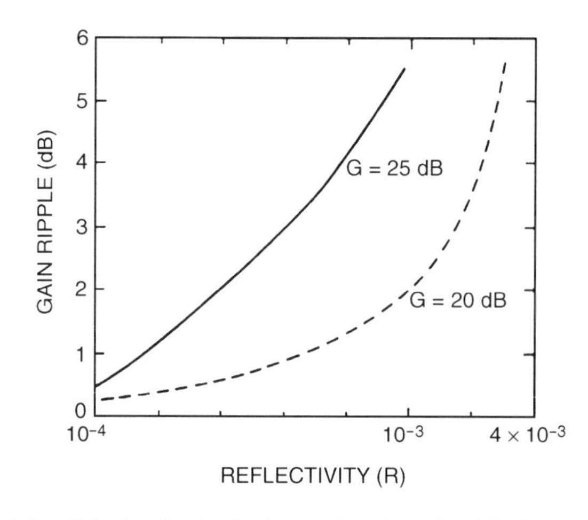

Fig. 8.3 Calculated gain ripple as a function of residual reflectivity.

8.2.1. LOW REFLECTIVITY COATINGS

A key factor for good performance characteristics (low gain ripple and low polarization selectivity) for TW amplifiers is very low facet reflectivity [7]. The reflectivity of cleaved facets can be reduced by dielectric coating.

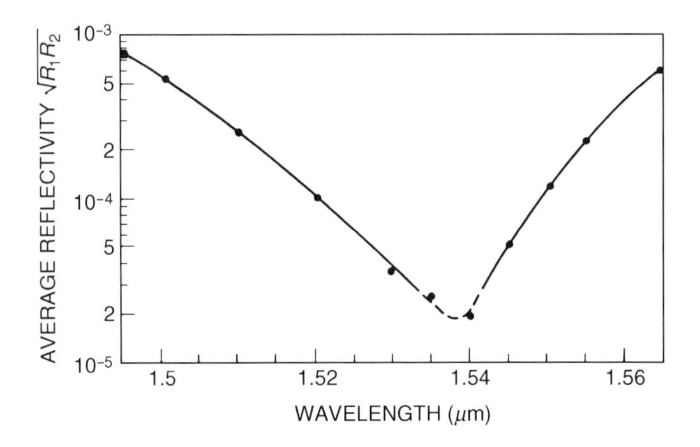

Fig. 8.4 The measured reflectivity is plotted as a function of wavelength.

For plane waves incident on an air interface from a medium of refractive index n, the reflectivity can be reduced to zero by coating the interface with a dielectric whose refractive index equals $n^{1/2}$ and whose thickness equals $\lambda/4$. However, the fundamental mode propagating in a waveguide is not a plane wave and therefore the $n^{1/2}$ law only provides a guideline for achieving very low ($\sim 10^{-4}$) facet reflectivity by dielectric coatings. In practice, very low facet reflectivities are obtained by monitoring the amplifier performance during the coating process. The effective reflectivity can then be estimated from the ripple at the Fabry-Perot mode spacings, caused by residual reflectivity, in the spontaneous emission spectrum. The result of such an experiment is shown in Fig. 8.4 [7]. The reflectivity is very low ($<10^{-4}$) only in a small range of wavelengths. Although laboratory experiments have been carried out using amplifiers that rely only on low reflectivity coatings for good performance, the critical nature of the thickness requirement and a limited wavelength range of good antireflection (AR) coating led to the investigation of alternate schemes as discussed subsequently.

8.2.2. BURIED FACET AMPLIFIERS

The principal feature of the buried facet (also known as the window structure) optical amplifiers relative to AR-coated cleaved facet devices is a polarization-independent reduction in mode reflectivity due to the buried facet, resulting in better control in achieving polarization-independent gain.

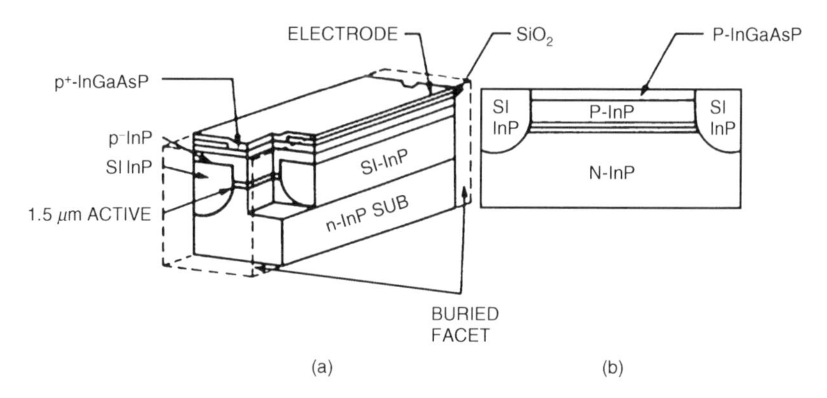

Fig. 8.5 Schematic of a buried facet optical amplifier.

A schematic cross-section of a buried-facet optical amplifier is shown in Fig. 8.5. Current confinement in this structure is provided by semi-insulating Fe-doped InP layers grown by the MOCVD growth technique. Fabrication of this device involves a procedure similar to that used for lasers. The first four layers are grown on a (100)-oriented n-InP substrate by MOCVD. These layers are (i) an n-InP buffer layer, (ii) an undoped InGaAsP ($\lambda \sim 1.55\ \mu$m) active layer, (iii) a p-InP cladding layer, and (iv) a p-InGaAsP ($\lambda \sim 1.3\ \mu$m) layer. Mesas are then etched on the wafer along the [110] direction with 15-μm-wide channels normal to the mesa direction using a SiO_2 mask. The mask is needed for buried-facet formation. Semi-insulating Fe-doped InP layers are then grown around the mesas by MOCVD with the oxide mask in place. The oxide mask and p-InGaAsP layers are removed and p-InP and p-InGaAsP ($\lambda \sim 1.3\ \mu$m) contact layers are then grown over the entire wafer by the MOCVD growth technique. The wafer is processed using standard methods and cleaved to produce 500-μm-long buried-facet chips with $\sim$7-μm-long buried facets at each end. Chip facets are then AR-coated using a single-layer film of ZrO_2.

Fabrication of cleaved-facet devices follows the procedure just described, except that the mesas are continuous with no channels separating them. This change is needed for defining the buried-facet regions. The semi-insulating layer, in both types of devices, provides current confinement and lateral index guiding. For buried-facet devices it also provides the buried-facet region.

The effective reflectivity of a buried facet decreases with increasing separation between the facet and the end of the active region. The effective

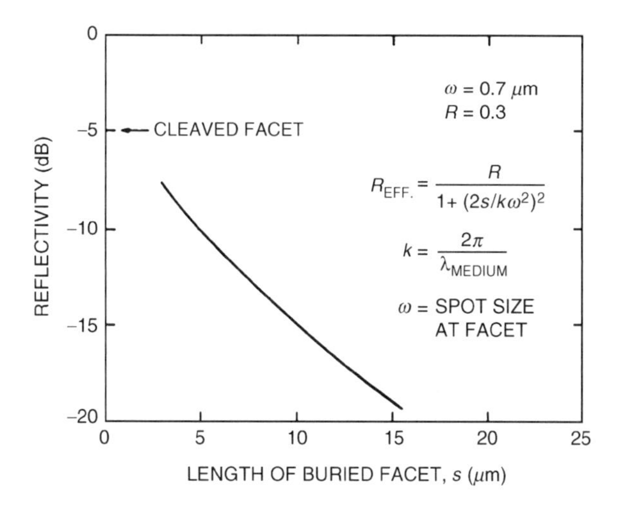

Fig. 8.6 Effective reflectivity of a buried-facet configuration plotted as a function of the length of the buried facet.

reflectivity (R_{eff}) of such a facet can be calculated by using a Gaussian beam approximation for the propagating optical mode. It is given by [8]

$$R_{\text{eff}} = R/(1 + (2S/kw^2)^2), \tag{8.1}$$

where R is the reflectivity of the cleaved facet, S is the length of the buried-facet region, $k \sim 2\pi/\lambda$, where λ is the optical wavelength in the medium, and w is the spot size at the facet. The calculated reflectivity using $w = 0.7$ μm and $R = 0.3$ for an amplifier operating near 1.55 μm is less than 10^{-2} for buried-facet lengths larger than $\sim$15 μm (Fig. 8.6).

Although increasing the length of the buried-facet region decreases the reflectivity, if the length is too long the beam emerging from the active region will strike the top metallized surface, producing multiple peaks in the far-field pattern, a feature not desirable for coupling into a single-mode fiber. The beam waist w of a Gaussian beam after traveling a distance z is given by the equation [9]

$$w^2(z) = w_0^2\left(1 + \left(\lambda z/\pi w_0^2\right)^2\right), \tag{8.2}$$

where w_0 is the spot size at the beam waist and λ is the wavelength in the medium. Because the active region is about 4 μm from the top surface of the chip, it follows from the preceding equations that the length of the buried-facet region must be less than 12 μm for single-lobed far-field operation.

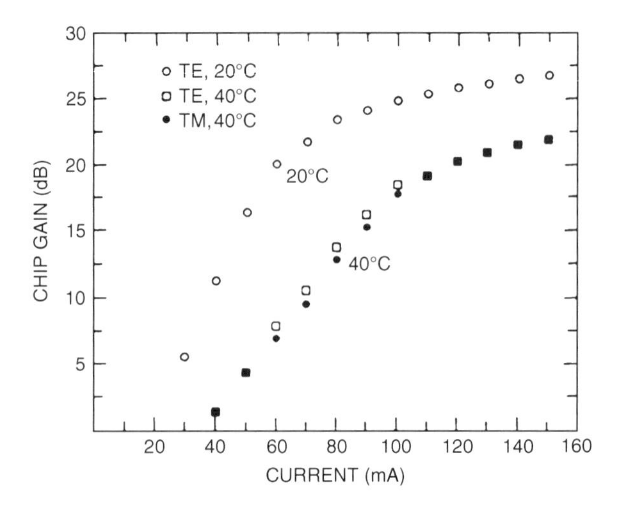

Fig. 8.7 Measured chip gain as a function of amplifier current.

The optical gain is determined by injecting light into the amplifier and measuring the output. The internal gain of an amplifier chip as a function of current at two different temperatures is shown in Fig. 8.7 [10]. Open circles and squares represent the gain for a linearly polarized incident light with the electric field parallel to the p-n junction in the amplifier chip (TE mode). Solid circles represent the measured gain for the TM mode at 40°C. Measurements were done for low input power (−40 dBm), so that the observed saturation is not due to gain saturation in the amplifier, but rather to carrier loss caused by Auger recombination. Note that the optical gain for the TE and TM input polarizations are nearly equal. Figure 8.8 shows the measured gain as a function of input wavelength for TE-polarized incident light. The modulation in the gain (gain ripple) with a periodicity of 0.7 nm is due to residual facet reflectivities. The measured gain ripple for this device is less than 1 dB. The estimated facet reflectivity from the measured gain ripple of 0.6 dB at 26 dB internal gain is 9×10^{-5}. The 3-dB bandwidth of the optical gain spectrum is 45 nm for this device. It has been shown that the gain ripple and polarization dependence of gain correlate well with the ripple and polarization dependence of the amplified spontaneous emission spectrum. Measurements of amplified spontaneous emission are much simpler to make than gain measurements, and provide a good estimate of the amplifier performance [8, 10].

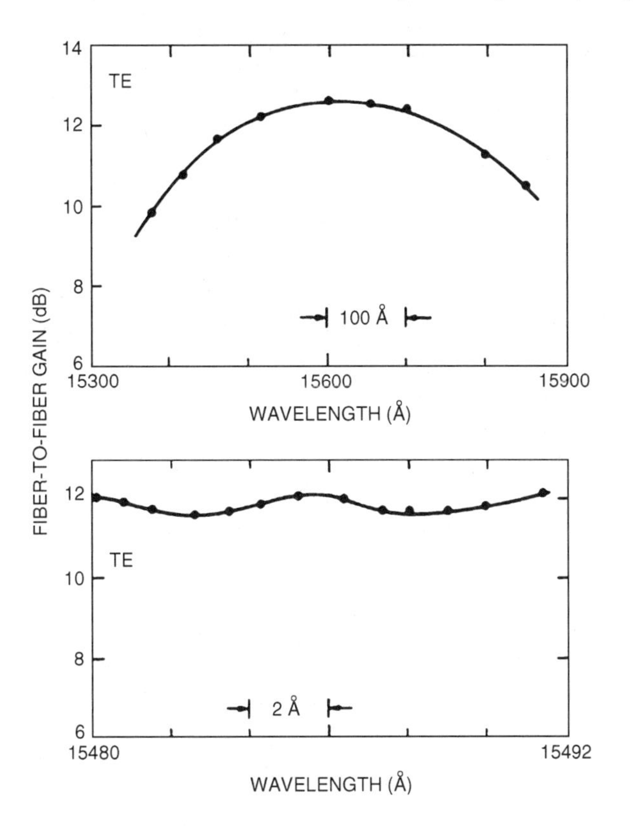

Fig. 8.8 Measured gain as a function of wavelength for the TE mode.

8.2.3. TILTED FACET AMPLIFIERS

Another way to suppress the resonant modes of the Fabry-Perot cavity is to slant the waveguide (gain region) from the cleaved facet so that the light incident on it internally does not couple back into the waveguide [11]. The process essentially decreases the effective reflectivity of the tilted facet relative to a normally cleaved facet. The reduction in reflectivity as a function of the tilt angle is shown in Fig. 8.9 for the fundamental mode of the waveguide.

The schematic of a tilted facet optical amplifier is shown in Fig. 8.10 [11]. Waveguiding along the junction plane is weaker in this device than that for the strongly index guided buried heterostructure device. Weak index guiding for the structure of Fig. 8.10 is provided by a dielectric

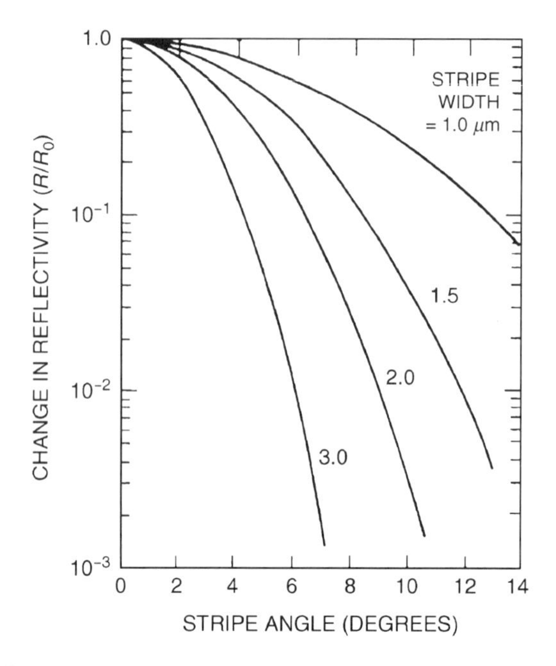

Fig. 8.9 Calculated change in reflectivity as a function of tilt angle of the facet.

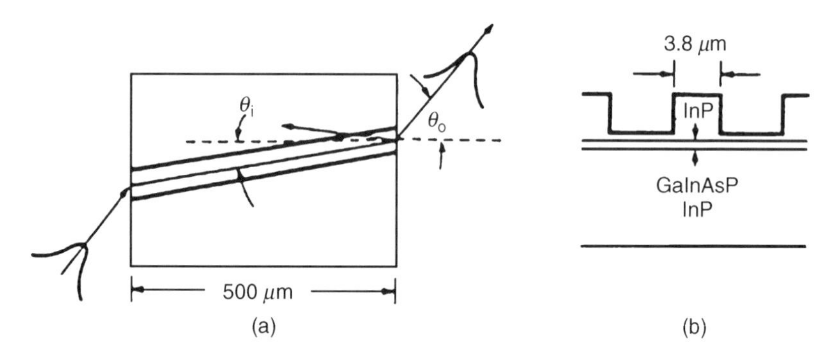

Fig. 8.10 Schematic of a tilted facet amplifier (Zah *et al.* [11]).

defined ridge. The fabrication of the device follows a procedure similar to that described previously.

The measured gain as a function of injection current for TM and TE polarized light for a tilted facet amplifier is shown in Fig. 8.11 [11]. Optical gains as high as 30 dB have been obtained using tilted facet amplifiers. Although the effective reflectivity of the fundamental mode decreases with increasing tilt of the waveguide, the effective reflectivity of the higher-order

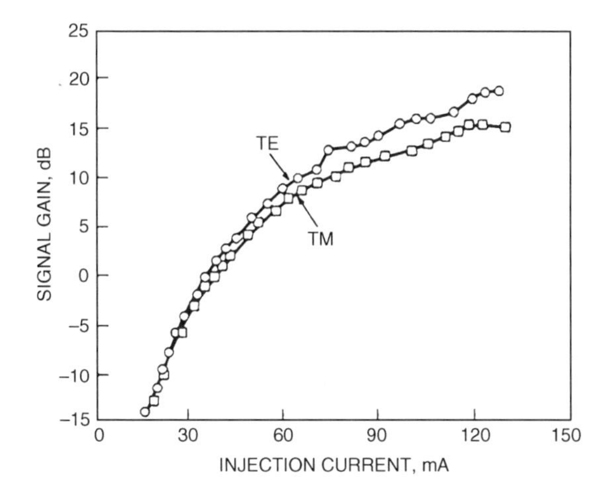

Fig. 8.11 Measured gain is plotted as a function of injection current (Zah *et al.* [11]).

modes increases. This may cause appearance of higher-order modes at the output (which may reduce fiber-coupled power significantly), especially for large ridge widths.

8.3. Integrated Amplifiers and Photonic Integrated Circuits

There have been a significant number of developments in the technology of integration of semiconductor lasers, amplifiers, and other related devices on the same chip. These chips allow higher levels of functionality than that achieved using single devices. For example, laser and electronic drive circuits have been integrated, serving as simple monolithic transmitters. The name photonic integrated circuits (PIC) is generally used when all the integrated components are photonic devices, e.g., lasers, detectors, amplifiers, modulators, and couplers [12, 13].

8.3.1. MQW AMPLIFIERS

For amplifier applications where the amplifier is integrated with another photonic device such as a laser, it is not necessary to have polarization-independent gain. Because multiquantum well (MQW) materials generally have higher gain than regular double heterostructure material, MQW active regions are generally used for amplifiers that are integrated in a

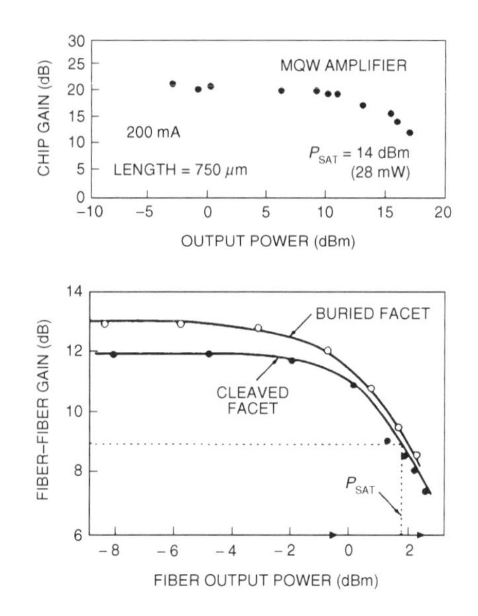

Fig. 8.12 Performance of optical amplifiers with regular and MQW active region.

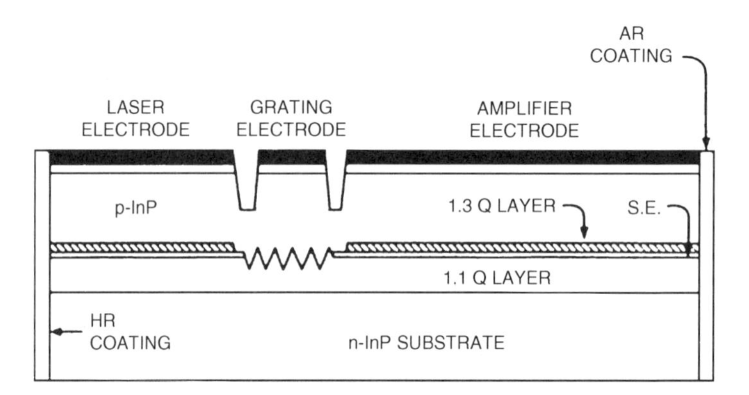

Fig. 8.13 Schematic of a photonic integrated circuit with a DBR (distributed Bragg reflector) laser and amplifier.

photonic circuit. The saturation output power—the power at which the gain decreases by 3 dB from its value at low power—is higher for MQW amplifiers. A comparison of regular and MQW amplifiers is shown in Fig. 8.12.

Semiconductor amplifiers are often integrated with tunable lasers to increase the output power. The schematic of such an integrated device is show in Fig. 8.13. The waveguide layer with a band gap of ~1.1 μm couples the laser and the amplifier devices. For the structure shown, both

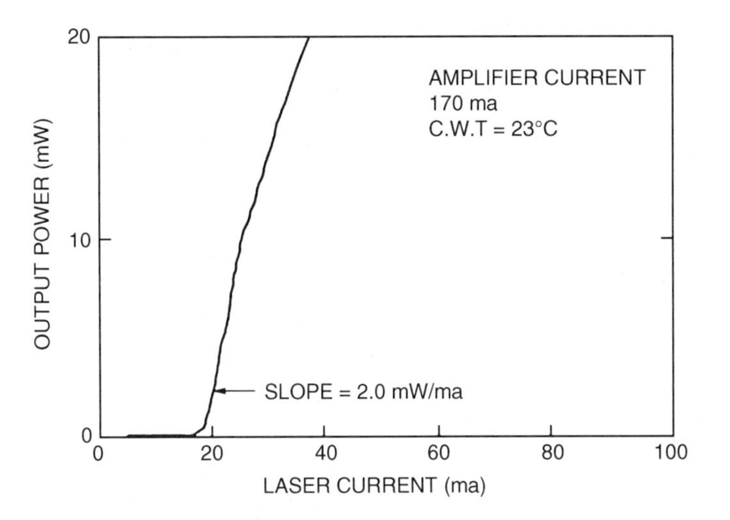

Fig. 8.14 Power output as a function of laser current of the device shown in Fig. 8.13.

the laser and the amplifier have a MQW active region. The light vs. current characteristics of the laser with the amplifier driven at 170 mA is shown in Fig. 8.14.

8.3.2. MULTICHANNEL WDM SOURCES WITH AMPLIFIERS

An alternative to single-channel very high speed (>20 Gb/s) data transmission for increasing transmission capacity is multichannel transmission using wavelength division multiplexing (WDM) technology. In a WDM system many (4, 8, 16 or 32) wavelengths carrying data are optically multiplexed and simultaneously transmitted through a single fiber. The received signal with many wavelengths is optically demultiplexed into separate channels, which are then processed electronically in a conventional form. Such a WDM system needs transmitters with many lasers at specific wavelengths. It is desirable to have all of these laser sources on a single chip for compactness and ease of fabrication, as in electronic integrated circuits.

Figure 8.15 shows the schematic of a photonic integrated circuit with multiple lasers for a WDM source [14]. This chip has 4 individually addressable DFB lasers, the outputs of which are combined using a waveguide-based multiplexer. Because the waveguide multiplexer has an optical loss of ~8 dB, the output of the chip is further amplified using

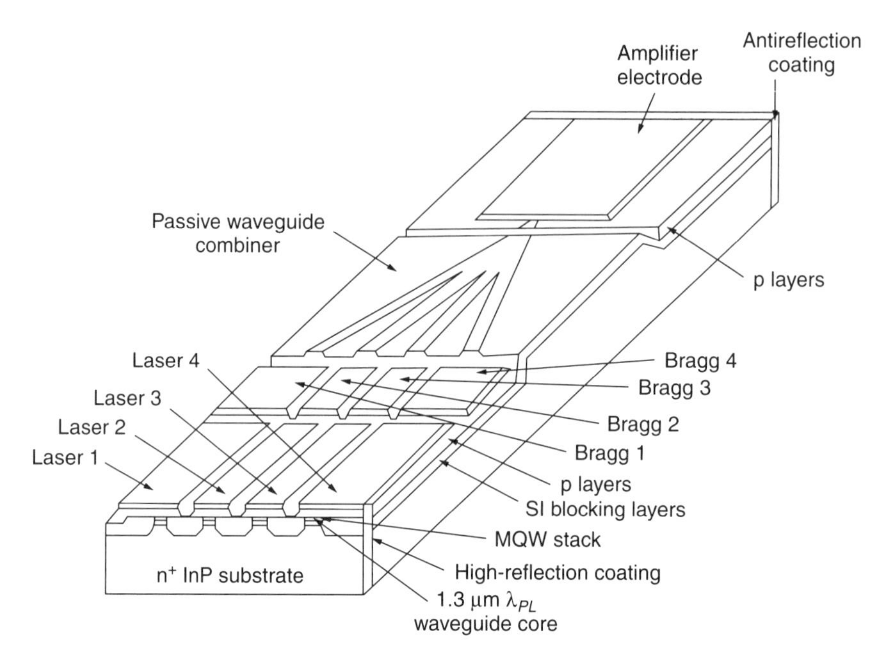

Fig. 8.15 Schematic of a photonic integrated circuit with multiple lasers for a WDM source.

a semiconductor amplifier. The laser output in the waveguide is TE polarized and hence an amplifier with a multiquantum well absorption region, which has a high saturation power, is integrated in this chip. Many similar devices that use SOA for output amplification have been fabricated [12, 13].

8.3.3. SPOT SIZE CONVERSION (SSC)

A typical amplifier diode has too wide a beam pattern ($\sim 30^\circ \times 40^\circ$) for good mode matching to a single-mode fiber. This results in a loss of power coupled to the fiber. Thus, an amplifier whose output spot size is expanded to match an optical fiber is an attractive device for low loss coupling to the fiber without a lens, and for wide alignment tolerances. Several researchers have reported such devices using lasers [16, 17]. Generally they involve producing a vertically and laterally tapered waveguide near the output facet of the laser. The tapering needs to be done in an adiabatic fashion so as to reduce the scattering losses. The schematic of a SSC laser is shown in Fig. 8.16 [17]. The amplifier designs are similar. The laser is fabricated using two MOCVD growth steps. The SSC section is about 200 μm long.

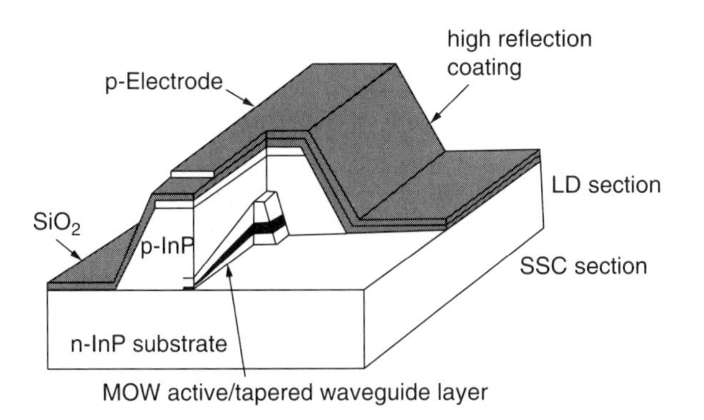

Fig. 8.16 Schematic of a spot size converter (SSC) laser (Yamazaki *et al.* [17]).

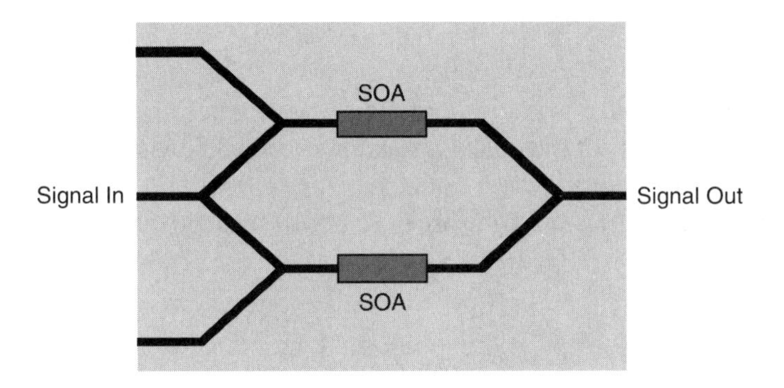

Fig. 8.17 Schematic of a Mach-Zehnder interferometer with integrated SOAs.

The waveguide thickness is narrowed along the cavity from 300 nm in the active region to ~100 nm in the region over the length of the SSC section. The laser emits near 1.3 μm, and has a multiquantum well active region and a laser section length of 300 μm. Beam divergences of 9° and 10° in the horizontal and vertical direction have been reported for SSC devices [17].

8.3.4. MACH-ZEHNDER INTERFEROMETER

Semiconductor optical amplifiers incorporated into Mach-Zehnder interferometer (MZI) or Michelsen interferometer configurations have been used for wavelength conversion and to demultiplex high-speed time division multiplexed optical signals [18, 19]. The schematic of the device is shown in Fig. 8.17. The input signal beam is split into two beams by an input y-junction, which propagate through semiconductor amplifiers positioned in the upper and lower arms of the interferometer. These beams then merge and interfere at an output y-junction. The phase of the interference is altered by injecting another signal in one of the other ports. This is discussed in the next section.

The ratio of the output (P_o) to the input signal (P_i) is given by

$$\frac{P_o}{P_i} = \frac{1}{8}\{G_1 + G_2 - 2\sqrt{G_1 G_2}\cos(\phi_1 - \phi_2)\} \tag{8.3}$$

where G_1, G_2 are the gains of the amplifiers and ϕ_1, ϕ_2 are phase changes induced by nonlinear effects in the two amplifiers. The phase difference, $\Delta\phi = \phi_1 - \phi_2$, at the output is given by

$$\Delta\phi = \frac{\alpha}{2P_s}\left\{-\frac{h\nu}{e}(I_1 - I_2) + \frac{P_c}{2}(G_1 - 1)\right\} \tag{8.4}$$

where $\alpha = \delta n_R/\delta n_{\mathrm{Im}}$ is the ratio of the carrier-induced change in the real and imaginary parts of the index, also known as the linewidth enhancement factor. P_c is the injected power in one of the amplifiers, P_s is the saturation power in the amplifier, and I_1, I_2 are the currents through the two amplifiers. For $I_1 \cong I_2$ and $G_1 \cong G_2$, Eq. (8.3) reduces to

$$\frac{P_o}{P_i} = \frac{G}{2}\sin^2\left(\frac{\Delta\phi}{2}\right) \tag{8.5}$$

$$\text{where } \Delta\phi = \frac{\alpha}{4}\frac{P_c}{P_s}(G - 1)$$

Using typical values $\alpha = 3$, $P_s = 10$ mW, and experimental saturated gains $G \cong 40$, we get $P_c \cong 1$ mW for $\Delta\phi = \pi$. Thus, the optical power that needs to be injected in one of the other ports for π phase shift can be easily achieved. There may be some thermally induced refractive index changes, but these are considerably smaller than the carrier-induced changes previously considered and can be ignored.

8.4. Functional Performance

8.4.1. WAVELENGTH CONVERSION

The semiconductor Optical Amplifier (SOA) has many potential applications in future all-optical transmission systems. Four wave mixing (FWM) in SOA has been proposed as a technique for performing wavelength conversion due to its high conversion efficiency and fast speed response in wavelength division multiplexing (WDM) [20–22]. Other mechanisms for wavelength conversion using SOA include the Mach-Zehnder modulator.

8.4.1.1. Four Wave Mixing

Semiconductor amplifiers are known to exhibit optical nonlinear effects, including four wave mixing. Four wave mixing is a process by which optical signals at different (but closely spaced) wavelengths can mix to produce new signals at other wavelengths, but with much lower power. In the FWM process, light at two frequencies, ω_0 and ω_1, is injected into the amplifier. These injected signals are generally referred to as pump and probe beams. The pump and probe beams emitting at frequencies ω_0 and ω_1 are obtained from two single wavelength distributed feedback (DFB) lasers. The pump signal is of higher power than the probe signal. Consider the case when both the pump and the probe signals are CW. The FWM signals at frequencies of $2\omega_0 - \omega_1$ and $2\omega_1 - \omega_0$ are measured by a spectrometer. The former has higher power if the pump signal strength is higher.

The measured FWM signal power as a function of the current in the SOA for two incident signals is shown in Fig. 8.18. Also plotted is the measured gain (fiber-to-fiber) as a function of the total current. The data is shown for the FWM signal at a frequency of $2\omega_0 - \omega_1$, where ω_0 and ω_1 are the frequencies of the pump and probe respectively.

Also shown is the measured gain of the same SOA for a low input signal.

Another parameter that affects the FWM power is the wavelength separation between the pump and the probe signal. The measured FWM power as a function of this wavelength separation is shown in Fig. 8.19. Also shown (solid line) is the calculated curve (described later) using typical material parameter values. Both the experimental results and calculation

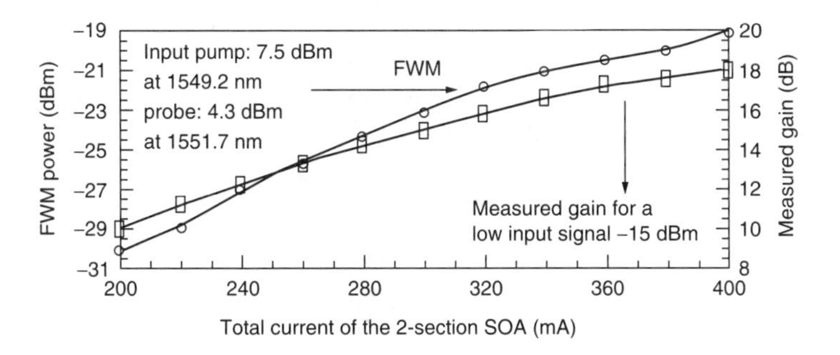

Fig. 8.18 Measured FWM signal as a function of current in the SOA.

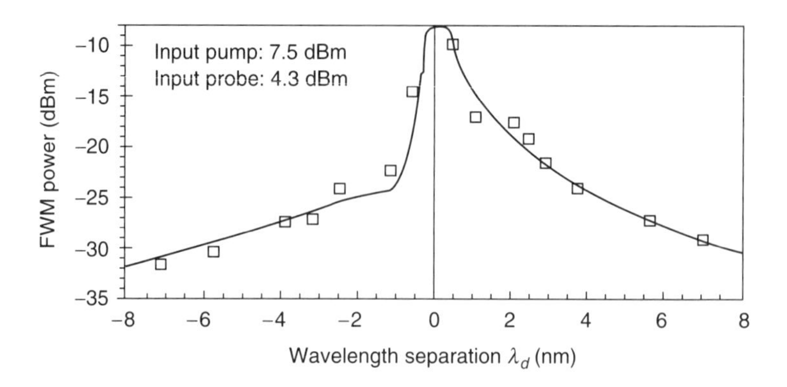

Fig. 8.19 Measured FWM power (circles) as a function of the wavelength separation. The solid line is the calculated result using typical parameter values.

show that the FWM power is high if the wavelength separation is small. It also shows that the curve is asymmetric—for a given $\lambda_d = \lambda_1 - \lambda_0$ FWM signal, power is higher for $\lambda_d > 0$ than for $\lambda_d < 0$.

The mechanisms responsible for FWM in a SOA are carrier density modulation and nonlinear gain [23]. The former has a time constant in the range of 0.1 to 0.5 ns and the latter can occur on a much faster time scale (~500 fs) [23]. Generally, $I_{\text{FWM}} << I_0, I_1$ where I_0, I_1 are the pump and probe power respectively. Under these assumptions, the equations for the electric field of the pump and probe are [23]

$$\begin{aligned}\frac{dE_0}{dZ} &= \frac{1}{2}g(1 - i\alpha)E_0 \\ \frac{dE_1}{dZ} &= \frac{1}{2}g(1 - i\alpha)E_1 \\ \frac{dE_{\text{FWM}}}{dZ} &= \frac{1}{2}g(1 - i\alpha)E_{\text{FWM}} - \frac{1}{2}g\sigma E_1^* E_0^2\end{aligned} \tag{8.6}$$

where
$$\sigma = \frac{1 - i\alpha}{1 + \frac{|E_0|^2}{I_s} - i2\pi \cdot f_d\tau_n} \cdot \frac{1}{I_s} + \frac{1 - i\beta}{1 - i2\pi \cdot f_d\tau_n} \cdot \frac{1}{I_n}$$

where $f_d = f_0 - f_1$ is the detuning frequency due to the wavelength separation between the pump and the probe. I_s, I_n, τ_s, τ_n are the saturation powers and the relaxation time constants for carrier density modulation and nonlinear gain, respectively. The quantity g is the material gain multiplied by the optical confinement factor, which is given by

$$g(z) = \frac{g_0}{1 + \frac{|E_0(z)|^2}{I_s}} \tag{8.7}$$

where g_0 is the unsaturated gain. The quantity α is the linewidth enhancement factor and β is the ratio of the real and imaginary parts of the refractive index change induced by gain nonlinearity. If $I_0(0) < I_s$, we can neglect the $|E_0|^2/I_s$ term in the denominator of σ. Also, for $g \cong g_0$, the equations for E_0 and E_1 are easily solved and substituting the solution into the equation for E_{FWM}, it reads as follows:

$$\frac{dE_{\mathrm{FWM}}}{dZ} = AE_{\mathrm{FWM}} - Be^{CZ} \tag{8.8}$$

where

$$A = \frac{1}{2}g(1 - i\alpha)$$

$$B = \frac{1}{2}g \cdot \left[\frac{1 - i\alpha}{1 - i2\pi \cdot f_d\tau_s} \cdot \frac{1}{I_s} + \frac{1 - i\beta}{1 - i2\pi \cdot f_d\tau_n} \cdot \frac{1}{I_n}\right] \cdot I_0(0) \cdot E_1(0)$$

$$C = \frac{1}{2}g(3 - i\alpha)$$

$I_0(0)$ and $E_1(0)$ are the pump intensity and the field of the probe at $Z = 0$, respectively. Using the boundary condition, $E_{\mathrm{FWM}}(Z = 0) = 0$, the solution of Eq. (8.8) is

$$E_{\mathrm{FWM}}(Z) = \frac{B}{A - C}(e^{CZ} - e^{AZ}) \tag{8.9}$$

The FWM power at $Z = L = L_1 + L_2$ is

$$I_{\mathrm{FWM}} = E_{\mathrm{FWM}}(L) \cdot E^*_{\mathrm{FWM}}(L) \tag{8.10}$$

$$I_{\mathrm{FWM}} = \frac{|B|^2}{g^2} \cdot f(G)$$

where $G = e^{gL}$ is the total gain of a 2-section SOA.

$$f(G) = G^3 - G$$

In Eq. (8.10), the absorption and reflection losses have been neglected. Although there are many simplifying assumptions, the equation is a good approximation. Figure 8.19 shows the calculated I_{FWM} (solid curve) using the typical parameter values: $\tau_s = 200$ ps, $\tau_n = 650$ fs, $I_s = 20$ mW, $I_n = 52I_s$, $\alpha = 5$, $\beta = -2.2$ [23], $I_0(0) = 7.5$ dBm, $I_1(0) = 4.3$ dBm, and $G = 8.0$ dB. The measured gains of the SOA for a low input signal and for the corresponding total input (pump and probe) are 18 dB and 0.8 dB respectively, so $G = 8.0$ dB is reasonable in the presence of saturation,

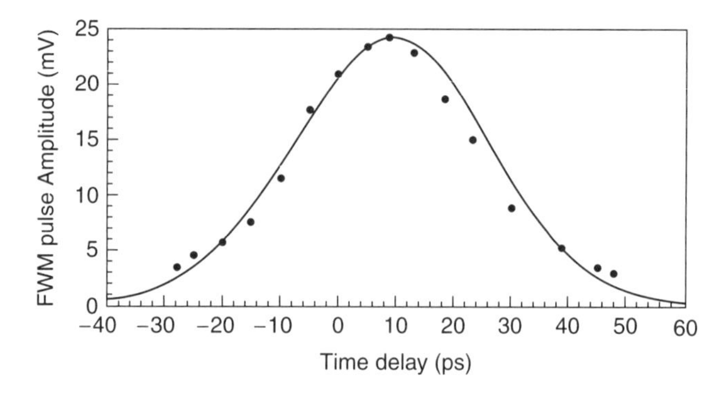

Fig. 8.20 The measured (circles) and calculated (solid line) FWM pulse amplitude as a function of the temporal position of the probe beam. The peak intensities of the pump and probe beams in the amplifier are 9 mW and 2 mW respectively.

when $g(z) = \frac{g_0}{1+\frac{|E_0(z)|^2}{I_s}}$. The FWM power is proportional to the $f(G)$. For $G > 5\,\text{dB}$, $f(G)$ varies as G^3. The FWM power is higher for SOA with high gain, which is consistent with the result shown in Fig. 8.18.

For certain transmission networks, signal add/drop capability is important. This can be carried out using FWM in a semiconductor amplifier. FWM can also be used for all-optical clock recovery and optical demultiplexing in high-speed transmission systems [24, 25]. For these systems both the pump and probe signals are pulsed. FWM experiments using pulsed pump and probe beams derived from gain-switched DFB lasers have been reported [26, 27]. For the results reported here, the widths of both pump and probe pulses were 33 ps and their repetition rate was 5.0 GHz. The gain saturation of the SOA takes place only when there is temporal overlap of the pump and probe pulses in the amplifier. In the experiment the temporal position of the probe pulse was varied by an optical delay line. The amplitude of the FWM beam (as measured using a sampling oscilloscope) as a function of the temporal position of the probe beam is shown in Fig. 8.20. The solid line is the calculated result using an analysis of FWM using pulsed light input as described subsequently.

The propagation of an electromagnetic wave along a traveling wave semiconductor optical amplifier is described by [26, 27]

$$\begin{aligned}\frac{\partial g}{\partial t} &= \frac{g_0 - g}{\tau_c} - \frac{g}{\tau_c}|E|^2 \\ \frac{\partial E}{\partial z} &= \frac{g}{2}(1 - i\alpha_c)E\end{aligned} \tag{8.11}$$

where $g(z, t)$ is the amplifier gain under saturation conditions and g_0 is the small signal unsaturated gain. The optical field normalized to $\sqrt{I_{\text{sat}}}$ (I_{sat} is the saturation intensity) is $E(z, t)$. τ_c is the carrier lifetime, α_c is the linewidth enhancement factor, and t is the time coordinate in the reference frame of the optical field.

The injected pulsed signals (assuming transform-limited Gaussian pulses) are represented by

$$\begin{aligned} E_0(t) &= A_0 \exp\left[-\frac{1}{2}\left(\frac{t}{t_0}\right)^2\right] \\ E_1(t) &= A_1 \exp\left[-\frac{1}{2}\left(\frac{(t+\phi)}{t_1}\right)^2\right] \end{aligned} \tag{8.12}$$

where E_0 is the pump signal at ω_0, E_1 is the probe signal at ω_1, ϕ is the delay time between the two pulses. For pump and probe pulses with 33 ps FWHM (full width at half maximum of intensity) the quantities t_0 and t_1 are 20 ps. The optical field at the input facet ($z = 0$) is

$$E_{\text{in}}(z = 0, t) = E_0 + E_1 \exp(i\Omega t), \quad \Omega = \omega_1 - \omega_0 \tag{8.13}$$

If we define an integrated gain $h(t) = \int_0^L g(z, t)dz$, then the preceding equations can be integrated to yield

$$\begin{aligned} \frac{dh}{dt} &= \frac{h_0 - h}{\tau_c} - (e^h - 1)\frac{|E_{\text{in}}(0, t)|^2}{\tau_c} \\ E_{\text{out}}(L, t) &= E_{\text{in}}(0, t) \exp\left[\frac{h}{2(1 - i\alpha_c)}\right] \end{aligned} \tag{8.14}$$

where $h_0 = g_0 L$, with L being the length of the SOA. If the probe signal E_1 is much smaller than the pump E_0, then

$$|E_{\text{in}}(0, t)|^2 \cong |E_0|^2 + E_0^* E_1 e^{i\Omega t} \quad \text{and} \quad |E_0|^2 \gg |E_0^* E_1| \tag{8.15}$$

so we can express $h(t)$ as the following:

$$h(t) = \bar{h}(t) + \delta h(t) e^{i\Omega t} \quad \text{with } |\bar{h}(t)| \gg |\delta h(t)| \tag{8.16}$$

where

$$\bar{h}(t) = -\ln\left[1 - \left(1 - e^{-h_0}\right) e^{-U_{\text{in}}(t)}\right] \quad \text{and} \quad U_{\text{in}}(t) = \frac{1}{\tau_c} \int_{-\infty}^{t} |E_0(x)|^2 dx$$

then the output optical field of SOA is [9]

$$E_{\mathrm{out}}(L,t) = E_0(L,t) + E_1(L,t)e^{i\Omega t} + E_2(L,t)e^{-i\Omega t} \tag{8.17}$$

where

$$\begin{aligned}
E_0(L,t) &= \exp\left[\frac{\bar{h}}{2(1-i\alpha_c)}\right]\left(1 + F_-(\Omega)|E_1|^2\right)E_0 \\
E_1(L,t) &= \exp\left[\frac{\bar{h}}{2(1-i\alpha_c)}\right]\left(1 + F_+(\Omega)|E_0|^2\right)E_1 \\
E_2(L,t) &= \exp\left[\frac{\bar{h}}{2(1-i\alpha_c)}\right]F_-(\Omega)E_0^2E_1^* \\
F_\pm(\Omega) &= -\frac{e^{\bar{h}}-1}{2}\left[\frac{1-i\alpha_c}{1+e^{\bar{h}}|E_0|^2 \pm i\Omega\tau_c} + \frac{1-i\beta_n}{1 \pm i\Omega\tau_n}K_n\right]
\end{aligned} \tag{8.18}$$

The preceding equations describe the FWM frequency product E_2 whose frequency is $w_2 = w_0 - \Omega = 2w_0 - w_1$. Other products are also created, but they are all much smaller in intensity than E_2 and are therefore neglected. The first term describes FWM due to carrier density pulsations and it represents an exact small signal solution. The second term accounts for the nonlinear gain effect of SOA, where K_n is the nonlinear gain coefficient.

The calculated FWM pulse amplitude is plotted in Fig. 8.20 using the following typical parameter values from references [23, 26–28]: $G_0 = 18$ dB, $t_0 = t_1 = 21$ ps, $\tau_c = 200$ ps, $\tau_n = 0.65$ ps, $\lambda_0 = 1548.3$ nm, $\lambda_1 = 1551.7$ nm, $\alpha_c = 5$, $\beta_n = -2.2$, $K_n = 2 \times 10^{-3}$. Figure 8.20 shows that the result of the calculation is consistent with the experimental data. The calculated FWM pulse width is shown in Fig. 8.21. Its value is nearly equal to $\frac{1}{\sqrt{3}}$ of the

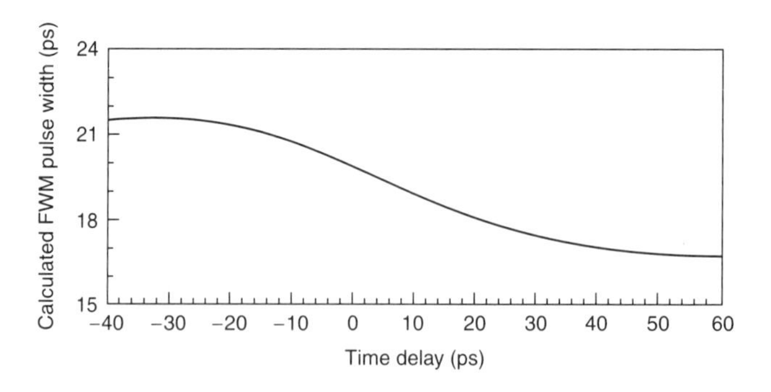

Fig. 8.21 Calculated FWM pulse width as a function of the temporal position of the pump pulse. The input pulse width (FWHM) is 33 ps for both the probe and the pump.

input pulse width (expected for a simple Gaussian model), with a slight modulation by the temporal mismatch of the input pulses. The measured FWM pulse was also found to be independent of the temporal mismatch with a value of 19 ± 3 ps.

Carrier density modulation and nonlinear gain effects (including dynamic carrier heating and spectral hole burning) in the semiconductor optical amplifier contribute to the generation of the FWM signal. When the wavelength separation λ_d between the pump and the probe beams is larger than 1 nm, nonlinear gain effects dominate in the generation of FWM. Because the effective time constant of the nonlinear gain effects is ~0.6 ps [23], which is significantly faster than the time constant for carrier density modulation, it has been suggested that the bandwidth of the FWM could reach as high as ~1 THz [29].

The calculated FWM efficiency for small signal modulation as a function of the modulation frequency is plotted in Figs. 8.22 and 8.23. The parameter values used are $\tau_c = 200$ ps, $\tau_n = 650$ fs, $P_s = 20$ mW, $P_n = 50P_s$, $\alpha = 5$, $\beta = -2.2$ [23, 26–28]. Figure 8.22 shows the bandwidth of FWM increases from 320 GHz to about 1.5 THz when the wavelength separation between the pump and the probe increases from 0.5 nm to 10 nm. This is because when the wavelength separation $\lambda_d = 0.5$ nm, both carrier density

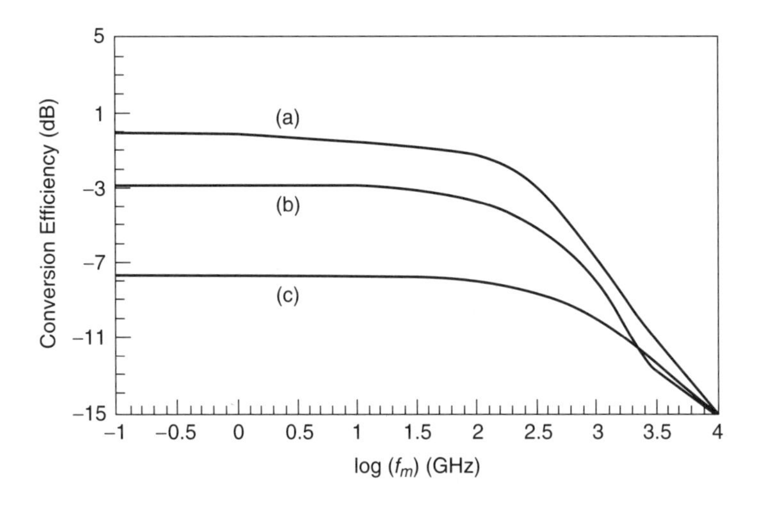

Fig. 8.22 Calculated frequency response of FWM conversion efficiency when $g_0L = 7$. (a) wavelength difference between the pump and the probe $\lambda_d = 0.5$ nm, FWM bandwidth = 320 GHz, (b) $\lambda_d = 2.5$ nm, FWM bandwidth = 460 GHz, (c) $\lambda_d = 10$ nm, FWM bandwidth = 1500 GHz. The x-axis is log(base 10) of the modulation frequency f_m in GHz.

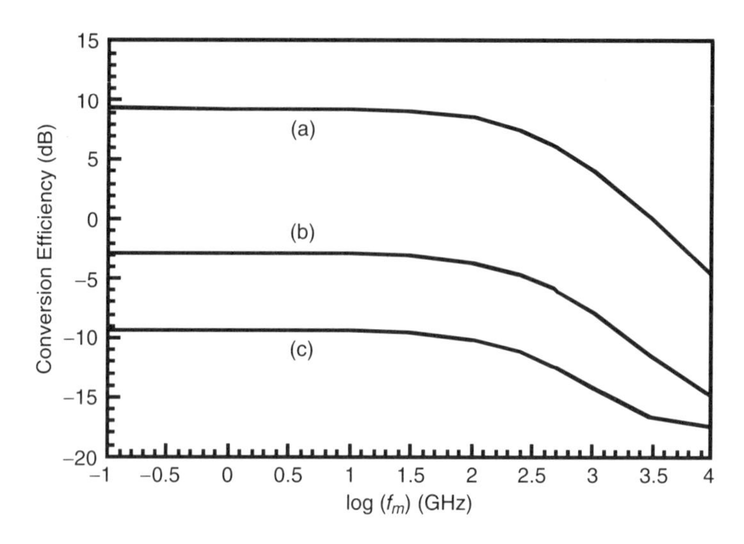

Fig. 8.23 Calculated frequency response of FWM conversion efficiency when $\lambda_d =$ 2.5 nm (a) $g_0L = 10$ in the 2-section SOA, FWM bandwidth = 440 GHz, (b) $g_0L = 7$, FWM bandwidth = 460 GHz, (c) $g_0L = 5$, FWM bandwidth = 470 GHz. The x-axis is log(base 10) of the modulation frequency f_m in GHz.

modulation and nonlinear gain effects contribute to the signal conversion from the probe to the FWM beam, while when $\lambda_d = 10$ nm, only the nonlinear gain effects contribute to the signal conversion.

Figure 8.23 shows the frequency response of the FWM conversion efficiency for different g_0L values, corresponding to different gains of the SOA. It shows that, although the absolute value of the FWM conversion efficiency depends strongly on the gain of the SOA, the bandwidth of FWM conversion efficiency decreases slightly when the g_0L value is doubled. However, because the FWM conversion efficiency increases with increasing gain (approximately as g^3), a high gain is necessary for applications, such as for optical demultiplexing and optical wavelength conversion. Thus, for large wavelength separation between the pump and the probe, where the nonlinear gain effects dominate the generation of the FWM, the bandwidth of the FWM could exceed 1 THz.

8.4.1.2. Cross-Phase Modulation

Cross-phase modulation (XPM) in a semiconductor optical amplifier (SOA) in an interferometric configuration is considered to be an important all-optical wavelength conversion scheme, because of its conversion efficiency, high extinction ratio, and low chirp characteristics [30, 31].

All XPM wavelength conversion devices require a CW probe source to provide the optical carrier onto which the incoming signal is to be encoded. Two methods have been used to provide this CW source: (i) Integration of a pre-amplifier SOA on the wavelength converter chip for boosting the received CW power from an off-chip source, and, (ii) integration of the very CW source (DFB laser) with the interferometer chip. Schematics of the two devices are shown in Fig. 8.24. The interferometer is a Mach-Zehnder interferometer (MZI) in both cases.

Both methods have the capability of enabling injection of large optical power into the MZI, which helps speed up interband carrier dynamics, and therefore can lead to higher speed device operation. The latter method

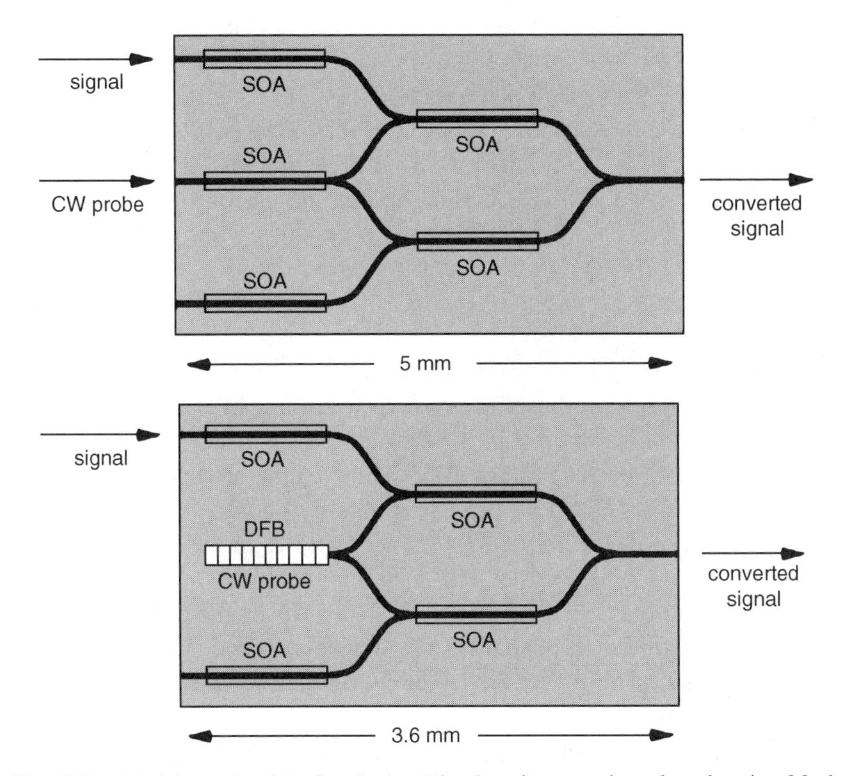

Fig. 8.24 (a) Schematic of the first device. Wavelength conversion takes place in a Mach-Zehnder interferometer. The CW probe supplying the converted wavelength is fed into the middle pre-amplifier SOA. Signal inputs are provided on either side for symmetry and redundancy reasons (only one is used). A signal pre-amplifier SOA compensates for variations in the input power. (b) Schematic of the second MZI wavelength converter. The CW source and input pre-amplifiers are monolithically integrated.

also guarantees the probe power to be delivered in TE polarization, which relaxes polarization-independence requirements of the device, and, more importantly, leads to a two port device (because an external connection for the CW source is obviously no longer needed), considerably simplifying packaging

The starting point for MZI device fabrication is a standard MOVPE-grown n-doped InP base wafer consisting of multi quantum well (MQW) active (amplifying) layers on top of a passive InGaAsP waveguide core. The quantum wells consist of an optimized compressively/tensile strained stack to provide polarization-independent gain. The quantum wells are etched away in the passive sections before waveguide formation. The confinement factor for the optical mode in the quantum wells is 5–10%, which is on the low side and not favorable for carrier speed, but necessary for obtaining low-loss transitions between the active and passive sections. After etching the waveguide pattern and optionally forming the DFB grating using phase-mask printing, blocking layers consisting of reversed p-n junctions are grown, in which slots are etched running on top of the active waveguide sections. This is done to confine current injection to the active regions only. After growth of the p-doped top and contact layers and gold contact formation, the wafer is cleaved into bars of devices. The facets are then coated with a SiOx layer of suitable thickness to suppress reflections.

Bit error rate curves for conversion of a PRBS of length $2^{23} - 1$ from 1559 nm to 1568 nm are shown in Fig. 8.25. When the MZI is optimized for maximum output extinction ratio (11 dB), a wavelength conversion penalty of 1.6 dB is measured at a BER of 10^{-9}. The eye diagram in the lower inset in the figure shows that this penalty is mainly caused by inter-symbol interference (ISI) due to rather slow gain dynamics in the SOA, which is driven with a moderate bias of 100 mA. The switching speed increases by adjusting the bias point of the MZI. This reduces the extinction ratio to 10 dB, but significantly enhances the switching speed, as shown by the eye in the upper inset of Fig. 8.25. The wavelength conversion penalty is reduced to 0.5 dB.

Wavelength-dependent conversion results are shown in Fig. 8.26. Conversion of signals in the 1550–1570 nm range, optimized for highest extinction ratios, gives similar penalties. The pattern length was $2^{23} - 1$, and the same settings for the MZI biases were used. 1 dB more input power (or a slightly higher pre-amp bias) was needed for input wavelengths of

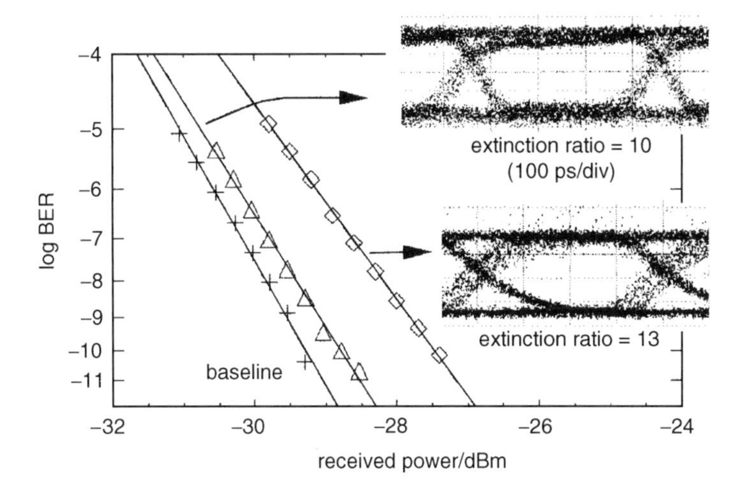

Fig. 8.25 BER curves for conversion from 1559 to 1568 nm, optimized for output extinction ratio (diamonds, lower eye diagram), and optimized for speed (triangles, upper eye diagram).

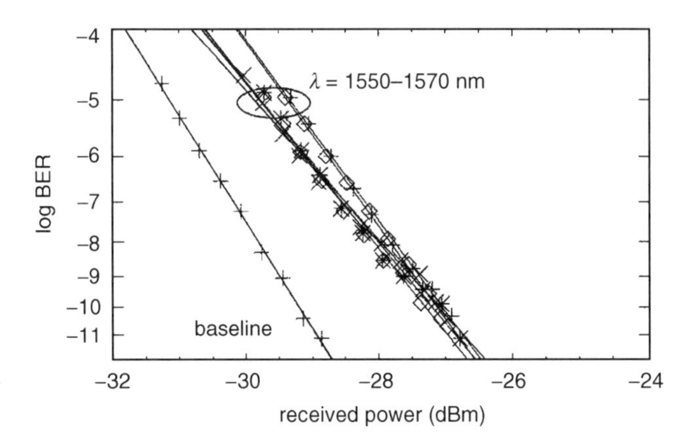

Fig. 8.26 Wavelength conversion of input signals in the 1550–1570 nm range to the output wavelength of 1568 nm, optimized for extinction ratio.

1550 and 1570 nm, to compensate for the roll-off at the extremes of the SOA gain spectrum.

The carrier lifetime in a SOA is an important parameter for high-speed operation of MZI devices. The measured lifetimes for various types of SOA devices are shown in Fig. 8.27. The times plotted are for 10% to 90% recovery. The horizontal axis shows bias current per unit device length.

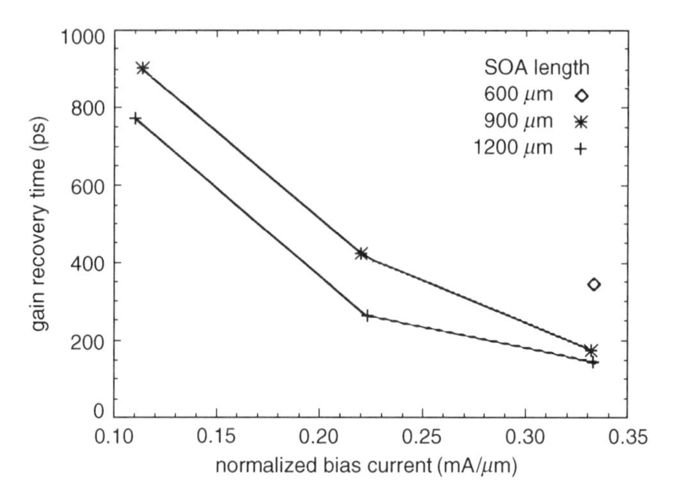

Fig. 8.27 Gain recovery times for cross-gain modulation in 600-, 900-, and 1200-μm-long SOAs, for bias currents up to 200, 300, and 400 mA, respectively. Bias currents are normalized with respect to the length of the devices.

Based on the data in this figure, with sufficiently high biases the devices with the longer amplifiers should be able to support high bit rates.

8.4.2. OPTICAL DEMULTIPLEXING

Among the techniques for optical demultiplexing are four wave mixing in semiconductor amplifiers, four wave mixing in fibers, and semiconductor Mach-Zehnder and Michelsen interferometers. All of these techniques need an optical clock. Four wave mixing in fibers does not lead to compact devices and hence it will not be discussed here.

8.4.2.1. Four Wave Mixing

An example of optical demultiplexing using four wave mixing (FWM) is shown Fig. 8.28. This technique allows any regularly spaced set of bits to be simultaneously extracted from the incoming data stream using a semiconductor optical amplifier (SOA) [32]. Consider the high-speed data (at a data rate B) at wavelength λ_1 and an optical clock (at a clock rate B/N) at wavelength λ_2 injected into an optical amplifier. The output of the amplifier will include four wave mixing signals at wavelengths $2\lambda_1 - \lambda_2$ and $2\lambda_2 - \lambda_1$. Typically, one of the FWM signals is filtered out using an optical filter at the exit port of the nonlinear element and represents the

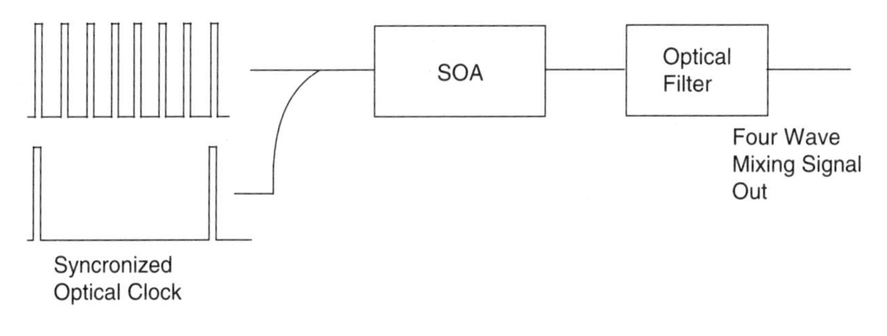

Fig. 8.28 Schematic of a four wave mixing based demultiplexer. The four wave mixing takes place in the SOA only when there is a temporal overlap between the signal and the clock pulse.

data signal before further processing. Thus, a signal representing the data every N_{th} bit time slot (where the mixing occurs with the clock signal) is generated by the four wave mixing process (Fig. 8.28). The clock signal can then be delayed by 1 additional bit each time to retrieve the original data in successive time slots as described following. A complete system would need N SOA elements in order to extract N data bits simultaneously. As shown in the previous section, the process can operate to speeds of 1 Tb/s.

8.4.2.2. Cross-Phase Modulation

Among the semiconductor devices suitable for high-speed all-optical demultiplexing are (i) Mach-Zehnder interferometer (MZI), and, (ii) Michelsen interferometer [18, 19]. These devices operate on the principle of a phase change caused by an incident optical clock signal (cross-phase modulation). This phase change can be adjusted so that the interferometer only yields output when a data signal overlaps a clock signal for demultiplexing applications. The schematic of a MZI device operating as a demultiplexer is shown in Fig. 8.29. An input y-junction splits the input signal beam into two beams, which propagate through semiconductor amplifiers positioned in the upper and lower arms of the interferometer. These beams then merge and interfere at an output y-junction. One of the amplifiers has the clock pulse incident on it. The clock frequency is $1/N$ times the signal bit rate. The clock pulse is absorbed by the amplifier, changing its carrier density, which induces an additional phase shift for the signal traveling through that amplifier arm. This changes the output of the device every N^{th} signal bit. The SOAs can be configured so that the phase difference between them is π (before the arrival of the clock pulse), so that the signals

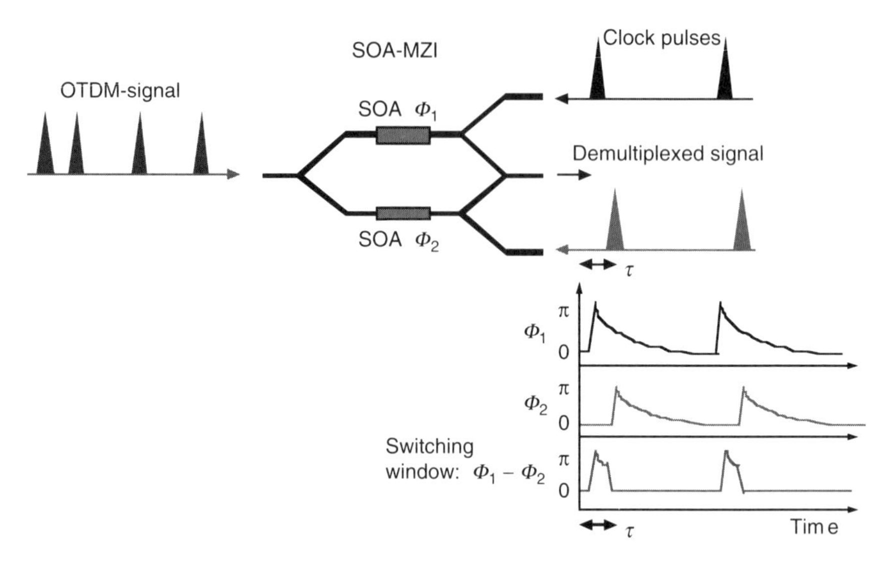

Fig. 8.29 Schematic of a Mach-Zehnder interferometer with integrated SOAs. The clock pulse introduces an additional phase shift in one of the arms of the interferometer. With two clock pulses a gate is produced for demultiplexing.

propagating through the two arms interfere destructively, resulting in no output. In the presence of the clock pulse, an additional π phase shift is introduced in one arm of the interferometer, which results in a nonzero output of every N^{th} bit. The device can be made to operate considerably faster using a differential phase shift scheme, which involves introducing a second delayed clock signal in the other arm of the interferometer. The principle of this operation is illustrated in Fig. 8.29. In the delayed pulse approach, the differential phase shift between the two arms is maintained over a short duration (1 bit period). As a result, a small gate is established for demultiplexing the high-speed optical time-division multiplexed (OTDM) signal. These devices have been used to demultiplex 80 and 160 Gb/s optical signals to 20 Gb/s signals. Figure 8.30 shows the experimental results of demultiplexing a 80 Gb/s signal into 8 channels at 10 Gb/s each [33].

8.4.3. *OPTICAL LOGIC*

Mach-Zehnder interferometers have been used to demonstrate several optical logic functions [34]. The schematic of a XOR operation is shown in Fig. 8.31. Consider the case when a clock signal (a series of 1s) is injected into the waveguide, which then splits into two and travels through both

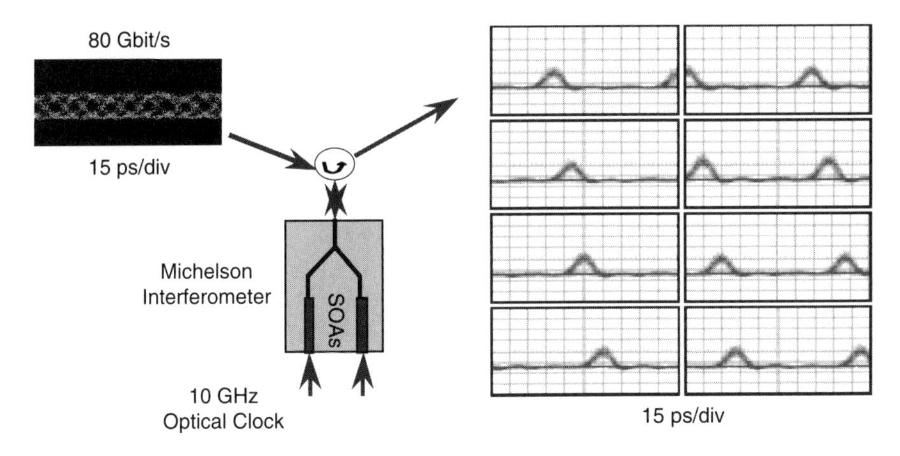

Fig. 8.30 Experimental results of demultiplexing a 80 Gb/s signal into 8 signals at 10 Gb/s each. All signals are RZ type.

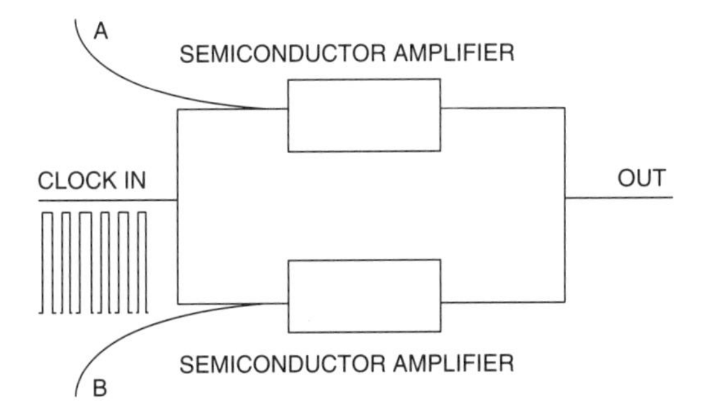

Fig. 8.31 Schematic of XOR function with MZ device. MZ is initially unbalanced $\Delta \times \phi\pi$. The clock shown here is a series of 1s. A CW clock may be used for lower speed operation. An example of XOR function obtained using the schematic (Fig. 8.31) is shown in the table.

amplifiers and interferes at the output. In the absence of any signal (A or B), the path lengths are such that the two interfering signals are π out of phase—i.e, the output is 0. If a signal of correct amplitude is now injected at the top amplifier ($A = 1$) and none at the bottom amplifier ($B = 0$), a differential π phase shift occurs (due to carrier-induced change in index of the amplifier), which causes "1" signal at the output. Similarly, when $A = 0$ and $B = 1$, the output is 1. Thus a XOR functionality shown in the Table 8.1 can be demonstrated using the schematic in Fig. 8.31. Other schemes are also feasible. For example, a simplification would be use of a CW signal instead of a clock (101010. . .) pattern. However, it may lead to slower

Table 8.1 **Logic XOR Operation**

A	*B*	*OUT(1 = CLOCK)*
1	0	1
0	1	1
1	1	0
0	0	0

operation. Other logic functions can also be performed by cascading and varying the inputs (A, B, A + B, etc.)

8.5. Summary and Future Challenges

Tremendous advances in InP-based semiconductor optical amplifiers (SOA) have occurred over the last decade. The advances in research and many technological innovations have led to improved designs of semiconductor amplifers. Although most lightwave systems use optical fiber amplifiers for signal amplification, SOAs are suitable for integration and can also be used as functional devices. These functional properties, such as wavelength conversion, optical demultiplexing, and, optical logic elements, make them attractive for all-optical network and optical time-division multiplexed systems.

The need for higher capacity is continuing to encourage research in WDM-based and OTDM-based transmission, both of which need optical demultiplexer and high-power tunable lasers. An important research area would continue to be the development of semiconductor amplifiers as Mach-Zehnder or Michelsen interferometers and as low-power amplifiers in integrated devices.

The amplifier-to-fiber coupling is also an important area of research. Recent development in spot-size converter-integrated amplifiers is quite impressive, but some more work, perhaps, needs to be done to make them easy to manufacture. This may require more process developments.

Amplifiers with better high-temperature performance comprise an important area of investigation. The InGaAsN-based system is a promising candidate for making gain regions near 1.3 μm with good high-temperature performance. New materials investigation is important.

Although WDM technology is currently being considered for increasing transmission capacity, the need for optical demultiplexers for OTDM systems and optical logic circuits for high data rate encryptors would remain. Hence, research on high-speed optical logic circuits is important. The capability to integrate many SOAs on the same chip with interconnecting waveguides is potentially important for these types of systems.

The vertically coupled amplifier is very attractive for two-dimensional arrays. Several important advances in this technology have occurred over the last few years. An important challenge is the fabrication of a device with characteristics superior to that of an edge-coupled amplifier.

Finally, many of the advances in semiconductor optical amplifier development would not have been possible without the advances in materials and processing technology. The challenges of much of the current semiconductor optical amplifier research are intimately linked with the challenges in materials growth, which include not only the investigation of new material systems but also improvements in existing technologies to make them more reproducible and predictable.

References

1. M. J. Coupland, K. G. Mambleton, and C. Hilsum, Phys. Lett., 7 (1963) 231.
2. J. W. Crowe and R. M. Graig, Jr., Appl. Phys. Lett., 4 (1964) 57.
3. W. F. Kosnocky and R. H. Cornely, IEEE, JQE QE-4 (1968) 225.
4. T. Saitoh and T. Mukai, In: Coherence, Amplification and Quantum Effects in Semiconductor Lasers, ed. by Y. Yamamoto (Wiley, New York, 1991) Chap. 7.
5. M. Nakamura and S. Tsuji, IEEE JQE QE-17 (1981) 994.
6. T. Saitoh and T. Mukai, J. Lightwave Technol., 6 (1988) 1656.
7. M. O'Mahony, J. Lightwave Technol., 5 (1988) 531.
8. N. K. Dutta, M. S. Lin, A. B. Piccirilli, and R. L. Brown, J. Appl. Phys., 67 (1990), 3943.
9. H. Kogelnik and T. Li, Proc. IEEE, 54 (1966) 1312.
10. M. S. Lin, A. B. Piccirilli, Y. Twu, and N. K. Dutta, Electron Lett., 25 (1989) 1378.
11. C. E. Zah, J. S. Osinski, C. Caneau, S. G. Menocal, L. A. Reith, J. Salzman, F. K. Shokoohi, and T. P. Lee, Electronic Lett., 23 (1987) 1990.
12. M. Dagenais, Ed., Integrated Optoelectronics (John Wiley, New York, 1995).
13. O. Wada Ed., Optoelectronic Integration: Physics Technology and Applications (Kluwar Academic, Amsterdam, 1994).
14. T. L. Koch, U. Koren, R. P. Gnall, F. S. Choa, F. Hernandez-Gil, C. A. Burrus, M. G. Young, M. Oron, and B. I. Miller, Electronics Lett., 25 (1989) 1621.

15. N. K. Dutta, J. Lopata, R. Logan, and T. Tanbun-Ek, Appl. Phys. Letts., 59 (1991) 1676–1678.
16. R. Y. Fang, D. Bertone, M. Meliga, I. Montrosset, G. Oliveti, and R. Paoletti IEEE Photonic Tech Lett., 9 (1997) 1084–1086.
17. H. Yamazaki, K. Kudo, T. Sasaki, and M. Yamaguchi, Proc. OECC '97 Seoul, Korea, paper 10C1-3 (1997) 440–441.
18. B. Mikklesen, S. L. Danielsen, C. Joregensen, and K. E. Stubkjaer, Proceedings ECOC, 97, Oslo, Norway, (1997) 2, 245–248.
19. S. L. Danielsen, P. B. Hansen, and K. E. Stubkjaer, IEEE JQE, 16, (1998) 2095–2108.
20. T. Durhuus, B. Mikkelsen, C. Joergensen, S. L. Danielsen, and K. E. Stubkjaer, J. Lightwave Technol., 14 (June 1996) 942–954.
21. A. Mecozzi, S. Scotti, and A. D'ottavi, IEEE. J. Quantum Electron., 31 (Apr. 1995) 689–699.
22. W. Shieh, E. Park, and A. E. Willner, IEEE Photonics Technol. Lett., 8 (Apr. 1996) 524–526.
23. K. Kikuchi, M. Kakui, C. E. Zah, and T. P. Lee, IEEE JQE, 28 (1992) 151.
24. O. Kamatani, S. Kawanishi, and M. Saruwatari, Electron. Lett., 30 (1994) 807.
25. S. Kawanishi, T. Morioka, O. Kamatani, H. Takara, J. M. Jacob, and M. Saruwatari, Electron. Lett., 30 (1994) 981.
26. G. P. Agrawal and N. A. Olsson. IEEE J. Quantum. Electron., 25 (1989) 11.
27. M. Shtaif and G. Eisenstein, Appl. Phys. Lett., 66 (1995) 1458.
28. J. Zhou, N. Park, J. W. Dawson, K. J. Vahala, M. A. Newkirk, and B. I. Miller, Appl. Phys. Lett., 63 (Aug. 1993) 1179.
29. C. Wu and N. K. Dutta, J. Appl Phys., 87 (2000) 2076.
30. K. E. Stubkjaer *et al.*, "Wavelength conversion devices and techniques," in Proc. 22nd European Conference on Optical Communication (ECOC '96), Oslo (Sept. 15–19, 1996) pp. 4.33–4.40 (Invited paper).
31. C. Joergensen, S. L. Danielsen, T. Durhuus, B. Mikkelsen, K. E. Stubkjaer, N. Vodjdani, F. Ratovelomanana, A. Enard, G. Glastre, D. Rondi, and R. Blondeau, IEEE Photon. Technol. Lett., 8 (Apr. 1996) 521.
32. N. K. Dutta, C. Wu, and H. Fan "Optical demultiplexing using semiconductor amplifiers, in SPIE Proceedings," 3945 (2000) 204–209.
33. A. Piccirilli *et al.*, MillCom 99 Atlantic City, NJ (April 99).
34. T. Fjelde, D. Wolfson, A. Kloch, B. Dagens, A. Coquelin, I. Guillemot, F. Gaborit, F. Poingt and M. Renaud, Electronics Letts., 36 (2000) 1863.

Chapter 9 | Optical Amplifiers for Terabit Capacity Lightwave Systems

Atul Srivastava
Yan Sun

Onetta Inc., 1195 Borregas Ave., Sunnyvale, CA 94089, USA

9.1. Introduction

The capacity of lightwave communication systems has undergone enormous growth during the last few years in response to huge capacity demand for data transmission. Laboratory demonstration of high-capacity tranmission exceeding 10 Tb/s capacity [1] has been recently reported, and terabit capacity commercial systems are now available from system vendors. The growth in transmission capacity of lightwave systems, both in laboratory experiments and commercial systems, is shown in Fig. 9.1. Here we will discuss only the transmission achieved in a multi-channel wavelength division multiplexed (WDM) transmission system, which is represented by blue triangles and black diamonds in the figure. The blue triangles represent laboratory results, while the black diamonds show progress in commercially deployed systems. The lines are merely guides for the eye. The solid lines were drawn with the data available nearly a year ago while the dashed line is added to include recently reported high-capacity experiments. The new data indicates that advances in the capacity of laboratory experiments since 1994, earlier estimated to be growing exponentially at a rate of 4 dB per year, have to be reevaluated because the current estimated rate lies nearly halfway between 2 and 4 dB per year. It also shows that the gap between laboratory demonstration and product availability has shrunk from 6 years in 1994 to less than 2 years at present.

WDM TECHNOLOGIES: PASSIVE
OPTICAL COMPONENTS
$35.00

ISBN: 0-12-225262-4

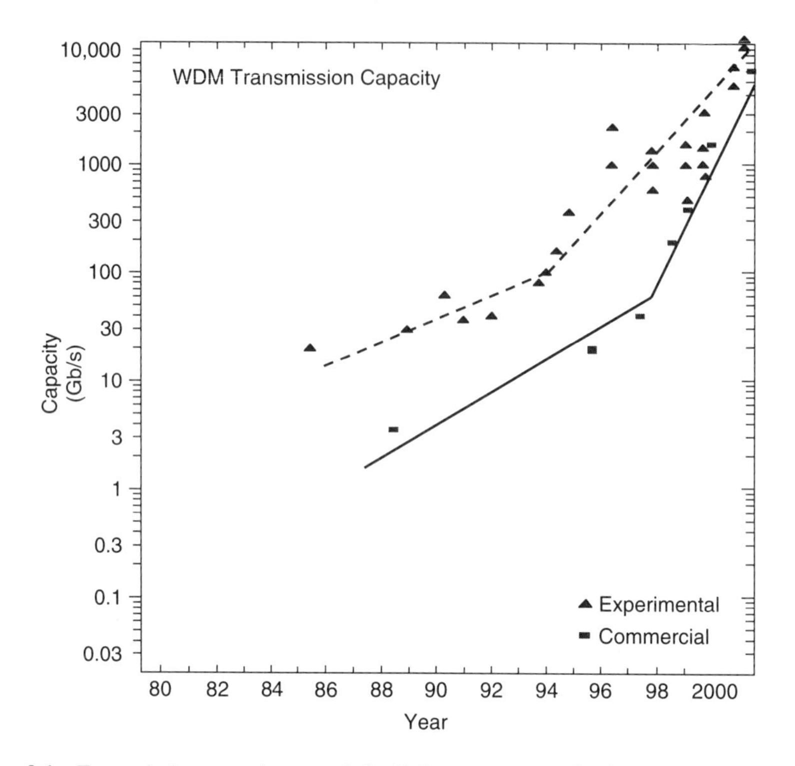

Fig. 9.1 Transmission capacity growth for lightwave communication systems. The blue triangles represent laboratory results while the black diamonds show progress in commercially deployed systems. The lines are merely guides for the eye.

These impressive strides in the capacity of lightwave systems have been made possible by several key technological advances in optical amplifiers, such as wide-bandwidth erbium-doped fiber amplifiers (EDFAs) [2]. Not only do EDFAs offer low noise, high output power, and wide bandwidth characteristics, they have evolved to become a subsystem, which incorporates a significant amount of loss in the mid-stage to accommodate dispersion compensation or optical add/drop modules needed in high-capacity systems. There has also been considerable progress in the understanding of the time-dependent behavior of the EDFA, which is key in the design of suitable amplifiers needed for dynamic networks.

In this chapter, following the introduction, we include a section covering the fundamentals of EDFAs and design considerations followed by system and network issues related to EDFAs. In the next section we describe recent advances in EDFAs, including wide-bandwidth conventional (C) band

1529–1562 nm EDFA, long (L) wavelength band 1570–1608 nm EDFA, and ultra-wideband EDFA that combines the C- and L-bands to provide an 84-nm bandwidth EDFA that was instrumental in the demonstration of the first terabit capacity WDM transmission over terrestrial distances. Next, we describe Raman/EDFA design of optical amplifiers, which combines the benefits of the Raman pre-amplifier, with its excellent noise characteristics, with EDFAs, with the very high power conversion efficiency needed for the power amplifier stage. The details of Raman amplifiers are covered in Chapter 10; here we limit the discussion to a couple of transmission experiments to illustrate the significant benefits of a Raman/EDFA design for terabit-capacity systems, in particular those carrying signals at 40 Gb/s line rates. We conclude this section by including a discussion on the issues related to EDFAs in dynamic WDM networks. After an introduction to gain dynamics of single EDFA, we describe the recently discovered phenomena of fast power transients in EDFA chains constituting optical networks and three schemes—pump control, link control, and laser control—suitable for controlling the signal channel power transients. In the following section we cover a related topic of semiconductor optical amplifiers, which have recently been demonstrated as in-line amplifiers for WDM applications. Again, the details of SOA technology are covered in Chapter 8; here we discuss the interesting aspects of WDM transmission related to SOAs in contrast to the EDFAs.

9.2. EDFAs for High-capacity Networks

In traditional optical communication systems, optoelectronic regenerators are used between terminals to convert signals from the optical to the electrical and then back to the optical domain. Since the first report in 1987 [3, 4], the erbium-doped fiber amplifier (EDFA) has revolutionized optical communications. Unlike optoelectronic regenerators, this optical amplifier does not need high-speed electronic circuitry and is transparent to data rate and format, which dramatically reduces cost. EDFAs also provide high gain, high power, and low noise figures. More importantly, all the optical signal channels can be amplified simultaneously within the EDFA in one fiber, thus enabling wavelength division multiplexing (WDM) technology.

In recent years, tremendous progress has been made in the development of EDFAs, including the erbium-doped fiber, semiconductor pump lasers, passive components, and splicing and assembly technology. Recent research in EDFAs has led to the development of an EDFA with a bandwidth of 84 nm. With these amplifiers, long-distance transmission at 1 terabit per second was achieved for the first time [5]. These developments have led to the application of EDFAs to commercial optical communication systems. The first field trial of a WDM optical communication system was carried out by AT&T in 1989 [6] and subsequently the first commercial WDM system was deployed in 1995. Since the first deployment, the capacity of WDM systems has been increasing at a very fast pace (see Fig. 9.1). Today, optical amplifiers and WDM technology are offering an unprecedented cost-effective means for meeting the ever-increasing demand for transport capacity, networking functionality, and operational flexibility.

9.2.1. BASICS CHARACTERISTICS OF EDFAs

EDFAs consist of erbium-doped fiber having a silica glass host core doped with active Er ions as the gain medium. Erbium-doped fiber is usually pumped by semiconductor lasers at 980 nm or 1480 nm.

Basic elements of an EDFA are shown schematically in Fig. 9.2. The gain medium in the amplifier is a specially fabricated optical fiber with its core doped with erbium (Er). The erbium-doped fiber is pumped by a semiconductor laser, which is coupled by using a wavelength selective coupler, also known as a WDM coupler, that combines the pump laser light with the signal light. The pump light propagates either in the same direction as the signal (co-propagation) or in the opposite direction (counter-propagation). Optical isolators are used to prevent oscillations and excess noise due to

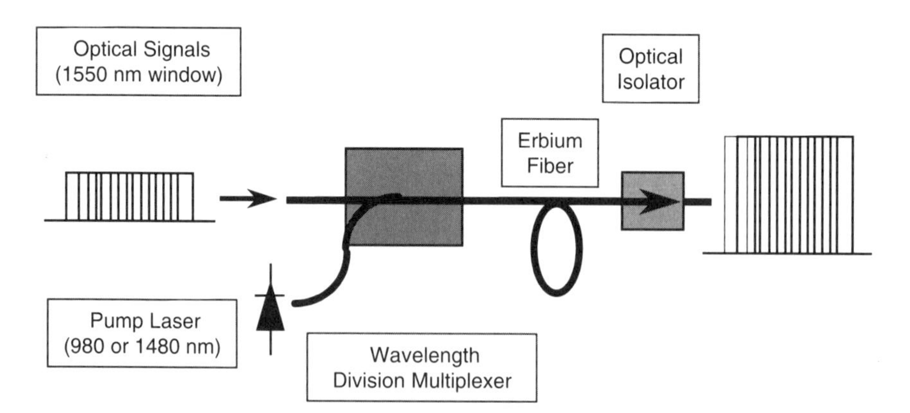

Fig. 9.2 Schematic diagram of an erbium-doped fiber amplifier.

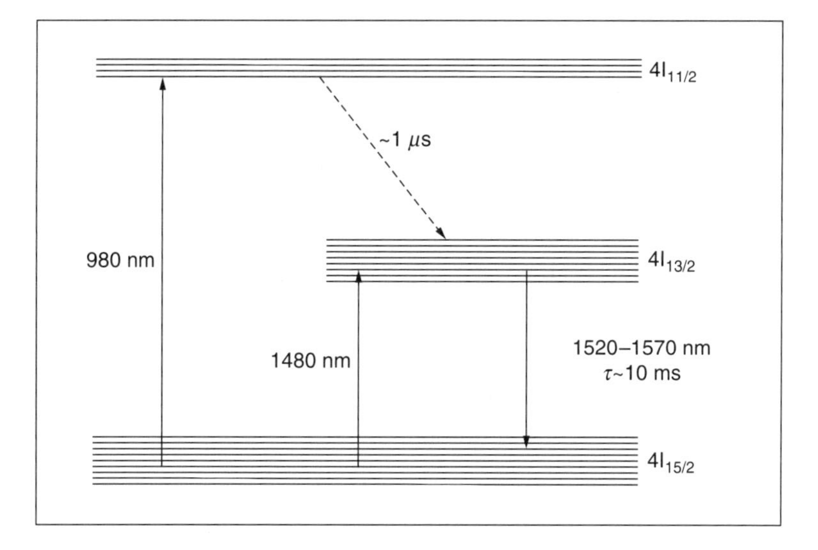

Fig. 9.3 Erbium ion energy-level scheme.

unwanted reflection in the assembly. More advanced architecture of an amplifier consists of mutiple stages designed to optimize the output power and noise characteristics while incorporating additional loss elements in the mid-stage.

The energy level scheme of the erbium ion and the associated spontaneous lifetime in the glass host are shown in Fig. 9.3. The atomic levels of Er-ions are broadened by local field variations at the microscopic level in the glass host. The light emission due to optical transitions from the first excited state ($^4I_{13/2}$) to the ground level are perfectly matched with the transmission window of silica transmission fiber (1525–1610 nm). The erbium ions can be excited to the upper energy levels by 980-nm or 1480-nm pumps. In both cases, it is the first excited state, $^4I_{13/2}$, that is responsible for the amplification of optical signals. The amplification is achieved by the signal photons causing stimulated emission from the first excited state. A three-level model can be used to describe the population of energy levels in the case of 980-nm pumping, while a two-level model usually suffices for the 1480-nm pumping case [7, 8]. Nearly complete inversion of Er ions can be achieved with 980-nm pumping, whereas due to the stimulated emission at the pump wavelength the inversion level is usually lower in the case of 1480-nm pumping [8]. A higher degree of inversion leads to lower noise level generated from the spontaneous emission process and is therefore highly desirable for the pre-amplifier stage. The quantum efficiency of the amplifier is higher for 1480 pumping due to better closer match between the signal and pump energies. The spontaneous lifetime of the metastable energy level ($^4I_{13/2}$) is about 10 ms, which is much slower than the signal bit rates of practical interest. These slow dynamics are responsible for the key advantage of EDFAs because of their negligible inter-symbol distortion and inter-channel crosstalk.

9.2.1.1. Amplifier Gain

The gain of the amplifier is defined as the ratio of the signal output and input powers. It is related to the gain factor $g(z)$, which changes along the length z of the erbium-doped fiber. The gain factor dependence on z arises due to pump depletion and gain saturation.

$$\mathrm{G} \equiv P_{\mathrm{out}}/P_{\mathrm{in}} = \exp\left(\int_0^L g(z)dz\right) \tag{9.1}$$

The gain factor $g(z)$ is a measure of the local growth of optical power $P(z)$.

$$g(z) = \frac{1}{P(z)}\frac{dP(z)}{dz} = \rho\Gamma(\sigma_{\mathrm{e}}N_2 - -\sigma_{\mathrm{a}}N_1) = \frac{g_0}{1 + P/P_{\mathrm{sat}}} \tag{9.2}$$

where N_2 and N_1 are the populations of the upper and lower energy levels of the active Er ions, ρ is the active ion density, and Γ is a confinement factor that is a measure of the overlap of the signal field with the doped core. σ_{e} and σ_{a} are the emission and absorption cross-sections of the two-level system at frequency ν. The equation on the right expresses the homogeneous saturation of the gain. When the signal power P reaches the saturation power P_{sat}, the gain reduces to half of the small signal gain g_0. The small signal gain factor g_0 is related to the pump power P_{p} as described by the following equation:

$$g_0 = \frac{\rho\Gamma\sigma_{\mathrm{e}}(P_{\mathrm{p}} - P_{\mathrm{TH}})}{P_{\mathrm{p}} + P_{\mathrm{TH}}\sigma_{\mathrm{e}}/\sigma_{\mathrm{a}}} \tag{9.3}$$

where the threshold pump power P_{TH} is given by

$$P_{\mathrm{TH}} = (\sigma_{\mathrm{a}}/\sigma_{\mathrm{e}})(h\nu_{\mathrm{p}}A/\Gamma_{\mathrm{p}}\tau\sigma_{\mathrm{p}}) \tag{9.4}$$

where τ, is the spontaneous lifetime of the upper energy level. A is the core area and h is Planck's constant. For pump powers lower than the threshold value, the gain is negative, and it reaches a maximum value $\rho\Gamma\sigma_{\mathrm{e}}$ for very

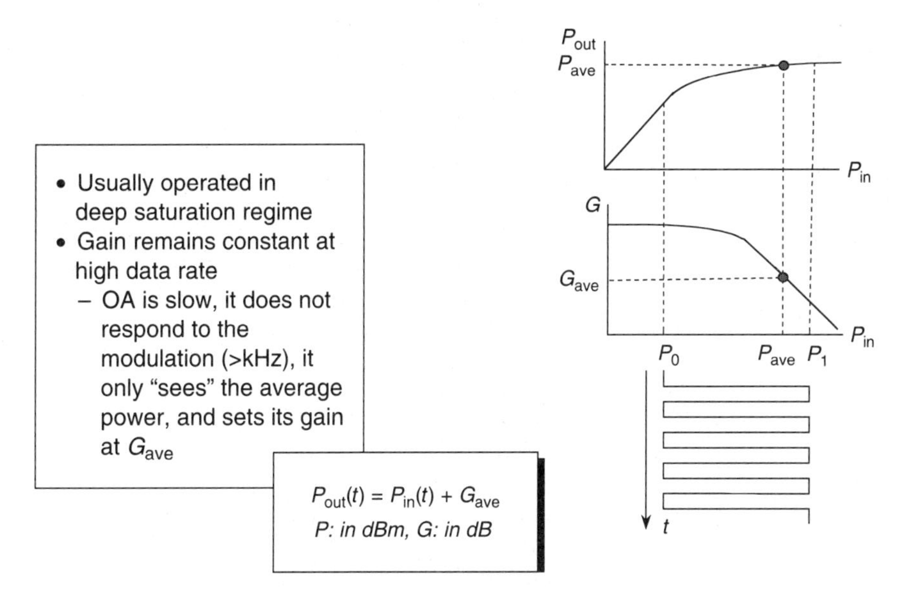

Fig. 9.4 Saturation characteristics of EDFA.

high pump powers. Γ_p is the pump confinement factor that is usually designed in the Er-doped fiber to maximize the overlap.

The amplifier saturation behavior is depicted in Fig. 9.4. For small input signal power the gain is constant and starts to decrease as the input power is increased further. In the saturation region, the output power is approximately proportional to the pump power. In this condition, the pump absorption from the ground state is balanced by stimulated emission from the first excited state induced by the signal. The balance shifts to higher signal power level as the pump power is increased. The output signal power at which the gain reduces to half of the small signal gain value is the saturation power P_{sat} and is given by

$$P_{sat} = \frac{h\nu A}{(\sigma_e + \sigma_a)\Gamma\tau}\left[1 + \frac{\sigma_a}{\sigma_e}\frac{P_p}{P_{TH}}\right] \tag{9.5}$$

For $P_p \gg P_{TH}$, P_{sat} is proportional to P_p/P_{TH}, while the preceding equations describe the gain saturation at a particular value of z. This local description is usually quite adequate in determining the amplifier saturation behavior because the saturation occurs primarily near the output end of the amplifier where the signal power is at its highest value.

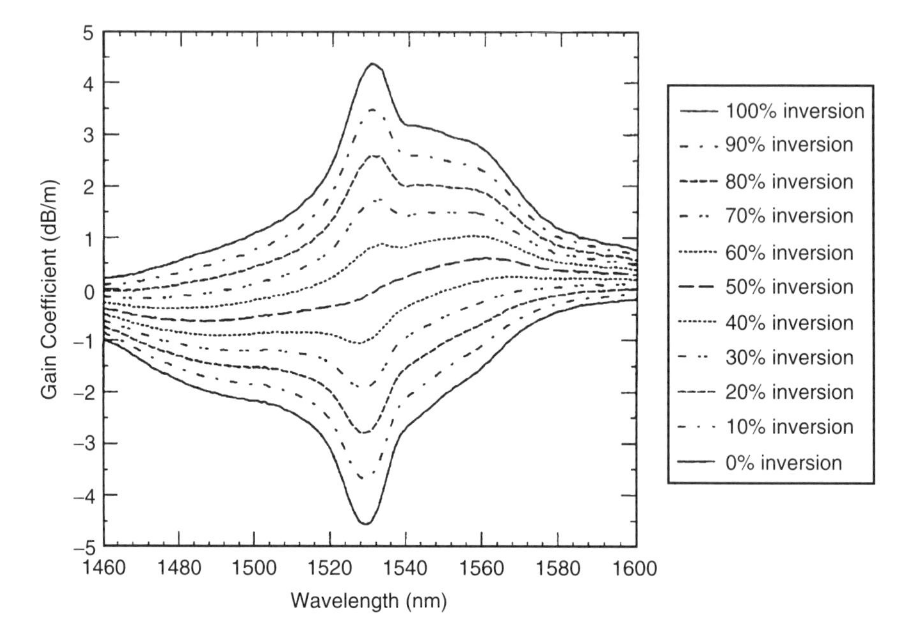

Fig. 9.5 The gain and loss coefficient spectra at different inversion levels for EDF with Al and Ge co-doping.

The gain and loss coefficient spectra are shown in Fig. 9.5 at different inversion levels for erbium-doped fiber with Al and Ge co-doping [9, 10]. Under a homogeneous broadening approximation, the overall gain spectrum of any piece of EDF always matches one of the curves after scaling, and does not depend on the details of pump power, signal power, and saturation level along the fiber [8]. The gain spectrum is very important for amplifier design.

9.2.1.2. Amplifier Noise

Amplification generates optical noise through the spontaneous emission process. The spontaneous emission arises when light emitted by spontaneous decay of ecited erbium ions is coupled to the optical fiber waveguide. The spontaneous emission is amplified by the gain in the medium and therefore increases with the gain G. The total amplified spontaneous emission noise power N_{ASE} at the amplifier output is given by

$$N_{\mathrm{ASE}} = 2n_{\mathrm{sp}}(G-1)h\nu B \tag{9.6}$$

where the factor of 2 takes into account the two modes of polarization supported by the fiber waveguide, B is the optical noise bandwidth, and n_{sp} is the spontaneous emission factor given by

$$n_{sp} = \sigma_e N_2 / (\sigma_e N_2 - \sigma_a N_1), \tag{9.7}$$

The spontaneous emission factor indicates the relative strengths of the spontaneous and stimulated emission process. It reaches its minimum value of 1 for complete inversion ($N_1 = 0$), when all the Er ions are in the excited state. The ASE power can be related to the noise figure (NF) of the amplifier, which is defined as the signal-to-noise ratio (SNR) corresponding to shot noise of the signal at the input divided by the SNR at the output. The NF is a measure of the degradation of the optical signal by the ASE noise added by the amplifier.

$$\text{NF} \equiv \text{SNR}_{\text{in}} / \text{SNR}_{\text{out}}, \tag{9.8}$$

which is given by

$$\text{NF} = N_{\text{ASE}} / h\nu BG = 2n_{sp}(G - 1)/G \tag{9.9}$$

As discussed earlier, for an EDFA with full inversion, the spontaneous emission factor reaches a value of unity.

A high inversion level provides a low noise figure, while a low inversion level yields high efficiency in the conversion of photons from pump to signal. To achieve both low noise figure and high efficiency, two or more gain stages are usually used, where the input stage is kept at a high inversion level and the output stage is kept at a low inversion level [11, 12]. An ASE filter is inserted in the middle stage to prevent gain saturation caused by the ASE peak around 1530 nm. For optical amplifiers with two or more gain stages, the overall noise figure is mainly decided by the high gain input stage and the output power is basically determined by the strongly saturated output stage. The passive components have minimal impact on noise figure and output power when they are in the mid-stage.

9.2.2. EDFA DESIGN CONSIDERATIONS

In recent years, the EDFA architecture has evolved significantly in order to provide the high level of performance required by the high-capacity WDM systems. The present generation of EDFAs needs to provide both high output power and low noise figure characteristics. In addition, for

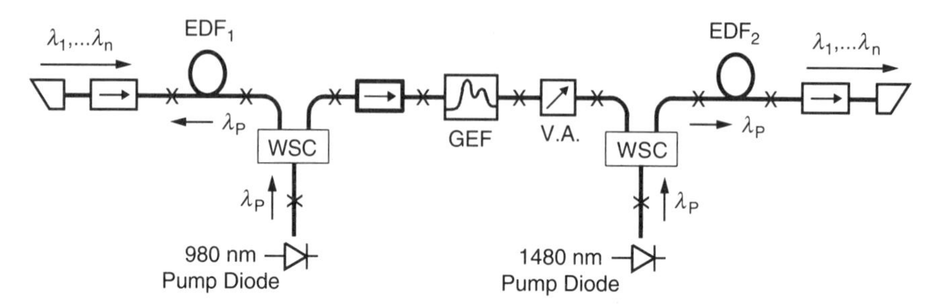

Fig. 9.6 A two-stage EDFA design with a mid-stage pumping configuration.

in-line applications, the amplifier design is required to accommodate 8–12 dB mid-stage loss, which is needed for dispersion compensation module or other elements such as optical add/drop multiplexer, etc. In this section we review the design considerations for the architecture of such EDFAs.

The amplifier design has to optimize the noise performance of the amplifier, while maintaining high output power and maintaining low cost and high reliability. While the first generation EDFAs were designed to have a single stage, the current requirements on the performance have led to a multi-stage design. Significant performance benefit can be derived from a basic two-stage architecture, shown schematically in Fig. 9.6. In order to maintain a low noise figure, the amplifier first stage is pumped by a 980-nm source. As discussed earlier, in order to achieve a low noise figure, the EDFA has to be inverted efficiently. The spontaneous emission generated near the EDFA input experiences the full gain of the amplifier, and therefore in order to minimize the noise figure it is essential to achieve the highest inversion near the front end of the amplifier. The most efficient pump for this purpose is a 980-nm source. The increased reliability of 980 pumps means they can be used for terrestrial applications. Using 1480-nm pumps, complete inversion of Erbium cannot be achieved because of the stimulated emission at the pump wavelength. Usually the noise figure of a 1480-nm pumped EDFA is 1dB higher than that obtained by a 980-nm pumped EDFA. For the power amplifier stage, however, it is the 1480-nm pump that is more suitable because higher pump conversion efficiency can be achieved. This is because of close photon energy match between the pump and the signal and also lower excited state absorption (ESA) as compared to 980 pumping. High gain efficiency in the Er-doped fiber reduces the pump power requirement and thereby enhances reliability.

In an EDFA, the erbium-doped fiber (EDF) is usually pumped by 980-nm or 1480-nm semiconductor lasers. In order to obtain a low noise figure, the first stage of EDFA is pumped using a 980-nm source, which can create a very high degree of inversion. The power stage in EDFA is pumped at 1480 nm, which provides high quantum conversion efficiency, a measure of the conversion of pump power to signal power. The quantum efficiency of the 980 nm pumped stage is poorer due to a greater mismatch of energies of the pump photons as compared to signal photons and a significant amount of excited state absorption. A three-level model can be used to describe a 980-nm pumped stage while a two-level model usually suffices for the 1480-nm pumped section [7, 8]. A key advantage of EDFAs in high data rate transmission systems is that the spontaneous lifetime of the metastable energy level ($^4I_{13/2}$) is about 10 ms, which is usually much slower than the time corresponding to signal bit rates of practical interest. As a result of the slow dynamics, inter-symbol distortion and inter-channel crosstalk are negligible.

The gain and loss coefficient spectra at different inversion levels for EDF with Al and Ge co-doping are given in Fig. 9.5. The inversion level at a point in Er fiber is expressed as the fractional percentage $N_2/(N_1 + N_2)$, where N_1 and N_2 are the populations of Er ions in the ground and excited states, respectively, and $(N_1 + N_2)$ is the total number of Er ions. Under a homogeneous broadening approximation, the overall gain spectrum of any piece of EDF always matches one of the curves after scaling, and does not depend on the details of pump power, signal power, and saturation level along the fiber. The derivation of a general two-level model that describes spectra and their dependence on EDF length and other parameters is given in ref. [8]. One key parameter in this model is the average fractional upper level density $\overline{N_2}(t)$, which is also called average inversion level [9], given by the average of N_2 over the length ℓ of EDF:

$$\overline{N_2}(t) = \frac{1}{\ell} \int_0^\ell N_2(z, t) dz \tag{9.10}$$

In the limit of strong inversion $\overline{N_2}$ takes its maximum value and the gain is highest. When the signal power becomes comparable to pump power, the EDFA is saturated and the level of invesion is reduced. The level of inversion and degree of saturation are closely related and are often used interchangeably. In the limit of low pump and low temperature, $\overline{N_2}(t) = 0$, where the absorption is the strongest. The gain spectra plotted in Fig. 9.5

are very useful in the study and design of EDFAs, such as locating the gain peak wavelengths at different inversion levels, finding out inversion level with wide flat gain range, and understanding relative gain variation at different wavelengths with changing inversion levels. The relative gain spectrum is only a function of the average inversion level for a given type of EDF, while the total integrated gain depends on the length of the EDF. There are small deviations in the gain spectrum from this model due to inhomogeneity in EDF that gives rise to spectral-hole burning, which is discussed in a subsequent section. Under the homogeneous approximation, however, if there are two EDFAs made of the same type of fiber but of different length, the instantaneous gain spectrum would be similar in shape scaled for the length if the instantaneous length averaged inversion level is the same.

A high inversion level provides a low noise figure while a low inversion level yields high efficiency in the conversion of photons from pump to signal [8]. To achieve both low noise figure and high efficiency, two or more gain stages are usually used, where the input stage is kept at a high inversion level and the output stage is kept at a low inversion level [11, 12]. Because the ASE power around the 1530 region can be high enough to cause saturation, an ASE filter can be added in the middle stage to block the ASE in this band [11]. These optical amplifiers have been successfully used in the early WDM optical networks [13].

For optical amplifiers with two or more gain stages, the overall noise figure is mainly decided by the high-gain input stage and the output power is basically determined by the strongly saturated output stage [11]. The passive components have minimal impact on noise figure and output power when they are in the mid-section of the amplifier. The noise figure of a two-stage EDFA is given by

$$NF = NF_1 + NF_2/L_1G_1 \tag{9.11}$$

where NF_1 and NF_2 are the noise figures of the two stages and L_1 and G_1 are the mid-stage loss and gain of the first stage, respectively. It can be seen from the noise figure formula (Eq. 9.11) that the overall noise figure of a two-stage amplifier is primarily determined by the noise figure of the first stage because the first stage gain is usually designed to be much larger than the middle-stage loss. For example, in a typical case, gain of the first stage $G_1 = 100$ and mid-stage loss $L_1 = 0.1$, the overall noise figure has only 10% contribution from the second-stage noise figure.

9.2.3. SYSTEM ISSUES

The EDFA is an optical amplifier that amplifies optical signals in the optical domain. Unlike the opto-electronic regenerators, used in earlier systems, the optical amplifier does not need high-speed electronic circuitry and is transparent to data rate and format, which dramatically reduces cost. All the optical signal channels can be amplified simultaneously within the EDFA, which provides high gain, high power, and low noise figure. In lightwave communication systems it can be used as a booster to increase the power of the transmitter, an in-line amplifier or a repeater to enhance the overall reach of the system, or a pre-amplifier to increase the receiver sensitivity. Due to slow gain dynamics in EDFAs, the high-speed data modulation characteristics of optical signals are preserved during the amplification process. Application of EDFAs as repeaters in WDM systems is particularly important because they offer a cost-effective means of faithfully amplifying all the signal wavelengths within the amplifier band simultaneously, thereby eliminating the need for costly opto-electronic regenerators. WDM-based lightwave systems were deployed by AT&T in 1996 for the first time. In present-day lightwave systems the EDFA has replaced the opto-electronic regenerator as the repeater of choice in both the sub-marine and terrestrial systems.

Currently, optical communication technology is moving from point-to-point systems to optical networking, where EDFAs can play a role at many places in WDM optical networks. The EDFAs are used mostly in optical transport line systems, which consist of a transmitter, fiber spans with in-line amplifiers, and a receiver. On the transmitter end, multiple optical channels are combined in an optical multiplexer, and the combined signal is amplified by a power amplifier before being launched into the first span of transmission fiber. At the receiver end, the incoming WDM signals are amplified by a pre-amplifier before being demultiplexed into individual channels that are fed into the respective receivers. Long-haul applications also require in-line repeater amplifiers to extend the total system reach.

9.2.3.1. Optical Signal-to-Noise Ratio

In an optically amplified system, channel power reaching the receiver at the end of the link is optically degraded by the accumulated amplified spontaneous emission (ASE) noise from the optical amplifiers in the chain. At the front end of the receiver, ASE noise is converted to electrical noise, primarily through signal-ASE beating, leading to bit-error-rate (BER) flooring.

System performance therefore places an important requirement on optical signal-to-noise ratio (OSNR) of each of the optical channels. OSNR therefore becomes the most important design parameter for an optically amplified system. Other optical parameters in system design consideration are channel power divergence, which is generated primarily due to the spectral gain non-uniformity in EDFAs (described in next section), and maximum channel power relative to the threshold levels of optical nonlinearities, such as self-phase modulation, cross-phase modulation, and four-photon mixing [14].

Although optical amplifiers are conventionally classified into power, in-line, and pre-amplifiers, state-of-the-art WDM systems require all three types of amplifiers to have low noise figure, high output power, and uniform gain spectrum. We will not distinguish these three types of amplifiers in the discussion presented in this section. The nominal OSNR for a 1.55-μm WDM system with N optical transmission spans can be given by the following formula [10]:

$$\mathrm{OSNR}_{\mathrm{nom}} = 58 + P_{\mathrm{out}} - 10\log_{10}(N_{\mathrm{ch}}) - L_{\mathrm{sp}} - NF - 10\log_{10}(N) \quad (9.12)$$

where OSNR is normalized to 0.1-nm bandwidth, P_{out} is the optical amplifier output power in dBm, N_{ch} is the number of WDM channels, L_{sp} is the fiber span loss in dB, and NF is the amplifier noise figure in dB. For simplicity, it has been assumed here that both optical gain and noise figure are uniform for all channels.

Equation (9.12) shows how various system parameters contribute to OSNR, for example, the OSNR can be increased by 1 dB by increasing the amplifier output power by 1 dB, or decreasing the noise figure by 1 dB, or reducing the span loss by 1 dB. This equation indicates that we can make trade-offs between number of channels and number of spans in designing a system. However, the trade-off may not be straightforward in a practical system because of the mutual dependence of some of the parameters. Other system requirements also impose additional constraints, for example, optical nonlinearities place an upper limit on channel power and this limit depends on the number of spans, fiber type, and data rate.

The simple formula in Eq. (9.12) highlights the importance of two key amplifier parameters: noise figure and output power. While it provides valuable guidelines for amplifier and system design, it is always necessary to simulate the OSNR evolution in a chain of amplifiers when designing a practical WDM system. The amplifier simulation is usually based on an accurate mathematical model of amplifier performance. Amplifier modeling is also a critical part of the end-to-end system transmission performance simulation that incorporates various linear and nonlinear transmission penalties.

9.2.3.2. Amplifier Gain Flatness

Amplifier gain flatness is another critical parameter for WDM system design. As the WDM channels traverse multiple EDFAs in a transmission system, the spectral gain nonuniformity adds up to create a divergence in channel powers. The worst WDM channel, the channel that consistently experiences the lowest amplifier gain, will have an OSNR value lower than the nominal value given in Eq. (9.12). The power deficit, which can be viewed as a form of penalty given rise to by amplifier gain nonuniformity, is a complicated function of individual amplifier gain shape [2], and correlation of the shapes of the amplifiers in the chain. The gain flatness is a parameter that can have a significant impact on the end-of-system OSNR. The penalty is especially severe for a long amplifier chain, as in the case of long-haul and ultra-long-haul applications.

The gain flatness affects system performance in multiple ways. Flat-gain amplifiers are essential to getting the system OSNR margin for routed channels and minimizing power divergence to allow practical implementation of networking on the optical layer. Wide bandwidth can enable either large channel spacing as a countermeasure of the filter bandwidth narrowing effect, or allow more optical channels for more flexibility in routing of traffic. Amplifier gain flatness, as discussed earlier, is critical to maintaining system performance under varied channel loading conditions caused by either network reconfiguration or partial failure.

Figure 9.7 shows how the OSNR penalty increases as a nonlinear function of the number of transmission spans for three cases: a ripple (flatness) of 1.0, 1.4, and 1.8 dB. The variations in signal strength may exceed the system margin and begin to increase the bit error rate (BER) if the SNR penalty exceeds 5 dB. When the gain spectrum starts with flatness of 1.8 dB, the OSNR penalty degrades by more than 5 dB after only 8 transmission spans. This OSNR penalty limits the reach of a WDM line system and requires signal regenerators at intervals of approximately 500 kilometers. These expensive devices convert the signals from the optical domain to the electrical domain, typically reshaping, retiming, and re-amplifying the signal before triggering lasers to convert the signal back from the electrical domain to the optical domain. In ultra-long-haul networks, carriers would like to increase the spacing between regenerators to several thousand

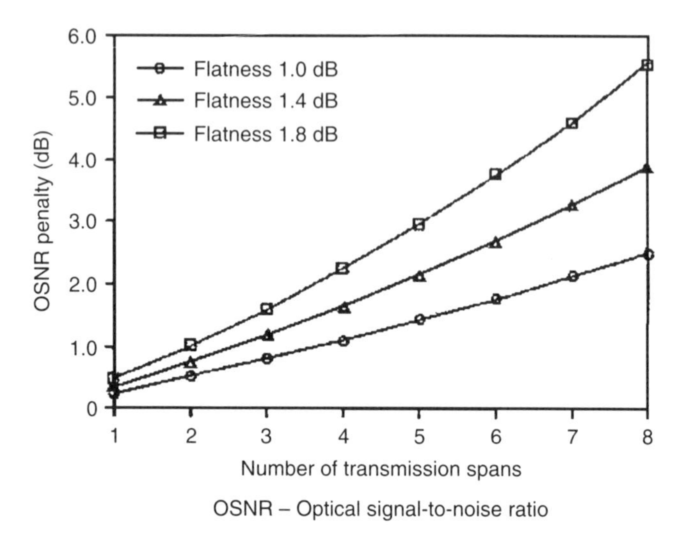

Fig. 9.7 Optical signal-to-noise ratio penalty vs. number of spans for different EDFA gain ripple [2].

kilometers, in which case the signal would have to pass through up to 50 amplifiers without electrical regeneration. These networks require EDFAs with excellent gain uniformity.

The impact of gain nonuniformity, which gives rise to channel power divergence in a chain of amplifiers, is, however, not limited to the OSNR penalty. While the weak channels see an OSNR penalty that limits the system performance, the strong channels continue to grow in power and may reach the nonlinear threshold, also limiting system performance. Additionally, large power divergence increases the total crosstalk from other WDM channels at the optical demultiplexer output. It is thus imperative to design and engineer optical amplifiers with the best gain flatness for WDM networking applications. The state-of-the-art optical amplifiers usually incorporate a gain equalization filter to provide uniform gain spectrum, as discussed in Section 9.3. To minimize the residual gain nonuniformity, it requires careful design, modeling, and engineering of the amplifiers in general, and the gain equalization filters in particular.

The gain equalization filters are optimized to flatten the gain spectrum of a fully loaded EDFA. But if a carrier wants to operate the system with fewer channels—for example, to reconfigure it dynamically—then the lower input power can decrease the EDFA's gain uniformity, thereby impairing the effectiveness of the GEF and increasing ripple in the network. Furthermore, as described later, spectral hole burning gives rise to channel-loading dependent changes in the gain spectrum of the EDFA by creating a dip in the region of the active channels. Spectral hole burning can create a gain spectrum for which the GEF was not optimized, making gain flattening very difficult. For all these reasons, future ultra-long-haul, dynamically re-configurable networks will require EDFAs with dynamic gain equalization. The gain spectrum of EDFA will be equalized by the use of a dynamically controlled filter having variable spectral loss characteristics. The dynamic gain equalizer can be controlled in a feedback loop in conjunction with an optical channel monitor to provide uniform channel powers or OSNR.

9.2.3.3. Amplifier Control

In an amplified system, optical amplifiers may not always operate at the gain value at which their performance, especially gain flatness, is optimized. Many factors contribute to this non-optimal operating condition. Among them is the fact that the span loss can be adjusted at system installation and maintained in the system's lifetime only to a finite range with respect to the value required by the amplifiers for optimal performance. As a result, amplifier gain will be tilted, and such tilt can have significant impact on system performance in ways similar to gain nonuniformity.

Gain tilt can, if not corrected, result in OSNR penalty and increased power divergence. Control of optical amplifier tilt is often necessary to extend the operational range of the amplifiers and compensate for loss tilt in the system due to, for example, fiber loss variation in the signal band. Control of amplifier gain tilt can be achieved by varying an internal optical attenuator [15, 16]. Implementation of such tilt control function requires a feedback signal that is derived from, for example, measured amplifier gain or channel power spectrum, and an algorithm that coordinates the measurement and adjustment functions. By changing the loss of the attenuator, the average inversion level [8] of the erbium-doped fiber can be adjusted, which affects the gain tilt in the EDFA gain spectrum.

9.2.4. DYNAMIC NETWORKS-RELATED ISSUES

Another important control function is amplifier power adjustment. In a WDM system, there is a need to adjust the total amplifier output as a function of the number of equipped channels. The total output power must be adjusted such that while the per-channel power is high enough to ensure sufficient OSNR at the end of the chain, it is low enough not to exceed the nonlinear threshold. In addition, per-channel power must be maintained within the receiver dynamic range as the system channel loading is changed. Such power adjustment has traditionally been achieved through a combination of channel monitoring and software-based pump power adjustment.

The recent advances in WDM optical networking have called for fast gain control to minimize the channel power excursion when a large number of channels are changed due to, for example, catastrophic partial system failure. Various techniques, as detailed in Section 9.4, have been demonstrated to stabilize amplifier gain, thereby achieving the goal of maintaining per-channel power. In addition to amplifier dynamics control, practical implementation in a system also requires a receiver design that can accommodate power change on a very short time scale.

9.3. Advances in EDFAs for High-Capacity Networks

9.3.1. WIDE-BANDWIDTH C-BAND EDFA

First-generation WDM systems utilized 8–10 nm of spectrum between 1540 and 1560 nm, where the gain of EDFAs is quite uniform. In this case the amplifier is operated at an inversion level on 70–80% (see Fig. 9.5). Another amplifier based on erbium-doped fluoride fiber (EDFFA) consisting of fluoro-zirconate was shown to have greater (24 nm) bandwidth [17, 18]. Unlike silica-based EDFAs, however, fluoride-based fiber is not a field-tested technology and there are concerns about its long-term reliability. Availability of gain equalization filters (GEFs) provides a way to increase the usable bandwidth in silica-based EDFAs. Several technologies have been studied to fabricate GEFs, including thin-film filters, long-period gratings [19], short-period gratings [20], silica waveguide structure [21], fused fibers, and acoustic filters [22].

Broadband EDFA having 40-nm bandwidth using long period gain-equalization filter has been demonstrated [23]. A similar design of EDFA with an optical bandwidth of 35 nm was used to demonstrate the transmission of 32 channels at 10 Gb/s over 8 spans of 80 km. [15]. The EDFAs reported in this work (Fig. 9.6) provided the first demonstration of two new features crucial for broadband, long-haul systems and optical networks. First, the EDFAs were gain flattened to provide flat gain over the full optical bandwidth of the erbium gain spectrum. The gain spectrum for an

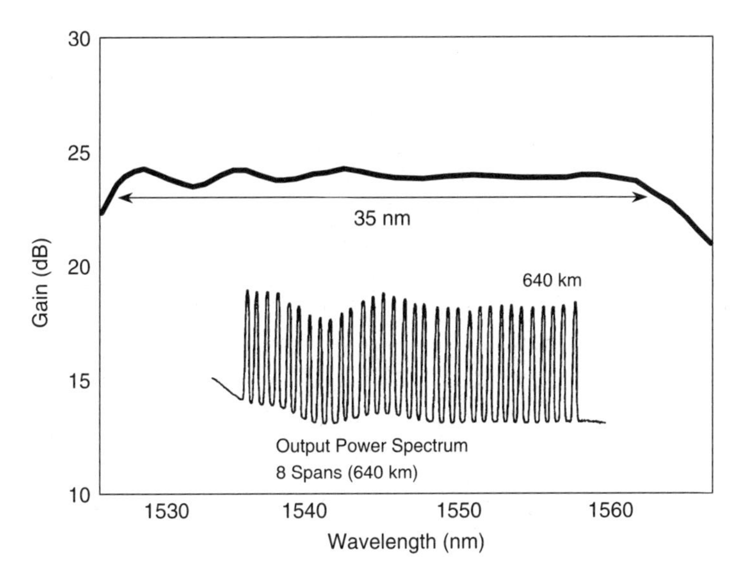

Fig. 9.8 Gain spectrum from the EDFA of Fig. 9.3. The gain is flattened with a long-period fiber grating equalization filter. The inset shows the channel power spectrum after transmission through 8×80 km spans and 8 EDFAs.

individual gain-flattened EDFA is shown in Fig. 9.8. Although the WDM channels in this experiment occupy an optical bandwidth of 25 nm, the EDFA's 1 dB optical bandwidth is 35 nm and the peak-to-peak gain variation over a 34-nm bandwidth is less than 0.6 dB or 2.5% of the gain. Second, with a mid-amplifier attenuator [16], the gain-flattened EDFA can be operated with this broad optical bandwidth in systems with a wide range of span losses. The attenuator can be adjusted to permit broadband, flat gain operation for a wide range of operating gain conditions, which is necessary to accommodate variations in span losses commonly encountered in practical transmission systems and multi-wavelength optical networks.

In the transmission experiment 32 channels with wavelengths ranging from 1531.4 nm (channel 1) to 1556.2 nm (channel 32) with 100 GHz channel spacing were transmitted. The signals were all modulated at 10 Gb/s ($2^{31} - 1$ PRBS) and after amplification the signals were launched into the transmission span. The transmission span consisted of four 80-km lengths of negative non-zero dispersion TrueWave (TW) fiber and four 80-km lengths of positive TrueWave fiber for a total transmission length of 640 km. This configuration results in a fiber span with low overall dispersion while maintaining finite local dispersion to reduce nonlinear effects. The positive

TW spans were chosen with dispersion slightly higher than 2.3 ps/nm-km at 1555 nm so that the zero-dispersion wavelength was below the short-wavelength signals in the 1530-nm range. The dispersion zero of the negative TW spans with dispersion of −2.3 ps/nm-km (at 1555 nm) was safely out of the range of the long-wavelength signals. The optical spectra after 640 km transmission is shown as an inset in Fig. 9.8. The signal-to-noise ratio (in a 0.1-nm bandwidth) is greater than 35 dB after the first amplifier, and remains greater than 23.7 dB after the eighth EDFA. The gain of the EDFAs operating in the system is 5 dB less than the ideal design gain. Nevertheless, with the mid-amplifier attenuators properly adjusted, the variation in channel powers after eight EDFAs and 140 dB of gain is only 4.9 dB, or 3.5% of the total gain. The channels were characterized for BER performance and achieved error rates below 10^{-9}.

9.3.2. L-BAND EDFA

In order to expand the optical bandwidth usage per fiber, the WDM systems have to be expanded beyond the conventional band or C-band (1525–1565 nm). The realization of EDFAs in the longer wavelength region [24] or L-band (1570–1610 nm) has doubled the usable transmission bandwidth. In addition to capacity, L-band EDFAs enable WDM system operation over different types of dispersion-shifted fiber (DSF) having low dispersion in the C-band. This is very significant because the deployment of non-zero dispersion-shifted fiber NZDSF now exceeds that of SMF. Dense WDM transmission over DSF/NZDSF was previously not feasible in the 1550-nm region due to the low values of dispersion, which result in unacceptably high levels of four wave mixing. The dispersion in fiber increases at longer wavelength and consequently the levels of FWM are reduced in the L-band.

In the long wavelength region, the EDF has nearly 0.2 dB/m gain coefficient for the inversion level between 20 to 30%. Even though at this inversion level the gain is much smaller than the gain at the highest peak in the C-band at 1532 nm, the gain shape in L-band is much more uniform as compared to that in C-band. This means that much less gain filtering is required in L-band EDFA. A comparison of the EDF gain spectra of C- and L-bands is shown in Fig. 9.9. The inversion level in EDF is optimized to provide the most bandwidth in both the cases. In order to achieve low noise figure operation, high gain in the first stage of EDFA is required. Due to small gain coefficient the EDF lengths exceeding 5 times that in a C-band EDFA are therefore necessary. Unfortunately, longer EDF

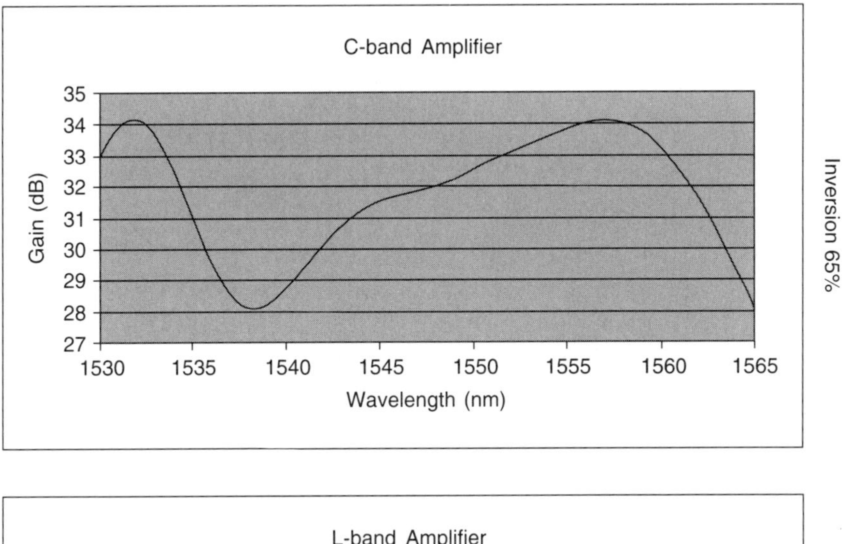

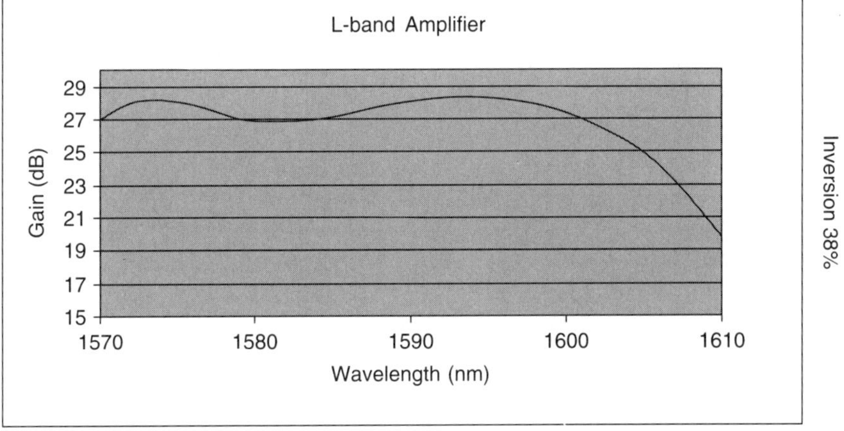

Fig. 9.9 A comparison of C- and L-bands and gain spectra.

length leads to larger background loss, which is detrimental to the noise figure. Recently, EDF optimized for low inversion L-band operation has been developed which can provide gain coefficient as high as 0.6 dB/m [25]. In new EDF larger overlap between the mode-field of the signals and the ion-doped core area is needed to increase the absorption, without producing concentration-quenching effect. The new EDF is designed to have longer cutoff wavelength around 1450 nm and has 2–3 times greater power overlap. The greater cutoff wavelength leads to smaller mode field diameter enabling better than 90% optical power confinement in the core area.

The new fiber has other advantages, such as lower background loss, increased tolerance to fiber bend, and higher pump efficiency [25]. In this type of fiber, power efficiency as high as 60% has been measured and 5-mm bend radius did not generate additional loss.

Several L-band transmission experiments at 10 Gb/s and higher rates in recent years [26–28] have demonstrated the feasibility of L-band EDFAs and WDM transmission over DSF/NZDSF. The first 64×10 Gb/s WDM transmission [29] with 50 GHz channel spacing over DSF verified that nonlinear effects such as four-photon mixing can be controlled in L-band.

EDFA used in the experiment had 25 dB design gain and 43.5 nm (1568.5 to 1612 nm) bandwidth. The noise figure in the L-band is ~6.5 dB. The amplifier has a high output power of 22 dBm, which is necessary to support a large number of channels. The amplifier incorporates an attenuator in a mid-section to correct the gain tilt caused by span loss variation. The mid-section can accommodate dispersion-compensating devices without affecting the gain flatness of the amplifier. The outputs of 64 external cavity lasers were combined using two 100-GHz waveguide grating routers. An attenuator/power meter was added at the output of each EDFA to control the power launched into the transmission span. The total power launched into fiber spans was between 15 and 16 dBm.

The transmission span consisted of five 100-km lengths of dispersion-shifted fiber having a total length of 500 km. The zero-dispersion wavelength of the spans were in the range 1546–1554 nm. The average value of span dispersion at channel 1 and channel 64 was 2.2 and 4.3 ps/nm-km, respectively. In the transmission band the overall dispersion was low while maintaining finite local dispersion to reduce nonlinear effects. The signal-to-noise ratio (in a 0.2-nm bandwidth) was greater than 32 dB after the first amplifier, and remains greater than 21 dB after the sixth EDFA. The gain of EDFAs operating in the system was 5 dB less than the design gain of the amplifier. With the mid-amplifier attenuators properly adjusted, the variation in channel powers after six EDFAs and 100 dB of gain was only 8.9 dB, or 8.9% of the total gain. After the fifth fiber span, the signals were amplified by an EDFA optical pre-amplifier and demultiplexed with a tunable optical bandpass filter with 0.25-nm 3 dB bandwidth. Subsequently, it was detected by a p-i-n detector and split for clock extraction and data. A DCF module having −310 ps/nm dispersion at 1555 nm was added in the mid-section of the preamplifier. The eye diagrams of all channels after transmission through 500 km are open and, as expected, exhibit little distortion due to chromatic dispersion. All channels achieved error rates below 10^{-9}. The power penalty was between 1.5 and 3.2 dB for 10^{-9}BER. The penalty could be due to nonlinear effects such as cross-phase modulation. Small channel separation (50 GHz) enhances these effects.

9.3.3. ULTRA-WIDE-BAND EDFA

Because the gain drops sharply on both sides of C-band at a 40–60% inversion, it is not practical to further increase the bandwidth with a GEF. However, a flat gain region between 1565 and 1615 (L-band) can be obtained at a much lower inversion level (−20 to −40%) [30]. The combined C- and L-Band amplifier can achieve bandwidth that is more than double that of the conventional band amplifier. The principle of such an amplifier is shown in Fig. 9.10. After the initial demonstration of the principle [30–32], much progress has been made in the understanding and design

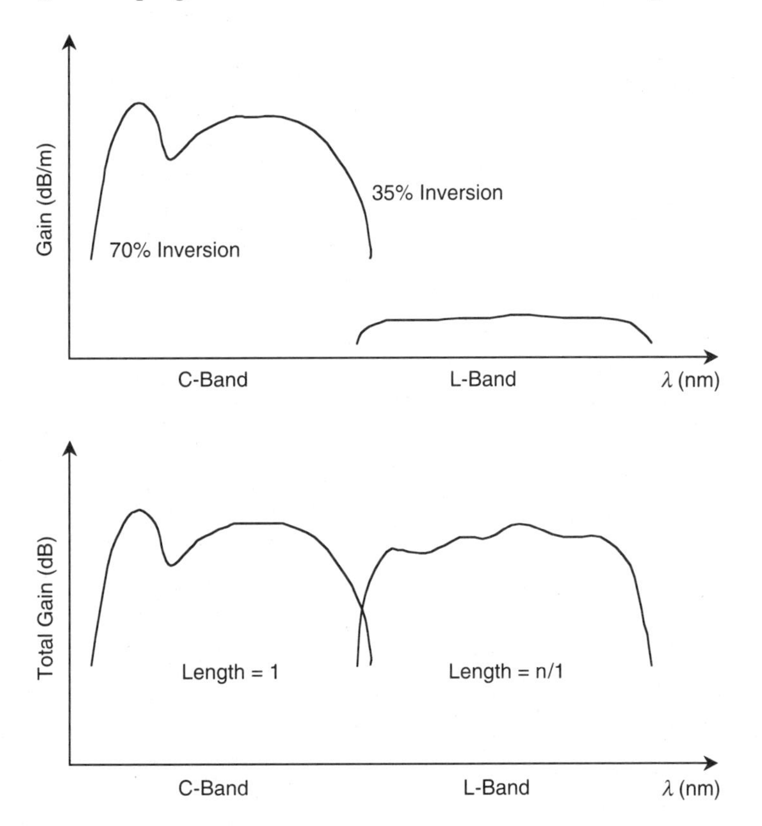

Fig. 9.10 Principle of combined C-band and L-band ultra-wide-band amplifier [2].

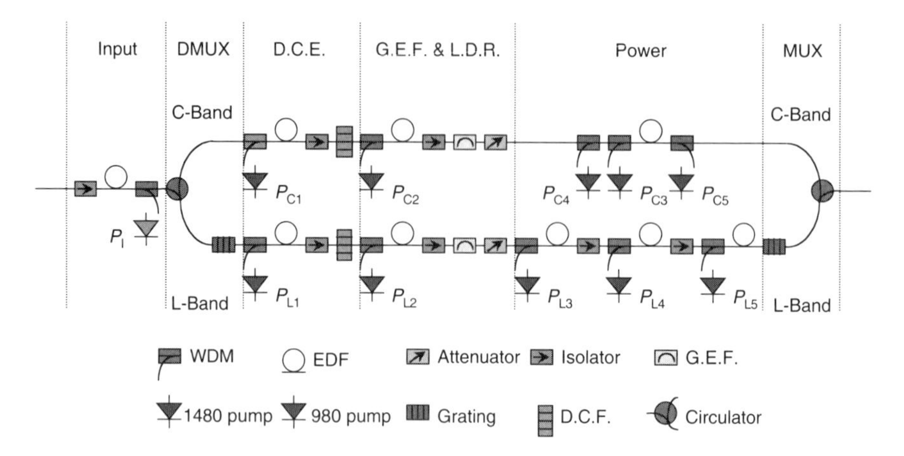

Fig. 9.11 Schematic of ultra-wide-band amplifier.

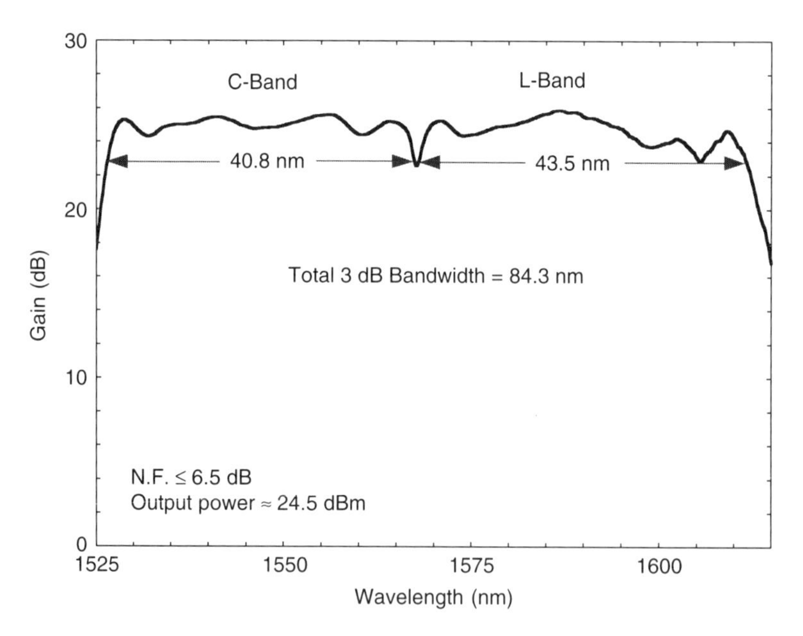

Fig. 9.12 Gain spectrum of ultra-wide-band amplifier [2].

of ultra-wide-band optical amplifiers with a split-band structure. A recent design shown schematically in Fig. 9.11 demonstrated a bandwidth of 84.3 nm. The gain and noise figure of the amplifier is shown in Fig. 9.12. Besides very wide bandwidth, this amplifier also provides power tilt control, which is realized by the variable attenuation, and dispersion compensation, which is realized by the dispersion compensation element after the second gain stages [33]. With this ultra-wide-band optical amplifier (UWBA), the first long-distance WDM transmission at 1 terabit capacity was demonstrated [5, 34, 35]. The experiment utilized 100 channels with 10 Gb/s line rate transmitted over four spans totaling 400 km of NZDSF. Sixty channels from 1546.6 nm to 1560.2 nm with 50 GHz channel spacing were transmitted in the C-band and 40 channels from 1569.4 nm to 1601.4 nm with 100 GHz spacing occupied the L-band. Error-free transmission of all 100 channels was demonstrated.

9.3.4. RAMAN/EDFAs

Raman effect in silica fiber has been intensively investigated in recent years. Stimulated Raman scattering transfers energy from the pump light to the signal via the excitation of vibrational modes in the constituent material. Measurement of Raman gain coefficient in silica fiber [36] has revealed that a significant amount of gain can be obtained at moderately high pump powers. The Raman gain is given by $G_r = \exp(P_p/2L_{eff}A_{eff})$, where P_p, L_{eff}, and A_{eff} are the pump power, effective length, and effective area, respectively. The Raman gain peak is offset by one Stokes shift in wavelength from the pump signal; thus Raman amplifiers can be implemented at any wavelength by selecting a suitable pump signal wavelength. The Stokes shift for silica fibers is approximately 13 THz, which corresponds to nearly 100 nm at 1550 nm. Raman gain spectrum is fairly uniform and has a 3 dB bandwidth of about 5 THz , corresponding to 40 nm in C-band.

Raman amplification can be achieved with low noise figure, which can be applied to enhance the system margin in WDM transmission systems. Unlike the EDFA, which requires a certain pump power to maintain high inversion level for low noise operation, the Raman amplifier can be inverted regardless of the pump level because the absorption of the signal photon to the upper virtual state is extremely small. The other significant advantage of Raman amplifiers is the ability to provide distributed gain in transmission fiber. The Raman gain, distributed over tens of kilometers of transmission fiber, effectively reduces the loss (L) of the fiber span and results in superior end-of-system OSNR. Demand for higher capacity and longer reach systems coupled with the availability of high-power pumps in the 1450 and 1480 nm wavelength region enabled the application of Raman amplifiers to WDM transmission systems. In addition, large deployment of smaller core area NZDSF in terrestrial networks has made it possible to obtain significant Raman gain with standard pump diodes. Several WDM transmission experiments have been reported, which used distributed Raman amplification in the transmission fiber to enhance the system OSNR. The Raman pumping is normally implemented in a counter-propagating configuration in order to avoid noise transfer from the Raman pump to the WDM signals. The counter-propagating pump configuration also efficiently suppresses any signal-pump-signal crosstalk that may occur if the Raman pump is depleted by the WDM channels.

The gain coefficient in the Raman amplifier is quite small and as a result gain medium consisting of tens of kilometer fiber length is needed. Because the power conversion efficiency in Raman amplifiers is $\sim$10%, which is several times smaller as compared to that in EDFAs (>60%), the most attractive design of low-noise figure amplifier is a hybrid configuration consisting of a Raman pre-amplifier stage followed by an EDFA power stage. In such a design, the gain of the Raman stage is kept below 16 dB in order to minimize inter-symbol interference arising from amplified double reflections of the signal from either discrete reflection points or from double Rayleigh scattering [37]. Noise figure improvement of 3–4 dB in such a hybrid design over the EDFA counterpart has been demonstrated [37, 38].

The enhanced margin derived from superior noise performance of Raman/EDFAs can be utilized in several ways, such as to increase the separation between amplifiers, to increase the overall reach of the transmission system, and to increase the spectral efficiency of transmission by reducing the channel separation or increasing the bit rate per channel. For example, a decrease in channel spacing to obtain higher spectral efficiency requires a reduction in launch power to avoid increased penalties from fiber nonlinearities, and must be accompanied by a noise figure reduction to maintain optical signal-to-noise (OSNR). Likewise, an increase in line rate requires a reduction in span noise figure to increase the OSNR at the receiver accordingly. Published results qualitatively confirm the outlined relation between channel spacing and noise figure for WDM systems limited by four wave mixing [39, 40]. In addition to superior noise performance, Raman amplification can provide gain in spectral regions beyond the C- and L-bands. As mentioned earlier, Raman gain curve is intrinsically quite uniform and broader gain spectra are naturally achievable in Raman amplifiers with gain-flattening filters. Alternatively, wide gain spectra may be obtained [41, 42] by the use of multi-wavelength pump sources, a technique applicable only for Raman amplifiers.

Three transmission experiments are described next, which show the benefits of incorporating Raman/EDFAs in WDM systems. The first experiment demonstrates that by the use of distributed Raman gain in a multi-span system the system margin can be enhanced by 4–5 dB. The second experiment shows that terabit capacity ultra-dense WDM transmission with 25 GHz channel spacing is made possible by the use of Raman gain because the launched power per channel can be lowered and nonlinear effects can be minimized. Finally, a transmission experiment at 40 Gb/s line rate covering C- and L-bands with overall capacity of 3.2 Tb/s is demonstrated by the use of Raman/EDFAs [43, 44]. More than 6 dB reduction in span noise figure obtained by distributed Raman amplification is expected to become essential in WDM transmission systems operating at 40 Gb/s. The improvement makes up for the 6 dB higher OSNR requirements for 40 Gb/s signals compared to 10 Gb/s signals. Thus, amplifier spacing used in today's 10 Gb/s WDM systems can be accommodated at 40 Gb/s by incorporating distributed Raman amplification.

A schematic of the first experiment [38] is shown in Fig. 9.13. Forty WDM channels modulated at 10 Gb/s in L-band with 50 GHz channel spacing were transmitted. Raman gain was provided at the input of each in-line amplifier by incorporating two polarization-multiplexed 1480-nm pump lasers with a wavelength selective coupler (shown in the lower section of Fig. 9.13). The DSF spans acted as the gain medium with the signal and pump propagating in opposite directions. The total Raman pump power in the fiber was 23.5 dBm, which resulted in a peak gain of 12 dB at 1585 nm. The total power launched into fiber spans was $\sim$15 dBm or -1dBm/ch. The transmission span consisted of five 120-km lengths of dispersion-shifted fiber, having a total length of 600 km. The eye diagrams of all channels after transmission through 600-km DSF are open and exhibit little distortion. All channels achieved error rates below 10^{-9}. The power penalty was between 2 and 4 dB for 10^{-9} BER. Without the use of Raman amplification, span lengths were restricted to 100-km length for similar BER performance. Thus the addition of Raman gain allows nearly 4 dB of additional span loss for error-free transmission.

In the second experiment, error-free transmission of 25-GHz-spaced 100 WDM channels at 10 Gb/s over 400 km of NZDS fiber was reported [45]. High spectral efficiency of 0.4 b/s/Hz was achieved by the use of distributed Raman gain. Raman pump power in the fiber was 22–23 dBm, which resulted in a peak gain of 10 dB at 1585 nm. Three fiber spans of positive NZDS fiber having zero dispersion wavelength in the range of

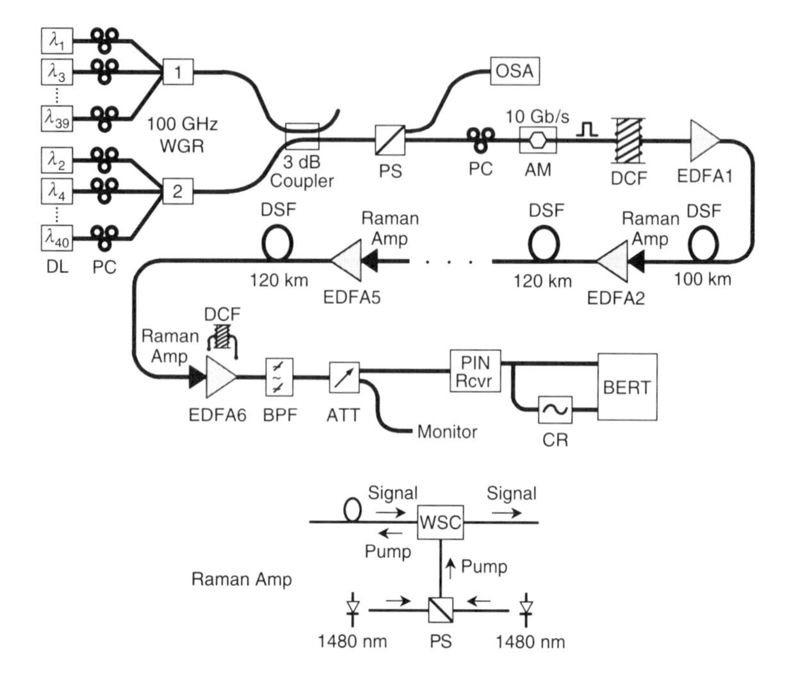

Fig. 9.13 A schematic of multi-span Raman gain-enhanced transmission experiment [38].

1508 to 1527 nm and lengths 125, 132, and 140 km, respectively, were used. The channel spectrum at the beginning and at the end of system is shown in Fig. 9.14. The eye diagrams of all the channels were open and, as expected, exhibited little distortion due to nonlinear effects. All channels achieved error rates below 10^{-9}. The power penalty at 10^{-9} BER was between 2.1 and 5.3 dB. The penalty can be attributed to the variation of OSNR due to some low power lasers and some due to the gain nonuniformity. Calculated value of OSNR 22 dB is in good agreement with the measured 21–25 dB.

Distributed Raman amplification was employed in two recent 40 Gb/s WDM transmission experiments, both achieving a spectral efficiency of 0.4 bit/s/Hz. In both experiments, distributed Raman amplification allowed the launch power into 100-km spans of TrueWave fiber to be as low as −1dBm/channel while maintaining sufficient optical signal-to-noise ratio at the output of the system for error-free performance of the 40 Gb/s WDM channels. In the first experiment 40 WDM channels were transmitted in the C-band over four 100-km spans of TrueWave fiber, using four hybrid

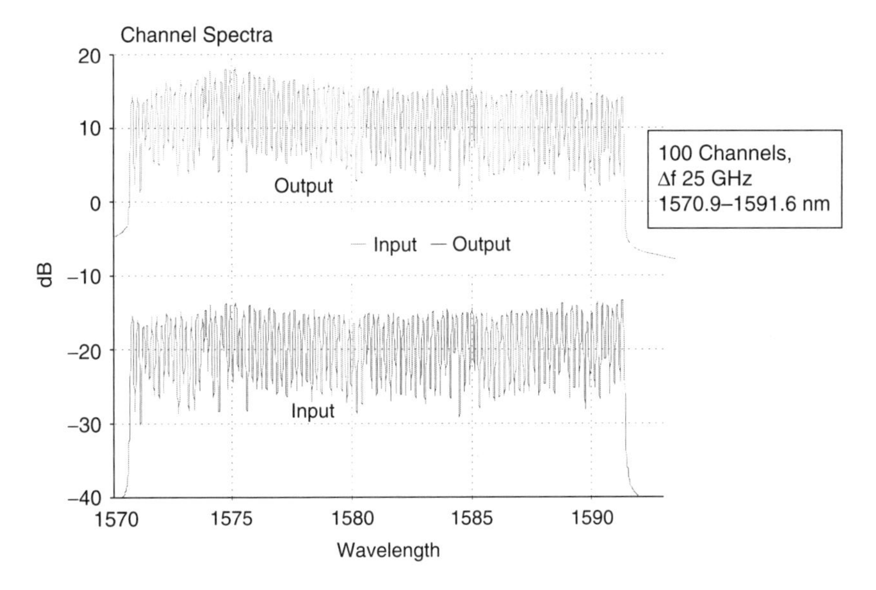

Fig. 9.14 Channel spectra of ultra-dense (25 GHz spaced) terabit capacity (100×10 Gb/s) WDM transmission [45].

Raman/erbium inline amplifiers [43]. A single Raman pump wavelength was sufficient to provide effective noise figures below 0 dB over the entire C-band, as shown in Fig. 9.15(a). For dual C- and L-band systems, at least two pump wavelengths are required to obtain adequate noise figures over the combined C- and L-bands. In the second experiment, 3.28 Tb/s were transmitted over three 100-km spans of non-zero dispersion-shifted fiber [44]. Two Raman pumps were used to achieve noise figures ranging from +1.5 dB in the lower end of the C-band to −1.7 dB in the L-band, as shown in Fig. 9.15(b). The two experiments illustrate one important complication that arises from using multiple Raman pumps in ultra broadband systems, namely that Raman pumps at lower wavelengths will be depleted by pumps at higher wavelengths. Pre-emphasis of the pumps must be used, and the noise figure of the WDM channels being pumped by the shorter wavelength Raman pump generally will be higher compared to the noise figure of the WDM channels being pumped by the longer wavelength Raman pump.

9.3.5. EDFAs FOR DYNAMIC NETWORKS

The lightwave networks are evolving from centrally planned provisioned circuits to one based on intelligent and dynamic network elements supporting point-and-click provisioning and simplified operations with reduced

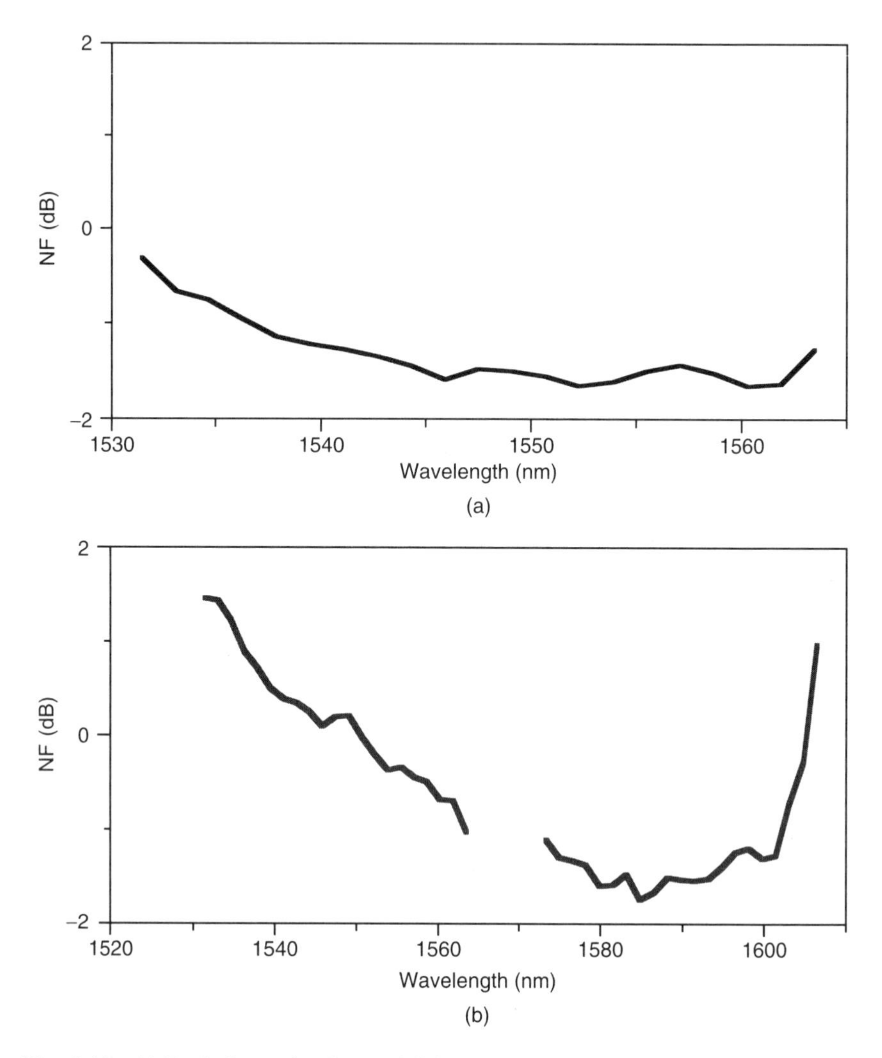

Fig. 9.15 (a) Equivalent noise figure of C-band Raman EDFA [43], (b) Equivalent noise figure of a combined C- and L-band Raman/EDFA [44].

capital and operational costs. Optical amplifiers are a key element of the present-day WDM systems. The WDM systems require high output power, low noise figure EDFAs with high level of gain flatness over a wide and well-managed bandwidth. In addition to high performance, the implementation of the new features such as dynamic wavelength provisioning and automated reconfiguration can lead to optical power transients in a conventional gain-controlled EDFA, which is usually operated under saturation. These signal transients can compromise the flexibility of network architecture and impact overall system performance. Should a network be reconfigured or a component fail, the number of WDM signals passing along a chain of amplifiers will change, thus causing the power of surviving channels to either increase or decrease. The service quality of surviving channels can be impaired through four mechanisms when channel loading changes. First, optical nonlinear effects in transmission fibers will occur or increase if the power excursions are large enough when signal channels are lost. For example, self-phase modulation (SPM) has been observed to affect the performance of the surviving channels [46]. Second, when channels are added, the optical power at the receiver can be reduced during the transient period, which would cause eye closure, and severe degradation of bit error rate. Third, optical SNR may be degraded due to the change of inversion level and therefore the change in gain spectrum during the transient period. Fourth, the received power at the receiver is varying, which requires that the threshold of the receiver be optimized at high speed, which can be a problem for certain receivers.

Conventional SONET and SDH ring networks have built-in protection switching and service restoration mechanisms, which have 50 ms response time. Alternatively, physical layer protection switching can be achieved in nearly 15 ms in an advanced WDM system. In both cases, transients will not be detected quickly enough to prevent system performance degradation as illustrated by the example in Fig. 9.16. The figure shows the signal power variation of a surviving channel in a 4-fiber span system when half of the channels are dropped and added (Fig. 9.16(a)) and the corresponding eye-diagram (Fig. 9.16(b)) showing bit-error-rate degradation to 1×10^{-4}. When the fast transient control is switched, the surviving channel power variation is restricted to within 0.5 dB. Under these conditions, the eye diagram shows significant improvement with BER $<1 \times 10^{-12}$. This example shows the necessity of fast transient control in a WDM network that guarantees quality of service.

9.3.5.1. Gain Dynamics of Single EDFAs

The speed of gain dynamics in a single EDFA is in general much faster than the spontaneous lifetime (10 ms) [47] because of the gain saturation effect. The time constant of gain recovery on single-stage amplifiers was measured to be between 110 and 340 μs [48]. The time constant of gain

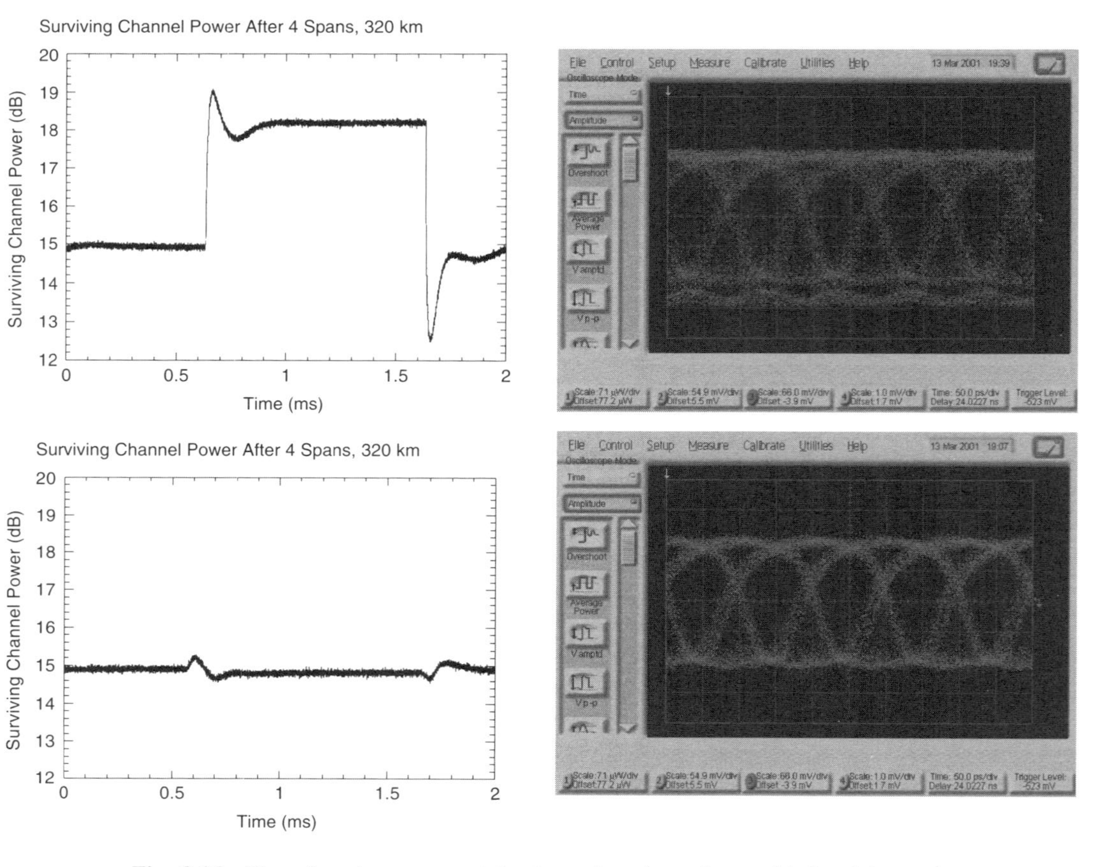

Fig. 9.16 Time-domain response at the time when channels are added and dropped.

dynamics is a function of the saturation caused by the pump power and the signal power. Present-day WDM systems with 40–100 channels require high-power EDFAs in which the saturation factor becomes higher leading to shorter transient time constant. In a recent report, the characteristic transient times were reported to be tens of microseconds in a two-stage EDFA [47]. The transient behavior of surviving channel power for the cases of one, four, and seven dropped channels, in an eight-channel system, is shown in Fig. 9.17. In the case of seven dropped channels, the transient time constant is nearly 52 μs. As can be seen, the transient becomes faster as the number of dropped channels decreases. The time constant decreases to 29 μs when only one out of eight channels is dropped. The rate equations [49] for the photons and the populations of the upper ($^4I_{13/2}$) and lower ($^4I_{15/2}$) states

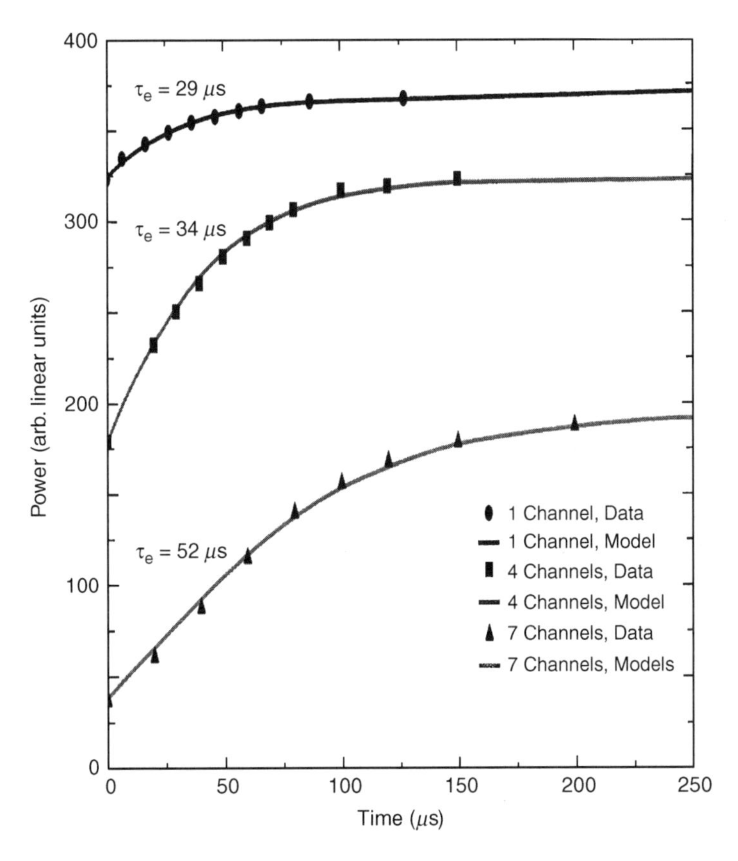

Fig. 9.17 Measured and calculated surviving power transients for the cases of 1, 4, 7 channels dropped out of 8 WDM channels.

can be used to derive the following approximate formula for the power transient behavior [50]:

$$P(t) = P(\infty)[P(0)/P(\infty)]^{\exp(-t/\tau_e)} \tag{9.13}$$

where $P(0)$ and $P(\infty)$ are the optical powers at time $t = 0$ and $t = \infty$, respectively. The characteristic time τ_e is the effective decay time of the upper level averaged over the fiber length. It is used as a fitting parameter to obtain best fit with the experimental data. The experimental data are in good agreement with the model for the transient response (Fig. 9.17). The model has been used to calculate the fractional power excursions in decibels of the surviving channels for the cases of one, four, and seven dropped channels. The times required to limit the power excursion to 1 dB are 18 and 8 μs when four or seven channels are dropped, respectively. As the EDFAs advance further to support larger number of WDM channels in lightwave networks, the transient times will fall below 10 μs. Dynamic gain control of the EDFAs with faster response times will be necessary to control the signal power transients.

A model of EDFA dynamics is needed to understand the transient behavior in large systems or networks. Recently, a simple model has been developed for characterizing the dynamic gain of an EDFA. The time-dependent gain is described by a single ordinary differential equation for an EDFA with an arbitrary number of signal channels with arbitrary power levels and propagation directions. Most previous EDFA models are represented by sets of coupled partial differential equations [51, 52], which can be solved only through iterative, computationally intensive numerical calculations, especially for multichannel WDM systems with counter-propagating pump or signals. The time-dependent partial differential equations can be dramatically reduced to a single ordinary differential equation. The mathematical details of the model are provided in ref. [53]. Here the simulation results from the model are compared with the measured time-dependent power excursions of surviving channels when one or more input channels to an EDFA are dropped. The structure of the two-stage EDFA used in the experiment [47] and simulation is shown in the inset of Fig. 9.18. The experimentally measured power of the surviving channel when one, four, or seven out of eight WDM channels are dropped are plotted in Fig. 9.18. In the simulation it is seen from the figure that the simulation results agree reasonably well with the experimental data without any fitting parameters. The exception is 0.9 dB difference at large t for the seven-channel drop case. This discrepancy is believed to arise from pump excited state absorption

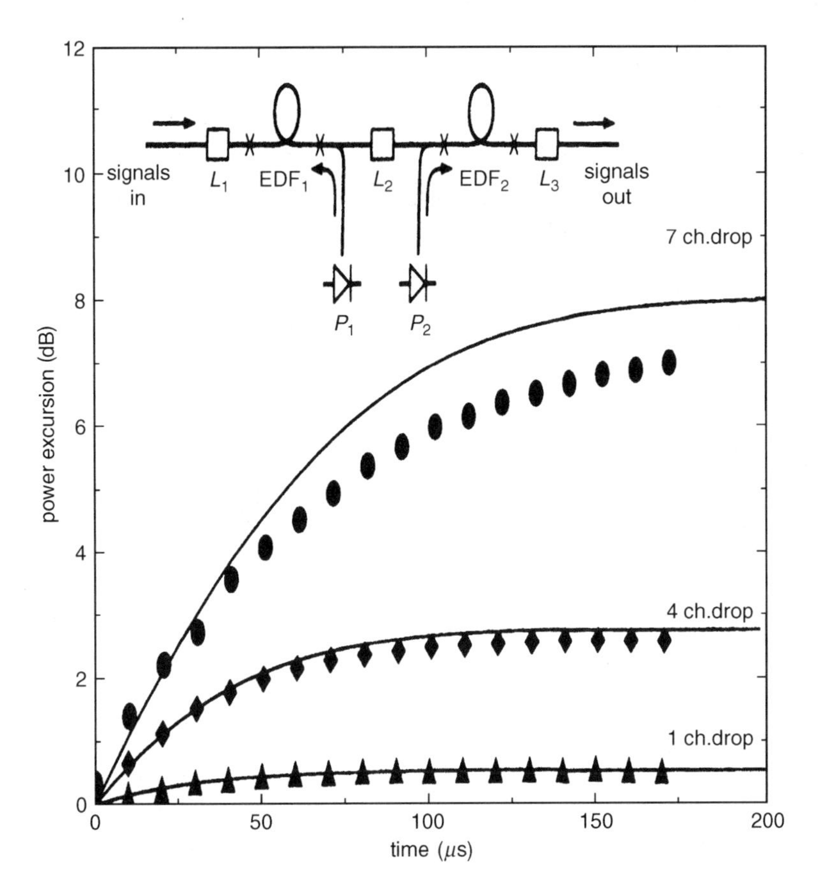

Fig. 9.18 Comparison between theory and experiment for output power excursions of surviving channels from a two-stage amplifier when one, four, seven of the input channels are dropped.

at high pump intensity. The model can be very useful in the study of power transients in amplified optical networks.

9.3.5.2. Fast Power Transients in EDFA Chains

In a recent work, the phenomena of fast power transients in EDFA chains was reported [54, 55]. The effect of dropped channels on surviving powers in an amplifier chain is illustrated in Fig. 9.19. When four out of eight WDM channels are suddenly lost, the output power of each EDFA in the chain drops by 3 dB and the power in each surviving channel then increases toward double the original channel power to conserve the saturated

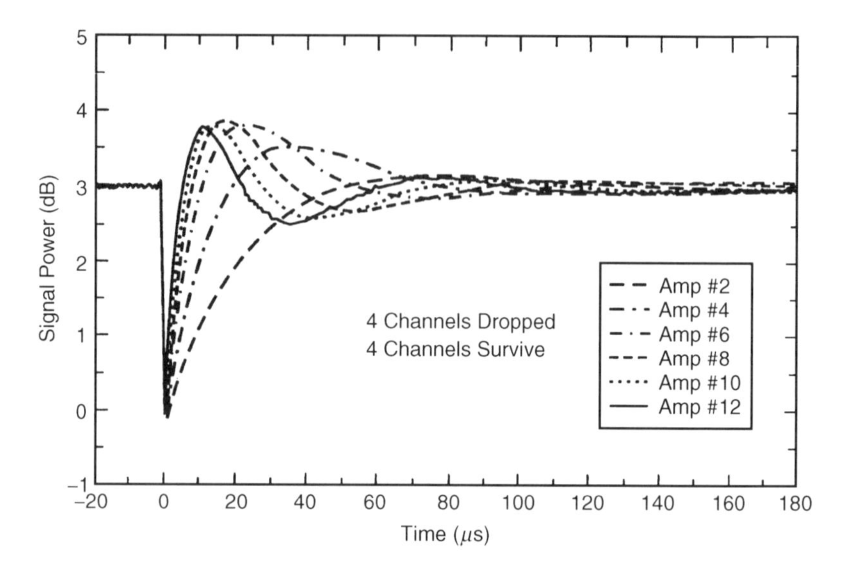

Fig. 9.19 Measured output power as a function of time after 0, 2, 4, 6, 8, 10, and 12 EDFAs (at time $t = 0$, 4 out of 8 WDM channels are dropped).

amplifier output power. Even though the gain dynamics of an individual EDFA are unchanged, the increase in channel power at the end of the system becomes faster for longer amplifier chains. Fast power transients result from the effects of the collective behavior in chains of amplifiers. The output of the first EDFA attenuated by the fiber span loss acts as the input to the second EDFA. Because both the output of the first EDFA and the gain of the second EDFA increase with time, the output power of the second amplifier increases at a faster rate. This cascading effect results in faster and faster transients as the number of amplifiers increase in the chain. To prevent performance penalties in a large-scale WDM optical network, surviving channel power excursions must be limited to certain values depending on the system margin. Take the MONET network as a example: the power swing should be within 0.5 dB when channels are added and 2 dB when channels are dropped [56]. In a chain consisting of 10 amplifiers, the response times required in order to limit the power excursions to 0.5 dB and 2 dB would be 0.85 and 3.75 μs, respectively. The response times are inversely proportional to the number of EDFAs in the transmission system.

The time response of EDFAs can be divided into three regions, the initial perturbation region, the intermediate oscillation region, and the final

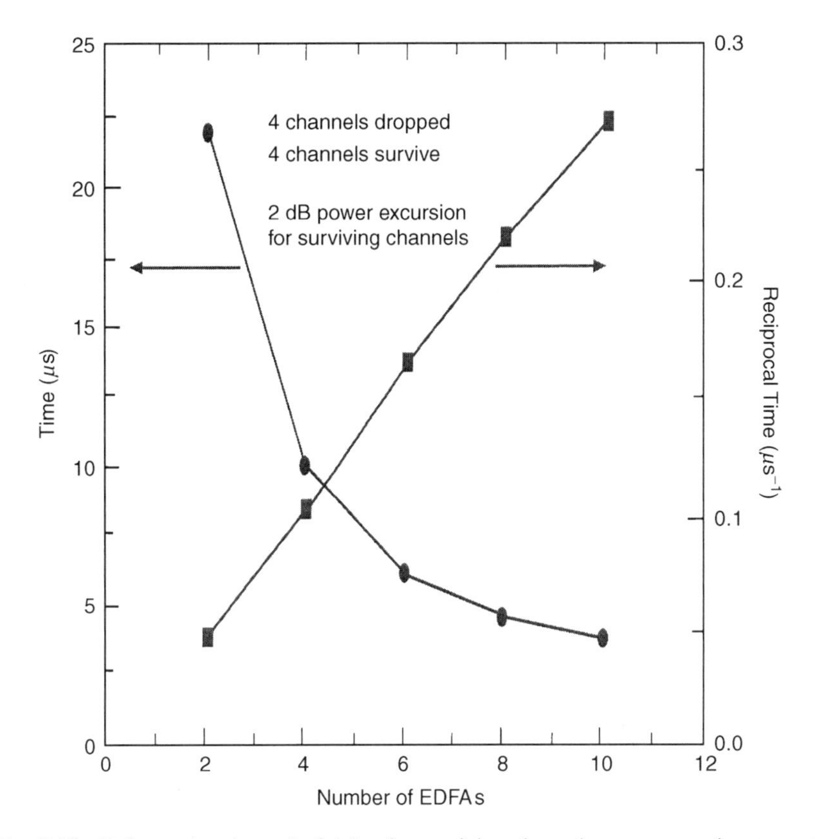

Fig. 9.20 Delay and reciprocal of delay for surviving channel power excursion to reach 2 dB after the loss of 4 out of 8 WDM channels.

steady state region. In the initial perturbation region, the gain of the EDFA increases linearly with time and the system gain and output power increase at a rate proportional to the number of EDFAs. The time delays for a channel power excursion of 2 dB are measured from Fig. 9.19 and the inverse of time delays, which are the power transient slope in the perturbation region, are plotted in Fig. 9.20. Assuming that the amplifiers operate under identical conditions, the rate of change of gain at each EDFA is the same and is proportional to the total lost signal power. The slope plotted in Fig. 9.20 therefore increases linearly with the number of EDFAs in the chain. These experimental results have been confirmed by modeling and numerical simulation from a dynamic model [8, 53].

In the intermediate region an overshoot spike can be observed after 2 EDFAs in Fig. 9.19. The first overshoot peak is the maximum power excursion because the oscillation peaks that follow are smaller than the first one. From the results of both experimental measurements and numerical simulation on a system with N EDFAs, the time to reach the peak is found to be inversely proportional to N and the slope to the peak is proportional to $N - 1$ [57]. This indicates that the overshoot peaks are bounded by a value determined by the dropped signal power and the operating condition of the EDFAs. These properties in the perturbation and oscillation regions can be used to predict the power excursions in large optical networks.

A study of dynamic behavior of L-band EDFA has been carried our recently [58]. In this work, transient response of the surviving channels in a two-stage L-band EDFA under different channel loading conditions was reported. The observed dynamic behavior in L-band is similar to that in C-band. However, the magnitude of the response time is very different. The response time constants as a function of the number of dropped channels under different saturation conditions is shown in Fig. 9.21. The time constants are about 105 μs and 260 μs when one and seven channels out of eight channels are dropped accordingly and the amplifier is well saturated. These values are about four to five times larger that that observed in a C-band EDFA. The difference can be explained by the different intrinsic saturation power in these two bands.

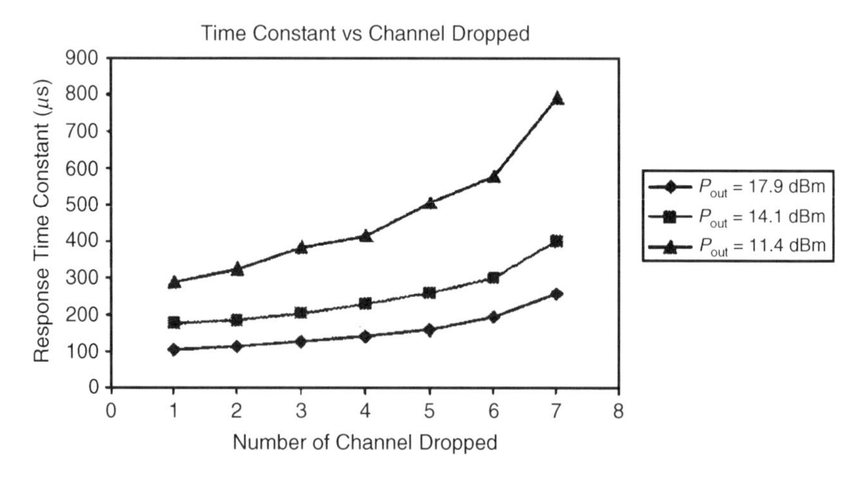

Fig. 9.21 Response time constant vs. the number of dropped channels under different saturation conditions in a two-stage L-band EDFA,

9.3.5.3. Channel Protection Schemes

As discussed earlier, channels in optical networks will suffer error bursts caused by signal power transients resulting from a line failure or a network reconfiguration. Such error bursts in surviving channels represent a service impairment that is absent in electronically switched networks and is unacceptable to service providers. The speed of power transients resulting from channel loading, and therefore the speed required to protect against such error bursts, is proportional to the number of amplifiers in the network and for large networks, can be extremely fast. Several schemes to protect against the fast power transients in amplified networks have been demonstrated in recent years.

Pump Control

The gain of an EDFA can be controlled by adjusting its pump current. Early reported work addressed pump control on time scales of the spontaneous lifetime in EDFAs [51]. One of the studies demonstrated low-frequency feed-forward compensation with a low-frequency control loop [48]. After the discovery of fast power transients, pump control on short time scales [59] was demonstrated to limit the power excursion of surviving channels. In the experiment, automatic pump control in a two-stage EDFA operating on a time scale of microseconds was demonstrated. The changes in the surviving channel power in the worst case of 7 channel drop/add in an 8-channel WDM system are shown in Fig. 9.22. In the absence of gain control, the change in surviving channel signal power exceeds 6 dB. When the pump control on both stages is active, the power excursion is less than 0.5 dB both for drop and add conditions. The control circuit acts to correct the pump power within 7–8 μs and this effectively limits the surviving channel power excursion.

Link Control

The pump control scheme previously described would require protection at every amplifier in the network. Another technique makes use of a control channel in the transmission band to control the gain of amplifiers. Earlier work demonstrated gain compensation in an EDFA at low frequencies (<1 kHz) using an idle compensation signal [60]. Recently, link control, which provides surviving channel protection against fast transients, has been demonstrated [46]. The scheme, as illustrated in Fig. 9.23, protects

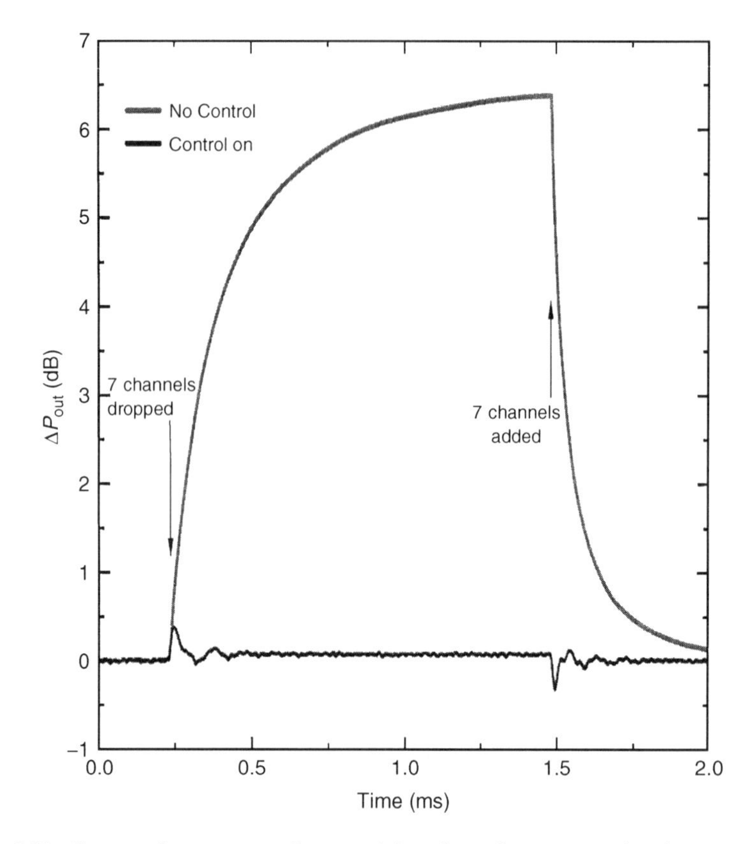

Fig. 9.22 Impact of pump control on surviving channel power transient in a two-stage EDFA when 7 out of 8 channels are dropped and added.

surviving channels on a link-by-link basis. A control channel is added before the first optical amplifier in a link (commonly the output amplifier of a network element). The control channel is stripped off at the next network element (commonly after its input amplifier) to prevent improper loading of downstream links. The power of the control channel is adjusted to hold constant the total power of the signal channels and the control channel at the input of the first amplifier. This will maintain constant loading of all EDFAs in the link.

The experimental demonstration of link control surviving channel protection is illustrated in Fig. 9.24. The experiment is set up with 7 signal channels and 1 control channel. A fast feedback circuit with a response time of 4 μs is used to adjust the line control channel's power to maintain total power constant. The signal channels and control channels are transmitted

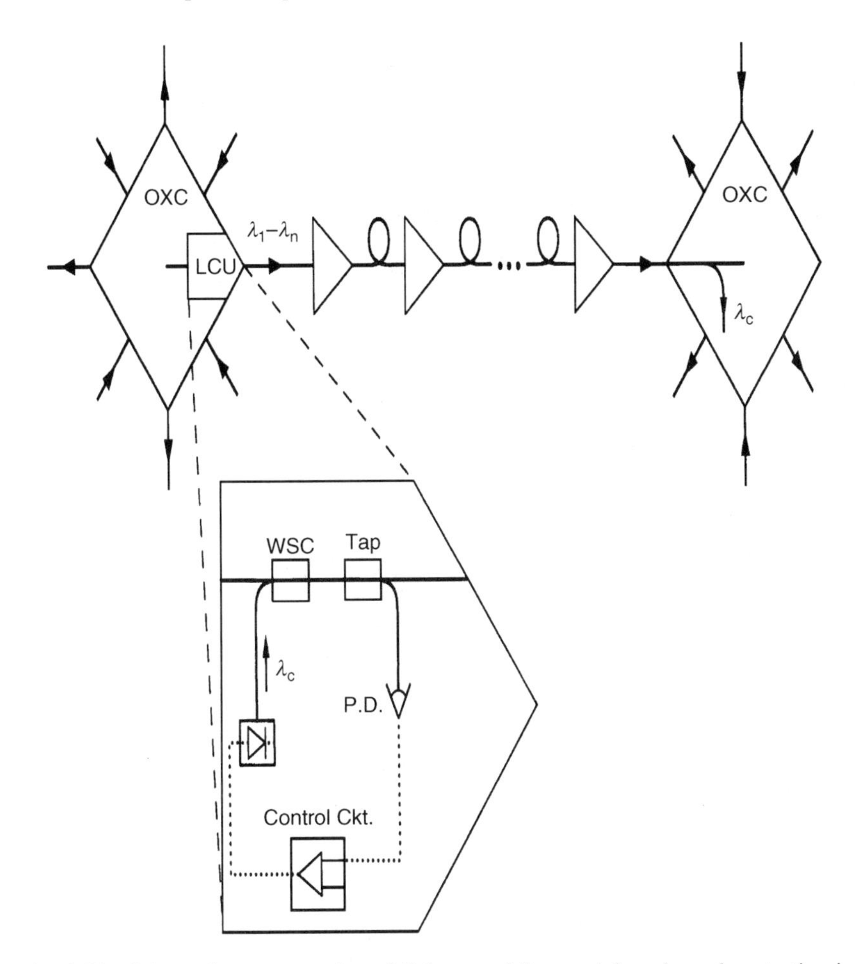

Fig. 9.23 Schematic representation of link control for surviving channel protection in optical networks.

through seven amplified spans of fiber and bit-error-rate performance of one of the signal channels is monitored. The measured results are summarized in Fig. 9.25. When 5 out of 7 signal channels are added/dropped at a rate of 1 kHz, the surviving channel suffers a power penalty exceeding 2 dB and a severe BER floor. An even worse BER floor is observed when 5 out of 6 channels are added/dropped, resulting from cross-saturation induced by change in channel loading. With the fast link control in operation, the power excursions are mitigated and BER penalties are reduced to a few tenths of a dB and error floors disappear.

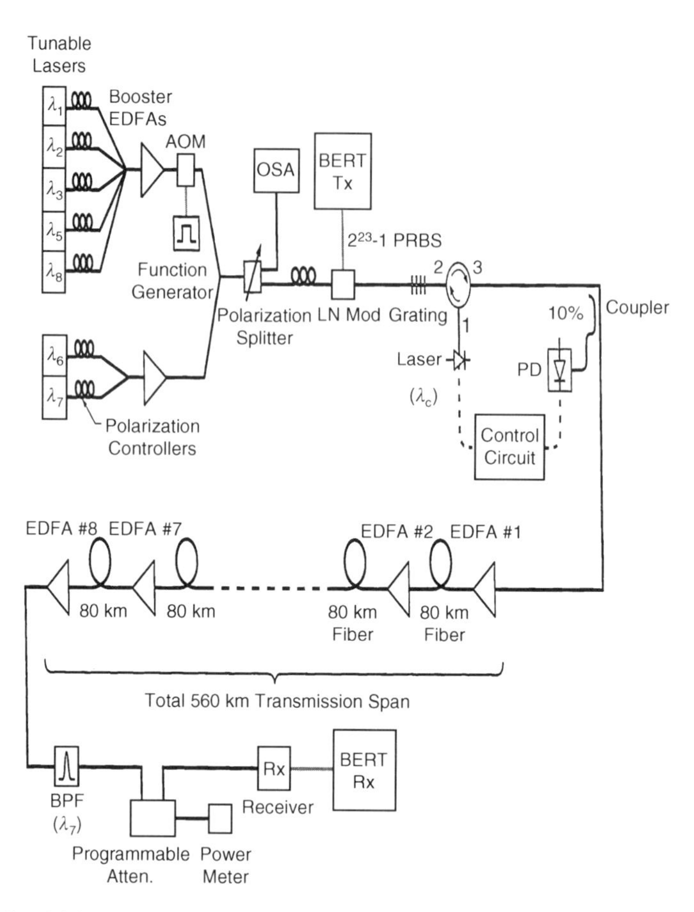

Fig. 9.24 Schematic of experimental setup for link control demonstration.

Laser Control

Laser automatic gain control has been extensively studied since it was experimentally demonstrated [61]. A new scheme for link control based on laser gain control has been proposed recently [62]. In this work, a compensating signal in the first amplifier is generated using an optical feedback laser loop and then propagates down the link. Stabilization is reached within a few tens of microseconds, and output power excursion after 6 EDFAs is reduced by more than an order of magnitude to a few tenths of a dB. For laser gain control, the speed is limited by laser relaxation oscillations [63], which are generally on the order of tens of microseconds or slower. Inhomogeneous broadening of EDFAs and the resulting spectral

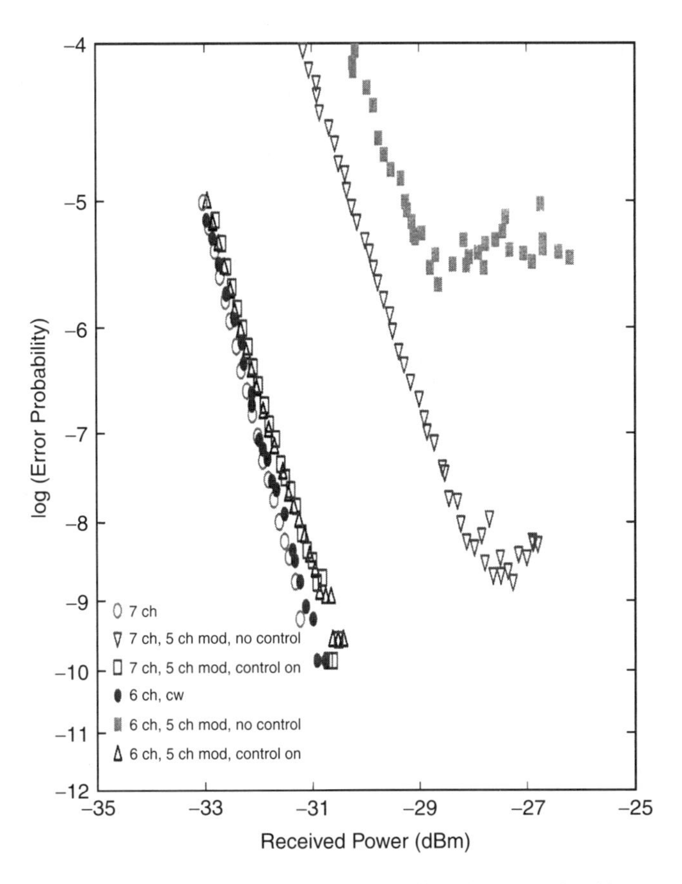

Fig. 9.25 Bit error rates measured for transmission of 6 and 7 channels without modulation and with modulation of channels 1, 2, 3, 5, and 8 on/off at 1 KHz with and without link control.

hole burning can cause gain variations at the signal wavelength, which will limit the extent of control from this technique. The same is true for link control scheme.

9.4. Semiconductor Optical Amplifiers for WDM systems

Erbium-doped fiber amplifiers (EDFA) are being widely used in optical communication systems. One major advantage of EDFA is its slow dynamics. For the bit rates of practical interest, the EDFA only sees the average power and provides constant gain to signal channels. The gain dynamic of a semiconductor optical amplifier (SOA), on the other hand, is much faster than that of an EDFA. The gain of a SOA changes quickly as the input power changes. As a result, the output power exhibits inter-modal distortion for the single-channel case and crosstalk distortion for WDM systems. So far, SOA has found very limited applications as in-line amplifiers in WDM optical communication systems.

We point out here that even with the inter-modal distortion and crosstalk in SOA, (1) accurate detection can be made with suitable adjustment of detection threshold level, and (2) the effect of gain fluctuations becomes smaller as the number of channels increases. To illustrate the first point, let's compare a SOA with an EDFA used in a WDM system with N channels. For EDFA, each channel takes an equal share of the total output power. It is more complicated for SOA. There are N possible values for bit "1," with the lowest possible value to be $1/N$ of the total power. If the detection threshold is set half of that, then accurate detection can be made, assuming there is enough dynamic range in the detector. In that sense, the only difference between a SOA and an EDFA is that the detection threshold of a SOA should be half of that of an EDFA.

The second property is the result of the statistics of the signal channels. Because the bit patterns of the WDM channels are random and are independent of each other, the total power is smooth in a WDM system with a large number of channels. This smoothed total effective power effect reduces input power variation to the SOAs and therefore the gain ripples. For an extremely large number of channels, the performance of a SOA approaches that of an EDFA in DWDM systems.

A major advantage of SOA is its large gain bandwidth. The gain spectrum of SOA is intrinsically flat and wide (80 nm). With the combination of several SOAs centered at different wavelengths, the whole transmission band of optical fiber can be covered. Therefore the optical amplifier will not be the limiting factor for the capacity of optical communication systems. Another key advantage of SOA is the potentially low manufacture cost. SOA with multiple gain stages can be integrated on the same chip during fabrication. The potential cost can be much lower than that of EDFA. There are other advantages such as small size and low power dissipation. Much more work needs to be done at the device and system level to bring SOA into commercial systems. The device issues include polarization dependence, nonlinear effects, noise figure and output power; while the system issues include nonlinear effects for bits with high power, amplifier cascading effects, and system control during the change of the number of channels.

However, with the key issue of crosstalk solved, the SOA can be used in metro networks first, where the required noise figure and output power are not stringent but low cost is the key. Upon improvements at both the device and system levels, the SOA can be expected to be used in long-distance optical communication systems and networks. As discussed earlier, a major advantage of the widely used erbium-doped fiber amplifier (EDFA) in present optical communication systems is its slow gain dynamics. An EDFA experiences only the average input power and provides constant gain to all the signal channels. The gain dynamic of a semiconductor optical amplifier (SOA) is much faster than that of an EDFA. The gain of a SOA, when operated under saturation, changes with the instantaneous input power and causes pulse distortion and crosstalk. However, the SOA has potential advantages in, among others, the coverage of the entire fiber transmission window, integration with other devices and platforms, and low cost.

In order to reduce the inter-modal distortion and crosstalk in SOAs, one can either operate a SOA in the small signal region or implement gain control schemes such as feed-forward, laser gain clamping, and pumping light injection. These schemes have been implemented in applications in single-channel system transmission [64, 65] and in WDM applications in loss compensation and system transmission [66–68]. However, for practical applications, optical amplifiers are desired to operate in the saturation region in order to achieve high output power and high efficiency. It is also important to look for economic gain control schemes with the least impact on noise figure and output power.

In a recent report, transmission of 32 dense WDM (DWDM) signal channels over 125 km of transmission fiber using 3 cascaded SOAs operated under saturation without local gain control. This demonstration was made possible because of the following three points. First, with suitable adjustment of the detection threshold level, accurate detection can be made under the presence of inter-modal distortion and crosstalk in a SOA. Second, the effect of power fluctuations due to gain crosstalk becomes smaller as the number of WDM channels increases [65]. And third, an external reservoir-channel was used to provide an effective means to further suppress the power fluctuations in the SOA. The use of reservoir-channel in optical networks (RON) with SOAs is effective to achieve high transmission quality.

The first point can be illustrated by comparing the output power of an EDFA with that of a SOA in a WDM system with N channels, as shown in Fig. 9.26. Here we assume that both EDFA and SOA are operated under

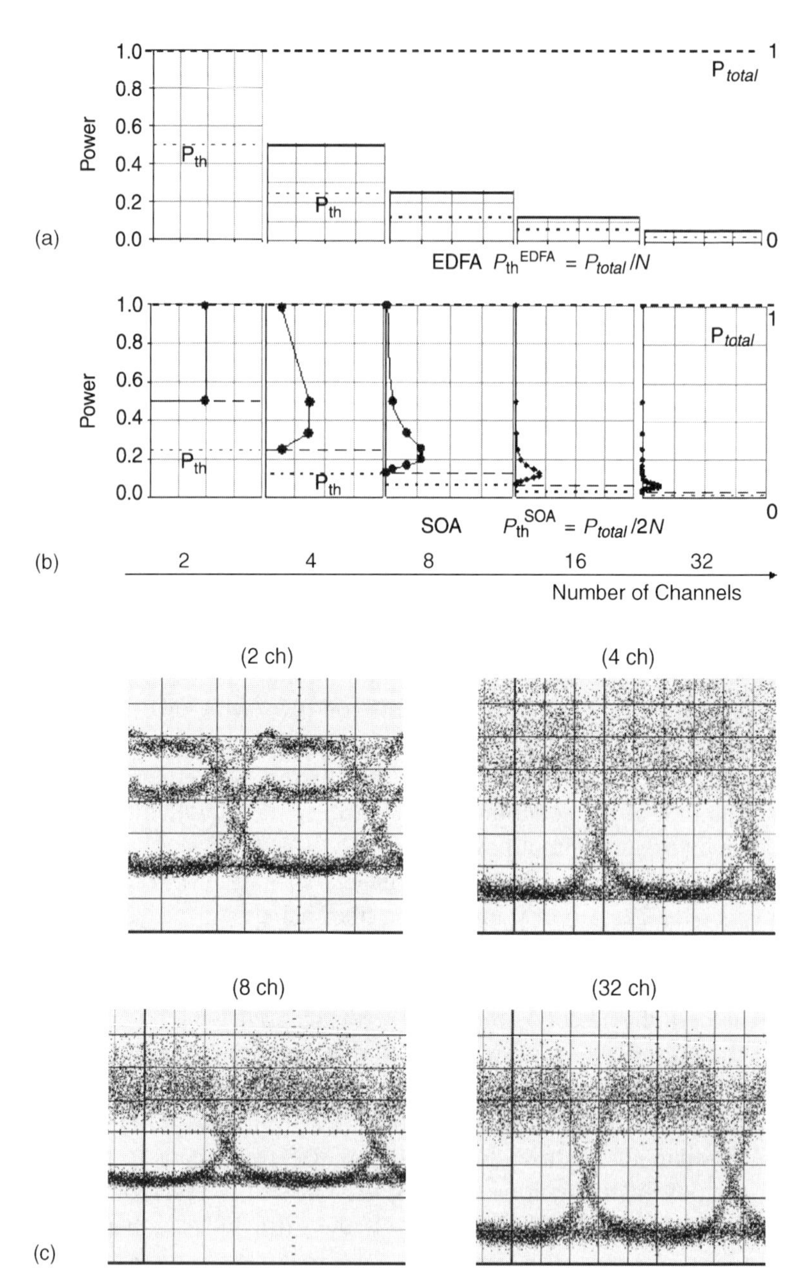

Fig. 9.26 Comparison of the performance of (a) EDFA and (b) SOA for different number of channels in WDM optical communication systems. (c) Eye diagrams showing the STEP Effect; a) 2, b) 4, c) 8, and d) 32 channels.

deep saturation and the total output power is fixed at P when there is input power. For an EDFA, each channel takes an equal share of the total output power P/N. As shown in Fig. 9.26(a), the bit "1"s have value $2P/N$ and the bit "0"s have a value 0, under the assumption that there are equal probabilities in RZ bit "1" and "0." The detection threshold should be set at P/N without noise taken into account. On the other hand, due to the cross-gain saturation in a SOA, there are N possible values for bit "1," with the lowest possible value being $1/N$ of the total power, as illustrated in Fig. 9.26(b). If the detection threshold is set at half of $2P/N$, accurate detection can be made, assuming that there are no nonlinear effects in the transmission system and that there is enough dynamic range and response speed in the detector. This implies that the detection threshold for a SOA should be half that for an EDFA, or 3 dB penalty.

The second point is related to the signal statistics [69]. Because the bit patterns of the WDM channels are random and are independent of each other, the statistics of the signal channels tend to smoothen the total power fluctuations in a system with large number of channels as shown in Fig. 9.26(b). The smoothed total effective power (STEP) effect is experimentally demonstrated in the eye diagrams at the output of a SOA, as shown in Fig. 9.26(c) with 2, 4, 8, and 32 signal channels, respectively. In a multi-span system with cascaded SOAs, the power fluctuations should become progressively smaller down the system if the dispersion is small. With large dispersion, relative shifts in the bits of different signal channels can cause further fluctuations in the input power for downstream SOAs and therefore cause further distortion in the system. In this case, the detection threshold needs to be adjusted to a lower level to accommodate the lowest possible value for bit "1." Alternatively, the relative bit shifts can be removed by complete dispersion compensation for all the signal channels.

To suppress power fluctuations, an unmodulated reservoir channel RON can be added. When the total input power is low, the RON shares more power as compared to when the input power is medium, thus reducing the power increment in the signal channels with bit "1." On the other hand, when the total input power is high, the RON will share less power and decrease the power reduction in bit "1." Therefore, the RON acts as a reservoir to suppress gain fluctuations in the SOAs.

The system experimental setup is schematically shown in Fig. 9.27. Thirty-two signal channels from 1534.95 nm to 1559.36 nm with 100 GHz spacing were combined with a waveguide router, simultaneously modulated

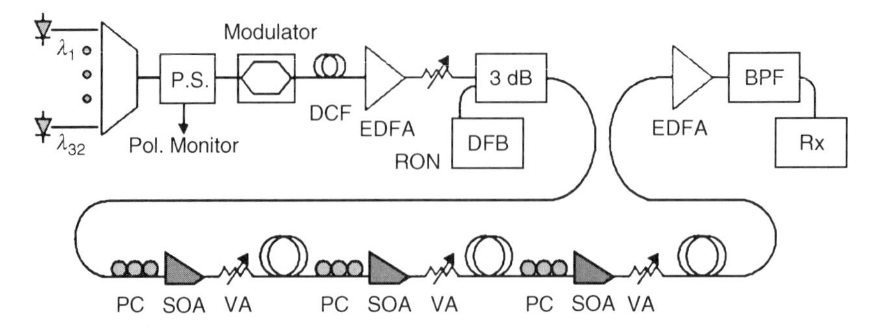

Fig. 9.27 Schematic of the transmission experiment. The total length of transmission fiber is 125 km.

with a single $LiNbO_3$ modulator, and de-correlated with DCF. The data rate was 2.5 Gbit/s and PRBS was $2^{31} - 1$. The reservoir channel RON was added at the input of the power booster SOA. The wavelength of the RON is at 1531.78 nm and the power is 13 dB higher than that of a single signal channel at the input. Equivalently, the RON takes 2.1 dB power margin.

In the transmission section, there were three spans of AllWave™ transmission fiber. The SOAs were single-stage, non-gain-clamped commercial devices. The total input power to each SOA was about −3.0 dBm and the gain was 15 dB, which is 5 dB below the small-signal gain. The output power of the SOAs was roughly 12 dBm and the noise figure was about 10 dB. Each transmission fiber span was between 41 and 42 km in length with around 9 dB loss. One variable attenuator was added after each SOA to make the total span loss to be at 15 dB. At the receiver end, an EDFA was used as the pre-amplifier and a band-pass-filter was placed before the detector.

Due to the intrinsic flat gain of SOA, there was only about 3 dB power tilt in the power spectrum at the end of the transmission section. Error-free transmission (BER 10^{-9}) for all the 32 channels with RON on was achieved with about 2 dB penalty, as shown in Fig. 9.28. In another experiment, without the RON turned on, we were able to obtain error-free transmission with 2 cascaded SOAs over 2 spans of transmission fiber totaling 85 km, by suitable adjustment of system parameters. Error-free transmission of 32×2.5 Gbit/s channels over 125 km of AllWave™ transmission fiber with three SOAs was achieved. A RON proves to be a simple, effective, and potentially economic method for the suppression of power fluctuations in

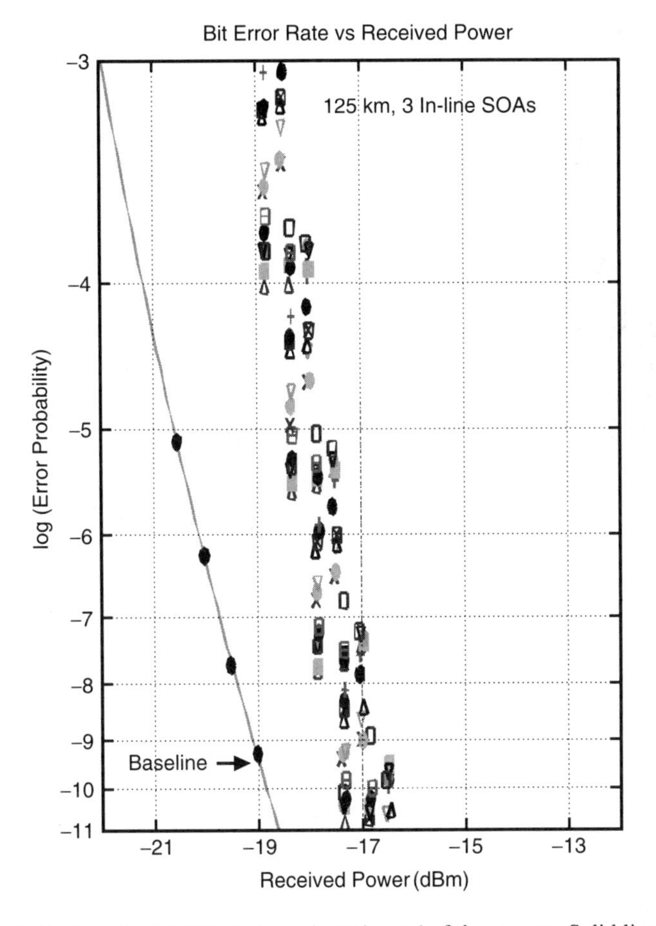

Fig. 9.28 BER data for 32 WDM channels at the end of the system. Solid lines through the data points are to guide the eye.

systems with cascaded SOAs. This experiment demonstrates the potential applications of the SOA for DWDM optical communications.

ACKNOWLEDGMENTS

Much of the work discussed in this paper was done at the Crawford Hill Laboratory of Lucent Technologies in Holmdel, New Jersey. The authors are grateful to all the colleagues for very fruitful collaborations. The authors are thankful to T. N. Nielsen for sharing some of his work on Raman amplifier based systems, which is included here.

References

1. K. Fukuchi *et al.*, "10.92 Tb/s (273 × 40 Gb/s) Triple-Band/Ultra-Dense WDM Optical Repeatered Transmission Experiment" PD 24, and S. Bigo *et al.*, "10.2 Tb/s (256 × 42.7 Gb/s PDM/WDM) Transmission over 100 km TeraLight Fiber with 1.28 bit/s/Hz Spectral Efficiency" PD-25, Proc. Optical Fiber Communications Conf., Anaheim, CA (March 2001).
2. Y. Sun, A. K. Srivastava, J. Zhou, and J. W. Sulhoff, "Optical Fiber Amplifiers for WDM Optical Networks," Bell Labs. Tech. J. (Jan–Mar 1999) 187–205.
3. R. J. Mears *et al.*, "Low noise erbium-doped fiber amplifier operating at 1.54 μm," Electron. Lett., 23 (September 10, 1987) 1026–1028.
4. E. Desurvire, J. R. Simpson, and P. C. Becker, "High gain erbium-doped traveling wave fiber amplifier," Optics Letters, 12 (November 1987) 888–890.
5. A. K. Srivastava, Y. Sun, J. W. Sulhoff, C. Wolf, M. Zirngibl, R. Monnard, A. R. Chraplyvy, A. A. Abramov, R. P. Espindola, T. A. Strasser, J. R. Pedrazzani, A. M. Vengsarkar, J. L. Zyskind, J. Zhou, D. A. Ferrand, P. F. Wysocki, J. B. Judkins, and Y. P. Li, "1 Tb/s transmission of 100 WDM 10 Gb/s channels over 400 km of TrueWave^{TM} fiber," OFC '98 Technical Digest, Postdeadline Papers, PD 10, San Jose, CA (February 22–27, 1998).
6. D. Fishman, J. A. Nagel, T. W. Cline, R. E. Tench, T. C. Pleiss, T. Miller, D. G. Coult, M. A. Milbrodt, P. D. Yeates, A. Chraplyvy, R. Tkach, A. B. Piccirilli, J. R. Simpson, and C. M. Miller, "A High Capacity Non-coherent FSK Lightwave Field Experiment using Er-Doped Fiber Optical Amplifiers," Photonic Technol. Lett., 2:9 (1990) 662.
7. C. R. Giles and E. Desurvire, "Modeling erbium-doped fiber amplifiers," J. Lightwave Technol., 9 (Feb. 1991) 271–283.
8. Y. Sun, J. L. Zyskind, and A. K. Srivastava, "Average Saturation Level, Modeling, and Physics of Erbium-Doped Fiber Amplifiers," IEEE J. Selected Areas in Quantum Elect., 3:4 (August 1997) 991–1007.
9. P. F. Wysocki, J. R. Simpson, and D. Lee, "Prediction of Gain Peak Wavelength for Er-Doped Fiber Amplifiers and Amplifier Chains," IEEE J. Photon. Technol. Lett., 6:9 (1994) 1098–1100.
10. J. L. Zyskind, J. A. Nagel, and H. D. Kidorf, "Erbium-Doped Fiber Amplifiers," in Optical Fiber Telecommunications, IIIB, I. P. Kaminow and T. L. Koch, Eds. (Academic Press, Amsterdam, 1997), 13–68.
11. R. G. Smart, J. L. Zyskind, and D. J. DiGiovanni, "Two-stage erbium-doped fiber amplifiers suitable for use in long-haul soliton systems," Electron. Lett., 30:1 (Jan. 1994) 50–52.
12. J.-M. P. Delavaux and J. A. Nagel, "Multi-stage erbium-doped fiber amplifier designs," J. Lightwave Technol., 13:5 (May 1995) 703–720.
13. A. R. Chraplyvy, J.-M. Delavaux, R. M. Derosier, G. A. Ferguson, D. A. Fishman, C. R. Giles, J. A. Nagel, B. M. Nyman, J. W. Sulhoff, R. E. Tench,

R. W. Tkach, and J. L. Zyskind, "1420-km Transmission of Sixteen 2.5-Gb/s Channels using Silica-Fiber-Based EDFA Repeaters," Photonic Technol. Lett., 6:11 (1994) 1371–1373.

14. F. Forgieri, R. W. Tkach, and A. R. Chraplyvy, "Fiber Nonlinearities and their Impact on Transmission Systems," in Optical Fiber Communications, Vol. IIIA, I. P. Kaminow and T. L. Koch, Eds. (Academic Press, Amsterdam, 1997) 196–264.
15. Y. Sun, J. B. Judkins, A. K. Srivastava, L. Garrett, J. L. Zyskind, J. W. Sulhoff, C. Wolf, R. M. Derosier, A. H. Gnauck, R. W. Tkach, J. Zhou, R. P. Espindola, A. M. Vengsarkar, and A. R. Chraplyvy, "Transmission of 32-WDM 10-Gb/s Channels Over 640 km Using Broad Band, Gain-Flattened Erbium-Doped Silica Fiber Amplifiers," IEEE Photon. Tech. Lett., 9:12 (December 1997) 1652–1654.
16. S. Kinoshita, Y. Sugaya, H. Onaka, M. Takeda, C. Ohshima, and T. Chikama, "Low-noise and wide dynamic range erbium-doped fiber amplifers with automatic level control for WDM transmission systems," Proc. Opt. Amplifiers and their Appl., Monterey, CA (July 11–13, 1996) 211–214.
17. S. Artigaud *et al.*, "Transmission of 16 × 10 Gb/s channels spanning 24 nm over 531 km of conventional single-mode fiber using 7 in-line fluoride-based EDFAs," OFC '96 Postdeadline Paper PD27 (1996).
18. M. Fukutoku *et al.*, "25 nm Bandwidth Optical Gain Equalization for 32-Channel WDM Transmission with a Lattice Type Optical Circuit," Optical Amplifiers and their Applications '96 (1996) 66.
19. A. M. Vengsarkar *et al.*, "Long-period fiber gratings based gain equalizers," Optics Lett. 21 (1996) 336.
20. R. Kashyap, R. Wyatt, and P. F. McKee, "Wavelength Flattened Saturated Erbium Amplifier using Multiple Side-tap Bragg Gratings," Elect. Lett. 29 (1993) 1025–1026.
21. Y. P. Li *et al.*, "Waveguide EDFA Gain Equalization Filter," Elect. Lett. 31 (1995) 2005–2006.
22. S. H. Yun, B. W. Lee, H. K. Kim, and B. Y. Kim, "Dynamic Erbium-Doped Fiber Amplifier with Automatic Gain Flattening," OFC'99 Post-deadline paper PD28 (1999).
23. P. Wysocki, J. B. Judkins, R. P. Espindola, M. Andrejco, and A. M. Vengsarkar, "Broadband Erbium-doped Fiber Amplifier flattened Beyond 40 nm Using Long Period Grating Filter," Photonic Technol. Lett., 9 (1997) 1343–1345.
24. H. Ono, M. Yamada, and Y. Ohishi, "Gain-flattened Er-doped fiber amplifier for a WDM signal in 1.57–1.60 λm wavelength region," IEEE Photonics Tech. Lett. 9 (1997) 596–598.
25. S. Ishikawa, "High Gain Per Unit Length Silica-Based Erbium Doped Fiber for 1580 nm Band Amplification," in Optical Amplifiers and Their Applications, 1998, 64–67.

26. Jinno *et al.*, Electron Lett. 1997
27. S. Aisawa, T. Sakamoto, J. Kani, S. Norimatsu, M. Yamada, and K. Oguchi, "Equally Spaced 8 × 10 Gb/s Optical Duobinary WDM Transmission in 1580-nm Band over 500 km of Dispersion-Shifted Fiber without Dispersion Compensation," OFC 1998 (Feb. 22–27, 1998) San Jose CA, 411–412.
28. Y. Yano, M. Yamashita, T. Suzuki, K. Kudo, M. Yamaguchi, and K. Emura, "640 Gb/s WDM Transmision over 400 km of Dispersion Shifted Fiber using 1.58 μm Band and Initial Chirp Optimization," ECOC 1998, (Sept. 20–24, 1998) Madrid, Spain 261–264.
29. A. K. Srivastava, Y. Sun, J. L. Zyskind, J. W. Sulhoff, C. Wolf, J. B. Judkins, J. Zhou, M. Zirngibl, R. P. Espindola, A. M. Vengsarkar, Y. P. Li, and A. R. Chraplyvy, "Error free transmission of 64 WDM 10 Gb/s channels over 520 km of TrueWave^{TM} fiber," Proc. ECOC '98, Madrid, Spain, (Sept. 20–24, 1998) 265.
30. Y. Sun, J. W. Sulhoff, A. K. Srivastava, J. L. Zyskind, C. Wolf, T. A. Strasser, J. R. Pedrazzani, J. B. Judkins, R. P. Espindola, A. M. Vengsarkar, and J. Zhou, "Ultra Wide Band Erbium-Doped Silica Fiber Amplifier with 80 nm of Bandwidth," Optical Amplifiers and their Applications, Postdeadline paper, PD2, Victoria B.C., Canada (July 1997).
31. M. Yamada *et al.*, Elect. Lett., 33:8 (April 1997) 710–711.
32. Y. Sun, J. W. Sulhoff, A. K. Srivastava, J. L. Zyskind, C. Wolf, T. A. Strasser, J. R. Pedrazzani, J. B. Judkins, R. P. Espindola, A. M. Vengsarkar, and, J. Zhou, "An 80 nm Ultra Wide Band EDFA Low Noise Figure and High Output Power," ECOC'97, Postdeadline paper, TH3C, Edinburgh, UK (Sept. 22, 1997) 69–72.
33. Y. Sun, J. W. Sulhoff, A. K. Srivastava, A. Abramov, T. A. Strasser, P. F. Wysocki, J. R. Pedrazzani, J. B. Judkins, R. P. Espindola, C. Wolf, J. L. Zyskind, A. M. Vengsarkar, and J. Zhou, "A Gain-Flattened Ultra Wide Band EDFA for High Capacity WDM Optical Communications Systems," Proc. ECOC '98, Madrid, Spain (Sept. 20–24, 1998) 53.
34. L. D. Garrett, A. H. Gnauck, F. Forghieri, V. Gusmeroli, and D. Scarano, "8 × 20 Gb/s – 315 km, 8 × 10 Gb/s – 480 km WDM Transmission over Conventional Fiber Using Multiple Broadband Fiber Gratings," OFC '98 Technical Digest, Postdeadline Papers, PD-18, San Jose, CA (Feb. 22–27, 1998).
35. S. Aisawa, T. Sakamoto, M. Fukui, J. Kani, M. Jinno, and K. Oguchi, "Ultra-wide Band, Long Distance WDM Transmission Demonstration: 1 Tb/s (50 × 20 Gb/s), 600 Km Transmission Using 1550 and 1580 nm Wavelength Bands," OFC '98 Technical Digest, Postdeadline Papers, PD 11, San Jose, CA (Feb. 22–27, 1998).
36. P. B. Hansen, L. Eskildsen, A. J. Stentz, T. A. Strasser, J. Judkins, J. J. Demarco, R. Pedrazzani, and D. J. DiGiovanni, "Rayleigh Scattering Limitations in Distributed Raman Preamplifiers," IEEE Photon Tech. Lett. 10:1 (Jan. 1998) 159–161.

37. L. D. Garrett, M. Eiselt, R. W. Tkach, V. Domini, R. Waarts, D. Gilner, and D. Mehuys, "Field Demonstration of Distributed Raman Amplification with 3.8 dB Q-Improvement for 5 × 120-km Transmission," IEEE Photon Tech. Lett. 13:2 (Feb. 2001) 157–159.
38. A. K. Srivastava Y. Sun, L. Zhang, J. W. Sulhoff, and C. Wolf., "System Margin Enhancement with Raman Gain in Multi-Span WDM Transmission," OFC'99 Paper FC2, San Diego, CA (Feb. 1999) 53–55.
39. H. Suzuki, J. Kani, H. Masuda, N. Takachio, K. Iwatsuki, Y. Tada, and M. Sumida, "25 GHz Spaced 1 Tb/s (100 × 10 Gb/s) Super Dense WDM Transmission in the C-Band over a Dispersion Shifted Cable Employing Distributed Raman Amplification," ECOC'99, PD2-4 Nice, France (Sept. 26–30, 1999).
40. Y. Yano *et al.*, ECOC'96, Paper ThB.3.1, Oslo Norway (1996).
41. K. Rottwitt and H. D. Kidorff, "A 92 nm Bandwidth Raman Amplifier," OFC '98 Technical Digest, Postdeadline Papers, PD-6, San Jose, CA (Feb. 22–27, 1998).
42. Y. Emori and S. Namiki, "100 nm Bandwidth Flat Gain Raman Amplifiers Pumped and Gain Equalized by 12-Wavelength Channel WDM High Power Laser Diodes," OFC'99 Postdeadline Paper PD-19, San Diego, CA (Feb. 1999).
43. T. N. Nielsen, A. J. Stentz, P. B. Hansen, Z. J. Chen, D. S. Vengsarkar, T. A. Strasser, K. Rottwitt, J. H. Park, S. Stultz, S. Cabot, K. S. Feder, P. S. Westbrook, and S. G. Kosinski, "1.6 Tb/s (40 × 40 Gb/s) Transmission over 4 × 100 km nonzero-dispersion fiber using hybrid Raman/Erbium-doped inline amplifiers," Postdeadline paper, ECOC'99 PD2-2, Nice, France (Sept. 26–30, 1999).
44. T. N. Nielsen, A. J. Stentz, K. Rottwitt, D. S. Vengsarkar, Z. J. Chen, P. B. Hansen, J. H. Park, K. S. Feder, T. A. Strasser, S. Cabot, S. Stultz, D. W. Peckham, L. Hsu, C. K. Kan, A. F. Judy, J. Sulhoff, S. Y. Park, L. E. Nelson, and L. Gruner-Nielsen, "3.28 Tb/s (82 × 40 Gb/s) Transmission over 3 × 100 km nonzero-dispersion fiber using dual C- and L-band hybrid Raman/ Erbium-doped inline amplifiers," Postdeadline paper, OFC'00 Paper PD-23, Baltimore, MD (Mar 7–10, 2000).
45. A. K. Srivastava, S. Radic, C. Wolf, J. C. Centanni, J. W. Sulhoff, K. Kantor, and Y. Sun, "Ultra-dense Terabit Capacity Transmission in L-Band," Postdeadline paper, OFC'00 Paper PD-27, Baltimore, MD (Mar 7–10, 2000).
46. A. K. Srivastava, J. L. Zyskind, Y. Sun, J. Ellson, G. Newsome, R. W. Tkach, A. R. Chraplyvy, J. W. Sulhoff, T. A. Strasser, C. Wolf, and J. R. Pedrazzani, "Fast-Link Control Protection of Surviving Channels in Multiwavelength Optical Networks," IEEE Photon. Technol. Lett., 9:11 (Dec. 1997) 1667–1669.
47. A. K. Srivastava, Y. Sun, J. L. Zyskind, and J. W. Sulhoff, "EDFA Transient Response to Channel Loss in WDM Transmission System," IEEE Photon. Technol. Lett. 9:3 (March 1997) 386–388.

48. C. R. Giles, E. Desurvire, and J. R. Simpson, "Transient Gain and Cross-talk in Erbium-doped Fiber Amplifier," Optics Letters, 14:16 (1989) 880–882.
49. E. Desurvire, "Analysis of Transient Gain Saturation and Recovery in Erbium-doped Fiber Amplifiers," IEEE Photon. Technol. Lett., 1 (1989) 196–199.
50. Y. Sun, J. L. Zyskind, A. K. Srivastava, and L. Zhang, "Analytical Formula for the Transient Response of Erbium-Doped Fiber Amplifier," Applied Optics, 38:9 (March 20, 1999) 1682–1685.
51. E. Desurvire, "Erbium-doped Fiber Amplifiers—Principles and Applications" (John Wiley, New York, 1994) 469–480.
52. P. R. Morkel and R. I. Laming, "Theoretical Modeling of Erbium-Doped Fiber Amplifiers with Excited State Absrorption," Opt. Lett., 14 (1989) 1062–1064.
53. Y. Sun, G. Luo, J. L. Zyskind, A. A. M. Saleh, A. K. Srivastava, and J. W. Sulhoff, "Model for Gain Dynamics in Erbium-doped Fiber Amplifiers," Electron. Lett., 32:16 (Aug. 1, 1996) 1490–1491.
54. J. L. Zyskind, Y. Sun, A. K. Srivastava, J. W. Sulhoff, A. L. Lucero, C. Wolf, and R. W. Tkach, "Fast Power Transients in Optically Amplified Multiwavelength Optical Networks," Proc. Optical Fiber Communications Conference, postdeadline paper PD 31 (1996).
55. Y. Sun, A. K. Srivastava, J. L. Zyskind, J. W. Sulhoff, C. Wolf, and R. W. Tkach, "Fast Power Transients in WDM Optical Networks with Cascaded EDFAs," Electron. Lett., 33 (1997) 313–314.
56. MONET 2nd Quarterly Report, Payable Milestone #2, 1995 (**query is incomplete**)
57. Y. Sun and A. K. Srivastava, "Dynamic Effects in Optically Amplifed Networks," Proc. Optical Amplifiers and their Applications, Victoria, Canada (July, 1997).
58. S. J. Shieh, J. W. Sulhoff, K. Kantor, Y. Sun, and A. K. Srivastava, "Dynamic Behavior in L-Band EDFA," OFC'00 Paper PD-27, Baltimore, MD (Mar 7–10, 2000).
59. A. K. Srivastava, Y. Sun, J. L. Zyskind, J. W. Sulhoff, C. Wolf, and R. W. Tkach, "Fast Gain Control in Erbium-doped Fiber Amplifiers," Proc. Optical Amplifiers and their Applications 1996, Postdeadline paper PDP4 (1996).
60. E. Desurvire, M. Zirngibl, H. M. Presby, and D. DiGiovanni, "Dynamic Gain Compensation in Saturated Erbium-Doped Fiber Amplifiers," IEEE Photon. Technol. Lett., 3 (May 1991) 453–455.
61. M. Zirngibl, "Gain Control in an Erbium-Doped Fiber Amplifier by an all-optical feedback loop," Electron. Lett. 27:7 (1991) 560–561.
62. J. L. Jackel and D. Richards, "All-Optical Stabilization of Cascaded Multi-Channel Erbium-doped Fiber amplifiers with changing numbers of channels," OFC '97 Technical Digest, Dallas, TX (Feb. 22–27, 1997) 84–85.

63. Luo, J. L. Zyskind, Y. Sun, A. K. Srivastava, J. W. Sulhoff, and M. A. Ali, "Relaxation Oscillations and Spectral Hole Burning in Laser Automatic Gain Control of EDFAs," OFC '97 Technical Digest, Dallas, TX (Feb. 22–27, 1997) 130–131. D. A. Ferrand, P. F. Wysocki, J. B. Judkins, and Y. P. Li, "1 Tb/s transmission of 100 WDM 10 Gb/s channels over 400 km of TrueWave^{TM} fiber," OFC '98 Technical Digest, Postdeadline Papers, PD 10, San Jose, CA (Feb. 22–27, 1998).
64. P. I. Kuindersma *et al.*, Proc. ECOC'96, 2, Oslo, Norway (Sept. 1996) 165–168.
65. G. Onishchukov *et al.*, Electron. Lett., 34 (Aug. 1998) 1597–1598.
66. M. Bachmann *et al.*, Electron. Lett., 32 (Oct. 1996) 2076–2078.
67. J. G. L. Jennen *et al.*, Proc. ECOC'98, 1, Madrid, Spain (Sept. 1998) 235–236.
68. K. P. Ho *et al.*, Electron. Lett., 32 (Nov. 1996) 2210–2211.
69. K. Inoue, J. Lightwave. Tech., 7 (July 1989) 1118–1124.

Chapter 10 | Fiber Raman Amplifier

Shu Namiki

Fitel Photonics Laboratory, Furukawa Electric Co., Ltd.
6 Yawata Kaigan Dori, Ichihara, Chiba 290-8555, Japan

10.1. Overview

The WDM transmission technologies have been constantly developed and improved in order to catch up to the explosive demands for high-capacity and global transmissions due to Internet growth. The EDFA (erbium-doped fiber amplifier)-based WDM transmission systems have finally begun to face their limit in terms of both bandwidth and noise. An optical amplifier besides EDFA is now sought in order to extend bandwidth and further reduce noise for higher capacity and longer distance transmissions. In this context, fiber Raman amplifiers are receiving much attention because of their adjustability of the gain band by choosing the proper pumping wavelength, and their low-noise nature due to an off-resonance amplification process that fits well in the distributed amplifier configuration.

In fact, fiber Raman amplifiers were already extensively studied before EDFA was developed in the late 1980s. However, because fiber Raman amplifiers needed more than 100 mW output power from the pump laser for most useful applications, fiber Raman amplifiers were not employed in the real world due to lack of compact and robust pump lasers at that time. Instead, EDFA, operating at pump powers less than 100 mW, could be pumped by compact laser diodes, so they were deployed in the real world. However, extensive studies in the 1970s to 1980s showed that the Raman amplification process could be used for optical transmission systems,

WDM TECHNOLOGIES: PASSIVE
OPTICAL COMPONENTS
$35.00

ISBN: 0-12-225262-4

leaving unsolved the issue of practical pumping sources. In the 1990s, with the invention of the so-called cascaded Raman resonator, the possibilities of Raman amplifiers being used in optical communications are again being explored, for better noise performance in a distributed amplifier configuration as well as viability in signal bands other than the EDFA band.

The explosive growth in Internet demands has boosted high output power requirements for pump laser diodes. As a result, in the late 1990s, 1480-nm pump lasers for EDFA became mature enough to achieve a few hundred milli-watt output into a single-mode fiber. Because the suitable pumping wavelength of a fiber Raman amplifier operating in 1550 nm, the same band as EDFA, is around 1440 nm, the same technologies for pumping EDFA could be used for a fiber Raman amplifier in a straightforward manner. Thus, the Internet demands, compelling the maturation of the pump laser diode, have given rise to the revisit of fiber Raman amplifiers for better systems performance. However, a fiber Raman amplifier has yet to be designed to fit well for WDM purposes.

Principles of Raman scattering have been elaborately covered in the literature [1, 2, 3]. The purpose of this chapter is to help readers grasp the general principles of fiber Raman amplifiers and refer to their applications in the most updated optical communications systems or WDM transmission systems. Section 10.2 presents an outlook on the principle of the Raman amplifier. Section 10.3 refers to physical properties with respect to gain and noise of Raman amplification in optical fiber. Section 10.4 develops arguments on how to design a fiber Raman amplifier specifically for WDM purposes, addressing a consistent design method as well as noise issues to be resolved. Section 10.5 reviews key components of the compose fiber Raman amplifier, and how they work in the real world. Section 10.6 investigates examples of system applications of fiber Raman amplifiers.

10.2. Principles

10.2.1. SPONTANEOUS RAMAN SCATTERING

Consider a scattering between a photon and a molecule or medium: As the photon is approaching a molecule, the electric field of the incident photon begins to apply to the molecule. Then, the photon leaves the molecule behind and goes away. The interaction will create a small distortion of the molecule, or a nonlinear polarization of the medium. As the photon moves away from the molecule, the force causing distortion of the molecule reduces. If the response of the molecule is instantaneous, the distorted portion of the molecule will retract back to the original equilibrium position as quickly as the incident photon moves away. Therefore, the molecule will stay "calm" after the photon is gone, as if nothing had happened in the first place.

However, if there is a delay in the response of the molecule to the electromagnetic field due to the incident photon, the distortion occurs after a certain delay once the incident photon enters. The distortion can also persist even though the photon is completely gone. This situation is similar to that of a molecule forcedly distorted and suddenly set free. It is easy to anticipate that the molecule will start vibrating or that a phonon will be created. The fact that the molecule starts vibrating implies there must be some energy input from outside to the molecule. In this case, the incident photon brings the energy necessary for the molecule to start vibrating. Therefore, for the sake of energy conservation, the photon must lose energy after the scattering. The frequency of the photon will red-shift by the equivalent amount of vibration energy of the molecule. In a classical description, the vibration of the molecule creates a temporal modulation of the refractive index to shift the frequency of incident light by the frequency of vibration. The frequency of the vibration is inversely proportional to the response time of the molecule. This is because the molecule moves at a common rate for both vibration and response.

In case the molecule is vibrating due to temperature before the photon interacts with the molecule, the photon can absorb the phonon and receive the energy equivalent to the vibration energy. In this case, the frequency of the photon will increase, or blue-shift. The red-shift is called "Stokes shift," while the blue-shift is called "anti-Stokes shift."

The frequency shift of light due to the scattering with molecules was discovered by C. V. Raman in 1928. This scattering is named "spontaneous Raman scattering" after the discoverer.

The process of Raman scattering can be described by using nonlinear polarization. Polarization induced by the input lightwave is written as

$$\mathbf{P} = N\rho E_{\text{in}} \tag{10.1}$$

where N is number of molecules in a unit volume, ρ is polarizability of molecule, and E_{in} is incident lightwave. Let us define $x(z, t)$ as the deviation from the equilibrium position of the molecule. The nonlinear polarizability ρ depends on the deviation of the molecule, x, and can be approximately expanded up to the first order as

$$\rho = \rho_0 + \rho_1 x(z, t). \tag{10.2}$$

The delayed response of the molecule can be described as a damped harmonic oscillator with a driving force.

$$\frac{d^2}{dt^2}x(z,t) + \Gamma \frac{d}{dt}x(z,t) + \omega_m^2 x(z,t) = \frac{\rho_1 \langle E_{\text{in}}^2(z,t) \rangle}{2M} \tag{10.3}$$

where Γ is the damping constant, ω_m is the resonance frequency of the molecule, and the right-hand side is the driving force due to incident lightwave divided by M, mass of the oscillator. As $x(z, t)$ oscillates, the polarizability also oscillates to modulate the incident lightwave. The nonlinear polarization can be calculated by using Eq. (10.1), substituting the solution of Eq. (10.3) into Eq. (10.2) (pp. 372–379 of [2]). Under this model, the Stokes-shifted lightwave hits the resonance and receives amplification from the response of the molecule. This process is called stimulated Raman scattering, or SRS for short. This is a four wave mixing process because the pump and Stokes waves are coupled with the incident waves in Eq. (10.1) through the force term in Eq. (10.3), creating amplification of the Stokes wave as well as absorption of the pump wave.

10.2.2. QUANTUM-MECHANICAL DESCRIPTION AND STIMULATED RAMAN SCATTERING (SRS)

Raman scattering in quantum-mechanical description is a process in which a pump photon at frequency ω_p is absorbed and a Stokes photon at frequency ω_s is emitted, while the molecule makes a transition from the ground state to an excited state, emitting an optical phonon. An energy diagram representing Raman scattering is depicted in Fig. 10.1. An electron of the molecule absorbs an incident pump photon, which is excited to the virtual intermediate states of the molecule. The electron is then quickly de-excited down to the molecule in the first energy state of the optical phonon. If a photon with Stokes-shifted frequency enters when the electron is de-excited, then the stimulated emission can occur. Stimulated Raman scattering is a process in which a Stokes-shifted photon receives gain when it enters into the molecules together with a pump photon. Stimulated Raman scattering was discovered by Woodbury and Ng in 1962.

The transition probability W_{fi} of Raman scattering is proportional to the number of pump photons and Stokes photons, as well as the transition rate

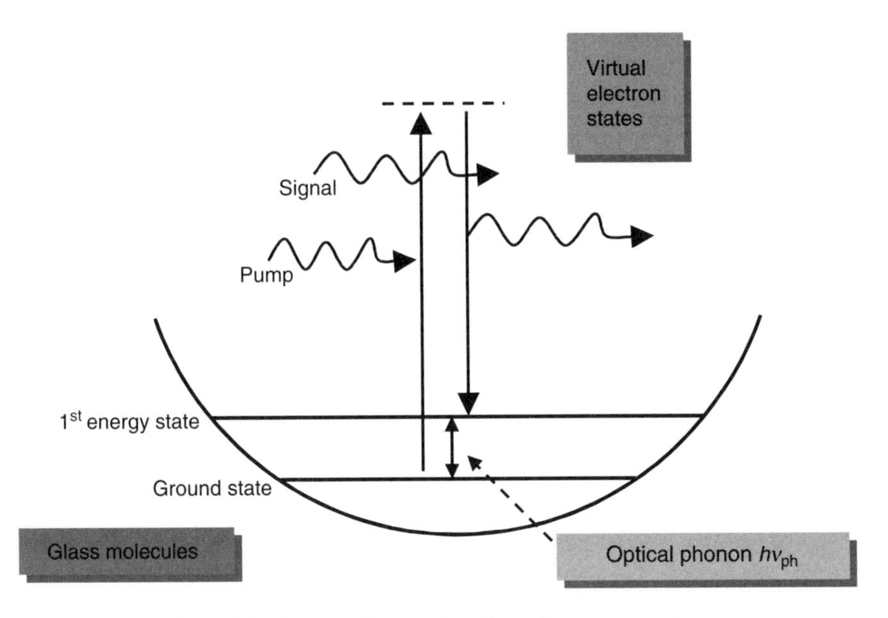

Fig. 10.1 Energy diagram describing Raman scattering.

of the molecule from ground to upper states due to incident electric field of photons, as follows:

$$
\begin{aligned}
W_{fi} &\propto |\langle f|K|i\rangle|^2 |\langle n_p - 1, n_s + 1|b_s^+ b_p|n_p, n_s\rangle|^2 \\
&= D n_p (n_s + 1) \qquad (10.4)
\end{aligned}
$$

where $|f\rangle$, $|i\rangle$, $|n_p, n_s\rangle$ are the excited and ground states of molecule, and the radiation field of n_p pump photons and n_s Stokes photons, respectively. b_s^+ and b_p are creation and annihilation operators for Stokes and pump photons, respectively. $\langle f|K|i\rangle$ is a compound matrix element with regard to the second-order perturbation describing the interaction between the molecule and electric field (pp. 143–146 of [1]). In this process, one pump photon is annihilated, while one Stokes photon is created. Note that, in the Raman medium, the direct electric dipole transition between $|f\rangle$ and $|i\rangle$ is forbidden and realized only through intermediate states having the opposite parity to $|f\rangle$ and $|i\rangle$. D is therefore a proportionality constant inherent in the material. The unity in the parenthesis, derived from the zero-point fluctuation, corresponds to spontaneous Raman scattering.

The evolution of Stokes photons along distance z can then be written as

$$\frac{d}{dz}n_s(z) = Cn_p(n_s + 1) - \alpha_s n_s \tag{10.5}$$

where C is a proportionality constant and α_s is the attenuation coefficient of the Stokes photon in the medium. It should be noted that the spontaneous Raman scattering term acts as a Langevin noise source because it has no correlation with the incident Stokes photons, but mixes in them. For a large enough number of Stokes photons and an almost constant number of pump photons, the number of Stokes photons grows exponentially as a function of distance, while in the absence of Stokes photons, the spontaneous term grows linearly with distance. Therefore, the intensity of stimulated Raman scattering is orders of magnitude larger than that of spontaneous Raman scattering, and stimulated Raman scattering can be used for amplification of Stokes photons.

10.2.3. RAMAN SCATTERING AND BRILLOUIN SCATTERING

In addition to conservation of energy, phase matching or conservation of momentum is necessary for a scattering process to be complete. For the Stokes process, the following conditions have to be met:

$$\hbar\omega_p - \hbar\omega_m = \hbar\omega_s \tag{10.6}$$

$$\mathbf{k}_p - \mathbf{k}_m = \mathbf{k}_s \tag{10.7}$$

where ω_p, ω_s, and ω_m, and $\mathbf{k}_p$, $\mathbf{k}_s$, and $\mathbf{k}_m$ are angular frequencies and wave vectors of pump photons, Stokes photons, and optical phonon, respectively. $\hbar$ is Planck's constant. For the anti-Stokes process, the minus signs on the left hand-side will be replaced with plus signs in Eqs. (10.6) and (10.7).

In general, there are two kinds of phonons in bulk or liquid mediums: optical and acoustic phonons. Raman scattering is the scattering of light with optical phonons, while Brillouin scattering is the scattering of light with acoustic phonons. For a diatomic chain model, the dispersion relation has two branches corresponding to optical and acoustic phonons. The dispersion curve for the optical phonon is relatively uniform over the first Brillouin zone so that the phase matching among pump photon, Stokes photon, and optical phonon can be easily achieved for any orientations of the wave vector of the pump photon with respect to that of Stokes photon. Therefore, the direction of the Stokes photon emitted through spontaneous

Raman scattering is almost isotropic. Also, stimulated Raman scattering can occur for both forward- and backward-pumping configurations (pp. 77–91 of [3]).

On the other hand, stimulated Brillouin scattering is known to occur only in a backward pumping configuration. The acoustic phonon has approximately a linear dispersion relation between frequency and wave number. Stimulated Brillouin scattering for the backward direction has the phase matching condition obeying this dispersion relation (pp. 331–337 of [2]). Also, stimulated Brillouin scattering is a process of Bragg reflection between incident light and index grating generated by acoustic phonons. In this regard, one can show the forward direction does not generate any frequency shift of light due to an acoustic phonon (pp. 88–91 of [3]).

Acoustic phonons in general have lower energy than optical phonons. For silica glass, the frequency of an optical phonon is approximately 13 THz, while that of an acoustic phonon is around 10 GHz [4]. The response of an optical phonon is much faster than that of an acoustic phonon.

10.3. Stimulated Raman Amplification in Optical Fiber

10.3.1. FIBER RAMAN AMPLIFIER AND ITS THEORETICAL MODEL

Figure 10.2 shows a general schematic of a fiber Raman amplifier. Because stimulated Raman scattering between Stokes and pump waves has to occur in a fiber Raman amplifier, the two waves have to be combined through a wavelength division multiplexing (WDM) coupler so that they co-exist in a common optical fiber. For amplification purposes, the frequency of the pump is chosen so that the signal would have the Stokes frequency with respect to the pump frequency.

Raman scattering in optical fiber was first observed by Stolen, Ippen, and Tynes in 1972 [5]. The Raman gain spectrum of optical fiber was first measured by Stolen and Ippen in 1973 [6]. Theoretical models of Raman scattering in optical fiber have been studied since [7–14]. Many reports on high stimulated Raman gain achieved in optical fiber were also published [15–18].

Converting Eq. (10.5) into power in single-mode fiber, and including the pump and neglecting the spontaneous emission term, the power evolutions

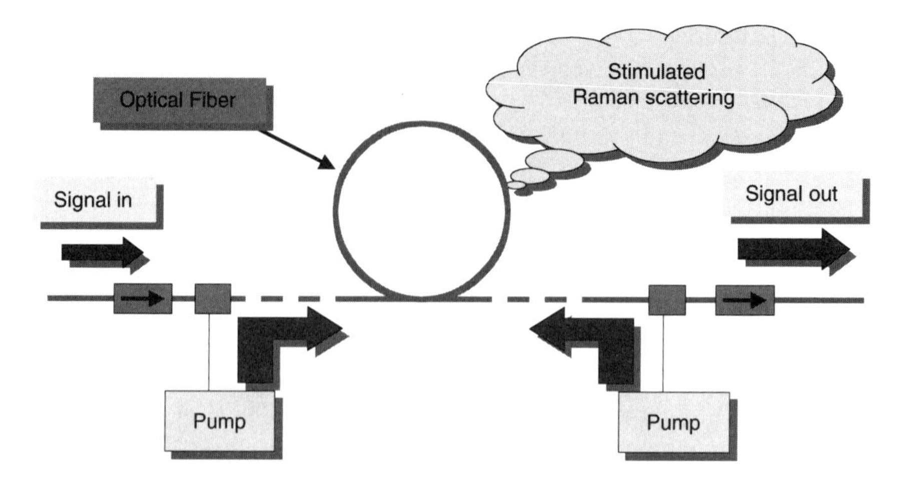

Fig. 10.2 General configuration of fiber Raman amplifier. S. Namiki and Y. Emori, "Ultrabroad-band Raman amplifiers pumped and gain equalized by wavelength-division-multiplexed highpower laser diodes," IEEE J. Selected Top Quantum Electron, vol. 7, No. 1, pp. 3–16, 2001. © 2001 IEEE.

of Stokes and pump waves along optical fiber are described as [4]

$$\frac{dP_s}{dz} = \frac{g_R}{\lambda_s A_{eff}} P_p P_s - \alpha_s P_s \tag{10.8}$$

$$\frac{dP_p}{dz} = -\frac{g_R}{\lambda_p A_{eff}} P_p P_s - \alpha_p P_p \tag{10.9}$$

where P_s and P_p are Stokes and pump powers respectively, g_R is Raman gain coefficient, λ_s and λ_p are the wavelength of Stokes and pump waves respectively, A_{eff} is the effective area of optical fiber, and α_s and α_p are attenuation coefficient of optical fiber at Stokes and pump frequencies, respectively. Effective area A_{eff} in Eqs. (10.8) and (10.9) is defined as

$$A_{eff} = \frac{\int da|u_s|^2 \int da|u_p|^2}{\int da|u_s|^2|u_p|^2} \tag{10.10}$$

where u_s and u_p are optical field distribution of Stokes and pump waves in the transverse plane, respectively, and $\int da$ denotes the integral over the entire transverse plane. It should be noted that A_{eff} depends on wavelength. Also, the total number of photons is conserved in the preceding coupled equations. Let us define $g_R/\lambda_s A_{eff}$ as Raman gain efficiency hereafter. Raman gain efficiency scales inversely with the wavelength and effective area. This scaling is the same as the nonlinear coefficient γ describing the self-phase modulation term in the nonlinear Schrödinger equation [4], which is defined as

$$\gamma \equiv \frac{2\pi n_2}{\lambda_s A_{eff}} \tag{10.11}$$

where n_2 is the nonlinear refractive index. As we have discussed in the previous sections, Raman scattering is a four wave mixing process with an imaginary part of the third-order nonlinearity. In other words, Raman scattering originates from a finite response time of the third-order nonlinearity [19]. Thus, Raman process is closely related to self-phase modulation or a real part of the third-order nonlinearity via the Kramers-Kronig relation [19–21]. In practice, however, it is not necessary to characterize A_{eff} alone; Raman gain efficiency needs to be measured as one property.

The first term on the right-hand side of Eq. (10.9) describes the effect of pump depletion, from which the gain saturation results. In cases where the Stokes signal is low enough that the amount of pump depletion is negligible as compared with the high pump power, one can solve Eqs. (10.8) and (10.9) to obtain the power of Stokes wave at distance z as

$$P_s(z) = P_s(0)\exp\left(\frac{g_R}{\lambda_s A_{eff}} P_p(0) L_{eff}(z) - \alpha_s z\right) \tag{10.12}$$

where L_{eff} is called the effective interaction length and defined as

$$L_{eff} = \frac{1 - \exp(-\alpha_p z)}{\alpha_p}. \tag{10.13}$$

In fact, the logarithmic gain of the Stokes wave is proportional to the term $P_p(0)L_{eff}$, which is the integration of the pump power over length. An exact solution of Eqs. (10.8) and (10.9) should also give the gain depending on the pump power integrated over length.

10.3.2. SPECTRA OF RAMAN GAIN EFFICIENCY FOR DIFFERENT FIBERS

In a small signal regime, one can measure the Raman gain efficiency by making use of Eq. (10.12). Figure 10.3 is the measurement setup for Raman gain efficiency. By comparing the signal output between the pump source turned on and off, the Raman gain efficiency is easily obtained. The on–off

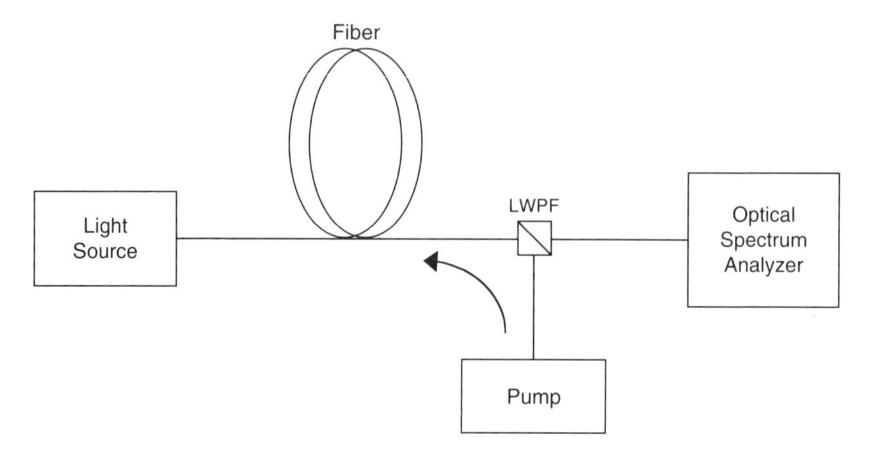

Fig. 10.3 Measurement setup for Raman gain efficiency. LWPF stands for long-wavelength path filter.

Raman gain is obtained by

$$G_R \equiv \frac{P_s(z;\, P_p = \text{on})}{P_s(z;\, P_p = \text{off})} = \exp\left(\frac{g_R}{\lambda_s A_{\text{eff}}} P_p(0) L_{\text{eff}}\right) \qquad (10.14)$$

Figure 10.4 shows the measured spectra of Raman gain efficiency for single-mode fiber (SMF), dispersion-shifted fiber (DSF), and dispersion-compensating fiber (DCF). Table 10.1 indicates the summary of characteristics of each fiber. Unlike molecules and crystals, the Raman gain spectra of optical fibers are broad. This reflects the amorphous nature of glass material: Silica (SiO_2) and Germania (GeO_2) glass have a continuum distribution of density of states for optical phonons. Two important differences among the three fibers can be pointed out—peak values of Raman gain efficiency and the spectral profile. The peak value depends on nonlinearity of the fiber. Because DCF has the smallest mode field diameter or effective area, it has the largest nonlinearity among the three kinds of fiber to have the highest peak value of Raman gain efficiency. Figure 10.5 shows the normalized Raman gain spectra for SMF, DSF, and DCF. The profile difference is due to the difference in the concentration of GeO_2.

Studies on vibrational modes such as bending, stretching, and rocking motions of vitreous SiO_2 and GeO_2 were extensively conducted to theoretically explain the spectral structures of Raman gain efficiency [22–24]. The spectral profile consequently depends on the concentration of GeO_2 [25, 26]. The peak value is proportional to the inverse effective area

Table 10.1 **Properties of SMF, DSF and DCF**

Parameter (@ 1550 nm)	α *[dB/km]*	*D [ps/nm/km]*	A_{eff} [μm^2]	γ *[1/km/W]*
Standard single-mode fiber	0.19	16.4	78.5	1.5
Dispersion-shifted fiber	0.20	−0.2	47.4	2.7
Dispersion-compensating fiber	0.57	−130	15.0	13.2

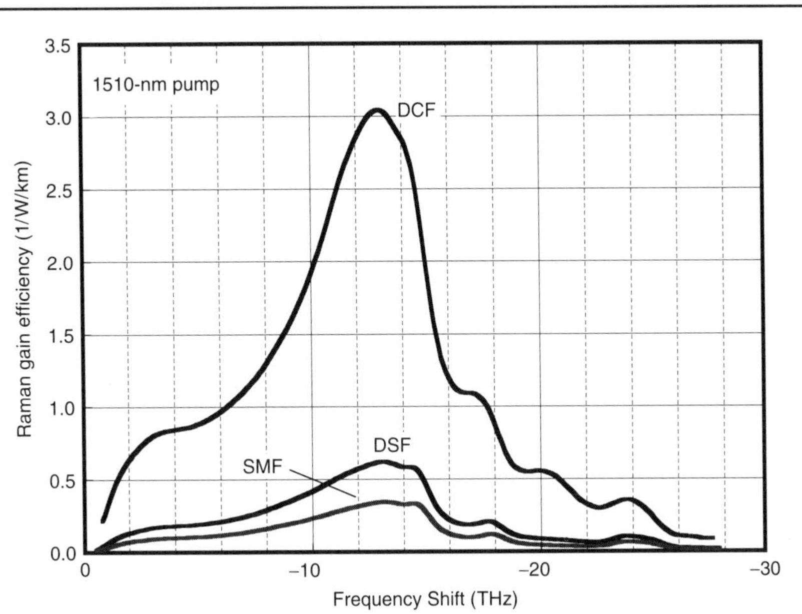

Fig. 10.4 Spectra of Raman gain efficiency for different fibers. S. Namiki and Y. Emori, "Ultrabroad-band Raman amplifiers pumped and gain equalized by wavelength-division-multiplexed highpower laser diodes," IEEE J. Selected Top Quantum Electron, vol. 7, No. 1, pp. 3–16, 2001. © 2001 IEEE.

according to Eq. (10.14) and the concentration of GeO_2 with respect to the host glass of silica [26, 27]. Detailed studies on Raman gain spectra for modern transmission fibers can be found, for example, in [28] and [29].

10.3.3. FLUORESCENCE LIFETIME

The Raman response time can be regarded as fluorescence lifetime in the context of optical amplifiers, which is estimated as 3–5 fs for silica [4, 19, 30]. For optical signals operating at a bit rate of 40 Gb/s or higher, this response time is almost instantaneous. This fast response nature of Raman amplifier is attributable to its process via virtual intermediate states.

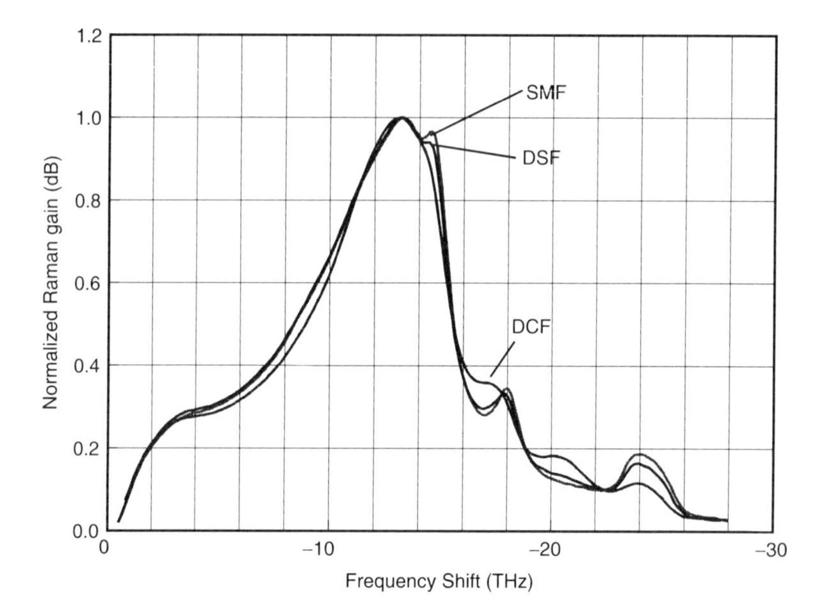

Fig. 10.5 Normalized Raman gain efficiency spectra for different fibers. S. Namiki and Y. Emori, "Ultrabroad-band Raman amplifiers pumped and gain equalized by wavelength-division-multiplexed highpower laser diodes," IEEE J. Selected Top Quantum Electron, vol. 7, No. 1, pp. 3–16, 2001. © 2001 IEEE.

10.3.4. POLARIZATION-DEPENDENT GAIN AND DEGREE OF POLARIZATION OF PUMP

Because the transition is quick, the Raman scattering process almost preserves the polarization states of pump and Stokes waves. Because of the phase-matching condition, the Raman scattering only occurs between waves in the same polarization states [6, 31–33]. Because the Raman interaction length is typically kilometers long, the states of polarization (SOP) of the pump and Stokes waves may become random with respect to each other through slightly birefringent segments of fiber. In the counter-pumping scheme, the pump and Stokes waves pass by each other so rapidly over distance that the relative orientation of the two polarization states will rapidly change and thus averages out to be completely random. However, this randomization process is strictly through the concatenation of local birefringent segments distributed over fiber. Therefore, a chance still exists for the polarization-dependent gain (PDG) to result. The co-pumping scheme has a better chance to have PDG because the pump and Stokes waves co-propagate with each other at velocities only different by the amount of chromatic dispersion. In order to avoid PDG, it is better to use

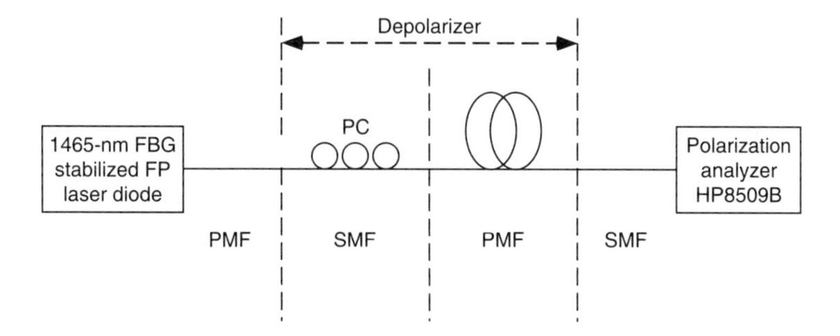

Fig. 10.6 Experimental setup of a depolarized pump LD. S. Namiki and Y. Emori, "Ultrabroad-band Raman amplifiers pumped and gain equalized by wavelength-division-multiplexed highpower laser diodes," IEEE J. Selected Top Quantum Electron, vol. 7, No. 1, pp. 3–16, 2001. © 2001 IEEE.

unpolarized pump sources. In the case where a laser diode is used for pumping, some depolarization technique has to be employed, as the output from the laser diode has a very clean linear polarization.

Figure 10.6 shows the experimental setup of a depolarized pump LD. Here, a 1465-nm Fabry-Perot laser diode (FP-LD) stabilized by a fiber Bragg grating (FBG) is depolarized through a depolarizer consisting of a polarization controller (PC) and a polarization-maintaining fiber (PMF). The measured value of the output degree of polarization (DOP) of the PMF depends on the PMF length and an input SOP of the PMF. In principle, the linearly polarized beam emitted from LD has to be incident to the PMF at exactly 45 degrees off of one of the eigen axes of the PMF. The PC was used here to minimize the risk of errors in aligning the incident angle. For practical use, the output of LD should be through a PMF pigtail and it should be fusion-spliced to the depolarizer PMF at 45 degrees. Any error from 45 degrees may result in partial incompleteness of depolarization. However, the depolarizer will be robust and stable.

Figure 10.7 shows the minimum DOP versus PMF length. As can be seen, the value of the DOP decreases to a few percents at 3 m, which is long enough with respect to the coherence length of the pump LD, and short enough to be low loss and low cost. Because this depolarizer is a Mach-Zehnder interferometer using slow and fast axes of PMF for the two arms with a delay, the DOP value as a function of PMF length (or delay) is a Fourier-transform of the optical output spectrum of the laser diode. Therefore, the optimum length of PMF depends on the optical spectrum of the laser diode.

Figure 10.8 shows the experimental setup for measuring PDG of Raman gain in DCF. The length of DCF was 2770 m with total dispersion of

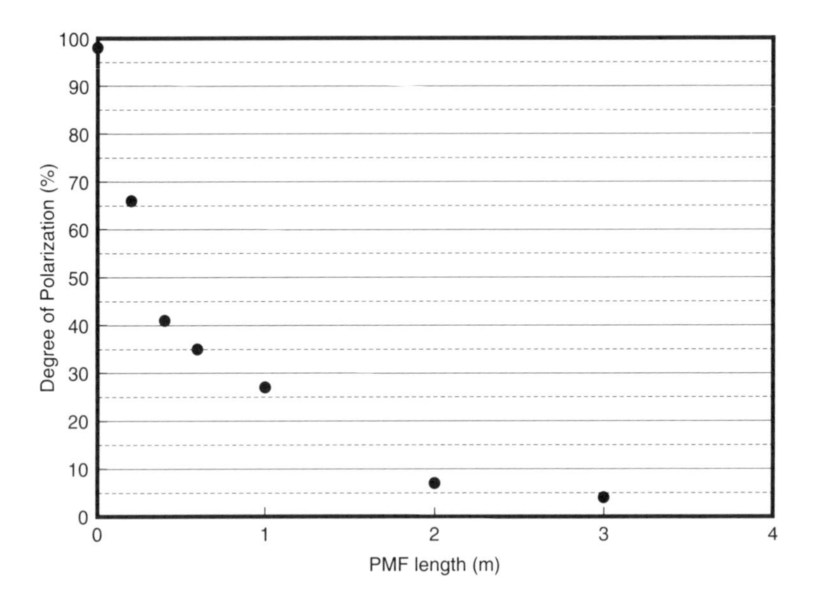

Fig. 10.7 Mimimum DOP versus PMF length [33]. S. Namiki and Y. Emori, "Ultrabroad-band Raman amplifiers pumped and gain equalized by wavelength-division-multiplexed highpower laser diodes," IEEE J. Selected Top Quantum Electron, vol. 7, No. 1, pp. 3–16, 2001. © 2001 IEEE.

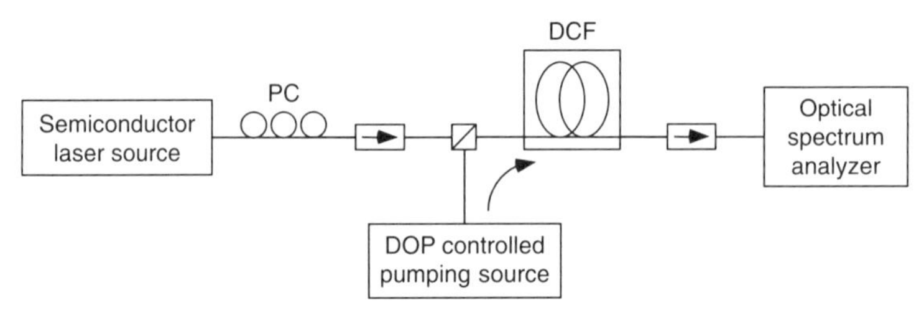

Fig. 10.8 Experimental setup for the measurement of PDG. S. Namiki and Y. Emori, "Ultrabroad-band Raman amplifiers pumped and gain equalized by wavelength-division-multiplexed highpower laser diodes," IEEE J. Selected Top Quantum Electron, vol. 7, No. 1, pp. 3–16, 2001. © 2001 IEEE.

−350 ps/nm. The DOP-controlled pump LD of 1465 nm and a semiconductor laser source of 1565 nm are used as the pump and signal, respectively. Co-propagating pumping is used, as larger PDG can be observed compared with counter-pumping. A SOP of the signal was controlled using PC for measurement of PDG. The Raman gain was measured to be 3 dB with pump power of 90 mW. Figure 10.9 shows the result of the measurement PDG versus DOP of the pump source. When the DOP is 7%, PDG is almost equal to the value of polarization-dependent loss of the passive components. This

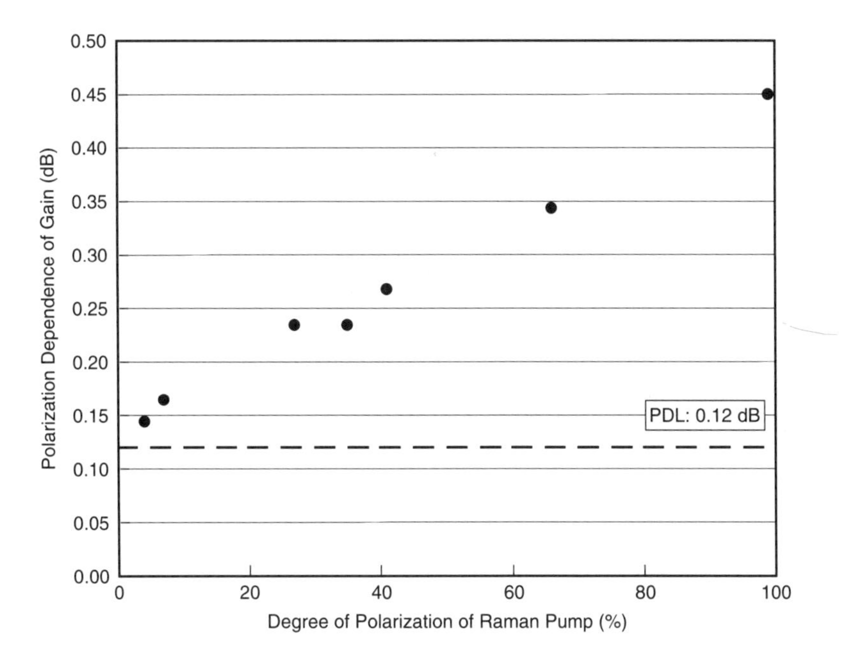

Fig. 10.9 Measured PDG value versus DOP of the pump source at Raman gain of 3 dB [33]. S. Namiki and Y. Emori, "Ultrabroad-band Raman amplifiers pumped and gain equalized by wavelength-division-multiplexed highpower laser diodes," IEEE J. Selected Top Quantum Electron, vol. 7, No. 1, pp. 3–16, 2001. © 2001 IEEE.

suggests that the PDG is sufficiently suppressed in this condition. It should be noted that the PDG value is proportional to the amount of Raman gain and dependent on fiber birefringence. Therefore, for larger Raman gain, DOP value has to be maintained strictly at a lower value.

10.3.5. GAIN SATURATION

As we discussed in previous sections, Raman response occurs almost instantaneously. Therefore, gain saturation impacts on the temporal shape of the amplified signal. Particularly for WDM transmissions in which multiple signals are simultaneously amplified, gain saturation can cause the so-called cross-gain modulation crosstalk among WDM signals when the pump is co-propagating.

Gain saturation in fiber Raman amplifiers occurs due to pump depletion, which is described in Eqs. (10.8) and (10.9). Because the saturation power of an optical amplifier is inversely proportional to the stimulated emission cross-section, and because Raman gain efficiency or scattering cross-section for an optical fiber is typically small, the saturation power of a

Raman amplifier is large. For example, as in Fig. 10.4, the peak Raman gain efficiency for DSF at pump wavelength of 1510 nm is 0.62/W/km. Typical loss coefficient of DSF is 0.2 dB/km in the 1.5 μm region. For simplicity, let us consider a small signal gain regime. The input pump power and fiber length are, say, 500 mW and 50 km, respectively. From Eq. (10.14), Raman gain is estimated as 6.7 dB, which is equivalent to the loss of 33.5-km DSF. Suppose that this Raman amplifier operates as a pre-amplifier attached at the very end of a fiber span. Then, the optical signal is usually amplified up to, say, 10 dBm. Consider that 10 dBm output from an optical amplifier with pump power of 500 mW. The power conversion efficiency in this case is merely 2%. Because gain saturation derives through pump depletion due to amplified signal power, the maximum power depleted due to amplifying the optical signal in this case is 10 mW out of 500 mW. Therefore, in a distributed pre-amplifier configuration, the saturation power can be said to be very high.

On the other hand, for discrete Raman amplifiers, small-signal net gain and output power are usually designed to achieve at least 20 dB and 20 dBm, respectively. In this case, Raman gain has to be larger than the passive loss due to fiber and other optical components because the gain saturation becomes larger when the power of the signal becomes comparable with that of the pump. Therefore, in order to increase the saturation power, one should increase the pump power while decreasing the length of the fiber. Using 9 or 10 km DSF, the input saturation power exceeding 2 dBm in an discrete amplifier with 20 dB small-signal gain was achieved using more than 1 W pump power [34].

The gain saturation impacts on a transient response in Raman amplifiers. Suppose one or a set of signal channels is suddenly added; the depth of gain saturation changes at the same time. However, because the interaction length of the Raman amplifier is kilometers in length, transient time from the initial to the saturated gain levels due to added signals can be 50 μsec long [35]. During this transient time, the gain is much larger than the steady state gain, and therefore burst errors of signal transmissions can occur.

10.3.6. NOISE

In this section, we discuss the noise sources in a Raman amplifier: quantum and thermal noise, double Rayleigh backscattering noise, pump-mediated noise, cross-gain saturation noise, and noise due to nonlinearity.

10.3.6.1. Quantum and Thermal Noise

Equation (10.5) includes the effect of spontaneous Raman scattering, which originates from zero-point fluctuation of the vacuum field. Because the Raman scattering process is very fast and goes through virtual intermediate states of material, it may be difficult for any classical noise to enter. Olsson and Hegarty experimentally observed that the spontaneous Raman scattering included in Eq. (10.5) was the predominant noise, with the spontaneous emission factor of almost unity [36]. This means that the Raman amplifier can operate at the minimum quantum limited noise level. The minimum quantum noise level is achieved when the inversion population is completely achieved. In the stimulated Raman scattering process as depicted in Fig. 10.1, there are no real upper energy levels that allow signal photons to be absorbed. In a sense, one may argue that when a pump photon is absorbed to excite an electron to its upper virtual intermediate states, the complete inversion is always effectively established. Mochizuki *et al.* showed that Smith's noise theory for a small-signal gain regime [7] agrees well with the experiment, and found that the counter-pumping scheme had larger spontaneous Raman scattering noise than the co-pumping scheme [37].

In addition to spontaneous Raman scattering, thermal noise can become important particularly for Stokes photons with smaller frequency shifts from the pump [38]. In optical amplifiers with real upper energy level, such as EDFA, the direct transition between lower and upper energy levels occurs, but a very small amount of black body radiation occurs because the optical frequencies are large enough. In the stimulated Raman scattering process in silica glass, however, optical phonons have only 13.5 THz of frequency, at which thermal noise can become non-negligible. In thermal equilibrium, the phonon occupancy obeys the Bose-Einstein distribution, which causes additive noise to Raman amplifiers [39, 40]. Taking into account both spontaneous Raman scattering noise and thermal noise, Eqs. (10.8) and (10.9) are thus rewritten as

$$\frac{dP_{\mathrm{s}}}{dz} = \frac{g_{\mathrm{R}}}{\lambda_{\mathrm{s}} A_{\mathrm{eff}}} P_{\mathrm{p}} P_{\mathrm{s}} - \alpha_{\mathrm{s}} P_{\mathrm{s}} + 2h\nu_{\mathrm{s}} \frac{g_{\mathrm{R}}}{\lambda_{\mathrm{s}} A_{\mathrm{eff}}} P_{\mathrm{p}} [1 + \Theta(\nu_{\mathrm{p}} - \nu_{\mathrm{s}}, T)] \Delta\nu_{\mathrm{s}} \tag{10.15}$$

$$\frac{dP_{\mathrm{p}}}{dz} = -\frac{g_{\mathrm{R}}}{\lambda_{\mathrm{p}} A_{\mathrm{eff}}} P_{\mathrm{p}} P_{\mathrm{s}} - \alpha_{\mathrm{p}} P_{\mathrm{p}} - 2h\nu_{\mathrm{p}} \frac{g_{\mathrm{R}}}{\lambda_{\mathrm{p}} A_{\mathrm{eff}}} P_{\mathrm{p}} [1 + \Theta(\nu_{\mathrm{p}} - \nu_{\mathrm{s}}, T)] \Delta\nu_{\mathrm{s}} \tag{10.16}$$

where h is Planck's constant, ν_s and ν_p are the frequencies of Stokes and pump waves, respectively, and $\Delta\nu_s$ is the bandwidth of interest. $\Theta(\nu_p - \nu_s, T)$ is the mean number of phonons in thermal equilibrium at temperature T, or the Bose-Einstein factor, as follows:

$$\Theta(\nu_p - \nu_s, T) = \frac{1}{\exp\left[\frac{h(\nu_p-\nu_s)}{kT}\right] - 1}. \tag{10.17}$$

The factor of 2 in the last term on the right-hand side of both Eqs. (10.15) and (10.16) results because during the propagation, polarization states of spontaneous emission and signal are mixed together to increase the amount of noise by the factor of 2. The last term on the right-hand side of Eq. (10.16) can be neglected for most cases where the pump depletion due to spontaneous emission is minor. This way of describing thermal noise is experimentally verified to produce excellent predictions [40]. Figure 10.10 plots simulation results based on a model extended from Eqs. (10.15) and (10.16) to include wavelength dependence to show the temperature dependence of gain and noise figure of a Raman amplifier that has 50 km DSF as its gain fiber, with pump power of 500 mW in a counter-pumping configuration. The total input power was -10 dBm over the wavelength range. It can be seen in Fig. 10.10 that noise increases for shorter wavelength due to the Bose-Einstein factor at room temperature.

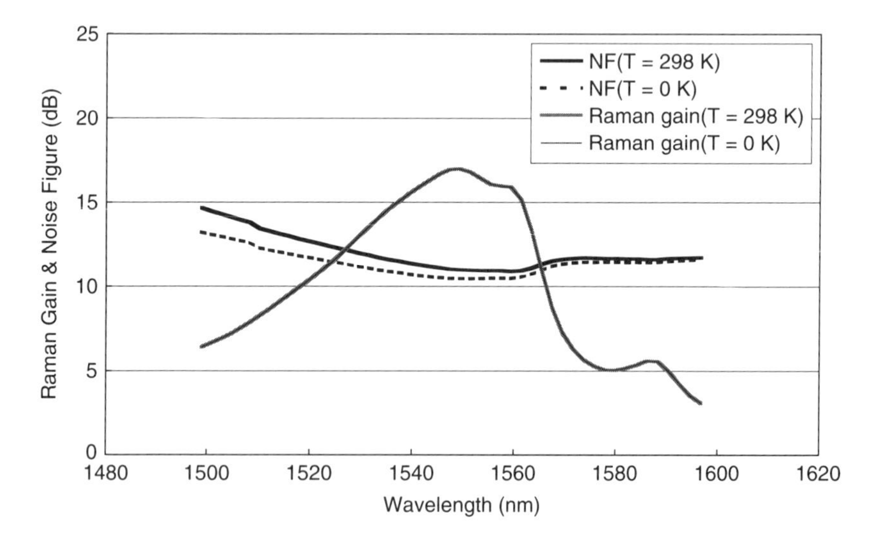

Fig. 10.10 Simulated temperature dependence of Raman gain and noise figure.

The noise figure spectrum at 0 K exhibits a wavelength dependence, which reflects on the loss spectrum of the fiber. (The detailed description of this extended model is discussed in Section 10.4.8.) This model assumes Raman gain efficiency is insensitive to temperature. In fact, the stimulated Raman scattering process is through temperature-insensitive phenomena. First, the nonlinear susceptibility of silica is quite uniform in the infrared wavelength region, which is well apart from the ultraviolet absorption, and hence insensitive to temperature. Second, the phonon energy scarcely depends on temperature, as the molecule vibrations of silica are well described by a harmonic oscillator model. (This has been verified experimentally [40, 41]; an experimental result will be presented in Section 10.4.10.4.)

10.3.6.2. Double Rayleigh Backscattering Noise

Rayleigh scattering is an elastic scattering of light with bound electrons, or glass molecules in optical fiber. In other words, it is the elastic version of Raman scattering. The typical value for the attenuation coefficient of optical fiber in 1550 nm band is −0.2 dB/km, which mainly derives from Rayleigh scattering. A small portion of Rayleigh-scattered light will be reflected to propagate backward through the optical fiber [42]. In optical fiber, the Rayleigh backscattering coefficient is typically −40 dB/km. If Rayleigh scattering occurs back and forth consecutively, it means that a small fraction of the lightwave is split and then mixed with the original one after a certain amount of relative time delay. This random interference may result in adverse effects to optical transmissions [43]. The amount of double Rayleigh scattering is very small for transmissions through passive optical fiber links. However, if there is a distributed gain over the optical fiber, then the small portion of Rayleigh scattered light is amplified and thus tends to persist. There are two mechanisms involved: one is that double Rayleigh-scattered signal causes the so-called multiple path interference (MPI), and the other is that spontaneous Raman-scattered light builds up due to the feedback through double Rayleigh scattering [44]. Therefore, in a distributed Raman amplifier, double Rayleigh backscattering has to be carefully estimated so as to avoid undesirable effects. Distributed Raman gain increases optical SNR of the system in the absense of Rayleigh scattering, while in the presence of Rayleigh scattering, the noise due to double Rayleigh backscattering increases with increasing Raman gain. To scrutinize how optical SNR changes with different Raman gains in the presence of double Rayleigh backscattering, one has to solve the following coupled rate equations:

$$\frac{dP_s^{\pm}}{dz} = \pm\frac{g_R}{\lambda_s A_{eff}} P_p P_s^{\pm} \mp \alpha_s P_s^{\pm} \pm \varepsilon_s P_s^{\mp} \pm 2h\nu_s \frac{g_R}{\lambda_s A_{eff}} \times P_p[1 + \Theta(\nu_p - \nu_s, T)]\Delta\nu_s \qquad (10.18)$$

where superscripts + and − denote forward and backward propagating lightwaves, and ε_s is the Rayleigh backscattering coefficient. We also need to know the evolution of the pump. For small signal regimes, one may use the so-called undepleted pump model where the pump is decreasing only through passive attenuation of fiber. Using the undepleted pump model, Hansen *et al.* calculated to find that there was an optimal Raman gain for the best achievable SNR, with experimental verification [44]. They found that the optimal launched pump power was about 980 and 550 mW for silica core fiber (SCF) and DSF, respectively, for which the detector sensitivity was improved by 7.0 and 6.2 dB, respectively. Because too much Raman gain causes a sharp decline of SNR, it may be a good idea to keep local Raman gain less than a certain level while the total Raman gain is large. Nissov *et al.* showed numerically that a bi-directional pumping scheme performs better than co- and counter-pumping schemes [45]. Lewis *et al.* have developed a modified time-domain extension method to characterize noise due to double Rayleigh backscattering [46]. Also, a direct temporal observation is proposed by Chinn [47].

10.3.6.3. Noise in Co-pumping Configuration

As we discussed in Section 10.2, a Raman amplifier can operate almost equally in both co- and counter-pumping schemes. Early studies in counter-pumped Raman amplification can be found in, for example, [15]. However, there are several issues involved with pumping orientation: The most important issues concern noise. Because the fluorescence lifetime of Raman gain is nearly instantaneous, the intensity noise of the pump will be imposed on the signal through Raman gain. For a co-pumping configuration, the difference in the group velocities between the pump and signal due to chromatic dispersion causes a "walk-off" and therefore an averaging effect in the noise transfer. On the other hand, for a counter-pumping configuration, the pump and signal pass by each other in opposite directions, resulting in the averaging-out of noise transfer. Therefore, the noise transfer functions for co- and counter-pumping configurations will have different bandwidths. A thorough analysis of the noise transfer has shown that for a non-dispersion shifted fiber (NDSF) Raman pump to create 10 dB gain, the counter- and co-pumping configurations may have the relative intensity noise (RIN) of −20 dB/Hz and −110 dB/Hz, respectively, in order to keep less than 0.1 dB penalty in Q-value [48].

If there is considerable gain saturation in the counter-pumping configuration, and if more than one signal is multiplexed in one fiber, the saturated gain is so instantaneous that the change in gain will have a temporal pattern reflecting the temporal structure of one signal, resulting in cross-talk to another signal. For the co-pumping configuration, the bandwidth of the cross-talk depends on the chromatic dispersion of the fiber, because of walk-off between signals and pump. This seldom happens for the co-pumping configuration [49, 50]. In the counter-pumping configuration, because the pump and signals are propagating in opposite directions, the rate of change in gain due to saturation is up-converted, so only low-frequency components of a signal will affect another signal.

10.3.6.4. Nonlinear Effects

The excursion of signal power in a distributed Raman amplifier system is different from that in a passive fiber link with lumped EDFAs. Nonlinear effects such as self-phase modulation (SPM), four wave mixing (FWM), cross-phase modulation (XPM), SRS of signals, all derive through the nonlinear phase shift. The cumulative nonlinear phase shift that the signal experiences in an optical fiber with the length of L is calculated by

$$\Delta\Phi_{\mathrm{NL}}(L) = \gamma \int_0^L P_{\mathrm{s}}(z)dz \tag{10.19}$$

where γ is the nonlinear coefficient of the fiber, and L is the length of the fiber. From Eq. (10.19), the nonlinear phase shift is proportional to the area underneath the signal power over distance. In the undepleted pump model, the excursion of signal power is derived from Eq. (10.12). Because Raman gain is distributed over kilometers, the nonlinear phase shift tends to be larger than for systems without Raman gain. Figure 10.11 shows the signal power excursions for co-, counter-, and bi-directional pumped Raman amplifiers in DSF. Pump powers for co- and counter-pumping configurations are both 200 mW, and pump power for the bi-directional pumping configuration is 100 mW for each pump. For the co-pumping configuration, the nonlinear phase shift increases remarkably by introducing Raman gain, while for the counter-pumping configuration, the nonlinear phase shift is not considerably increased because the signal is amplified when the power

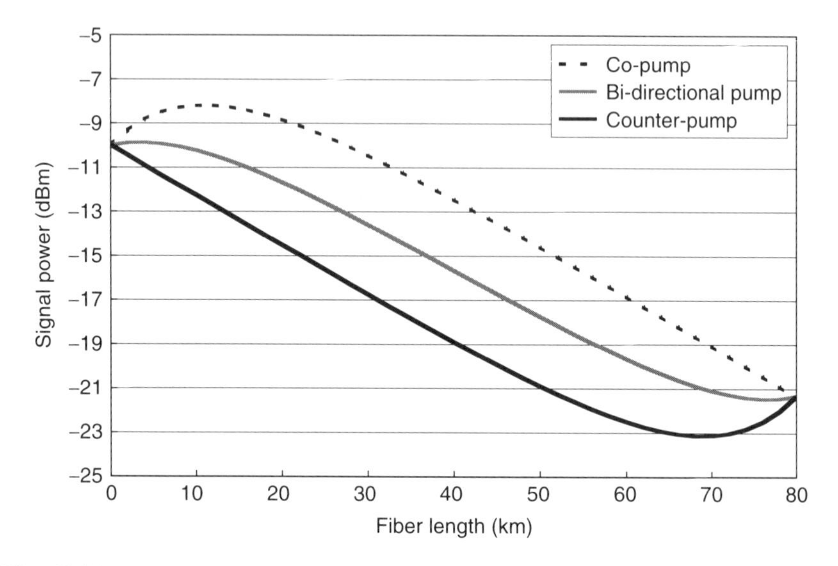

Fig. 10.11 Signal power excursion in co-, counter-, and bi-directionally pumped Raman amplifiers, in DSF.

level is low enough. On the other hand, a co-pumped Raman amplifier better improves the OSNR by reducing spontaneous emission noise than does a counter-pumped Raman amplifier. Therefore, by appropriately designing the Raman amplifier, one can improve the system OSNR by means of Raman gain that allows reducing the input signal level and hence nonlinear effects [51–55].

10.4. Raman Amplifiers for WDM Transmissions

10.4.1. APPLICATION OF FIBER RAMAN AMPLIFIERS TO OPTICAL COMMUNICATIONS

Stolen and Ippen were motivated to scrutinize fiber Raman amplifiers for communication purposes [6]. In 1983, Hasegawa proposed using fiber Raman amplifiers for optical soliton communications [20]. In 1985, Aoki *et al.* [56] and Hegarty *et al.* [57] reported the first demonstration of NRZ transmissions. Almost at the same time, Mollenauer *et al.* achieved the first experimental demonstration of optical soliton propagation in long fiber with the loss compensated by Raman gain [58, 59]. The first WDM transmission through a LD pumped Raman amplifier was demonstrated by Edagawa *et al.* [60].

From the late 1980s through the 1990s, because of a relatively low necessity for pumping power, diode-pumped EDFAs were successfully developed and widely deployed in commercial systems, while Raman amplifiers could not find practical pumping sources. It was in repeaterless optical transmission systems that Raman amplifiers found their potentials of performing better than EDFAs. Nakagawa *et al.* first demonstrated that the Raman amplifier is effective in extending the transmission distance in repeaterless transmissions [61]. Hansen *et al.* have investigated the detector sensitivity improvement and capacity upgrades by Raman gain [62]. They found that for silica core fiber with 28.1 dB gain by 1 W pump at 1.45 μm, the detector sensitivity improved by more than 7 dB for 10 Gbit/s signals, allowing a system upgrade from a single channel at 2.5 Gbit/s to 4 WDM channels at 10 Gbit/s.

Nissov *et al.* demonstrated 100 Gbit/s (10×10 Gbit/s) WDM transmission over 7200 km using distributed Raman amplification [63]. This demonstration was indeed decisive in proving that a distributed Raman amplifier system is superior to lumped EDFA systems in terms of amplified spontaneous emission (ASE) noise and hence performance. Since this demonstration, there have been reported a number of ultra-high capacity WDM transmission systems using fiber Raman amplifiers. Most of the high-capacity systems reported, especially for terrestrial systems, have used a hybrid combination of EDFA and Raman amplifiers [64–66]. This could be attributed to an optimization with respect to double Rayleigh backscattering noise and achievable Raman gain with a limited pump power.

The ultimate pursuit for high capacity has resulted in the achievement of more than 10 Tbit/s transmissions by making use of Raman amplifiers [67, 68]. 80 Gbit/s at 0.8 bit/s/Hz spectral efficiency over 120 km NDSF was achieved with Raman amplification [69]. The use of Raman amplification has also enabled achievement of multi-Tbit/s WDM transmissions over transoceanic distances [70, 71]. It is also noteworthy that 1 Tbit/s over more than 8000 km using all-Raman distributed amplifier repeaters was achieved [72].

10.4.2. ADVANTAGES OF DISTRIBUTED AMPLIFICATION

The fundamental quantum noise, or amplified spontaneous emission (ASE) noise, is a dominant noise in a cascaded optical amplifier chain. In quantum mechanics, loss invites the entrance of additional quantum noise that originates from the zero-point fluctuation [73]. In a cascaded optical amplifier chain, as the signal experiences loss in the optical fiber, it loses SNR at the detector because the shot noise power scales linearly with the signal intensity while the signal power scales quadratically with the signal intensity [74, 75]. In other words, in order to maintain SNR, one should keep intensity of signal at a certain level by somehow combating the loss of fiber. In this regard, a distributed optical amplifier can perform better than a lumped one in terms of noise, because distributed amplifiers can retain the signal level throughout transmission distance, while lumped ones must allow the signal level to become much lower at the output end of a fiber span. Of course, one could increase the signal level in order to achieve a desirable SNR. However, too high signal level causes system performance to deteriorate due to adverse nonlinear effects.

A simple picture of the systems design of optical transmissions is to optimize the signal level diagram, that is, the optical signal level versus distance. When a lumped EDFA chain is used in a system, the signal level decreases with increasing distance in the optical fiber span due to attenuation and is regenerated at a point where one of the EDFAs is located. On the other hand, in a distributed amplifier-based system, the signal level does not monotonically decrease but is maintained within a smaller excursion. Figure 10.12 shows this difference. In general, system designers suffer from a dilemma between nonlinear effects and excess noise. Too high a signal level may result in a penalty due to nonlinear effects, while too low a signal level may result in insufficient SNR. Likewise, the signal amplified by a distributed amplifier better fits in the transmission window, avoiding both

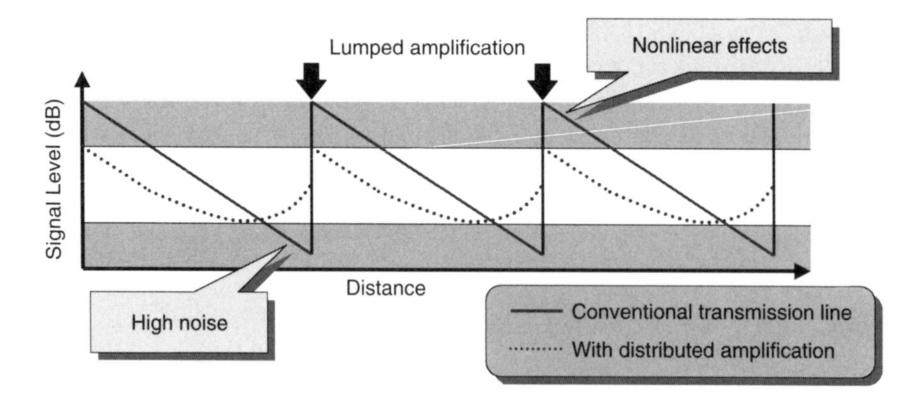

Fig. 10.12 Comparison of signal level diagram between lumped and distributed amplifiers. S. Namiki and Y. Emori, "Ultrabroad-band Raman amplifiers pumped and gain equalized by wavelength-division-multiplexed highpower laser diodes," IEEE J. Selected Top Quantum Electron, vol. 7, No. 1, pp. 3–16, 2001. © 2001 IEEE.

high noise and nonlinearity, and hence performs better than one amplified by a lumped amplifier. To avoid double Rayleigh backscattering noise, the Raman gain sometimes cannot be large enough to compensate for the span loss. In this case, EDFA has to jointly be used, as assumed in Fig. 10.12.

Rottwitt *et al.* proposed so-called second-order pumping to better achieve a constant signal level over distance [76]. In a conventional Raman amplifier, the pump attenuates due to passive loss of optical fiber, and so does the gain, which limits the distance range of distributed gain. If the pump is also amplified through Raman amplification, then the pump can persist over a longer distance than in the former case. In this way, one can design a better signal excursion for the system. In this technique, the orientations of the first- and second-order pumps are important parameters to determine the signal level excursions.

As was discussed in Section 10.3, fiber Raman amplifiers are an excellent distributed amplifier. This is why the fiber Raman amplifier is receiving renewed attention for WDM transmissions.

10.4.3. COMPARISON WITH EDFA

It is of interest to compare Raman amplifier with EDFA. Even though the general schemes resemble each other except for the gain fiber, their fundamental characteristics are in good contrast. One characteristic difference is the stimulated emission processes through either virtual or real energy states. The fluorescence lifetime of Raman is in femtoseconds, while that of EDFA is in milliseconds. The state of polarization of the lightwave is maintained during stimulated Raman scattering, while it is almost completely depolarized in EDFA. Raman amplification is inherently of low noise regardless of pump level because the absorption of the signal photon to the upper virtual state is extremely small or just the Rayleigh scattering loss; while in EDFA a certain pump level has to be maintained in order to achieve a sufficient population inversion for low noise operation. This is part of the reason distributed EDFAs are not very attractive in spite of extensive studies in the early 1990s [77–81]. The temperature dependence of the Raman gain spectrum is very small because of the off-resonant process and the harmonic nature of optical phonons [40, 41], while that of EDFA is noticeable due to the temperature sensitivity of the local electric field created by glass molecules surrounding erbium ions [82, 83]. The temperature dependence was discussed in Section 10.3.6.1.

Another characteristic difference between Raman amplifier and EDFA is either weak or strong interactions. Because the Raman gain coefficient is considerably smaller than the gain coefficient of EDFA, the interaction lengths of the amplifiers are very different. A Raman amplifier needs tens of kilometers of fiber, while EDFA uses only tens of meters of erbium-doped fiber (EDF), for comparable and sufficient amount of gain. Therefore, the Raman amplifier is, in nature, a distributed amplifier rather than a lumped one. A drawback is then low efficiency. However, efficiency may not be an important issue in distributed amplifiers. If the Raman amplifier had high efficiency, the pump would be depleted so rapidly in a short distance that it might not be a "distributed" amplifier any more. As we have discussed in the previous section, a distributed amplifier is superior in terms of noise performance. Also, because of the low efficiency, most of the pump power is lost due to passive attenuation of the optical fiber rather than energy transfer to the signal. This fact makes the Raman amplifier almost a linear amplifier with high saturation power. For a Raman amplifier with a very fast fluorescence lifetime, the linearity is essential to WDM application. In contrast, EDFA is efficient enough to have a low saturation power, which means EDFA is a nonlinear amplifier. However, because EDFA has a very long fluorescence lifetime, it is also viable for WDM transmissions. Thus, both Raman amplifier and EDFA are good for WDM transmissions.

10.4.4. DISCRETE AND HYBRID RAMAN AMPLIFIERS

Besides the advantages due to distributed amplification, another merit of the Raman amplifier is that any gain band can be realized by proper choice of pump wavelength. One of the main purposes of discrete Raman amplifiers is to realize an amplifier operating in different windows than EDFA. There have been many efforts to develop discrete Raman amplifiers operating in 1.3 [84–87], 1.52 [88], and 1.65 μm [89, 90] bands. Because the interaction length of the Raman amplifier is typically orders of magnitude longer than that of EDFA, nonlinearity, saturation, and double Rayleigh backscattering may become serious issues. However, by optimizing the length of the gain fiber [91] and using a two-stage structure [92], one may be able to design discrete Raman amplifiers that are good for signal transmissions. Many successful transmission experiments using discrete Raman amplifiers have been reported [93–95].

For systems faster than 10 Gbit/s, dispersion compensation is of critical importance. However, dispersion-compensating fibers (DCF) generally have large losses. Raman amplification in DCF is an effective means to compensate for the loss in DCF [96–98]. It was demonstrated that the use of Raman gain in DCF improved noise figure of an inline EDFA with DCF as a mid-stage device by 1.5 dB [99]. As can be seen from Fig. 10.4, DCF is an excellent fiber from the aspect of Raman amplification efficiency [28, 33].

Because of increasing demands on bandwidth, many efforts have been made to extend the signal band beyond the EDFA gain range. Masuda *et al.* proposed a supplementary use of Raman amplifiers to extend the EDFA band [100–102] and also to flatten the composite gain of EDFA and Raman amplifiers [103]. Kani and Jinno proposed a similar approach to Masuda's but used thulium-doped fiber amplifier (TDFA) instead of EDFA for the wider operation of the S-band [104]. Bayart *et al.* tried to achieve the plausibly widest signal bandwidth of 17.7 THz by a combination of EDFA, TDFA, and Raman amplifiers [105]. Fukuchi *et al.* achieved more than 10-Tbit/s WDM transmission by using EDFA, TDFA, and Raman amplifiers [68].

There are also attempts to increase the signal bandwidth by using a Raman amplifier alone. Rottwitt *et al.* proposed the use of two pump wavelengths, 1453 and 1495 nm, to achieve a 92-nm lossless bandwidth window for 45-km-long DSF [106], while Lewis *et al.* have achieved 114-nm bandwidth of Raman amplification by simultaneously launching 1423, 1455, and 1480-nm pump lasers [107]. Emori *et al.*, invented an efficient technique called "WDM pumping," which enables us to realize a sufficient, wideband, and flat gain in Raman amplifiers [97, 99, 108].

10.4.5. WDM PUMPING

As we shall discuss in subsequent sections, 1480-nm pump laser diodes for EDFA have matured [109] since completion of their development [110]. The idea of "WDM pumping" is to prepare a set of pump lasers operating at different wavelengths combined through WDM couplers into a single fiber to realize a composite Raman gain. The composite Raman gain created by different wavelengths of pumps can be so designed as to be arbitrarily large, wide, and flat without using any gain-flattening filters.

For stable and efficient WDM pumping, pump lasers ought to have narrow and stable lasing spectra. Koyanagi *et al.* have achieved a very high power from fiber Bragg grating (FBG) stabilized pump laser diodes [111]. Tanaka *et al.* developed integrated imbalance Mach-Zehnder interferometers (MZI) on a planar lightwave circuit (PLC) based on the flame hydrolysis deposition (FHD) method [112]. Figure 10.13 shows the schematic

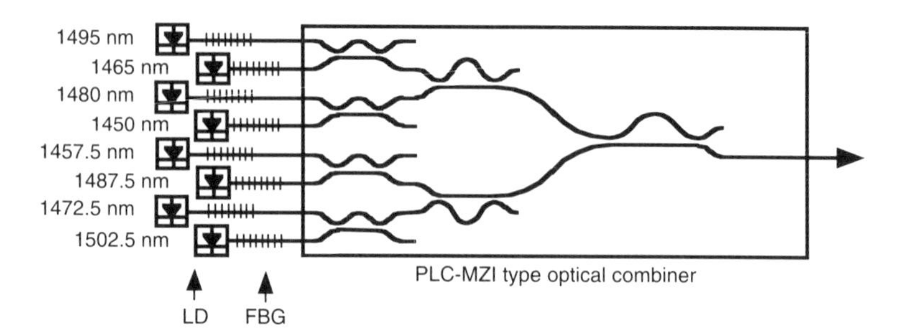

Fig. 10.13 Schematic diagram of WDM pumping [112]. S. Namiki and Y. Emori, "Ultrabroad-band Raman amplifiers pumped and gain equalized by wavelength-division-multiplexed highpower laser diodes," IEEE J. Selected Top Quantum Electron, vol. 7, No. 1, pp. 3–16, 2001. © 2001 IEEE.

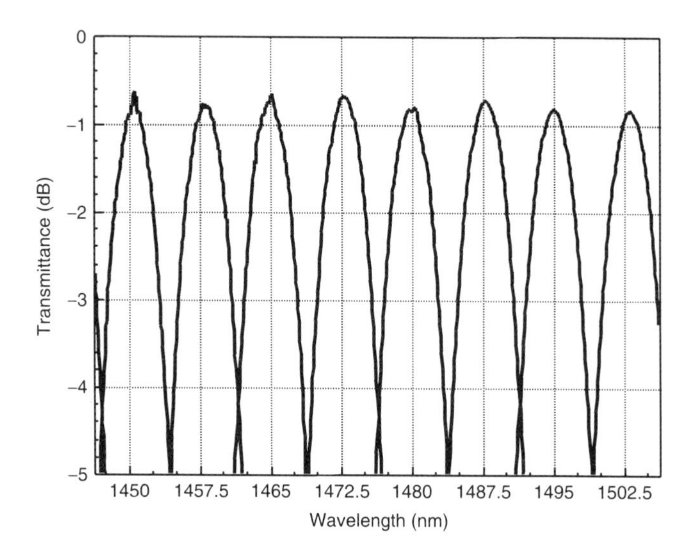

Fig. 10.14 Transmission spectra of a PLC-MZI-WDM coupler [112]. S. Namiki and Y. Emori, "Ultrabroad-band Raman amplifiers pumped and gain equalized by wavelength-division-multiplexed highpower laser diodes," IEEE J. Selected Top Quantum Electron, vol. 7, No. 1, pp. 3–16, 2001. © 2001 IEEE.

diagram of WDM pumping. Figures 10.14 and 10.15 show the transmission spectra of a PLC-MZI-WDM coupler, and pump laser output spectra without and with FBG stabilization, respectively. It is found from Figs. 10.14 and 10.15 that the FBG-stabilized spectrum fits in the transmission window of the coupler. Figures 10.16 and 10.17 show the combined output optical spectrum and light-current curve of the WDM pump lasers, respectively. The average insertion loss in this case was only 1.2 dB, and the maximum combined output power exceeded 1 Watt.

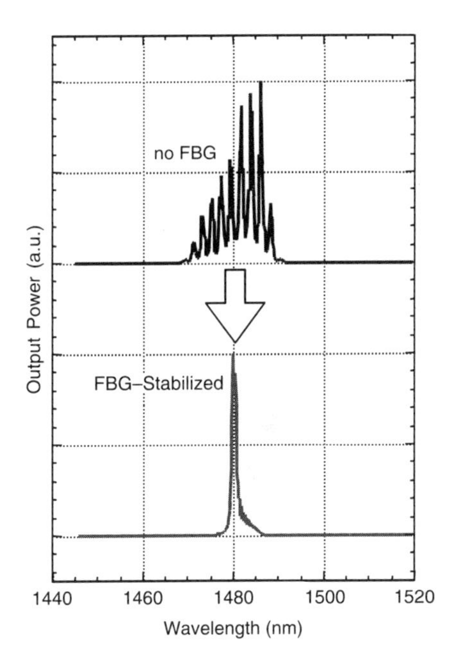

Fig. 10.15 Output spectra of a FBG-stabilized pump laser diode and the pump laser diode without FBG. S. Namiki and Y. Emori, "Ultrabroad-band Raman amplifiers pumped and gain equalized by wavelength-division-multiplexed highpower laser diodes," IEEE J. Selected Top Quantum Electron, vol. 7, No. 1, pp. 3–16, 2001. © 2001 IEEE.

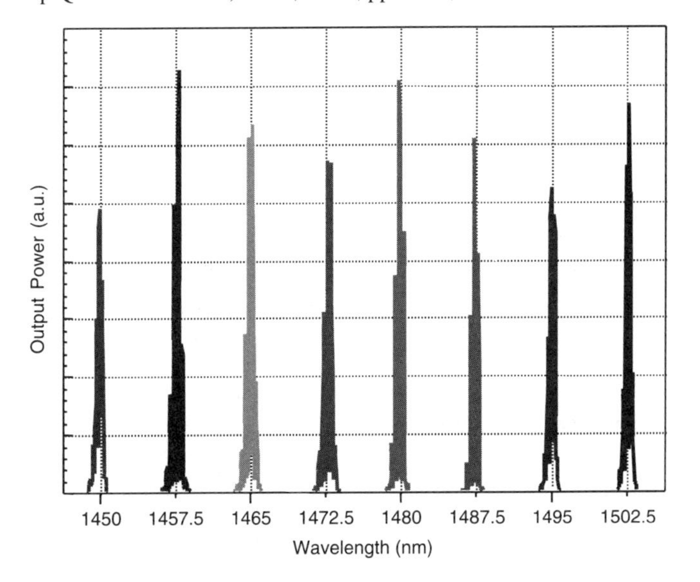

Fig. 10.16 Output spectrum of WDM pumping. S. Namiki and Y. Emori, "Ultrabroad-band Raman amplifiers pumped and gain equalized by wavelength-division-multiplexed highpower laser diodes," IEEE J. Selected Top Quantum Electron, vol. 7, No. 1, pp. 3–16, 2001. © 2001 IEEE.

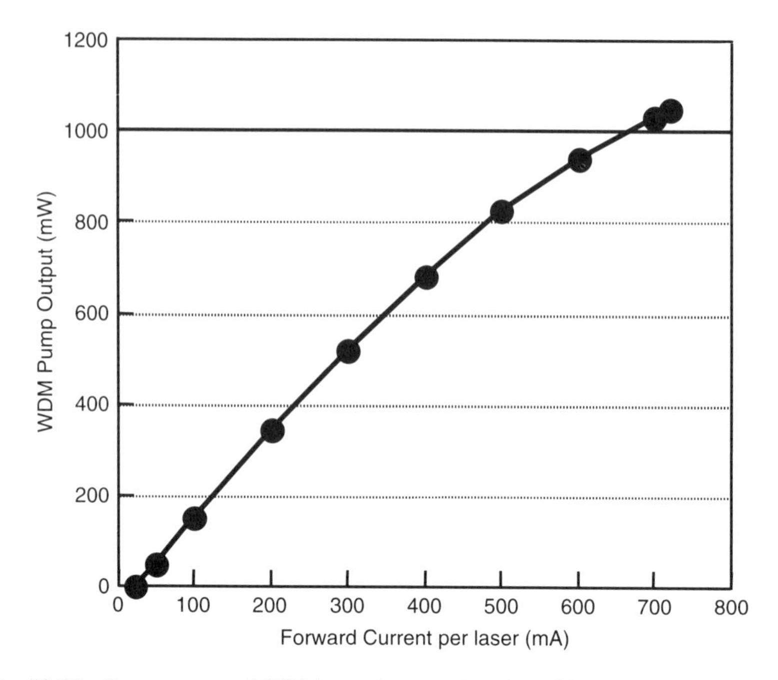

Fig. 10.17 Output power of WDM pumping as a function of forward current per laser.

Other different WDM coupler types, such as dielectric thin-film filter type, fused-fiber coupler type, and so on, can also be used for the same purpose. There may be an optimal choice of these alternatives, depending upon the allocation of wavelengths and powers of pump lasers to be multiplexed. Using this technique, one can realize both sufficiently high power output and broad bandwidth at the same time. The larger the number of WDM pumping channels, the wider and more powerful Raman amplifiers with various gain shapes can be designed. This technique is also good for high-power EDFA [113].

10.4.6. SUPERPOSITION RULE

Figure 10.18 shows the principle of the WDM-pumped Raman amplification process. Because the Raman shift is due to the excitation of optical phonon in the glass, the wavelength at which Raman gain takes the maximum value is solely determined by the pump wavelength. If two pump waves at different wavelengths are launched into a common fiber at the same time, then the two processes shown in Fig. 10.18 occur, creating a wide composite gain.

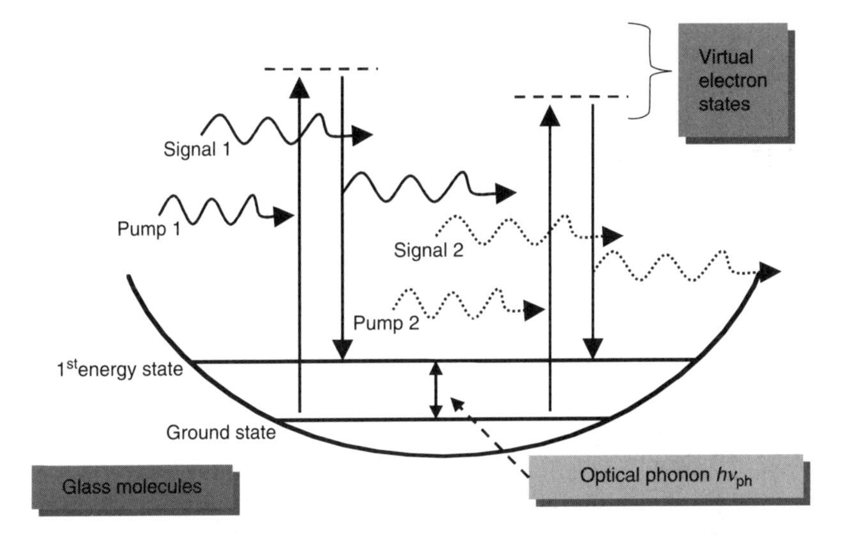

Fig. 10.18 Principle of WDM-pumped Raman amplifier. S. Namiki and Y. Emori, "Ultrabroad-band Raman amplifiers pumped and gain equalized by wavelength-division-multiplexed highpower laser diodes," IEEE J. Selected Top Quantum Electron, vol. 7, No. 1, pp. 3–16, 2001. © 2001 IEEE.

The gain profile of a WDM-pumped Raman amplifier is thus expressed as a logarithmic superposition of the gain profiles pumped by their respective pumping wavelengths. The shape of Raman gain pumped by a single wavelength is independent of the pumping wavelength as long as the peak value is normalized to be almost unity and the horizontal axis is expressed by the frequency shift from the pump. A possible gain profile of any pumping wavelength can be predicted from a gain profile pumped at a wavelength, though the magnitude of Raman gain is not determined only by the pumping wavelength. However, the predicted gain profile is sure to be realized by an appropriate pump power level unless it is beyond the practical limit. Therefore, the pumping wavelength allocation of a WDM-pumped Raman amplifier can be determined by adjusting the magnitude and position of the respective gain profile so as to make the superposed profile a desirable one. Figure 10.19 illustrates the determination process of pumping wavelengths. The parameter "effective gain" corresponds to the magnitude of each gain profile in the right graph or a weighting factor.

When the pumps are combined via dielectric filters, there is no restriction to choose the pumping wavelengths. In case of Mach-Zehnder interferometer (MZI) type wavelength multiplexer, the pumping wavelengths have to be on a grid with an equal frequency interval. The pumping wavelengths are determined so that the gain flatness is as good as possible. The number

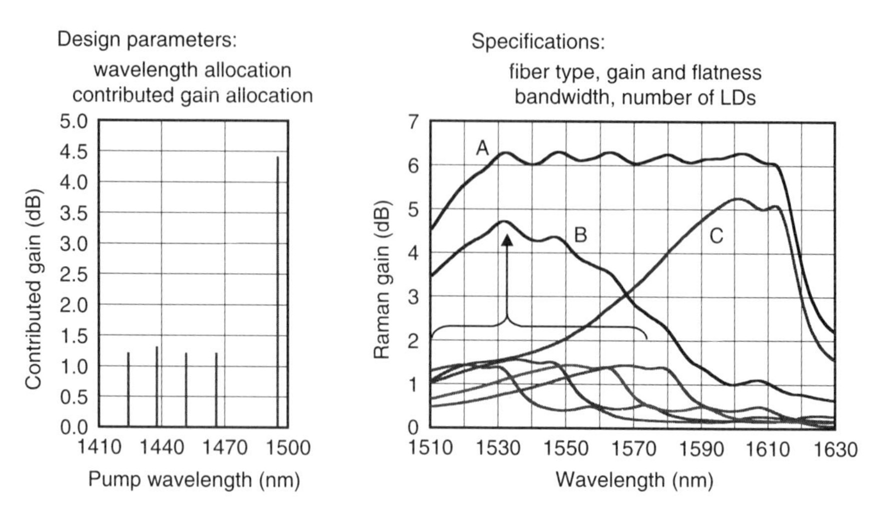

Fig. 10.19 Design of composite Raman gain by WDM pumping.

of pumping wavelengths reflects on the granularity of composite gain. Therefore, the larger number of pumping wavelengths creates a flatter and larger composite gain [108].

A guideline to achieve a flat composite gain can be discussed as follows. Closely looking into the Raman gain spectra in Fig. 10.4, we find the profiles asymmetric around the gain peak. The spectra have a smoother and less steep curve toward smaller frequency shift from the peak. In order to create a flat-top gain spectral shape, a set of many small gain spectra that are laterally shifted with respect to each other along the wavelength axis, is necessary to compose an opposite smooth gain slope (labeled as curve B in Fig. 10.19) to the smoother gain slope on the short wavelength side from the peak (labeled curve C in Fig. 10.19) [114]. This guideline is actually applied to the examples discussed in Sections 4.10.1 through 4.10.3.

10.4.7. ACCOUNTING FOR PUMP-TO-PUMP RAMAN INTERACTION

The required pump power for a certain Raman gain is affected by several factors, such as Raman gain coefficient, polarization effect, fiber length, fiber loss at pump wavelength, pump depletion, and so on. In the case of WDM pumping, the wavelength dependence of these factors is one of the principal parameters for determining the pump power allocation. For example, Raman gain coefficient at a shorter wavelength is larger than at a

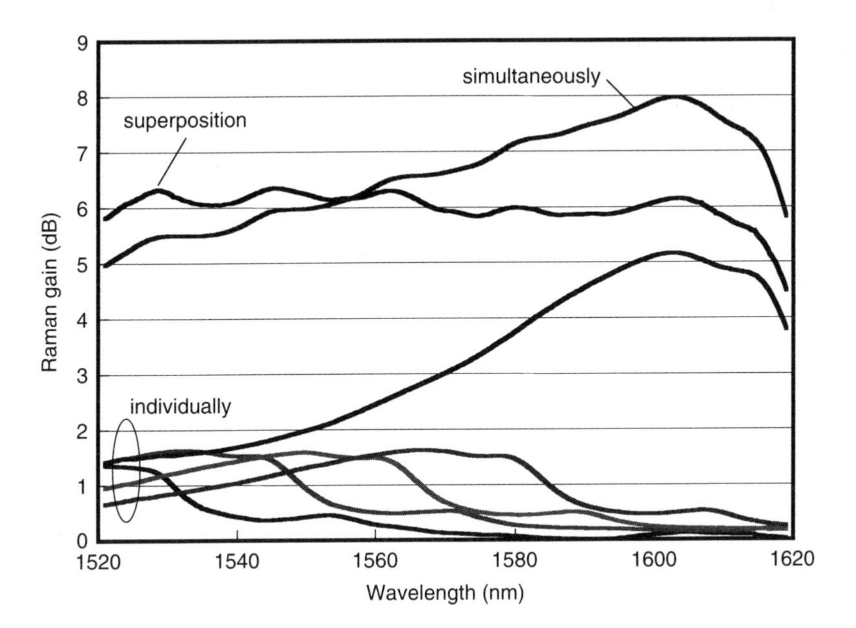

Fig. 10.20 The gain tilt caused by pump-to-pump Raman interaction. The fiber used is 25 km DSF. Pump wavelengths are 1420, 1435, 1450, 1465, and 1495 nm with fiber-launched powers of 61, 55, 48, 47, and 142 mW, respectively. S. Namiki and Y. Emori, "Ultrabroad-band Raman amplifiers pumped and gain equalized by wavelength-division-multiplexed highpower laser diodes," IEEE J. Selected Top Quantum Electron, vol. 7, No. 1, pp. 3–16, 2001. © 2001 IEEE.

longer one, while fiber loss is generally larger at a shorter wavelength due to Rayleigh scattering. Another important effect is pump-to-pump Raman interaction. It increases the required pumping power at shorter wavelengths, because a longer-wavelength pump absorbs the energy of a shorter one. Figure 10.20 shows the gain tilt caused by pump-to-pump Raman interaction. There are the measured gain profiles individually pumped by each channel and simultaneously pumped by all channels. Numerical superposition of individually pumped gain profiles is also drawn in the figure. When the pumps with different wavelength are simultaneously launched into a fiber, the gain tilt results.

10.4.8. MODELING

Let us now consider how we describe WDM-pumped Raman amplifiers by a theoretical model. Basically, we need to extend Eq. (10.18) for the multi-wavelength case in which all the pumps and signals are interacting with each other through Raman gain. Several models have been proposed [108, 115–118]. Although these are similar at a glance, detailed approximations are different. In this section, let us review the model developed in [108, 117], where the total photon number is conserved and the wavelength dependence of the Raman gain coefficient and effective area is taken into account.

The evolution of WDM pumps, WDM signals, and spontaneous Raman scattering noise are expressed in terms of the following set of equations [108, 117]:

$$\begin{aligned}\frac{dP_\nu^\pm}{dz} &= -\alpha_\nu P_\nu^\pm + \varepsilon_\nu P_\nu^\mp + P_\nu^\pm \sum_{\mu>\nu} \frac{g_{\mu\nu}}{A_\mu}(P_\mu^+ + P_\mu^-) \\ &\quad + 2h\nu \sum_{\mu>\nu} \frac{g_{\mu\nu}}{A_\mu}(P_\mu^+ + P_\mu^-)[1 + \Theta(\mu - \nu, T)]\Delta\mu \\ &\quad - P_\nu^\pm \sum_{\mu<\nu} \frac{\nu}{\mu}\frac{g_{\nu\mu}}{A_\nu}(P_\mu^+ + P_\mu^-) \\ &\quad - 4h\nu P_\nu^\pm \sum_{\mu<\nu} \frac{g_{\nu\mu}}{A_\nu}[1 + \Theta(\nu - \mu, T)]\Delta\mu \qquad (10.20)\end{aligned}$$

where subscripts μ and ν denote optical frequencies, superscripts + and − denote forward- and backward-propagating waves, respectively, P_ν is optical power within infinitesimal bandwidth $\Delta\nu$ around ν, α_ν is attenuation coefficient, ε_ν is Rayleigh backscattering coefficient, A_ν is effective area of optical fiber at frequency μ, $g_{\mu\nu}$ is Raman gain parameter at frequency μ due to pump at frequency ν, h is Planck's constant, k is Boltzmann constant, T is temperature, and $\Delta\mu$ is the bandwidth of each frequency component around frequency μ. We note that g_R/λ_s in Eq. (10.18) is related to $g_{\mu\nu}$ in Eq. (10.20). In this equation, we include pump-to-pump, pump-to-signal, and signal-to-signal Raman interactions, pump depletions due to Raman energy transfer, Rayleigh backscattering, fiber loss, spontaneous Raman emission noise, and thermal noise. We have put a factor of 2 in the noise terms, the fourth and sixth terms, in Eq. (10.20). This is because spontaneous emission and thermal noise are not correlated with signal; although the signal is polarized, the states of polarization (SOP) of signal and noise are scrambled and mixed together in the process of propagating through optical fiber.

Table 10.2 **Peak Values of Raman Gain Efficiency, $g_{\mu\nu}/A_\nu$ (1/W/km)**

Pump Wavelength (μm)	***1.420***	***1.510***
Standard single-mode fiber	0.43	0.35
Dispersion-shifted fiber	0.76	0.62

For design of composite Raman gain, loss and gain terms are important. And because the noise associated with stimulated Raman scattering is low and saturation power is high, the noise terms are relatively small and unimportant. Rayleigh-backscattered light does not grow significantly in the case of moderate pump power level, which is always the case for systems of interest. We found that the frequency dependence of effective area A_ν plays an important role for accurate design of composite Raman gain, especially when the pump band is wide. It is difficult to describe its dependence for various kinds of fibers in general terms. Also, as discussed in Eq. (10.10), rigorously speaking, the effective area depends on both ν and μ. For the present purpose, however, we do not have to directly measure the effective area A_ν. Instead, the property $g_{\mu\nu}/A_\nu$ has to be measured with respect to wavelength for each fiber under consideration. Table 10.2 shows some exemplary measured peak values of $g_{\mu\nu}/A_\nu$ at different wavelengths for SMF and DSF, from which we find the significance of accounting for its wavelength dependence in the case of wide pump band. We also note that the normalized spectral profile of $g_{\mu\nu}$ scarcely depends on pump wavelength. Accordingly, for thorough designs, one needs to numerically solve Eq. (10.20) while knowing the peak values for $g_{\mu\nu}/A_\nu$ at pertinent pump wavelengths. Figure 10.21 shows the simulated pump power evolution along distance when 5 different wavelengths with 100 mW of each power level are simultaneously launched into the fiber. From Fig. 10.21, it is seen that the longest wavelength acquires gain while the shortest one suffers from the largest attenuation due to energy transfer to longer wavelengths via SRS.

The composite Raman gain in this case is expressed as the logarithmic sum of each Raman gain created by each pump wavelength, with a weighting factor. The weighting factor is proportional to the area underneath the curve with respect to distance in Fig. 10.21, which is calculated by integrating the pump power level over distance. In this regard, the length of fiber is not important for the gain profile itself, or rather it only determines the amount of weighting factors by setting the upper limit of the integral

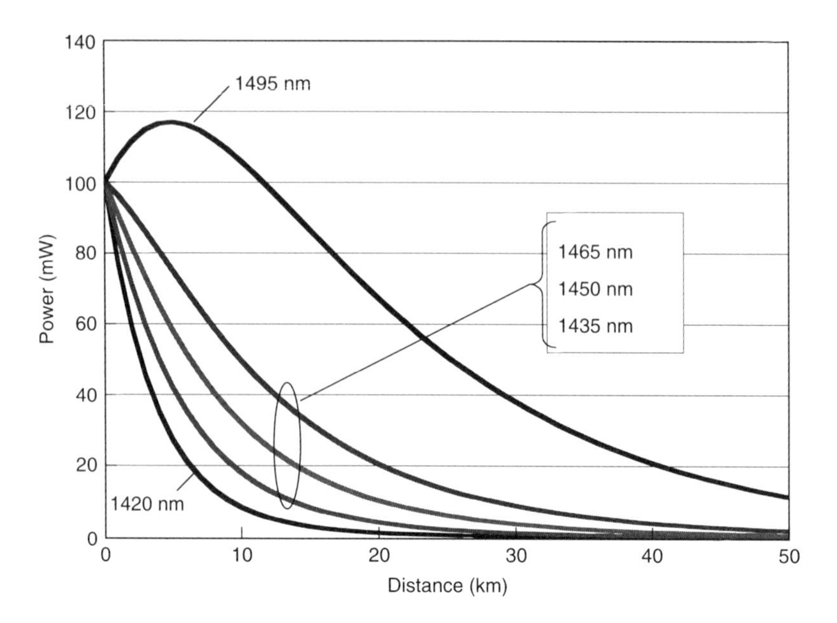

Fig. 10.21 Pump power excursion over propagation distance in case of simultaneous input of WDM pumping into DSF. S. Namiki and Y. Emori, "Ultrabroad-band Raman amplifiers pumped and gain equalized by wavelength-division-multiplexed highpower laser diodes," IEEE J. Selected Top Quantum Electron, vol. 7, No. 1, pp. 3–16, 2001. © 2001 IEEE.

to calculate the area. Likewise, the design procedure can be simplified: (1) One determines pump wavelengths with which a desired composite Raman gain can be obtained by adding in logarithmic scale individual Raman gain spectra shifted by the respective wavelength differences with adequate weighting factors. (2) One predicts how much power should be launched in order to realize the weighting factors through numerical simulations based on Eq. (10.20). Figure 10.22 shows the comparison between the composite Raman gains calculated with weighting factors and experimentally realized by adjusting pump powers, where 11 pump channels are used. In this comparison, the simulated profile was simply created by adding the individual gain profiles weighted by appropriate factors. The agreement is superb. We have also verified Eq. (10.20) to predict both composite Raman gain and NF at given pump input. Figures 10.23 and 10.24 show the comparison between the experiment and simulation results for the same case. Figure 10.23 shows an excellent agreement except for a slight discrepancy in the middle of the spectra. We have confirmed that this was due to polarization-dependent gain; one pump channel had an imperfect depolarization of the output. Actually, the spectral profile is perfectly

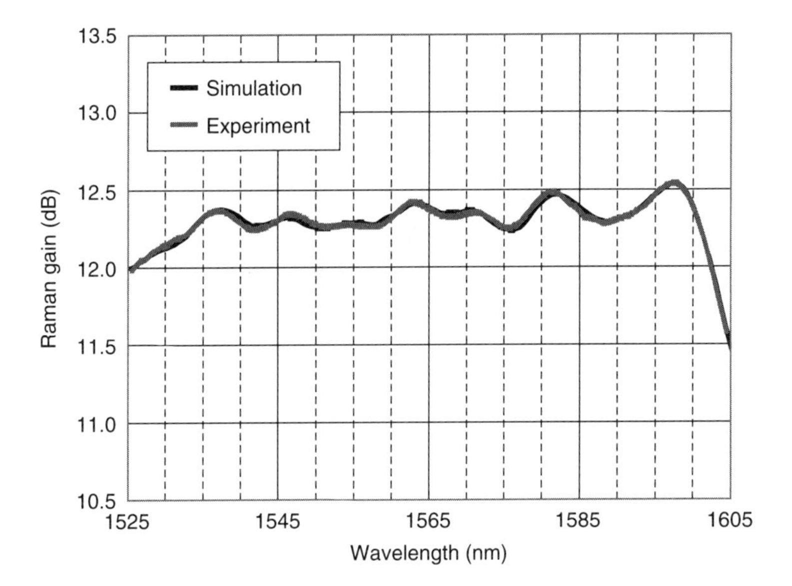

Fig. 10.22 Comparison between the composite Raman gains calculated with weighting factors and experimentally realized by adjusting pump powers. The fiber used was 25-km DSF. S. Namiki and Y. Emori, "Ultrabroad-band Raman amplifiers pumped and gain equalized by wavelength-division-multiplexed highpower laser diodes," IEEE J. Selected Top Quantum Electron, vol. 7, No. 1, pp. 3–16, 2001. © 2001 IEEE.

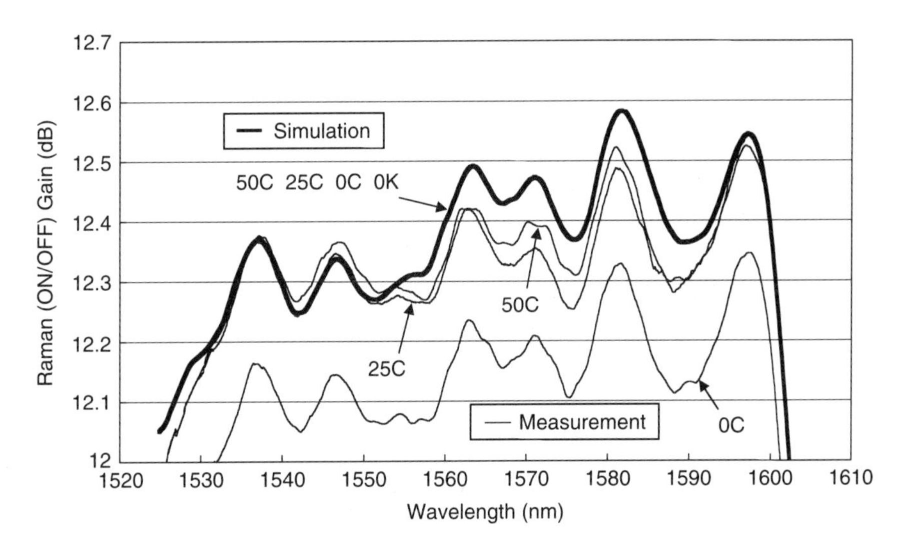

Fig. 10.23 Comparison of measured composite gain with numerical simulation. The measurements were carried out at 0, 25, and 50°C, and the simulations were performed for 0 K, 0, 25, and 50°C.

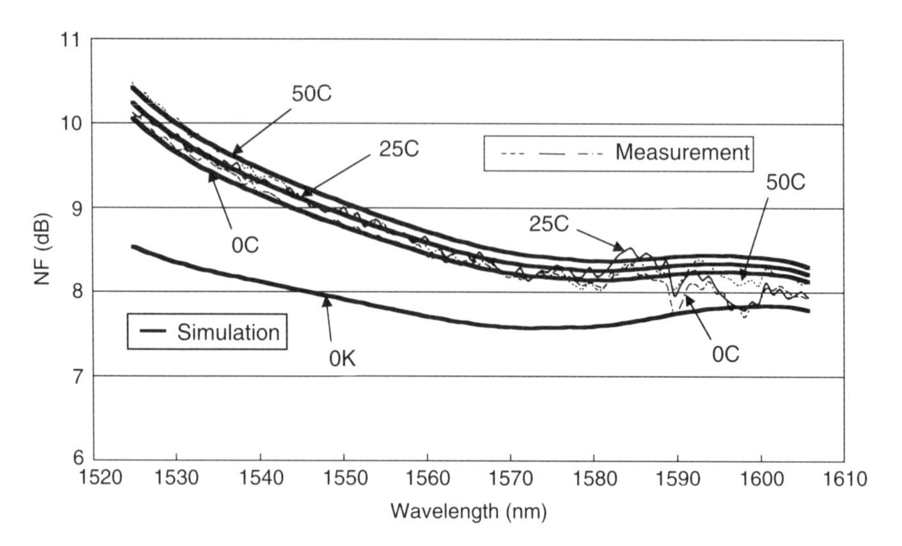

Fig. 10.24 Comparison of measured composite NF with numerical simulation at 0, 25, and 50°C, and a simulation at 0 K.

explained by the superposition rule as shown in Fig. 10.22. Figure 10.24 shows an excellent agreement between the NF values obtained by simulation and experiment. Further discussions of temperature dependence and wavelength-dependent noise will be developed in Section 10.4.10.4.

10.4.9. GAIN SATURATION

The gain saturation in a WDM-pumped Raman amplifier occurs also through pump depletion. However, in this case, a saturating signal is depleting more than one pump in accordance with a different amount of gain from each pump at a wavelength. Pump-to-pump energy transfer is also involved in the pump depletion process, resulting in a wavelength dependence of gain saturation. Depending on the wavelength of the saturating signal, gain tilt due to saturation may vary [119].

10.4.10. SOME SPECIFIC DESIGN ISSUES

This section looks at some specific design issues to be considered for WDM-pumped Raman amplifiers. First we discuss some design examples, followed by some noise-related issues.

10.4.10.1. Ultra-wideband Operation [120]

Let us take a look at the demonstration of 100-nm-bandwidth Raman amplifiers using a 12-wavelength-channel WDM pump laser diode unit. The gain flatness is less than ±0.5 dB, which is achieved through an asymmetric channel allocation of pump and without any gain equalization filters. Because the Raman gain shift is approximately 110 nm in a 1550-nm signal window, we note that the bandwidth of 100 nm is the broadest limit of bandwidth in which cross-talk between pump and signal can be easily taken out.

Figure 10.25 shows the schematic diagram of our 12-wavelength-channel WDM high-power pump LD unit. We use high-power GRIN-SCH strained layer MQW BH laser diode modules. Each channel has polarization-multiplexed LDs using PBC. Each LD is stabilized by a fiber Bragg grating (FBG) to have a narrow and stable spectrum for low insertion loss through the WDM coupler. The WDM coupler is a silica-based PLC comprising eleven Mach-Zehnder interferometers (MZI).

Figure 10.26 shows the output spectrum of the pumping unit. The insertion loss through PBC and WDM was typically 2.5 dB. The maximum total output power was more than 2.2 W. The channel spacing in the short-wavelength section (1405–1457.5 nm) is 1 THz (~7.5 nm), while that in the long one (1465–1510 nm) is 2 THz (~15 nm).

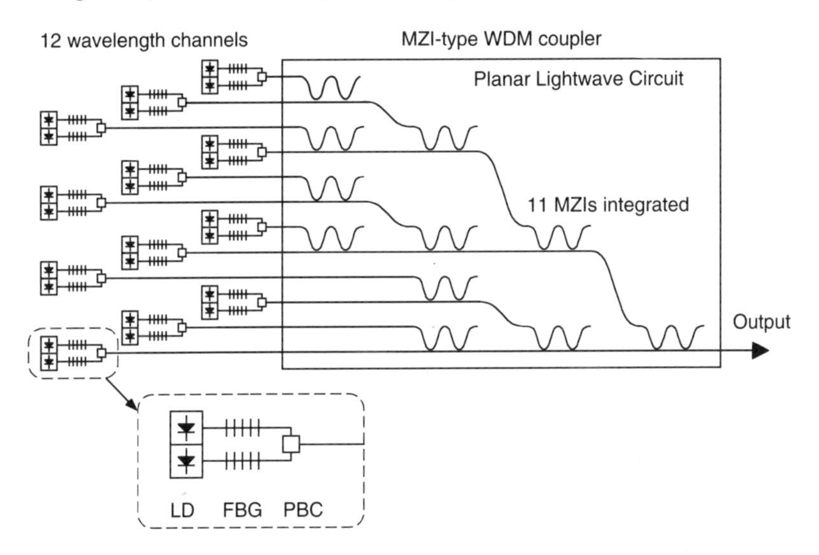

Fig. 10.25 The schematic diagram of our 12-wavelength-channel WDM high-power pump LD unit [120]. S. Namiki and Y. Emori, "Ultrabroad-band Raman amplifiers pumped and gain equalized by wavelength-division-multiplexed highpower laser diodes," IEEE J. Selected Top Quantum Electron, vol. 7, No. 1, pp. 3–16, 2001. © 2001 IEEE.

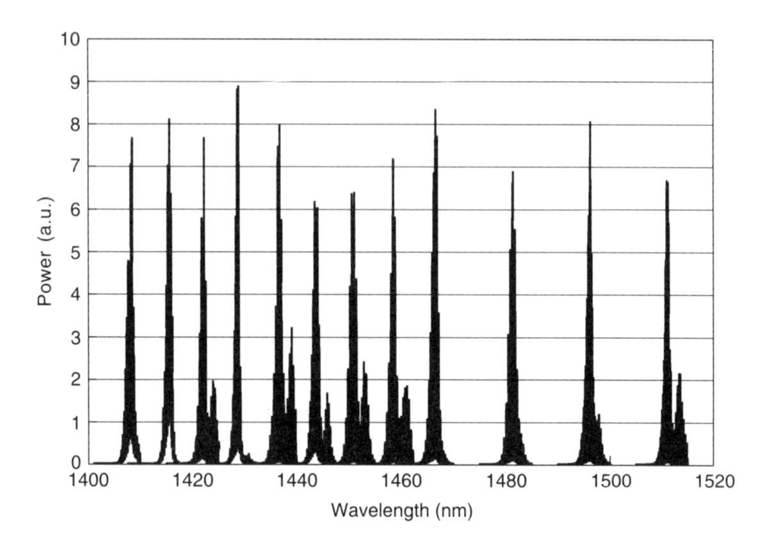

Fig. 10.26 The output spectrum of the pumping unit [120]. S. Namiki and Y. Emori, "Ultrabroad-band Raman amplifiers pumped and gain equalized by wavelength-division-multiplexed highpower laser diodes," IEEE J. Selected Top Quantum Electron, vol. 7, No. 1, pp. 3–16, 2001. © 2001 IEEE.

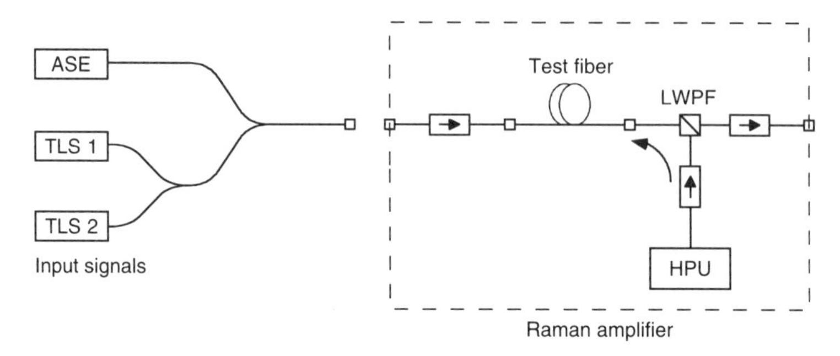

Fig. 10.27 The experimental setup of the broadband Raman amplifiers. HPU stands for high-power WDM pumping unit [120]. S. Namiki and Y. Emori, "Ultrabroad-band Raman amplifiers pumped and gain equalized by wavelength-division-multiplexed highpower laser diodes," IEEE J. Selected Top Quantum Electron, vol. 7, No. 1, pp. 3–16, 2001. © 2001 IEEE.

Figure 10.27 shows the experimental setup of the broadband Raman amplifiers. Counter-propagating pumping is employed to avoid the pump-induced amplitude noise. The total insertion loss of the isolators and long-wavelength pass filter (LWPF) along the signal path is 1.6 dB at 1550 nm. The loss variation over the 1520–1620 nm range is less than 1 dB. We use a broadband ASE light source as the input signal, as well as two tunable laser sources (TLS) used as probe signals to measure the gain profile.

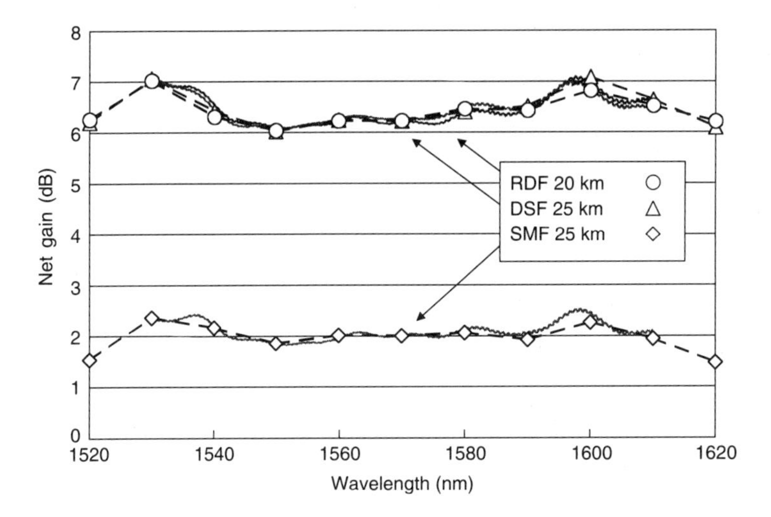

Fig. 10.28 Net gain profile of a 25 km SMF, a 25 km DSF, and a 20 km RDF. Transmission loss of each test fiber is about 5 dB [120]. S. Namiki and Y. Emori, "Ultrabroad-band Raman amplifiers pumped and gain equalized by wavelength-division-multiplexed highpower laser diodes," IEEE J. Selected Top Quantum Electron, vol. 7, No. 1, pp. 3–16, 2001. © 2001 IEEE.

We have tested three different types of fiber as Raman media—SMF, DSF, and reverse dispersion fiber (RDF). Figure 10.28 shows the net gain profile of a 25-km SMF, a 25-km DSF, and a 20-km RDF, and launched pump power of each channel is shown in Fig. 10.29. The 100-nm net gain bandwidth with ±0.5 dB variations is obtained for all types of fibers. The average gain is 2 dB for SMF, 6.5 dB for DSF and RDF. The nonlinear refractive indices divided by the effective areas are from small to large in the order of SMF, DSF, and RDF. Because the Raman scattering cross-section is larger for fibers with larger nonlinearity, RDF exhibits the highest pump-to-signal efficiency and the largest contrast in pump power distribution over channels due to pump-to-pump energy transfer. When the pump power in a longer wavelength becomes larger, the signal gain created by the pump at a shorter wavelength decreases. Hence, the maximum average gain is determined by the maximum pump power at shorter wavelengths as long as the gain has to be equalized by adjusting pump power distribution only.

10.4.10.2. Upgradeable Operation

In most cases, the number of WDM channels is gradually increased in the system due to the gradual increase in bandwidth demand. In this context, the operating bands are also upgraded accordingly. For example, EDFAs

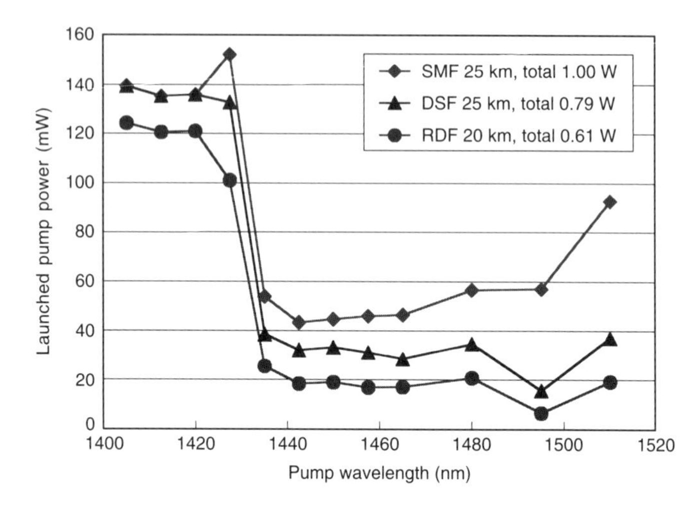

Fig. 10.29 Launched pump power of each channel [120]. S. Namiki and Y. Emori, "Ultrabroad-band Raman amplifiers pumped and gain equalized by wavelength-division-multiplexed highpower laser diodes," IEEE J. Selected Top Quantum Electron, vol. 7, No. 1, pp. 3–16, 2001. © 2001 IEEE.

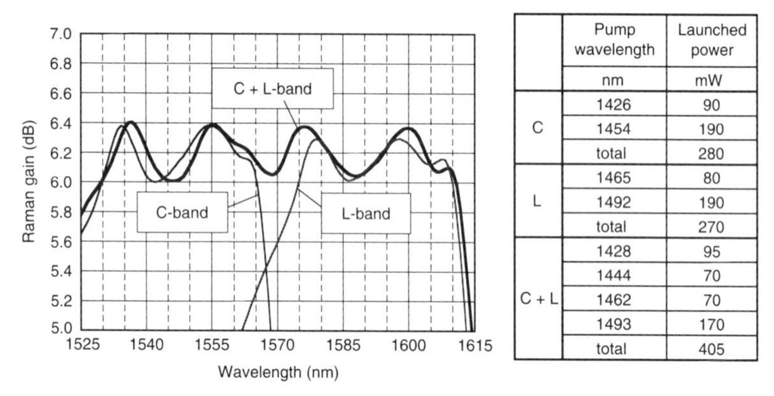

	Pump wavelength	Launched power
	nm	mW
C	1426	90
	1454	190
	total	280
L	1465	80
	1492	190
	total	270
C + L	1428	95
	1444	70
	1462	70
	1493	170
	total	405

Fig. 10.30 Calculated Raman gain spectra of C-, L- and C-L-band Raman amplifiers with optimal sets of pumps [108]. The fiber assumed is 50-km SMF. S. Namiki and Y. Emori, "Ultrabroad-band Raman amplifiers pumped and gain equalized by wavelength-division-multiplexed highpower laser diodes," IEEE J. Selected Top Quantum Electron, vol. 7, No. 1, pp. 3–16, 2001. © 2001 IEEE.

operating in C-band are first deployed and then those operating in L-band are additionally deployed later when needed. Similarly, it is desirable for Raman amplifiers to be upgradeable from only C- to simultaneous C- and L-bands operation.

Figure 10.30 shows the gain curves of practical examples of C-, L-, and C-L-band Raman amplifiers. A reasonable amount of gain and flatness are

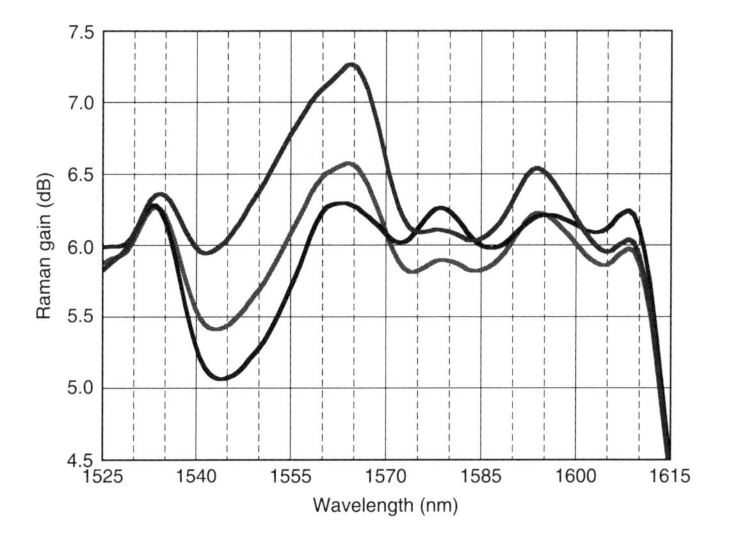

Fig. 10.31 Calculated Raman gain spectra of optimized C-L-band operations using the fixed set of pumps designed for individual C- and L-bands in Fig. 10.30 [108]. The fiber assumed is 50-km SMF. S. Namiki and Y. Emori, "Ultrabroad-band Raman amplifiers pumped and gain equalized by wavelength-division-multiplexed highpower laser diodes," IEEE J. Selected Top Quantum Electron, vol. 7, No. 1, pp. 3–16, 2001. © 2001 IEEE.

achieved with a reasonable number of pump lasers with commercially available amount of output powers. However, by comparing pump wavelengths for each case, they are all slightly different from each other. Each of C- and L-band Raman amplifiers cannot be used in the simultaneous operation of C- and L-bands. This means that this system is not upgradeable and that they have to be wholly replaced when upgraded from either C- or L-band operation to the simultaneous operation of C- and L-bands. Figure 10.31 shows the optimal Raman gain achievable with simultaneous operation of C- and L-band Raman amplifiers instead of C-L-band Raman amplifier. The flatness is apparently not acceptable in regard with the performances of the other amplifiers.

We interestingly found that providing an ad-hoc pump channel between the two shortest wavelengths can make the system upgradeable. The ad-hoc channel is not used in individual band operations, but only turned on for C-L-band operation. This derives from the asymmetric profile of Raman gain spectra, as was discussed in Section 10.3. The simultaneous operation of C- and L-band Raman amplifiers does not realize an asymmetric allocation of pump wavelengths. By adding an ad-hoc pump channel, an asymmetrically allocated pump channel can only be realized for respective C-, L-, and C-L-bands. Figure 10.32 shows an example in which an ad-hoc

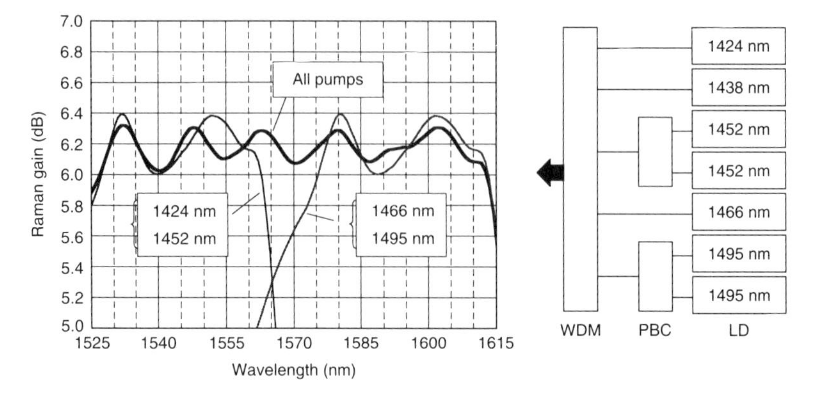

Fig. 10.32 Calculated Raman gain spectra of an upgradeable Raman amplifier system with an ad-hoc pump channel [108]. The fiber assumed is 50-km SMF. S. Namiki and Y. Emori, "Ultrabroad-band Raman amplifiers pumped and gain equalized by wavelength-division-multiplexed highpower laser diodes," IEEE J. Selected Top Quantum Electron, vol. 7, No. 1, pp. 3–16, 2001. © 2001 IEEE.

pump channel is adopted. It is clear that this scheme provides an upgradeable solution with a reasonable flatness maintained for all cases.

10.4.10.3. Ultra-flat and Dynamic Gain-tilt Operation [114]

In this section, we consider a WDM pumping scheme where wavelength-multiplexed laser diodes on a 1-THz spaced grid are used, and demonstrate it is capable of 0.1-dB Raman gain flatness over the C- plus L-band (1527–1607 nm) without any gain-flattening filters. By using this pumping scheme, we also demonstrate a smooth gain tilt to compensate for inter-signal Raman tilt.

Figure 10.33 shows a simulated broadband Raman gain profile illustrating how the superposition of one-wavelength-pumped gain profiles shifted by the pump frequency difference can create a flattened gain profile. In this simulation, we have used adequate 11 channels among the 16 frequencies spaced by 1 THz between each other. We assume the use of DSF for the Raman fiber.

In Fig. 10.33, the curves A, B, and C represent the total Raman gain of 11 channels, the sum of 9 channels in the shorter wavelength range, and that of 2 in the longer range, respectively. The 1-THz-spaced gain curves having almost same and small peak levels are added so as to obtain a smooth slope like curve B. Because curve C has an opposite slope of curve B, the total gain profile A is flattened. This example demonstrates that 0.1-dB flatness can be achieved by using this scheme.

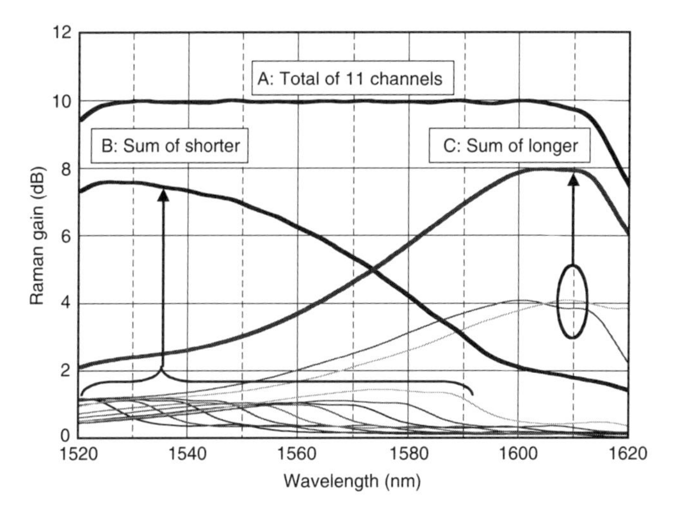

Fig. 10.33 A simulated broadband Raman gain profile using 11 1-THz-spaced pump channels [108]. The fiber is assumed to be DSF. S. Namiki and Y. Emori, "Ultrabroadband Raman amplifiers pumped and gain equalized by wavelength-division-multiplexed highpower laser diodes," IEEE J. Selected Top Quantum Electron, vol. 7, No. 1, pp. 3–16, 2001. © 2001 IEEE.

We prepare a pumping unit comprising 12 wavelength channels in the range of 212.2–199.3 THz (1412.5–1504.5 nm) on a 1-THz-spaced grid, where the source of each channel is a high-power GRIN-SCH strained layer MQW BH laser diode module.

Figure 10.34 shows the measured Raman gain profile and the launched pump power of each channel. In this experiment, a 25-km DSF is used as the Raman fiber, with a counter-propagating pump configuration. The gain flatness is about 0.1 dB over the wavelength range of the C- plus L-band (1527–1607 nm). Although the prepared unit has different absolute pumping frequencies from those in Fig. 10.33, the obtained flatness is as small as simulated in the previous section. This is because the frequency interval was comparable to realize the almost same flatness. The average Raman gain is almost 10.5 dB, which can compensate for approximately 50-km transmission fiber. Because of the pump-to-pump interaction, the launched powers of lower frequency pumps are made smaller in order to flatten the gain profile. In other words, the maximum available Raman gain in this scheme is determined by the maximum output power of higher frequency pumps.

This WDM pumping scheme can create a very flexible Raman gain profile. A smooth tilted profile can also be formed as shown in Fig. 10.35, where the gain by the lower frequency pumps is decreased and the others are adjusted so as to shape various slanted straight profiles. Inter-signal

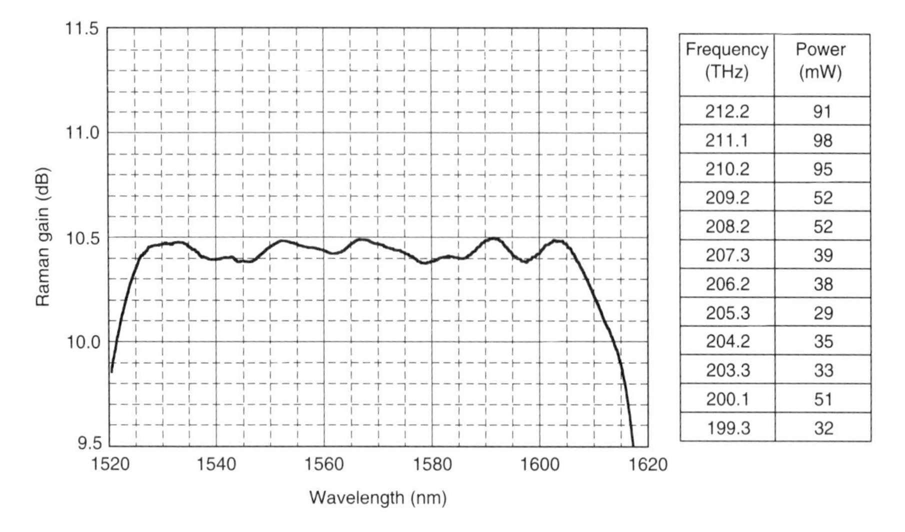

Frequency (THz)	Power (mW)
212.2	91
211.1	98
210.2	95
209.2	52
208.2	52
207.3	39
206.2	38
205.3	29
204.2	35
203.3	33
200.1	51
199.3	32

Fig. 10.34 The measured Raman gain profile and the launched pump power of each channel [108]. The fiber is 25-km DSF. S. Namiki and Y. Emori, "Ultrabroad-band Raman amplifiers pumped and gain equalized by wavelength-division-multiplexed highpower laser diodes," IEEE J. Selected Top Quantum Electron, vol. 7, No. 1, pp. 3–16, 2001. © 2001 IEEE.

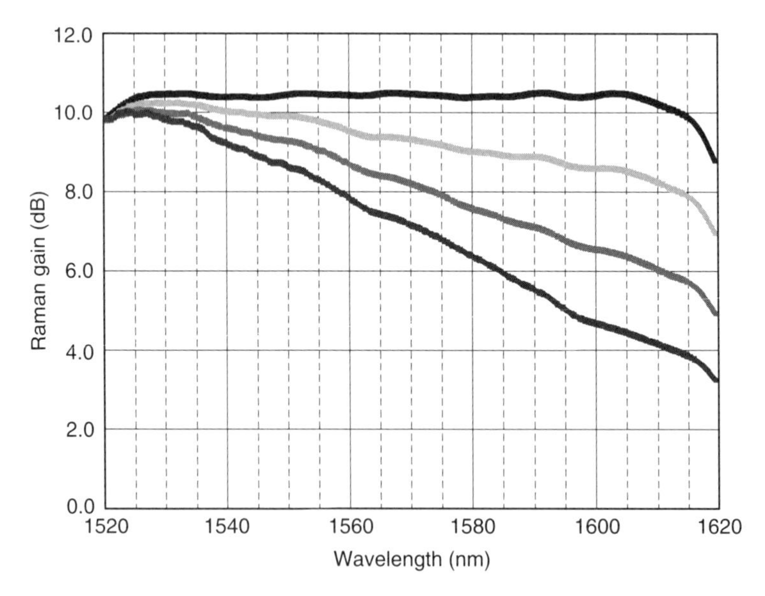

Fig. 10.35 Experimental demonstration of various slanted straight profiles [108]. The fiber is 25-km DSF. S. Namiki and Y. Emori, "Ultrabroad-band Raman amplifiers pumped and gain equalized by wavelength-division-multiplexed highpower laser diodes," IEEE J. Selected Top Quantum Electron, vol. 7, No. 1, pp. 3–16, 2001. © 2001 IEEE.

Raman tilts in broadband WDM transmissions [121] can be canceled by using this pumping scheme.

10.4.10.4. Temperature Stability and Wavelength-dependent Noise

Figure 10.36 shows the temperature dependence on Raman gain and noise figure in the broadband Raman amplifier [41]. The launched pump power of each wavelength channel is maintained through the experiment and the only amplifier fiber is put into a thermal chamber. The variation due to temperature is less than 0.5 dB for both gain and noise figure. The small reduction of Raman gain at 0°C may be derived from micro-bending loss of pumps. The temperature stability of the Raman gain is magnificent. The simulation model (Eq. (10.20)) was used to accurately predict the temperature dependence as shown in Figs. 10.23 and 10.24.

NF values increase with decreasing wavelength; this is due to three effects. First, thermal noise is larger for smaller frequency shift, as discussed in Eq. (10.15). Second, pump-to-pump energy transfer causes pumps at shorter wavelengths to attenuate more quickly, as shown in Fig. 10.21. (In a counter-pumping configuration, because the signals at shorter wavelength

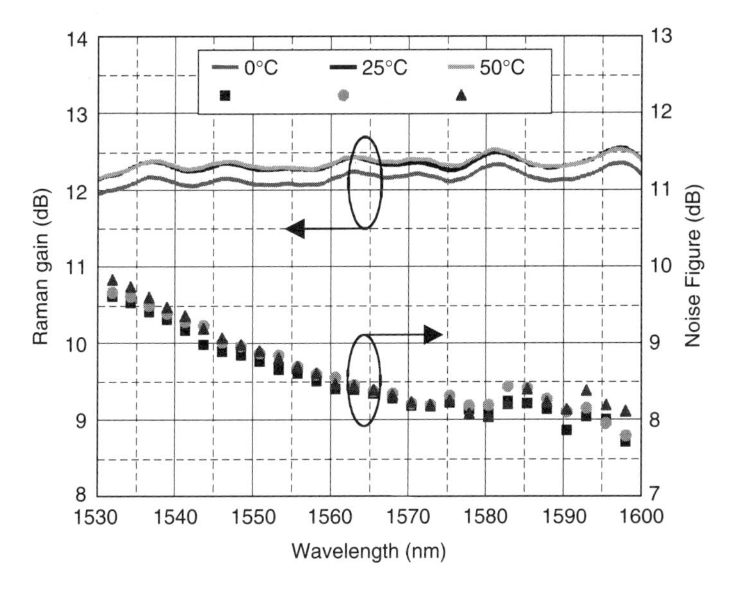

Fig. 10.36 The temperature dependence on Raman gain and noise figure [108]. The fiber is 25-km DSF. S. Namiki and Y. Emori, "Ultrabroad-band Raman amplifiers pumped and gain equalized by wavelength-division-multiplexed highpower laser diodes," IEEE J. Selected Top Quantum Electron, vol. 7, No. 1, pp. 3–16, 2001. © 2001 IEEE.

suffer from larger attenuation before they acquire larger gain than those at longer wavelength, they acquire higher noise than signals at longer wavelength [122].) Third, the loss spectrum of fiber has larger values for shorter wavelength because of the λ^{-4} scaling of Rayleigh scattering, which causes the similar dependence of NF spectrum.

10.4.10.5. Noise Issues in WDM Pumping

As we have already contextually discussed in part, there are several noise-related issues particularly taken care of in the configuration of WDM pumping. They are (1) wavelength-dependent spontaneous emission noise, (2) wavelength-dependent double Rayleigh backscattering noise, (3) cross-gain saturation noise, and (4) FWM generation among pumps.

Double Rayleigh backscattering noise (DRBS) in a WDM-pumped Raman amplifier may be smaller than a Raman amplifier pumped by a very high power single-wavelength laser with a gain-flattening filter. Because DRBS noise critically grows near a threshold, the amount of DRBS noise depends on gain and hence wavelength [46]. Therefore, Raman amplifiers gain-equalized by WDM pumping may be desirable in terms of DRBS.

When the wavelength range of WDM pumping contains the zero dispersion wavelength of the gain fiber, FWM sidebands of WDM pumps can be generated [123]. The FWM sidebands may overlap the signals and thus degrade performance. The countermeasures for this adverse phenomenon are difficult because it is an inherent issue in fiber and operating band, neither of which can be replaced. However, for a totally new system in which an optical fiber cable is newly deployed, this issue has to be resolved by optimizing the design of the optical fiber.

10.5. Key Components and Raman Pump Unit

This section discusses some practical aspects of a Raman pumping unit. First we examine some design issues regarding the choice of gain fibers. Second, we refer to the technologies of 14XX pump lasers. We also review pump combiners briefly, followed by the application of heat-pipe for the heat sink of Raman pumping units.

10.5.1. GAIN FIBER

Figures 10.4 and 10.5 imply that the gain shape is almost insensitive to production errors of optical fiber as the difference in shape among SMF, DSF, and even DCF is not so remarkable in spite of the huge difference of fiber structures. Therefore, the main property may be the magnitude of Raman gain efficiency. As we discussed in Section 10.3, the high gain efficiency means low saturation power and a less distributed nature. Therefore, there is a design trade-off of efficiency versus linearity and long distribution of gain. An ideal fiber for Raman gain may be a fiber having very low loss and high Raman gain at the same time. In a low-loss fiber, most of the loss is due to Rayleigh scattering, which is closely related to the Raman scattering cross-section. Low-loss fiber always has low Raman gain efficiency. For example, pure silica core fiber has very low loss characteristics and very low Raman gain efficiency. It is known that Germania has a higher Raman gain coefficient than silica. It should be noted that the transmission characteristics of fiber also depend on the core material. The fiber with high Germania concentration has a smaller core size to keep the single-mode operation in general, which tends to restrict the dispersion value. Also, Raman gain efficiency is closely related to Kerr nonlinearity. Fiber with higher Raman gain efficiency always has a large nonlinear coefficient. Therefore, to design a fiber link based on distributed Raman amplification, one should perform thorough optimizations depending on system requirements [124, 125].

10.5.2. LASER DIODE

This section reviews pump laser diodes that are commonly used in WDM-pumped Raman amplifiers. First, Fig. 10.37 shows the appearance of 14XX-nm pump laser diode. The package is very thin (less than 8-mm high) and

Fig. 10.37 Photograph of 14XX-nm pump laser diode.

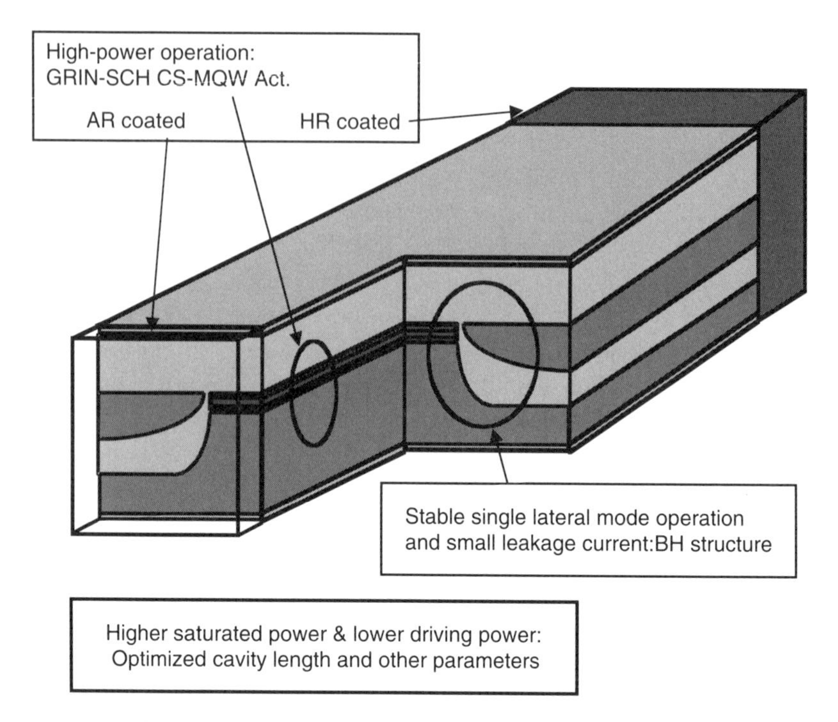

Fig. 10.38 Structure of 14XX-nm pump laser diode chip.

compact 14-pin butterfly package, carrying a laser chip, a back-facet monitor PIN photodiode, coupling optics, a thermistor sensor, and a thermoelectric cooler inside. Figure 10.38 shows the structure of a high-power 14XX-nm pump laser diode chip. The chip structure is graded index, separate confinement hetero-structure, buried hetero-structure (GRIN-SCH-BH) with compressively strained multiple quantum well (MQW) active layer. There are many design parameters for the laser diode [126]; but once these technologies become the state-of-the-art technology, the remaining issues are mainly the optimization of cavity length of the chip with respect to the targeted operating conditions such as power and current [127]. Figure 10.39 shows front-facet light output power of laser diode chip versus forward current for a different cavity length. From the figure, it is evident that the operating power and current increase with increasing cavity length. For different purposes, we can choose the most suitable cavity lengths. Figure 10.40 shows the cross-sectional schematic of packaging. An optimized two-lens coupling optics is employed to achieve near 90% coupling efficiency [110]. The high-power FBG-stabilized 14XX-nm pump lasers

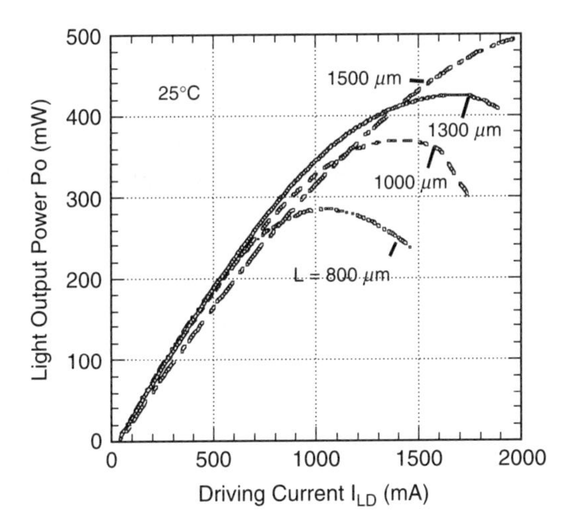

Fig. 10.39 Front-facet light output power of laser diode chip versus forward current for different cavity length.

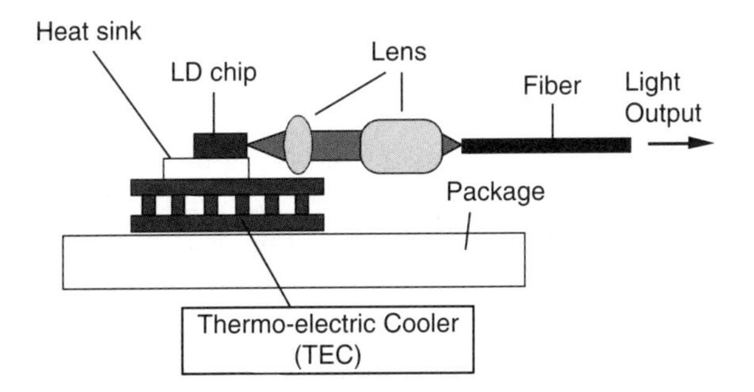

Fig. 10.40 The cross-sectional schematic of laser module packaging.

were achieved based on these technologies [111]. The reliability has been well established, mainly through their widespread field use for pumping EDFA.

Other laser diode chip structures have been reported to achieve 1 Watt output [128, 129].

10.5.3. CASCADED RAMAN RESONATOR

The so-called cascaded Raman resonator is a technology that can achieve watt-class output in the 1480-nm band [130]. This technology is based on high-power cladding pumped fiber laser [131] as a pump launched

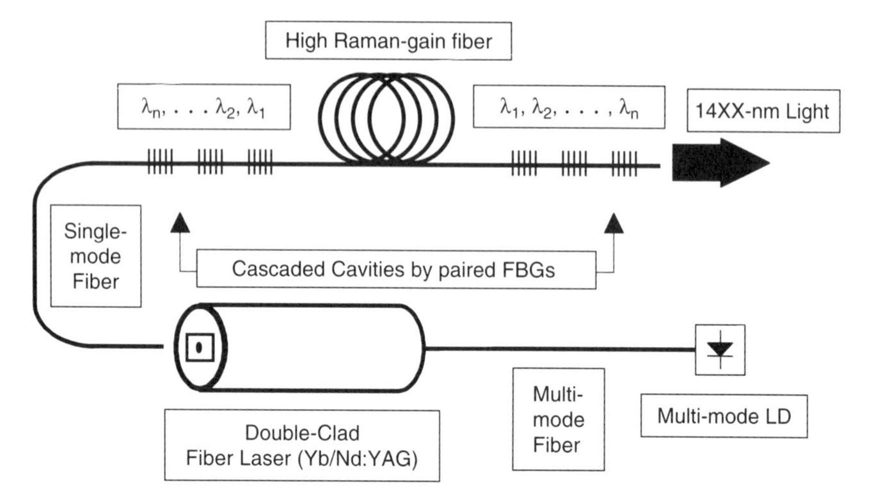

Fig. 10.41 The schematic diagram of the cascaded Raman resonator.

into a cascaded Raman resonator in which a multiple of Stokes waves are generated. Figure 10.41 shows the schematic diagram of the cascaded Raman resonator. Many pioneering experiments using this technology were performed [132].

Unlike diode pumps, this laser can generate multi-watt output at a single center wavelength. Therefore, output power by itself is sufficient for Raman amplifiers. However, some drawbacks are higher noise than laser diodes, less efficiency, and large volume so that it does not fit very well in the WDM pumping scheme. A more important issue of this technology is the fact that the reliability is not quite established yet, or that it has not been widely used in real field telecommunications.

In order to increase efficiency and available lasing wavelength range, Dianov *et al.* have reported the use of phosphosilicate glass fiber for the cascaded Raman resonator [133]. Other efforts have been made to operate more than two lasing wavelengths in a common cascaded Raman resonator, by which one can achieve a compact but multi-wavelength pumping to achieve a wide flat gain spectrum in a Raman amplifier [134, 135].

10.5.4. PUMP COMBINERS

There are two methods for efficiently combining two pump lasers into a single-mode fiber. One is polarization division multiplexing (PDM), in which two orthogonally polarized beams are combined through a polarization beam combiner (PBC). The other is wavelength division multiplexing (WDM), in which two lasers operating at different wavelengths are combined through a WDM coupler. PDM allows only two lasers to be combined, while WDM allows in principle an arbitrary number of pump wavelengths to be combined in a cascaded configuration of WDM couplers. PDM at the same center wavelength is effective for increasing launched power at the center wavelength and depolarizing the pump output, while WDM of many center wavelengths is effective for extending bandwidth, improving gain flatness, and increasing magnitude of composite gain of the Raman amplifier.

PBC uses either a polarization beam splitter prism or a birefringent crystal such as rutile. Figure 10.42 shows the photograph of a commercial PBC. The body size of this device is 6 mm H, 32 mm L, and 11 mm D. There are several technologies to realize WDM couplers; one is to use a dielectric thin-film interference filter, another is to use a fiber-fused coupler technology. Figure 10.43 shows the photograph of a WDM coupler device based on dielectric thin-film interference filter technologies. The body size of this device is 36 mm L, and ϕ5.5 mm. WDM couplers based on both dielectric thin-film filter and fiber-fused coupler are so-called discrete couplers. In the case of many pump channels, it is advantageous to integrate WDM couplers in one chip. Planar lightwave circuit (PLC) technologies are good for this purpose, as was discussed in Sections 10.4.5 and 10.4.10.1 [97, 112, 120]. Figure 10.44 shows the photograph of a PLC-MZI-WDM coupler for 15 wavelength inputs. Even though it contains 14 imbalanced MZI interferometers, the body size is as compact as 9.5 mm H, 115 mm L, and 7.9 mm D.

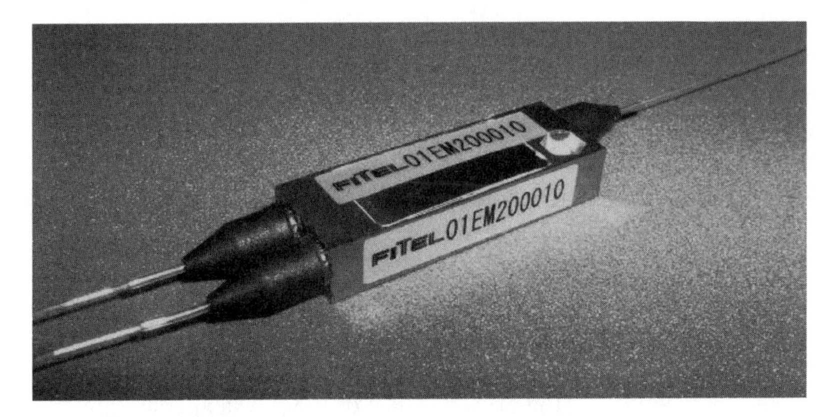

Fig. 10.42 Photograph of polarization beam combiner (PBC).

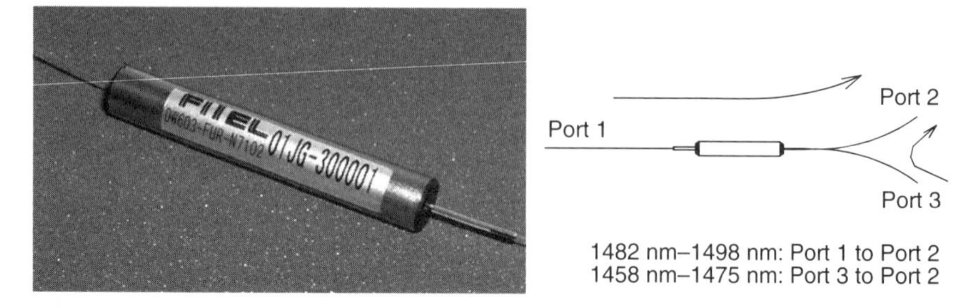

Fig. 10.43 Photograph of WDM coupler using a dielectric thin-film interference filter.

Fig. 10.44 Photograph of integrated 15-wavelength PLC-MZI-WDM coupler.

10.5.5. RAMAN PUMP UNITS

For distributed Raman amplifiers, the gain fiber is the deployed transmission fibers. Therefore, when we install a distributed Raman amplifier, we install a Raman pump unit that includes pump lasers, PBC, WDM couplers, isolators, tap couplers for monitoring, monitor photodiodes, signal-pump-WDM coupler, drive and monitoring circuits, and heat sink, etc. Figure 10.45 depicts a schematic diagram of a Raman pump unit, in which 5 pump wavelength channels are used.

An important issue for the design of a Raman pump unit is thermal management. Because the pump lasers are driven relatively hard as compared with those built in EDFA, the power consumption as well as thermal dissipation becomes an issue. In order to enhance thermal dissipation, a larger heat sink is inevitable. However, because of the physical constraints from the system, the size of the Raman pump unit, in particular the height, is

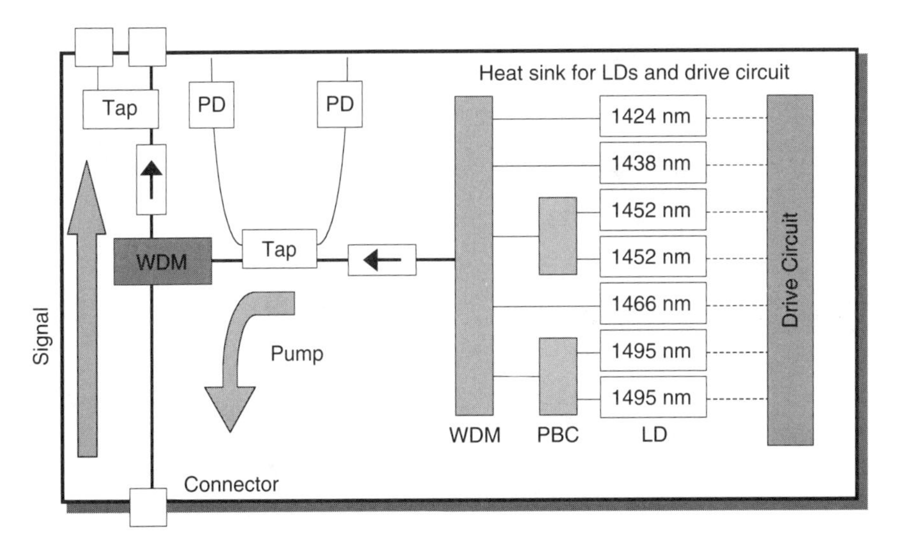

Fig. 10.45 Schematic diagram of Raman pumping unit.

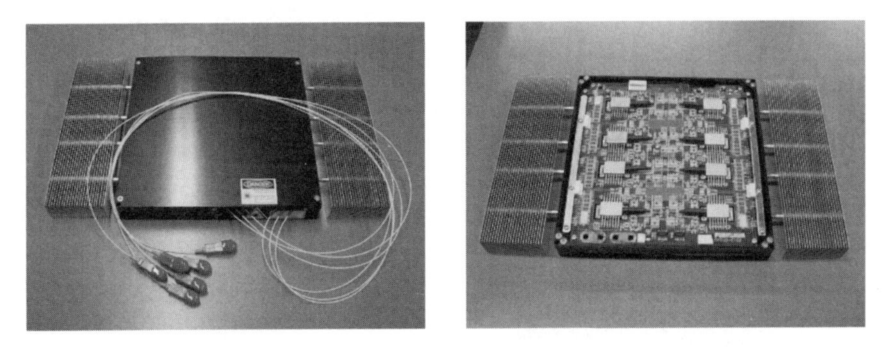

Fig. 10.46 Photograph of Raman pumping unit with heat pipe heat sink.

usually limited. When enough space for a heat sink is not available, the use of "heat pipe" is effective to transport the heat from one place to another. Heat pipe is a copper-sealed tube in which the pressure is very low, but containing a tiny amount of water. The water inside evaporates in the operating temperature range to create a current of molecules transporting heat from one end to the other. The effective thermal conductivity is 100 times as large as diamond. Figure 10.46 shows the photograph of a Raman pump unit equipped with heat pipes. This unit can carry up to eight pump lasers and is only one inch thick, yet has enough heat dissipation. Figure 10.47 compares the temperature increase with and without heat pipes. For the air

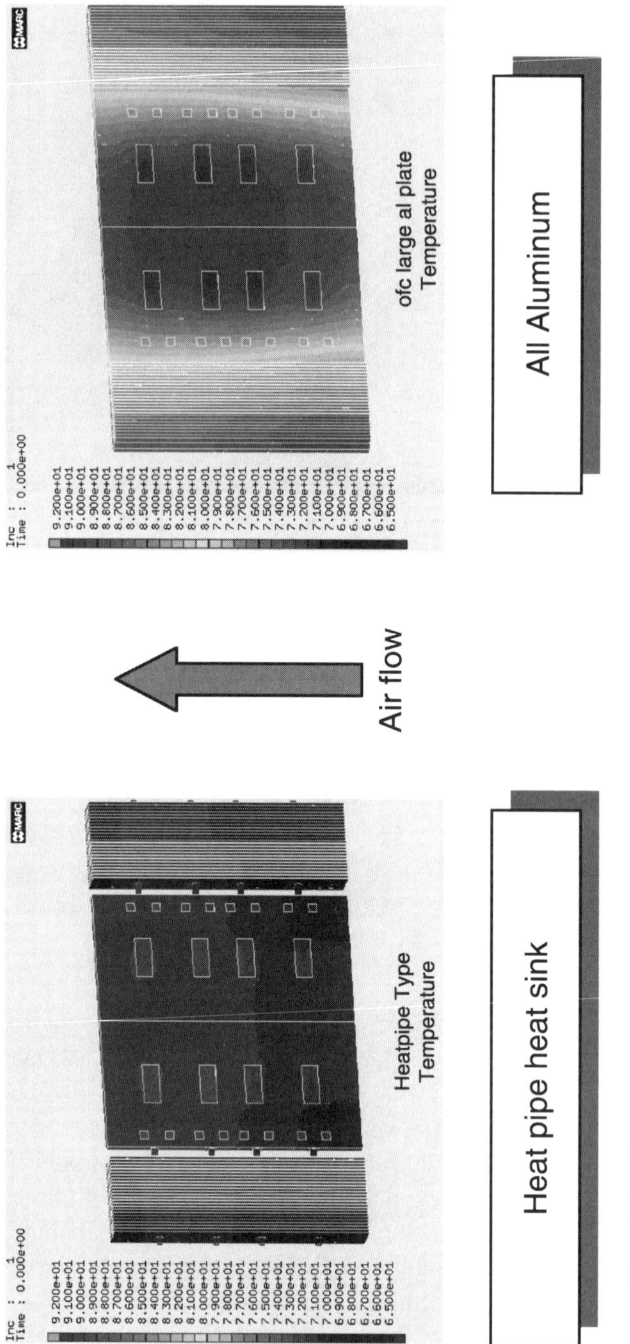

Fig. 10.47 Simulated comparison between the temperature increase of heat sink with and without heat pipe. (*See color plate*).

flow of 1 m/s, the heat dissipation per laser of 14 W including the power consumed by the driving FETs, and the ambient temperature of 50°C, the highest temperature of heat-pipe heat sink is 17.3°C lower than that of normal aluminum heat sink with the same size.

10.6. System Applications

We have already discussed the remarkable performances of Raman amplifiers for system applications, especially in Section 10.4.1. This section further examines some of the references regarding optical transmission experiments using Raman amplifiers. Here we attempt to categorize the applications of Raman amplifiers into the following items: (1) reports emphasizing improved OSNR, (2) reports pursuing low nonlinearity, (3) reports on extending repeater spacing, including festoon and repeaterless undersea systems, (4) reports on the so-called "pseudo-linear" transmissions assisted by Raman amplification, and (5) reports on field trials.

Improvement in OSNR, Q-factor, detector sensitivities, and non-regenerated transmission distances by means of Raman amplification is found in [136–139], in which the importance of distributed Raman amplifiers is studied especially for a higher bit rate such as 40 Gbit/s. For example, it is reported that a 40-Gbit/s system using distributed Raman amplifiers with a repeater span of 80 km performs as well as an EDFA-based system without using a Raman amplifier with an EDFA repeater span of 40 km [139].

As discussed in Section 10.3.6.4, distributed Raman amplifiers are effective in reducing nonlinearity such as FWM, SPM, and XPM by spending the increased power budget margin [51–55]. It is therefore effective to achieve an ultra-dense allocation of WDM channels [140, 141].

Longer repeater spans are realized in long-haul high-capacity WDM systems assisted by distributed Raman amplifiers. For example, repeater spans of 140 km [142] and 100 km [143] were achieved in 10 Gbit/s $\times$ 128-, and 80-ch WDM systems, respectively. There are recent similar results based on a 40-Gbit/s bit rate; 32 $\times$ 40 Gbit/s over 1000 km with 160-km Raman-amplified repeater spans [144], and 77 $\times$ 42.7 Gbit/s over 1200 km with 100-km Raman-amplified repeater spans [145]. It is important for next-generation WDM systems to handle optical networking. 52 $\times$ 12.3 Gbit/s with second-generation forward error correction (FEC) over 3600 km with 100-km Raman-amplified repeater span carrying periodic 100% add/drop access was demonstrated in [146].

Studies on festoon-type undersea optical transmissions show the effectiveness of distributed Raman amplifiers [147–149]. The repeater spacing of such systems is usually 200 to 300 km. Also, in repeaterless undersea systems, distributed Raman amplifiers play an indispensable role [150, 151]. In transoceanic systems, transmission bandwidth can be increased by means of WDM pumping [152]. Also, management and optimization of transmission fibers are of critical importance, as they determine the interplay between nonlinearity and dispersion, as well as Raman gain efficiency [153].

The so-called "pseudo-linear" transmission is one of the pursuits toward next-generation ultra-high-speed transmissions [154]. In this transmission technology, distributed Raman amplifiers also play an important role, as they better achieve linear transmissions than lumped EDFAs. Transmissions of 160 Gbit/s over 400 km [155] and 320 Gbit/s over 200 km [156] have been achieved by this technology.

After discussing many rationales for Raman amplifiers to further boost the transmission performances over EDFA-based WDM technologies, we should significantly note some experimental reports from the real field trials. In the postdeadline paper session of the Optical Communication Conference (OFC) held in Baltimore 2000, two field transmission experiments using distributed Raman amplifiers were reported; one was all-optical network field trial with 10 Gb/s DWDM transmission over Qwest's 2410 km NZ-DSF terrestrial route [157], and the other was a field demonstration of distributed Raman amplification with 3.8 dB Q-improvement in AT&T's 5×120 km transmission fiber loop between Freehold and Jamesburg, NJ [158, 159]. Also, published in March 2001, was 160 Gbit/s transmission over Deutche Telecom's 116 km field-installed fiber connecting Darmstadt, Gross Gersu, and Bensheim, Germany [160]. In the postdeadline paper session of OFC 2001, 3.2 Tbit/s field trial (80×40 Gbit/s) over WorldCom's 3×82-km standard SMF between Richardson and Dallas using FEC, Raman, and tunable dispersion compensation, was reported [161].

There is always an uncertainty when a new technology is actually applied in the field, although plenty of beautiful experiments have been demonstrated beforehand in laboratories. For Raman amplifiers, one such concern may be related to the high power operation of the pumps. In the mentioned field trials, the actual launched pump power ranges from 250 mW to 950 mW. As long as these reports say, the effectiveness of Raman amplifiers is confirmed in real field fiber cables. In particular, field-deployed cables contain many excess losses due to splicing or connectors; Raman amplification still produced 3.8 dB Q-improvement at an optimum pump power [159].

10.7. Concluding Remarks

We have discussed the issues related to fiber Raman amplifiers for telecommunication purposes as many as possible, ranging from fundamental principles through practice of fiber Raman amplifiers. Even though the principle of Raman amplifier was demonstrated by Stolen and Ippen back in 1973, the widespread applications of Raman amplifiers to the real optical transmission systems are still forthcoming; it is expected that many efforts have to yet to be done for adapting this technology to the society. However, it is convincing that Raman amplifiers will play significant roles in next generation WDM network systems.

ACKNOWLEDGMENTS

The author would like to thank Drs. Yoshihiro Emori and Misao Sakano for invaluable discussions, Yoshio Tashiro and Souko Kado for proofreading, Junji Yoshida for providing figures of pump laser, Tomohiko Kimura and Dr. Kanji Tanaka for providing photographs of pump combiners, and Ryosei Kunifuji and Masami Ikeda for assistance of references. He is also indebted to Drs. K. Ohkubo and Y. Suzuki, Mr. T. Kikuta, and Mr. A. Hasemi for their encouragement.

References

1. Y. R. Shen, The Principles of Nonlinear Optics (Wiley, New York, 1984).
2. R. W. Boyd, Nonlinear Optics (Academic Press, San Diego, 1992).
3. E. G. Sauter, Nonlinear Optics (Wiley, New York, 1996).
4. G. P. Aggrawal, Nonlinear Fiberoptics, 3rd ed. (Academic Press, San Diego, 2001).
5. R. H. Stolen, E. P. Ippen, and A. R. Tynes, Appl. Phys. Lett., 20 (1972) 62.
6. R. H. Stolen and E. P. Ippen, Appl. Phys. Lett., 22 (1973) 276.
7. R. G. Smith, Applied Optics, 11 (1972) 2489.
8. J. Auyeung and A. Yariv, IEEE J. Quantum Electron., QE-14 (1978) 347.
9. K. Mochizuki, IEEE J. Lightwave Technol., LT-3 (1985) 688.
10. M. L. Dakks and P. Melman, IEEE J. Lightwave Technol., LT-3 (1985) 806.
11. E. Desurvire, M. J. F. Digonnet, and H. J. Shaw, IEEE J. Lightwave Technology, LT-4 (1986) 426.
12. T. Nakashima, S. Seikai, M. Nakazawa, and Y. Negishi, IEEE J. Lightwave Technol., LT-4 (1986) 1267.
13. Y. Aoki, IEEE J. Lightwave Technol., 6 (1988) 1225.
14. Y. Aoki, Optical and Quantum Electronics, 21 (1989) S89.
15. C. Lin and R. H. Stolen, Appl. Phys. Lett., 29 (1976) 428.

16. Y. Aoki, S. Kishida, H. Honmou, K. Washio, and M. Sugimoto, Electron. Lett., 19 (1983) 620.
17. K. Nakamura, M. Kimura, S. Yoshida, T. Hidaka, and Y. Mitsuhashi, IEEE J. Lightwave Technol., LT-2 (1984) 379.
18. M. Nakazawa, Appl. Phys. Lett., 46 (1985) 628.
19. R. H. Stolen, J. P. Gordon, W. J. Tomlinson, and H. A. Haus, J. Opt. Soc. Am., B 6 (1989) 1159.
20. A. Hasegawa, Opt. Lett., 8 (1983) 650.
21. A. Hasegawa, Applied Optics, 23 (1984) 3302.
22. F. L. Galeener and G. Lucovsky, Phys. Rev. Lett., 37 (1976) 1474.
23. F. L. Galeener, Phys. Rev. B, 19 (1979) 4292.
24. N. Shibata, M. Horiguchi, and T. Edahiro, J. Non-Crystalline Solids, 45 (1981) 115.
25. F. L. Galeener, J. C. Mikkelsen, Jr., R. H. Geils, and W. J. Mosby, Appl. Phys. Lett., 32 (1978) 34.
26. S. T. Davey, D. L. Williams, B. J. Ainslie, W. J. M. Rothwell, and B. Wakefield, IEE Proc., 136 (1989) 301.
27. V. L. da Silva and J. R. Simpson, "Comparison of Raman efficiencies in optical fibers," in Optical Fiber Communication Conference, OSA Technical Digest Series (Optical Society of America, Washington, D.C., 1994) paper WK13, p. 136.
28. Y. Akasaka, I. Morita, M.-C. Ho, M. E. Marhic, and L. G. Kazovsky, "Characteristics of optical fibers for discrete Raman amplifiers" in 26th European Conference on Optical Communications, Proceedings, Vol. 1 (1999) 288.
29. C. Fludger, A. Maroney, N. Jolly, and R. Mears, "An analysis of the improvements in OSNR from distributed Raman amplifiers using modern transmission fibres," in Optical Fiber Communication Conference, OSA Technical Digest Series (Optical Society of America, Washington, D.C., 2000), paper FF2.
30. A. K. Atieh, P. Myslinski, J. Chrostowski, and P. Galko, J. Lightwave Technol., 17 (1999) 216.
31. R. H. Stolen, IEEE J. Quantum Electron., QE-15 (1979) 1157.
32. D. J. Dougherty, F. X. Kärtner, H. A. Haus, and E. P. Ippen, Opt. Lett., 20 (1995) 31.
33. Y. Emori, S. Matsushita, and S. Namiki, "Cost-effective depolarized diode pump unit designed for C-band flat-gain Raman amplifiers to control EDFA gain profile," in Optical Fiber Communication Conference, OSA Technical Digest Series (Optical Society of America, Washington, D.C., 2000), paper FF4.
34. S. A. E. Lewis, S. V. Chernikov, and J. R. Taylor, Electron. Lett., 35 (1999) 923.
35. C.-J. Chen and W. S. Wong, Electron. Lett., 37 (2001) 371.

36. N. A. Olsson and J. Hegarty, IEEE J. Lightwave Technol., LT-4 (1986) 396.
37. K. Mochizuki, N. Edagawa, and Y. Iwamoto, IEEE J. Lightwave Technol., LT-4 (1986) 1328.
38. K. Rottwitt, M. Nissov, and F. Kerfoot, "Detailed analysis of Raman amplifiers for long-haul transmission," in Optical Fiber Communication Conference, OSA Technical Digest Series (Optical Society of America, Washington, D.C., 1998) paper TuG1, p. 30.
39. R. H. Stolen, S. E. Miller, and A. G. Chynoweth, eds., Nonlinear Properties of Optical Fibers: Optical Fiber Telecommunications (Academic Press, New York, 1979), Chap. 5, pp. 127–133.
40. S. A. E. Lewis, S. V. Chernikov, and J. R. Taylor, Opt. Lett., 24 (1999) 1823.
41. S. Namiki and Y. Emori, "Broadband Raman amplifiers design and practice," in Optical Amplifiers and Their Applications, OSA Technical Digest Series (Optical Society of America, Washington, D.C., 2000), paper OMB2, p. 7.
42. A. H. Hartog and M. P. Gold, IEEE J. Lightwave Technol., LT-2 (1984) 76.
43. See, for example, P. Wan and J. Conradi, IEEE J. Lightwave Technol., 14 (1996) 268.
44. P. B. Hansen, L. Eskildsen, A. J. Stentz, T. A. Strasser, J. Judkins, J. J. DeMarco, R. Pedrazzani, and D. J. DiGiovanni, IEEE Photon. Technol. Lett., 10 (1998) 159.
45. M. Nissov, K. Rottwitt, H. D. Kidorf, and M. X. Ma, Electron. Lett., 35 (1999) 997.
46. S. A. E. Lewis, S. V. Chernikov, and J. R. Taylor, IEEE Photon. Technol. Lett., 12 (2000) 528.
47. S. R. Chinn, IEEE Photon. Technol. Lett., 11 (1999) 1632.
48. C. R. S. Fludger, V. Handerek, and R. J. Mears, Electron. Lett., 37 (2001) 15.
49. F. Forghieri, R. W. Tkach, and A. R. Chraplyvy, "Bandwidth of cross talk in Raman amplifiers," in Optical Fiber Communication Conference, OSA Technical Digest Series (Optical Society of America, Washington, D.C., 1994), paper FC6, p. 294.
50. J. S. Wey, D. L. Butler, M. F. Van Leeuwen, L. G. Joneckis, and J. Goldhar, IEEE Photon. Technol. Lett., 11 (1999) 1417.
51. P. B. Hansen, A. Stentz, T. N. Nielsen, R. Espindola, L. E. Nelson, and A. A. Abramov, "Dense wavelength-division multiplexed transmission in "Zero-Dispersion" DSF by means of hybrid Raman/Erbium-doped fiber amplifiers," in Optical Fiber Communication Conference, OSA Technical Digest Series (Optical Society of America, Washington, D.C., 1999), paper PD8.
52. N. Takachio, H. Suzuki, H. Masuda, and M. Koga, "32 $\times$ 10 Gb/s distributed Raman amplification transmission with 50 GHz channel spacing in the

zero-dispersion region over 640 km of 1.55-mm dispersion-shifted fiber," in Optical Fiber Communication Conference, OSA Technical Digest Series (Optical Society of America, Washington, D.C., 1999), paper PD9.
53. H. Suzuki, N. Takachio, H. Masuda, and M. Koga, Electron. Lett., 35 (1999) 1175.
54. H. Kawakami, Y. Miyamoto, K. Yonenaga, and H. Toba, "Highly efficient distributed Raman amplification system in a zero-dispersion-flattened transmission line," in Optical Amplifiers and Their Applications, OSA Technical Digest Series (Optical Society of America, Washington, D.C., 1999), paper ThB5.
55. M. E. Marhic, Y. Akasaka, K. Y. Wong, F. S. Yang, and K. G. Kazovsky, Electron. Lett., 36 (2000) 1637.
56. Y. Aoki, S. Kishida, K. Washio, and K. Minemura, Electron. Lett., 21 (1985) 191.
57. J. Hegarty, N. A. Olsson, and L. Goldner, Electron. Lett., 21 (1985) 290.
58. L. F. Mollenauer, R. H. Stolen, and M. N. Islam, Opt. Lett., 10 (1985) 229.
59. L. F. Mollenauer, J. P. Gordon, and M. N. Islam, IEEE J. Quantum Electron., QE-22 (1986) 157.
60. N. Edagawa, K. Mochizuki, and Y. Iwamoto, Electron. Lett., 23 (1987) 196.
61. J. Nakagawa, K. Shimizu, T. Mizuochi, K. Takano, K. Motoshima, and T. Kitayama, "10 Gbit/s–270 km non-repeated optical transmission experiment with high receiver sensitivity," in 21st European Conference on Optical Communications, Proceedings, Vol. 1 (1995) p. 601.
62. P. B. Hansen, L. Eskildsen, S. G. Grubb, A. J. Stentz, T. A. Strasser, J. Judkins, J. J. DeMarco, R. Pedrazzani, and D. J. DiGiovanni, IEEE Photon. Technol. Lett., 9 (1997) 262.
63. M. Nissov, C. R. Davidson, K. Rottwitt, R. Menges, P. C. Corbett, D. Innis, and N. S. Bergano, "100 Gbit/s (10×10Gb/s) WDM transmission over 7200 km using distributed Raman amplification," in 23rd European Conference on Optical Communications, Proceedings, Vol. 5, Post-deadline papers (1997) p. 9.
64. T. N. Nielsen, A. J. Stentz, K. Rottwitt, D. S. Vengsarkar, Z. J. Chen, P. B. Hansen, J. H. Park, K. S. Feder, S. Cabot, S. Stulz, D. W. Peckham, L. Hsu, C. K. Kan, A. F. Judy, S. Y. Park, L. E. Nelson, and L. Grüner-Nielsen, IEEE Photon. Techno. Lett., 12 (2000) 1079.
65. S. Bigo, E. Lach, Y. Frignac, D. Hamoir, P. Sillard, W. Idler, S. Gauchard, A. Bertaina, S. Borne, L. Lorcy, N. Torabi, B. Franz, P. Nouchi, P. Guenot, L. Fleury, G. Wien, G. Le Ber, R. Fritschi, B. Junginger, M. Kaiser, D. Bayart, G. Veith, J.-P. Hamaide, and J.-L. Beylat, Electron. Lett., 37 (2001) 448.
66. S. Bigo, A. Bertaina, Y. Frignac, S. Borne, L. Lorcy, D. Hamoir, D. Bayart, J.-P. Hamaide, W. Idler, E. Lach, B. Franz, G. Veith, P. Sillard, L. Fleury,

P. Guenot, and P. Nouchi, "5.12 Tbit/s (128 × 40 Gbit/s WDM) transmission over 3 × 100 km of teralight fibre," in 26th European Conference on Optical Communications, Proceedings, Vol. 5, Post-deadline paper, PD1-2 (2000).
67. S. Bigo *et al.*, "10.2 Tbit/s (256 × 42.7 Gbit/s PDM/WDM) transmission over 100 km Teralight fiber with 1.28 bit/s/Hz spectral efficieny," in Optical Fiber Communication Conference, OSA Technical Digest Series (Optical Society of America, Washington, D.C., 2001), paper PD25.
68. K. Fukuchi, T. Kasamatsu, M. Morie, R. Ohhira, T. Ito, K. Sekiya, D. Ogasahara, and T. Ono, "10.92 Tb/s (273 × 40 Gb/s) triple-band/ultra-dense WDM optical-repeatered transmission experiment," in Optical Fiber Communication Conference, OSA Technical Digest Series (Optical Society of America, Washington, D.C., 2001), paper PD24.
69. W. S. Lee, Y. Zhu, B. Shaw, D. Watley, C. Scahill, J. Homan, C. Fludger, M. Jones, and A. Hadjifotiou, "2.56 Tbit/s capacity, 0.8 b/Hz.s DWDM transmission over 120 km NDSF using polarization-bit-interleaved 80 Gbit/s OTDM signal," in Optical Fiber Communication Conference, OSA Technical Digest Series (Optical Society of America, Washington, D.C., 2001), paper TuU1.
70. T. Naito, N. Shimojoh, T. Tanaka, H. Nakamoto, M. Doi, T. Ueki, and M. Suyama, "1 Terabit/s WDM transmission over 10,000 km," in 25th European Conference on Optical Communications, Proceedings, Vol. 5, Post-deadline paper, PD2-1 (1999).
71. T. Tanaka, N. Shimojoh, T. Naito, H. Nakamoto, I. Yokota, T. Ueki, A. Sugiyama, and M. Suyama, "2.1-Terabit/s WDM transmission over 7,221 km with 80-km repeater spacing," in 26th European Conference on Optical Communications, Proceedings, Vol. 5, Post-deadline paper, PD1-8 (2000).
72. H. Nakamoto, T. Tanaka, N. Shimojoh, T. Naito, I. Yokota, A. Sugiyama, T. Ueki, and M. Suyama, "1.05 Tbit/s WDM transmission over 8,186 km using distributed Raman amplifier repeaters," in Optical Fiber Communication Conference, OSA Technical Digest Series (Optical Society of America, Washington, D.C., 2001), paper TuF6.
73. H. A. Haus and J. A. Mullen, Phys. Rev., 128 (1962) 2407.
74. A. Yariv, Optical Electronics, 4th ed. (Oxford University Press, Oxford, 1991).
75. A. Yariv, Opt. Lett., 15 (1990) 1064.
76. K. Rottwitt, A. Stentz, T. Nielsen, P. Hansen, K. Feder, and K. Walker, "Transparent 80 km Bi-directionally pumped distributed Raman amplifier with second order pumping," in 25th European Conference on Optical Communications, Proceedings, Vol. 2 (1999) 144.
77. P. Urquhart and T. J. Whitley, Appl. Opt., 29 (1990) 3503.
78. G. R. Walker, D. M. Spirit, D. L. Williams, and S. T. Davey, Electron. Lett., 27 (1991) 1390.
79. D. N. Chen and E. Desurvire, IEEE Photon. Technol. Lett., 4 (1992) 52.

80. K. Rottwitt, A. Bjarklev, O. Lumholt, J. H. Povlsen, and T. P. Rasmussen, Electron. Lett., 28 (1992) 287.
81. M. Nissov, H. N. Poulsen, R. J. Pedersen, B. F. Jørgensen, M. A. Newhouse, and A. J. Antos, Electron. Lett., 32 (1996) 1905.
82. N. Kagi, A. Oyobe, and K. Nakamura, "Temperature Dependence of the Gain in Erbium Doped Fibers" IEEE J. Lightwave Technol., 9 (1991) 261–265.
83. T. Tsuda, M. Miyazawa, K. Nishiyama, T. Ota, K. Mizuno, Y. Mimura, Y. Tashiro, Y. Emori, and S. Namiki, "Gain-flattening filters with autonomous temperature stabilization of Erbium gain," in Proc. Optical Amplifiers and Their Applications, paper OWA4 (2000) 173–175.
84. A. J. Stentz, "Progress on Raman amplifiers," in Optical Fiber Communication Conference, OSA Technical Digest Series (Optical Society of America, Washington, D.C., 1997), paper FA1.
85. E. M. Dianov, A. A. Abramov, M. M. Bubnov, A. V. Shipulin, A. M. Prokhorov, S. L. Semjonov, A. G. Schebunjaev, G. G. Devjatykh, A. N. Guryanov, and V. F. Khopin, Optical Fiber Technology, 1 (1995) 236.
86. E. M. Dianov, M. V. Grekov, I. A. Bufetov, V. M Mashinsky, O. D. Sazhin, A. M. Prokhorov, G. G. Devyatykh, A. N. Guryanov, and V. F. Khopin, Electron. Lett., 34 (1998) 669.
87. D. V. Gapontsev, S. V. Chernikov, and J. R. Taylor, Opt. Communications, 166 (1999) 85.
88. J. Kani, M Jinno, and K. Oguchi, Electron. Lett., 34 (1998) 1745.
89. T. Horiguchi, T. Sato, and Y. Koyamada, IEEE Photon. Technol. Lett., 4 (1992) 64.
90. H. Masuda, S. Kawai, K.-I. Suzuki, and K. Aida, Electron. Lett., 34 (1998) 2339.
91. F. Koch, S. V. Chernikov, S. A. E. Lewis, and J. R. Taylor, Electron. Lett., 36 (2000) 347 .
92. A. J. Stentz, S. G. Grubb, C. E. Headley III, J. R. Simpson, T. Strasser, and N. Park, "Raman amplifier with improved system performance," in Optical Fiber Communication Conference, OSA Technical Digest Series (Optical Society of America, Washington, D.C., 1996), paper TuD3, p. 16.
93. T. N. Nielsen, P. B. Hansen, A. J. Stentz, V. M. Aquaro, J. R. Pedrazzani, A. A. Abramov, and R. P. Espindola, IEEE Photon. Technol. Lett., 10 (1998) 1492.
94. J. Bromage, J.-C. Bouteiller, J. J. Thiele, K. Brar, J. H. Park, C. Headley, L. E. Nelson, Y. Qian, J. DeMarco, S. Stulz, L. Leng, B. Zhu, and B. J. Eggleton, "S-band all-Raman amplifiers for 40 $\times$ 10 Gb/s transmission over 6 $\times$ 100 km of non-zero dispersion fiber," in Optical Fiber Communication Conference, OSA Technical Digest Series (Optical Society of America, Washington, D.C., 2001), postdeadline Paper PD4.

95. A. B. Puc, M. W. Chbat, J. D. Henrie, N. A. Weaver, H. Kim, A. Kaminski, A. Rahman, and H. A. Fèvrier, "Long-haul WDM NRZ transmission at 10.7 Gb/s in S-band using cascade of lumped Raman amplifiers," in Optical Fiber Communication Conference, OSA Technical Digest Series (Optical Society of America, Washington, D.C., 2001), postdeadline Paper PD39.
96. P. B. Hansen, G. Jacobovitz-Velselka, L. Grüner-Nielsen, and A. J. Stentz, Electron. Lett., 34 (1998) 1136.
97. Y. Emori, Y. Akasaka, and S. Namiki, Electron. Lett., 34 (1998) 2145.
98. S. A. E. Lewis, S. V. Chernikov, and J. R. Taylor, Electron. Lett., 36 (2000) 1355.
99. Y. Emori, Y. Akasaka, and S. Namiki, "Less than 4.7 dB noise figure broadband in-line EDFA with a Raman amplified –1300 ps/nm DCF pumped by multi-channel WDM laser diodes," in Optical Amplifiers and their Applications, OSA Technical Digest Series (Optical Society of America, Washington, D.C., 1998), postdeadline Paper PD3.
100. H. Masuda, S. Kawai, K. Suzuki, and K. Aida, IEEE Photon. Technol. Lett., 10 (1998) 516.
101. H. Masuda, S. Kawai, and K. Aida, Electron. Lett., 34 (1998) 1342.
102. H. Masuda, S. Kawai, and K. Aida, Electron. Lett., 35 (1999) 411.
103. H. Masuda and S. Kawai, IEEE Photon. Technol. Lett., 11 (1999) 647.
104. J. Kani and M. Jinno, Electron. Lett., 35 (1999) 1004.
105. D. Bayart, P. Baniel, A. Bergonzo, J.-Y. Boniort, P. Bousselet, L. Gasca, D. Hamoir, F. Leplingard, A. Le Sauze, P. Nouchi, F. Roy, and P. Sillard, Electron. Lett., 36 (2000) 1569.
106. K. Rottwitt and H. D. Kidorf, "A 92 nm bandwidth Raman amplifier," in Optical Fiber Communication Conference, OSA Technical Digest Series (Optical Society of America, Washington, D.C., 1998), postdeadline Paper PD6.
107. S. A. E. Lewis, S. V. Chernikov, and J. R. Taylor, Electron. Lett., 35 (1999) 1761.
108. S. Namiki and Y. Emori, IEEE J. Selected Topics in Quantum Electron., 7 (2001) 3.
109. T. Kimura, N. Tsukiji, A. Iketani, N. Kimura, H. Murata, and Y. Ikegami, "High temperature operation quarter watt 1480 nm pump LD module," in Optical Amplifiers and their Applications, OSA Technical Digest Series (Optical Society of America, Washington, D.C., 1999), paper ThD12.
110. S. Namiki, Y. Ikegami, Y. Shirasaka, and I. Oh-ishi, "Highly coupled high power pump laser modules," in Optical Amplifiers and their Applications, OSA Technical Digest Series (Optical Society of America, Washington, D.C., 1993), paper MD5.
111. S. Koyanagi, A. Mugino, T. Aikiyo, and Y. Ikegami, "The ultra high-power 1480 nm pump laser diode module with fiber Bragg grating," in Optical

Amplifiers and their Applications, OSA Technical Digest Series (Optical Society of America, Washington, D.C., 1998), paper MC2.
112. K. Tanaka, K. Iwashita, Y. Tashiro, S. Namiki, and S. Ozawa, "Low-loss integrated Mach-Zehnder interferometer-type eight-wavelength multiplexer for 1480-nm band pumping," in Optical Fiber Communication Conference, OSA Technical Digest Series (Optical Society of America, Washington, D.C., 1999), paper TuH5.
113. Y. Tashiro, S. Koyanagi, K. Aiso, and S. Namiki, "1.5 W erbium doped fiber amplifier pumped by the wavelength division-multiplexed 1480 nm laser diodes with fiber Bragg grating," in Optical Amplifiers and their Applications, OSA Technical Digest Series (Optical Society of America, Washington, D.C., 1998), paper WC2.
114. Y. Emori, S. Matsushita, and S. Namiki, "1-THz-spaced multi-wavelength pumping for broadband Raman amplifiers," in 26th European Conference on Optical Communications, Proceedings, Vol. 2, Paper 4.4.2 (2000) 73.
115. H. Kidorf, K. Rottwitt, M. Nissov, M. Ma, and E. Rabarijaona, IEEE Photon. Technol. Lett., 11 (1999) 530.
116. B. Min, W. J. Lee, and N. Park, IEEE Photon. Technol. Lett., 12 (2000) 1486.
117. M. Sakano, A. Aikawa, and S. Namiki, internal memo, Furukawa Electric Co., Ltd. (2001).
118. M. Achtenhagen, T. G. Chang, B. Nyman, and A. Hardy, Appl. Phys. Lett., 78 (2001) 1322.
119. S. A. E. Lewis, S. V. Chernikov, and J. R. Taylor, Electron. Lett., 35 (1999) 1178.
120. Y. Emori, K. Tanaka, and S. Namiki, Electron. Lett., 35 (1999) 1355.
121. S. Bigo, S. Gauchard, A. Bertaina, and J.-P. Hamaide, IEEE Photon. Technol. Lett., 11 (1999) 671.
122. C. R. S. Fludger, V. Handerek, and R. J. Mears, "Fundamental noise limits in broadband Raman amplifiers," in Optical Fiber Communication Conference, OSA Technical Digest Series (Optical Society of America, Washington, D.C., 2001), paper MA5.
123. R. E. Neuhauser, P. M. Krummrich, H. Bock, and C. Glingener, "Impact of nonlinear pump interactions on broadband distributed Raman amplification," in Optical Fiber Communication Conference, OSA Technical Digest Series (Optical Society of America, Washington, D.C., 2001), paper MA4.
124. R. Hainberger, J. Kumasako, K. Nakamura, T. Terahara, and H. Onaka, "Optimum span configuration of Raman-amplified dispersion-managed fibers," in Optical Fiber Communication Conference, OSA Technical Digest Series (Optical Society of America, Washington, D.C., 2001), paper MI5.
125. T. Okuno, T. Tsuzaki, and M. Nishimura, "Novel lossless optical transmission line with distributed Raman amplification," in 26th European Conference on Optical Communications, Proceedings, Vol. 2, Paper 4.4.3 (2000) 75.

126. A. Kasukawa, Chapter 3, WDM Technologies: Active Optical Components, (A. K. Dutta, N. K. Dutta, and M. Fujiwara, Eds.) (Academic Press, Boston, 2002).
127. J. Yoshida, N. Tsukiji, A. Nakai, T. Fukushima, and A. Kasukawa, "Highly Reliable High Power 1480 nm Pump Lasers," in Testing, Reliability, and Application of Optoelectronic Devices (SPIE, Society of Photo-Optical Instrumentation Engineers, Bellingham, WA, 2001) Proceedings of SPIE, Vol. 4285, 146–158.
128. A. Mathur, M. Ziari, and V. Dominic, "Record 1 Watt fiber-coupled-power 1480 nm diode laser pump for Raman and erbium doped fiber amplification," in Optical Fiber Communication Conference, OSA Technical Digest Series (Optical Society of America, Washington, D.C., 2000), postdeadline Paper PD15.
129. D. Garbuzov, R. Menna, A. Komissarov, M. Maiorov, V. Khalfin, A. Tsekoun, S. Todorov, and J. Connolly, "1400–1480 nm ridge-waveguide pump lasers with 1 Watt CW output power for EDFA and Raman amplification," in Optical Fiber Communication Conference, OSA Technical Digest Series (Optical Society of America, Washington, D.C., 2001), postdeadline Paper PD18.
130. S. G. Grubb, S. T. A, W. Y. Cheung, W. A. Reed, V. Mizrahi, T. Erdogan, P. J. Lemaire, and A. M. Vengsarkar, "High-power 1.48 μm cascaded Raman laser in germanosilicate fibers," in Optical Amplifiers and their Applications, OSA Technical Digest Series (Optical Society of America, Washington, D.C., 1995), paper SaA4.
131. H. Po, J. D. Cao, B. M. Lliberte, R. A. Minns, R. F. Robinson, B. H. Rockney, R. R. Tricca, and Y. H. Zhang, Electron. Lett., 29 (1993) 1500.
132. A. J. Stentz, "Raman and cladding-pumped fiber amplifiers and lasers," in Optical Fiber Communication Conference, OSA Technical Digest Series (Optical Society of America, Washington, D.C., 1999), tutorial paper ThR.
133. E. M. Dianov, M. V. Grekov, I. A. Bufetov, S. A. Vasiliev, O. I. Medvedkov, V. G. Plotnichenko, V. V. Koltashev, A. V. Belov, M. M. Bubnov, S. L. Semjonov, and A. M. Prokhorov, Electron. Lett., 33 (1997) 1542.
134. D. I. Chang, D. S. Lim, M. Y. Jeon, Hak Kyu Lee, K. H. Kim, and T. Park, Electron. Lett., 36 (2000) 1356.
135. M. D. Mermelstein, C. Headly, J.-C. Bouteiller, P. Steinvurzel, C. Horn, K. Feder, and B. J. Eggloton, "A high-efficiency power-stable three-wavelength configurable Raman fiber laser," in Optical Fiber Communication Conference, OSA Technical Digest Series (Optical Society of America, Washington, D.C., 2001), postdeadline paper PD3.
136. T. Matsuda, M. Murakami, and T. Imai, Electron. Lett., 37 (2001) 237.
137. R. Ohhira, Y. Yano, A. Noda, Y. Suzuki, C. Kurioka, M. Tachigori, S. Moribayashi, K. Fukuchi, T. Ono, and T. Suzaki, "40 Gbit/s × 8-ch NRZ WDM transmission experiment over 80 km × 5-span using distributed

Raman amplification in RDF," in 25th European Conference on Optical Communications, Proceedings, Vol. 2 (1999) 176.

138. I. Morita, K. Tanaka, N. Edagawa, and M. Suzuki, Electron. Lett., 36 (2000) 2084.
139. I. Morita, K. Tanaka, and N. Edagawa, Electron. Lett., 37 (2001) 507.
140. H. Suzuki, J. Kani, H. Masuda, N. Takachio, K. Iwatsuki, Y. Tada, and M. Sumida, IEEE Photon. Technol. Lett., 12 (2000) 903.
141. A. K. Srivastava, S. Radic, C. Wolf, J. C. Centanni, J. W. Sulhoff, K. Kantor, and Y. Sun, "Ultra-dense terabit capacity WDM transmission in L-band," in Optical Fiber Communication Conference, OSA Technical Digest Series (Optical Society of America, Washington, D.C., 2000), postdeadline paper PD27.
142. T. Terahara, T. Hoshida, J. Kumasako, and H. Onaka, "128 × 10.66 Gbit/s transmission over 840-km standard SMF with 140-k optical repeater spacing (30.4-dB loss) employing dual-band distributed Raman amplification," in Optical Fiber Communication Conference, OSA Technical Digest Series (Optical Society of America, Washington, D.C., 2000), postdeadline paper PD28.
143. B. Zhu, P. B. Hansen, L. Leng, S. Stulz, T. N. Nielsen, C. Doerr, A. J. Stentz, D. S. Vengsarkar, Z. J. Chen, D. W. Peckham, and L. Gruner-Nielsen, Electron. Lett., 36 (2000) 1860.
144. Y. Zhu, W. S. Lee, C. Scahill, C. Fludger, D. Watley, M. Jones, J. Homan, B. Shaw, and A. Hadjifotiou, Electron. Lett., 37 (2001) 43.
145. B. Zhu, L. Leng, L. E. Nelson, Y. Qian, S. Stulz, C. Doerr, L. Stulz, S. Chandrasekar, S. Radic, D. Vengsarkar, Z. Chen, J. Park, K. Feder, H. Thiele, J. Bromage, L. Gruner-Nielsen, and S. Knudsen, "3.08 Tb/s (77 × 42.4 Gb/s) transmission over 1200 km of non-zero dispersion shifted fiber with 100-km spans using C- and L-band distributed Raman amplification," in Optical Fiber Communication Conference, OSA Technical Digest Series (Optical Society of America, Washington, D.C., 2001), postdeadline paper PD23.
146. I. Haxell, M. Ding, A. Akhtar, H. Wang, and P. Farrugia, "52 × 12.3 Gbit/s DWDM transmission over 3600 km of True Wave fiber with 100 km amplifier spans," in Optical Amplifiers and their Applications, OSA Technical Digest Series (Optical Society of America, Washington, D.C., 2000), postdeadline paper PD5.
147. M. Ma, H. D. Kidorf, K. Rottwitt, F. W. Kerfoot, III, and C. R. Davidson, IEEE Photon. Technol. Lett., 10 (1998) 893.
148. T. Miyakawa, N. Edagawa, and M. Suzuki, "210 Gbit/s (10.7 Gbit/s × 21 WDM) transmission over 1200 km with 200 km repeater spacing for the festoon undersea cable system," in Optical Fiber Communication Conference, OSA Technical Digest Series (Optical Society of America, Washington, D.C., 2000), paper FC5.

149. J.-P. Blondel, F. Boubal, E. Brandon, L. Buet, L. Labrunie, P. Le Roux, and D. Toullier, "Network application and system demonstration of WDM systems with very large spans (Error-free 32 × 10 Gbit/s 750 km transmission over 3 amplified spans of 250 km)," in Optical Fiber Communication Conference, OSA Technical Digest Series (Optical Society of America, Washington, D.C., 2000), postdeadline paper PD31.
150. T. Miyakawa, I. Morita, K. Tanaka, H. Sakata, and N. Edagawa, "2.56 Tbit/s (40 Gbit/s × 64 WDM) unrepeatered 230 km transmission with 0.8 bit/s/Hz spectral efficiency using low-noise fiber Raman amplifier and 170 μm^2-A_{eff} fiber," in Optical Fiber Communication Conference, OSA Technical Digest Series (Optical Society of America, Washington, D.C., 2001), postdeadline paper PD26.
151. E. Brandon, J.-P. Blondel, F. Boubal, L. Buet, V. Harvard, A. Hugbart, L. Labrunie, L. Le Roux, D. Toullier, and R. Uhel, "1.28 Tbit/s (32 × 40 Gbits) unrepeatered transmission over 250 km," in 26th European Conference on Optical Communications, Proceedings, Vol. 2, Paper 10.1.4 (2000) 21.
152. F. Boubal, E. Brandon, L. Buet, V. Havard, L. Labrunie, P. Le Roux, and J.-P. Blondel, "Broadband (32 nm) 640 Gbit/s unrepeatered transmission over 300 km with distributed dual wavelength Raman pre-amplification," in Optical Amplifiers and their Applications, OSA Technical Digest Series (Optical Society of America, Washington, D.C., 2000), postdeadline paper PD8.
153. K. Shimizu, K. Kinjo, N. Suzuki, K. Ishida, S. Kajiya, K. Motoshima, and Y. Kobayashi, "Fiber-effective-area managed fiber lines with distributed Raman amplification in 1.28-Tb/s (32 × 40 Gbit/s), 202 km unrepeatered transmission," in Optical Fiber Communication Conference, OSA Technical Digest Series (Optical Society of America, Washington, D.C., 2001), paper TuU2.
154. B. Mikkelsen, G. Raybon, and R.-J. Essiambre, "160 Gbit/s TDM transmission systems," in 26th European Conference on Optical Communications, Proceedings, Vol. 2 (2000) 176.
155. G. Raybon, B. Mikkelsen, B. Zhu, R.-J. Essiambre, S. Stulz, A. Stentz, and L. Nelson, "160 Gb/s TDM transmission over record length of 400 km fiber (4 × 100 km) using distributed Raman amplification only," in Optical Amplifiers and their Applications, OSA Technical Digest Series (Optical Society of America, Washington, D.C., 2000), postdeadline paper PD1.
156. G. Raybon, B. Mikkelsen, R.-J. Essiambre, A. J. Stentz, T. N. Nielsen, D. W. Peckham, L. Hsu, L. Gruner-Nielsen, K. Dreyer, and J. E. Johnson, "320 Gbit/s single-channel pseudo-linear transmission over 200 km of nonzero-dispersion fiber," in Optical Fiber Communication Conference, OSA Technical Digest Series (Optical Society of America, Washington, D.C., 2000), postdeadline paper PD29.

157. I. Haxell, N. Robinson, A. Akhtar, M. Ding, and R. Haigh, "2410 km all-optical network field trial with 10 Gb/s DWDM transmission," in Optical Fiber Communication Conference, OSA Technical Digest Series (Optical Society of America, Washington, D.C., 2000), postdeadline paper PD41.
158. L. D. Garrett, M. Eiselt, R. W. Tkach, V. Dominic, R. Waarts, D. Giltner, and D. Mehuys, "Field demonstration of distributed Raman amplification with 3.8 dB Q-improvement for 5 × 120 km transmission," in Optical Fiber Communication Conference, OSA Technical Digest Series (Optical Society of America, Washington, D.C., 2000), postdeadline paper PD42.
159. L. D. Garrett, M. Eiselt, R. W. Tkach, V. Dominic, R. Waarts, D. Giltner, and D. Mehuys, IEEE Photon. Technol. Lett., 13 (2001) 157.
160. U. Feiste, R. Ludwig, C. Schubert, J. Berger, C. Schmidt, H. G. Weber, B. Schmauss, A. Munk, B. Buchold, D. Briggmann, F. Kueppers, and F. Rumpf, Electron. Lett., 37 (2001) 443.
161. D. Chen, S. Wheeler, d. Nguyen, A. Färbert, A. Schöpflin, A. Richter, C.-J. Weiske, K. Kotten, P. M. Krummrich, A. Schex, and C. Glingener, "3.2 Tb/s field trial (80 × 40 Gb/s) over 3 × 82 km SSMF using FEC, Raman and tunable dispersion compensation," in Optical Fiber Communication Conference, OSA Technical Digest Series (Optical Society of America, Washington, D.C., 2001), postdeadline paper PD36.

Part 4 Critical Technologies

Chapter 11 | Semiconductor Monolithic Circuit

Yuzo Yoshikuni

NTT Photonics Laboratories, 3-1 Morinosato-wakamiya, Atsugishi, Kanagawa, 243-0198 Japan

11.1. Introduction

The explosive growth of Internet traffic makes the development of integrated photonic devices for next-generation photonic systems an important issue. The enormous number of devices that will be needed in future optical systems means that integrated devices will be indispensable if we want to make such systems at a reasonable cost and in a reasonable space. Monolithic integration on semiconductor substrates is attractive because it allows the semiconductor devices for optical systems, such as lasers, modulators, and photodetectors, to be included on a single chip. Starting from a simple integrated device containing a DFB laser and an EA modulator, various kinds of integrated photonic devices have been developed by using hybrid and monolithic technologies.

Historically, semiconductor integrated optical devices were developed for active devices based on semiconductor lasers. Fabrication technologies for semiconductor lasers were extended to make integrated devices such as multi-wavelength lasers, tunable lasers, and modulator-integrated lasers. Research showed that semiconductor materials also had attractive features for passive devices. The ability to control the refractive index in semiconductor materials in a wide range by changing material compositions has led to compact optical circuits; the bending radius can be reduced by making the refractive index difference between the core and cladding

WDM TECHNOLOGIES: PASSIVE
OPTICAL COMPONENTS
$35.00

ISBN: 0-12-225262-4

materials in waveguides large. The availability of a large refractive index in semiconductor devices has also contributed to the development of compact interferometers due to the enlarged optical path length. In spite of these advantages, the application of semiconductor materials to passive devices had been limited due to relatively large propagation and coupling losses, polarization dependence, and so on. Recent progress in fabrication technologies and device design, however, has made it possible to extend semiconductor monolithic technologies to passive devices. Fabrication technologies, such as crystal growth and dry etching, enable fabrication of waveguides with uniform and smooth interfaces, which significantly contributes to reducing propagation losses in waveguides. They also enable the precise control of waveguide thickness that is needed in order to fabricate polarization-insensitive waveguides. In addition, the recent invention of the arrayed waveguide grating (AWG) and the multi-mode interference (MMI) coupler enables us to make optical devices with significantly improved fabrication tolerance. With these technologies, we can extend the application of semiconductor materials to passive optical devices and, consequently, to complicated monolithic circuits containing both passive and active devices. This chapter describes basic technologies supporting semiconductor monolithic circuits and then introduces some typical examples of the semiconductor monolithic devices reported so far.

11.2. Semiconductor Materials for Optical Integrated Devices

Semiconductor active optical devices, which emit, amplify, or detect light, mostly use band-to-band transition of electrons. In this context, a semiconductor material for active optical devices should have band-gap energy that corresponds to the operation wavelength of light. A direct bandgap is also required in order to obtain a large transition probability. For these reasons, unlike electrical ICs, most semiconductor optical devices are made from III-V compound semiconductors. In compound semiconductors, we can obtain the desired band-gap energy or wavelength by changing the composition of the compound. This flexibility enables us to make optical devices that can emit, absorb, or amplify light at arbitrary wavelengths.

Semiconductor optical devices are fabricated by sequential growth of semiconductor layers with different compositions. Unlike glass materials

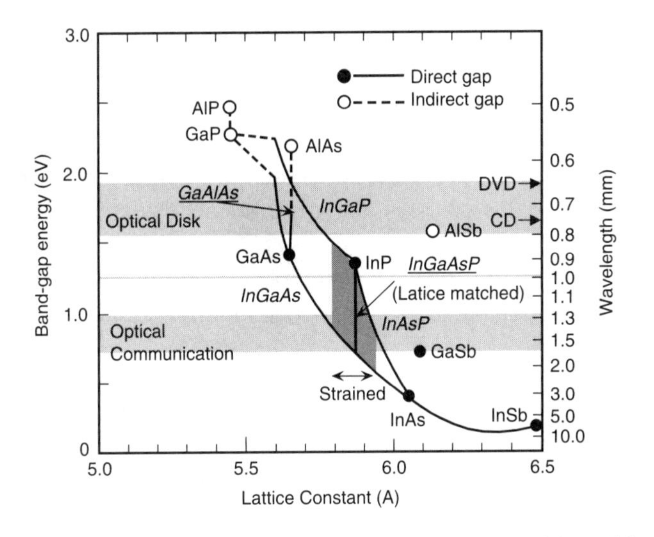

Fig. 11.1 Summarized characteristics of III-V semiconductor materials used for semiconductor monolithic circuits. The horizontal axis represents the lattice constant of crystals. The left vertical axis represents the band-gap energy of materials, and the right axis represents light wavelength corresponding to the band-gap energy. Curves show the characteristics of ternary or quaternary mixed crystal, obtained by interpolation.

such as silica, semiconductor monolithic devices are made of a single crystal. Although optical devices contain many semiconductor layers of differing compositions, the entire device forms a single crystal. In this context, lattice matching among different layers is important; each layer should have the same lattice constant, i.e., the same distance between neighboring unit cells, as the substrate crystal. This condition limits the wavelength range covered by a semiconductor device fabricated on a certain substrate material. Figure 11.1 summarizes the band-gap energy and consequent operation wavelength possibly obtainable in III-V compound materials as a function of lattice constant.

There are two principal substrate materials, indium phosphide (InP) and gallium arsenide (GaAs), for optical semiconductors. Optical devices on GaAs substrate use the ternary compound GaAlAs, which is a mixture of GaAs and AlAs. Because the lattice constant of GaAs and AlAs is almost the same, about 5.6 Å, GaAlAs with any composition can be grown on GaAs substrate. This fortunate coincidence makes fabrication of GaAs-based optical devices easy and is accelerating development of semiconductor optical devices. Light-emitting and light-detecting devices for wavelengths between about 0.7 and 0.9 μm can be fabricated with GaAlAs on GaAs substrates by adjusting the composition of the compound. Although band-gap energy of up to about 0.6 μm can be obtained on GaAs, compounds for wavelength shorter than about 0.7 μm have an indirect bandgap and are not suitable for optical devices. Quaternary compounds like GaAlInP cover this wavelength range. GaAs-based optical devices are mainly used in optical storage, lasers for compact disks at 780 nm, and for digital versatile disks (DVD) at 650 nm. They are seldom used for optical communication, with the exception of Mach-Zehnder-type modulators, which do not need optical absorption.

Optical devices on InP substrate use the quaternary compound InGaAsP, a mixture of InP, GaAs, and InAs. Thanks to two independent parameters, the ratio of In to Ga and that of P to As, the band-gap wavelength and the lattice constant can be independently adjusted. Thus, quaternary compounds of InGaAsP with arbitrary band-gap energies can be fabricated on InP substrates with a perfect lattice match to the substrates. The quaternary compounds of InGaAsP cover the wavelength range between about 1.0 and 1.7 μm, which includes all wavelengths used in optical fiber communications. By adjusting the composition, quaternary compounds of InGaAsP can be used either for active waveguides or for transparent passive waveguides. A compound whose band-gap wavelength is longer than the light wavelength exhibits strong light absorption or amplification, depending on the carrier density, and can be used for active devices. On the other hand, a compound whose band-gap wavelength is shorter than the light wavelength is transparent to the light and can be used for passive devices.

Controllability of the bandgap is also important in passive devices for obtaining materials with arbitrary refractive index. Waveguides in passive devices require materials with different refractive indexes for the core and cladding regions. Semiconductor materials with different band-gap wavelength are used for this purpose. The refractive index in a semiconductor strongly depends on the relative location of the light wavelength to the band-gap wavelength. Passive devices use semiconductor materials with band-gap wavelength longer than the operating wavelength to avoid band-to-band absorption. There are several empirical models describing the relationships between the material composition, the band-gap energy, and the refractive index in InP-based quaternary materials [1, 2]. Here we summarize these relationships as obtained by the modified single-effective-oscillation (MSEO) method [3]. Using some empirical parameters, the refractive index n can be approximated by,

$$n = \sqrt{1 + \frac{E_d}{E_0} + \frac{E_d \cdot E^2}{E_0^3} + \frac{\eta}{\pi} \cdot E^4 \cdot \ln\left(\frac{2 \cdot E_0^2 - E_0^2 - E_g^2 - E^2}{E_g^2 - E^2}\right)} \tag{11.1}$$

with

$$\begin{aligned} E_0 &= 3.391 - 1.652 \cdot y + 0.863 \cdot y^2 - 0.123 \cdot y \\ E_d &= 28.91 - 9.278 \cdot y + 5.626 \cdot y^2 \\ \eta &= \frac{\pi \cdot E_d}{2 \cdot E_0^3 \cdot \left(E_0^2 - E_g^2\right)} \end{aligned} \tag{11.2}$$

where $E_g = 1.24/\lambda_g$ is the band-gap energy, $E = 1.24/\lambda$ is the photon energy of the light, and y is the ratio of the amount of As to P in the compound, which is given by

$$y = \frac{0.72 - \sqrt{0.72 - 0.48(1.35 - E_g)}}{0.24}. \tag{11.3}$$

Figure 11.2 shows the refractive index variation in InGaAsP as a function of the light wavelength. The curves represent the band-gap wavelength denoted at the top of the figure. Note that the calculated refractive index at wavelengths close to or shorter than the bandgap is not accurate due to the limitation of the approximation. The refractive index is large at wavelengths close to the band-gap wavelength. It decreases rapidly at longer wavelength. Typically, optical waveguides in a semiconductor material use InP as a cladding material and InGaAsP as a core material. The composition of the core material is chosen from a band-gap wavelength shorter than the light wavelength. For example, compounds with a bandgap between about 1.0 and 1.5 μm can be used for a waveguide in a 1.55-μm optical communication band. The relative difference between the refractive indexes of the core and cladding is larger for longer band-gap wavelength and is about 1.5% for a core with a 1.0 μm bandgap and about 10% for 1.5 μm. A waveguide with a large index difference has strong light confinement. This leads to the possibility of a small bending radius with low bending loss, though the coupling loss to optical fibers is large and stripe width has to be extremely narrow to obtain single-mode waveguides. A waveguide with a small index difference, on the other hand, allows a wider stripe width and consequently a larger optical spot size, which makes fabrication easier and reduces optical fiber coupling loss.

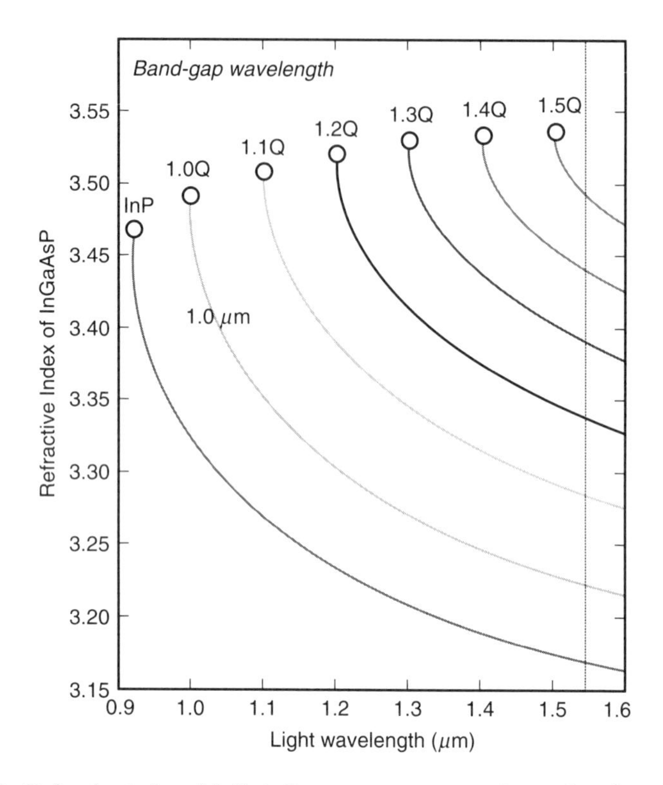

Fig. 11.2 Refractive index of InGaAsP quaternary compounds as a function of the light wavelength. Curves correspond to different composition in the compound. The labels on each curve, e.g., 1.0 Q, means quaternary material with the band-gap wavelength of 1.0 μm.

11.3. Semiconductor Waveguide Structure for Optical Integrated Devices

In addition to the variability of material compositions, semiconductor optical waveguides offer flexibility in selecting the waveguide structure. Figure 11.3 illustrates three typical waveguide structures used in many semiconductor monolithic circuits: the buried heterostructure, the ridge structure, and the deep-ridge structure. Semiconductor monolithic circuits typically use one of the three, depending on various factors. A small bending radius with negligible excess loss is requested in most applications to make sophisticated circuits and compact chips. In this context, a waveguide structure with strong optical confinement and consequently large refractive index difference is required for lateral direction. Strong optical confinement,

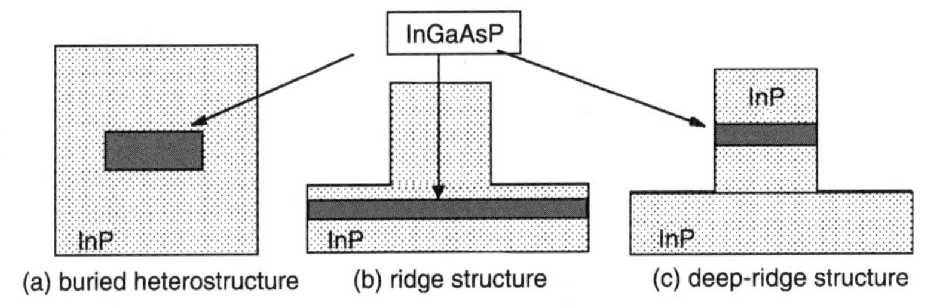

Fig. 11.3 Typical waveguide structures used in semiconductor monolithic circuits.

however, tends to support multi-mode operation, which causes a complicated problem due to the interference between modes. Polarization dependence is a severe issue in semiconductor waveguides. As the polarization of the light signal is uncertain and unstable after the light is transmitted through fiber, most devices in optical communication systems require polarization-independent operation. The polarization dependence of waveguide parameters causes polarization dependence of device characteristics. A semiconductor waveguide typically has an asymmetric structure for directions parallel (TE) and perpendicular (TM) to the substrate. This causes a dependence of the waveguide parameters on light polarization in the lateral and vertical directions. In what follows, the characteristics of the three waveguide structures are summarized.

(1) Buried heterostructure (BH): A BH waveguide consists of a core made of a longer band-gap wavelength compound semiconductor surrounded by cladding made of a shorter band-gap wavelength compound semiconductor. BH waveguides are fabricated in a sequential process that involves etching of the waveguide and regrowth of the cladding material. Most semiconductor active devices such as lasers and amplifiers use a BH structure for waveguiding, mainly for the reliability advantage. Passive waveguides with a BH structure, therefore, have the significant advantage of compatibility with active devices with respect to both the waveguide structure and fabrication process. The complicated fabrication process with regrowth, however, has limited BH application to passive waveguides. The minimum bending radius is relatively large due to weak optical confinement. A typical BH waveguide shows large birefringence due to asymmetry in the thickness and width of the core.

Figure 11.4 shows the effective refractive index of a BH waveguide as a function of waveguide width. Calculated optical fields are shown on the right. The calculation was carried out by the finite difference method (FDM)

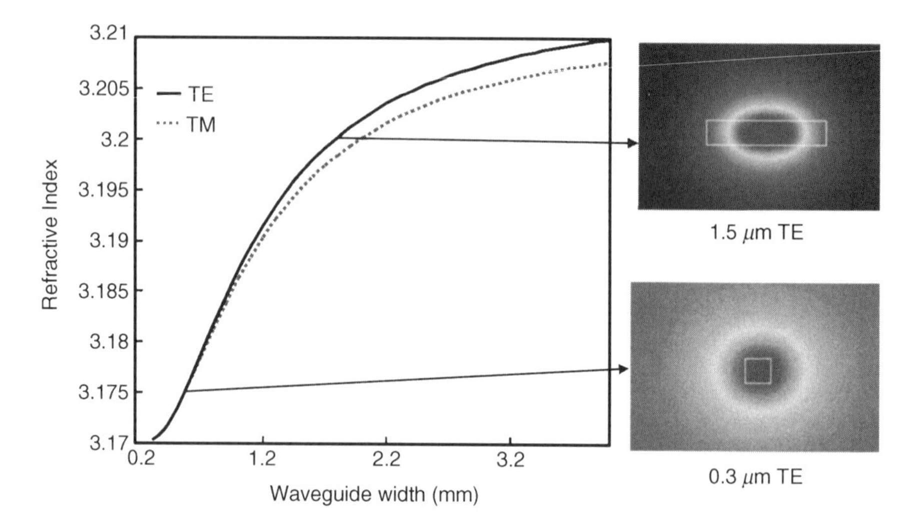

Fig. 11.4 Optical modes in buried heterostructure semiconductor waveguides calculated by the finite difference method. The left figure shows the equivalent refractive index as a function of the waveguide width. Photographs on the right show optical field profiles in the waveguide. (*See color plate*).

for a core layer containing 1.05 μm band-gap materials with a thickness of 0.3 μm and a cladding layer containing InP. A typical waveguide with width of 1 to 2 μm has a larger refractive index for TE polarization, reflecting the fact that the optical field extends in the lateral direction. This birefringence decreases for a narrower waveguide. The refractive index does not depend on the polarization at a width of 0.3 μm, as expected from the symmetry of the waveguide core.

(2) Ridge waveguide structure: A ridge waveguide consists of narrow ridge stripe formed on a two-dimensional slab waveguide. A significant advantage of the ridge waveguide is its simple fabrication process. Optical confinement in the lateral direction is weak due to small difference of the refractive index under the ridge stripe. The minimum bending radius is therefore large compared to a deep-ridge waveguide. This asymmetry in optical confinements causes inherent birefringence. Figure 11.5 shows the effective refractive index of a ridge waveguide as a function of waveguide width. Calculated optical fields are shown on the right. The calculation was carried out by the finite difference method (FDM) for a core layer containing 1.05 μm band-gap materials with thickness of 0.3 μm and a cladding layer containing InP. The refractive index for TE is larger than TE

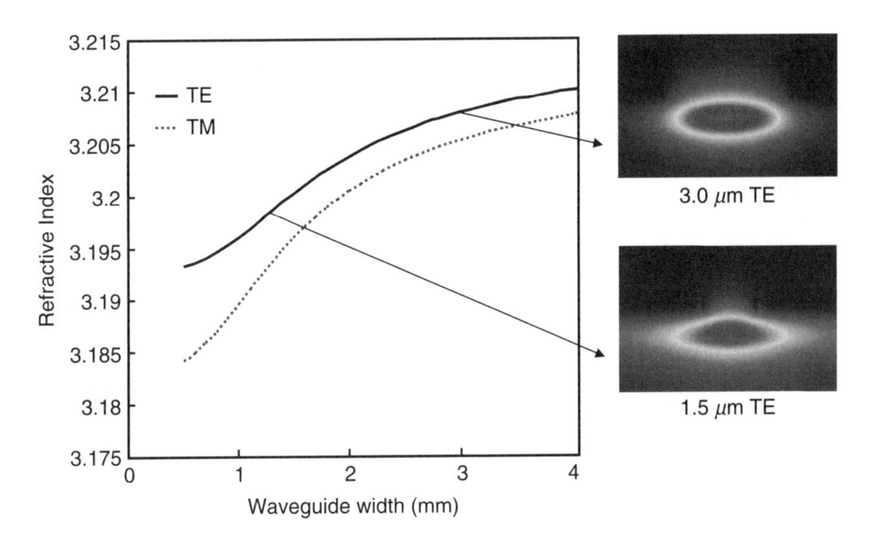

Fig. 11.5 Optical modes in a ridge-structure semiconductor waveguide calculated by the finite difference method. The left figure shows the equivalent refractive index as a function of waveguide width. Photographs on the right show optical field profiles in the waveguide. (*See color plate*).

for all widths. Even in narrow-stripe waveguides, the optical field extends out of the ridge due to the weak confinement.

(3) Deep-ridge waveguide: The deep-ridge waveguide consists of a deeply etched stripe containing the waveguide core. Optical confinement in the lateral direction is very strong, implying that a very small bending radius is possible with negligible bending loss. Although a typical deep-ridge waveguide has large birefringence, polarization-independent design is possible. Figure 11.6 shows the effective refractive index of BH waveguide as a function of the waveguide width. Calculated optical fields are shown on the right. The calculation was carried out by the finite difference method (FDM) for a core layer containing 1.05-μm band-gap materials with thickness of 0.3 μm and a cladding layer containing InP.

For wide waveguide, the optical field is flat and consequently, the refractive index is larger for TE than TM. By reducing the waveguide width, the field becomes circular and consequently the birefringence becomes small. The refractive index is polarization independent for the width of about 2.1 μ. The optical field of this non-birefringent waveguide resembles that of typical active devices. This feature brings a high coupling efficiency if this is connected to active devices.

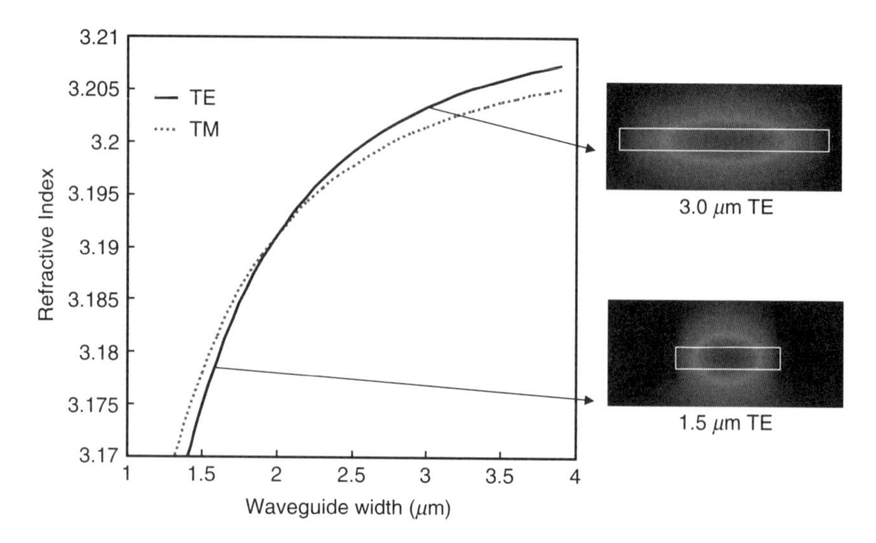

Fig. 11.6 Optical modes in a deep ridge semiconductor waveguide calculated by the finite difference method. The left figure shows the equivalent refractive index as a function of waveguide width. Photographs on the right show optical field profiles in the waveguide. (*See color plate*).

The waveguide structure for a monolithic circuit is chosen on the basis of various factors: minimum bending radius, compatibility with devices integrated in the circuit, polarization sensitivities, waveguide loss, and so on. In simple circuits, including semiconductor lasers such as modulator-integrated laser arrays, BH waveguides are normally used because of their compatibility with lasers. Deep-ridge waveguides are mainly used for complicated optical circuits such as arrayed waveguide gratings, because of the extremely small bending radius.

11.4. Semiconductor Arrayed Waveguide Grating (AWG)

Obviously, the significant motivation for developing optical integrated circuits is the progress being made in wavelength division multiplexing (WDM) technology. WDM technology enables ultra-wide-band transmission with 1 Tbit/s capacity and provides various attractive new functions for optical networks. The large number of devices needed in a WDM system, however, means that integrated optical devices are indispensable for practical implementation.

An important device in WDM systems is an optical filter, which provides multiplexing and demultiplexing functions. There are various types of optical filters, such as multi-layer interference filters, diffraction gratings [4, 5], fiber gratings [6], and array waveguide gratings (AWGs) [7]. The AWG, which is also called a PHASAR (phased array grating) or WGR (waveguide grating router), has significant advantages over the others. An AWG filter is a planar device that can be fabricated with processes similar to those used for electrical ICs. It is also a multi-channel filter that eliminates the cumulative losses that possibly occur if single-channel filters are serially connected. In addition, the multi-beam interference nature of an AWG possibly eases the distractive effect of the phase error in a waveguide. Finally, in principle, an AWG provides no reflection for input. These features are especially important in semiconductor monolithic circuits, making the AWG a key component for them.

The advantages of semiconductor-based AWGs are small device size and potential for integration with other semiconductor devices, such as photodiodes, semiconductor optical amplifiers (SOAs), and laser diodes. The channel crosstalk of a semiconductor-based AWG is about $-30\,\mathrm{dB}$, which may be of practical worth. Current research on semiconductor-based AWGs targets the improvement of characteristics and enhancement of functionality by integration with other semiconductor devices. These targets encompass increasing the maximum channel number, reducing the channel spacing, reducing crosstalk, improving miniaturization, reducing polarization sensitivity, reducing insertion loss, reducing temperature sensitivity [8], and improving the tuning characteristics. Research has also concentrated on monolithic integration with photodiodes, semiconductor optical amplifiers, laser diodes, modulators, optical switches, and so on.

11.4.1. STRUCTURE AND OPERATION PRINCIPLE OF AWGs

As shown in Figure 11.7, an AWG consists of input and output waveguides and two focusing slab waveguides connected by a waveguide array. The light from the input waveguide diffracts in the first slab waveguide, where it is divided and starts to propagate through the waveguides in the array. The array consists of tens of waveguides and the path length is longer for the outer waveguides. The path length difference between two adjacent waveguides is constant, ΔL. The light is injected into the second slab waveguide, where it is focused in the focal plane and collected by a waveguide.

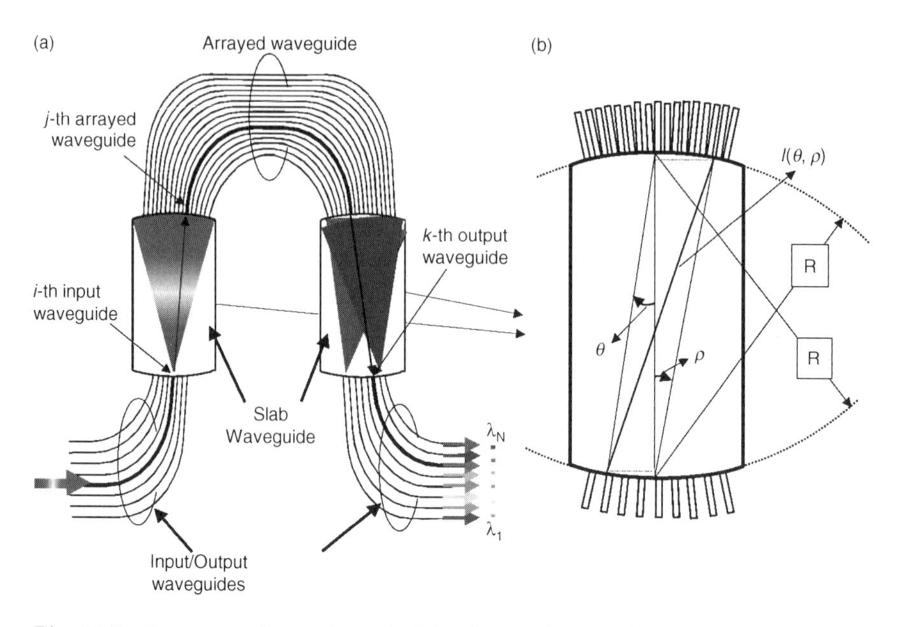

Fig. 11.7 Structure and operation principle of arrayed waveguide gratings (AWGs). The left figure shows the structure of an AWG. The right figure is an enlargement of the structure of the slab waveguide in AWG. A simple geometric analysis gives us an expression for distance between two points on the circles.

The position of the focal point varies according to wavelength because of the interference between lights traveling through paths of different length.

Let us start by investigating the light propagation in the slab waveguide using Fig. 11.7(b). The slab waveguide contains two circular interfaces connected to input/output and arrayed waveguides. The center of the circle locates on the other circle as shown in the figure. The light from a input waveguide is diffracted at the slab interface and is coupled to many waveguides in the array. The phase of the light in a waveguide in array depends on the distance between the input and the output waveguide. A simple geometric analysis gives the distance between two waveguides at an angler positions of θ and ρ on the output circles, $l(\theta, \rho)$ as,

$$l(\theta, \rho) = R\sqrt{(\sin(\theta) + \sin(\rho))^2 + (\cos(\theta) + \cos(\rho) - 1)^2}$$
$$\approx R(1 + \theta\rho), \tag{11.4}$$

where R is the radius of the circles.

The input light passes through the input slab, the arrayed waveguides, and the output slab. The propagation in the device changes the phase of the light depending on the optical path length. In the output slab, lights pass through different paths, interfering with one another, and the constrictive interference makes a focus at an arbitrary position on the output slab.

Let us consider the optical path depicted by the thick line in Fig. 11.7(a), where light comes through the i-th input port, propagates through the j-th waveguide in the arrayed-waveguide, and exits from the k-th output port. The total optical path length l_{ijk} is given by

$$\begin{aligned} l_{ijk} &= R(1+\theta_i \cdot j\theta_{\rm d}) \cdot n_{\rm s} + (L_0 - j \cdot \Delta L) \cdot n_{\rm g} + R(1 - j\theta_{\rm d} \cdot \theta_k) \cdot n_{\rm s} \\ &= 2Rn_{\rm s} + L_0 \cdot n_{\rm g} + [(\theta_i - \theta_k)d_{\rm a} \cdot n_{\rm s} - \Delta L \cdot n_{\rm g}] \cdot j \qquad (11.5) \end{aligned}$$

where $\theta_{i(k)}$ is the angular positions of the $i(k)$-th waveguide in the input (output) slab, $\theta_{\rm d}$ the angular difference in the adjacent waveguides in the arrayed waveguide, $d_{\rm a} = R\theta_{\rm d}$ the separation of the arrayed waveguides at the slab-array interface, L_0 the path length of the 0-th waveguide in the arrayed waveguide, ΔL the path length difference between adjacent waveguides in the array, and $n_{\rm s}$ ($n_{\rm g}$) is the effective refractive index in the slab (arrayed) waveguide.

As the input light is distributed to any waveguide in the array, there are many paths connecting the input and the output waveguides. Therefore, the light focuses at a point where the optical path lengths for all possible j are the same. This condition is satisfied, if the path length difference expressed by the square bracketed term in Eq. (11.5) equals the wavelength λ multiplied by an integer number m;

$$(\theta_k - \theta_i)d_{\rm a}n_{\rm s} + \Delta L n_{\rm g} = m\lambda, \qquad (11.6)$$

here m determines the grating order of the AWG. If there are no path length differences in the arrayed waveguide, i.e., if $\Delta L = 0$, the equation simply expresses the lens operation. Light injected at θ_i on the input slab focuses at $\theta_k = -\theta_i$, which means that the optical field at the input slab makes an inverted image at the output slab. The path length difference, ΔL, shifts focus position depending on the light wavelength. The solution of Eq. (11.6) for θ_k,

$$\theta_k = (m\lambda - \Delta L \cdot n_{\rm g})/d_{\rm a}n_{\rm s} + \theta_i, \qquad (11.7)$$

determines the focus position on the output slab. The center wavelengths λ_0^m, that is, the peak wavelength of the transmission for an m-th grating order with $\theta_k = \theta_i$, is determined by

$$\lambda_0^m = \Delta L \cdot n_{\rm g}/m, \qquad (11.8)$$

and the free spectral range (FSR) is given by

$$\Delta\lambda_{\mathrm{FSR}} = \Delta L \cdot n_{\mathrm{g}}\left(\frac{1}{m} - \frac{1}{m+1}\right) \approx \frac{\lambda_0^{m^2}}{\Delta L \cdot n_{\mathrm{g}}}. \tag{11.9}$$

The focus position on the output slab shifts depending on wavelength λ around λ_0^m as

$$\theta_k = \left(\lambda - \lambda_0^m\right) m / d_{\mathrm{a}} n_{\mathrm{s}} + \theta_i. \tag{11.10}$$

Accordingly, the light from the input port at θ_i focuses at θ_k depending on the light wavelength and consequently couples to the output port at the position.

Figure 11.8 shows the light propagation in the output slab waveguide in an AWG, calculated by the beam propagation method (BPM), assuming the light comes from the center of the input waveguides and with a wavelength of λ_0^m. The light comes from the arrayed waveguides (left side of the figure). The phase of the light at the slab interface is the same for all waveguides in the array. In this case, the light focuses at the center of the output circle and couples to the center waveguide. Two focusing points on either side of the center focusing point correspond to different order diffractions,

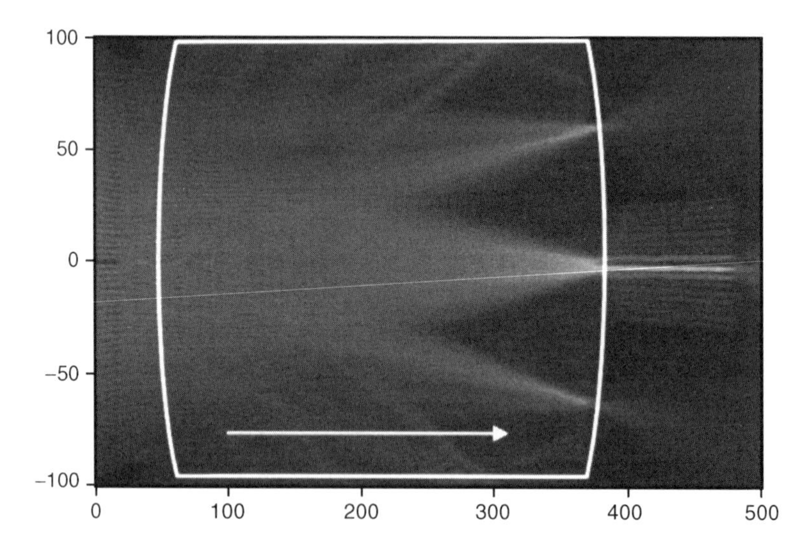

Fig. 11.8 Simulation of light propagation in the output slab waveguide of an AWG calculated by the beam propagation method (BPM). Light wavelength λ is the center in the FSR, i.e., $\lambda = \lambda_0^m$. (*See color plate*).

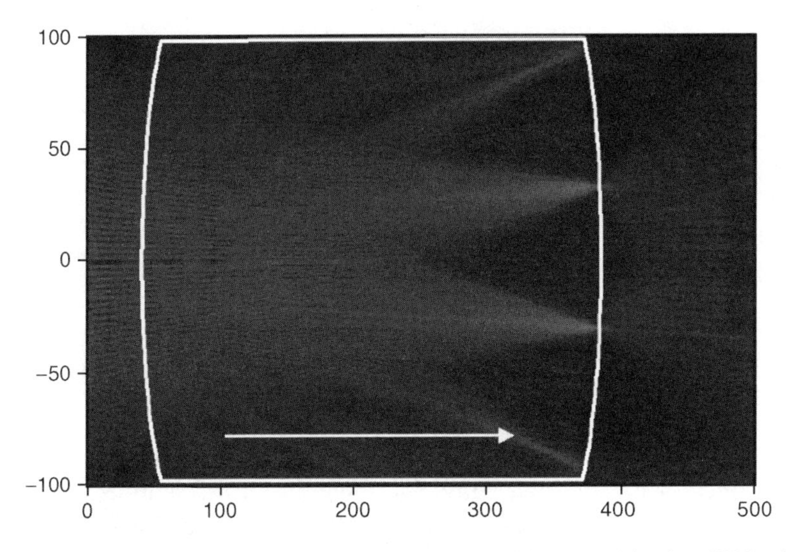

Fig. 11.9 Simulation of light propagation in the output slab waveguide of an AWG calculated by the beam propagation method (BPM). Light wavelength λ is the edge in the FSR, i.e., $\lambda = (\lambda_0^m + \lambda_0^{m+1})/2$. (*See color plate*).

i.e., $m+1$ and $m-1$. Although these side diffractions may cause significant loss, careful design of the width of the arrayed waveguide and FSR can minimize it.

Figure 11.9 shows the same simulation for light at the wavelength of the edge of the FSR. The path length difference in the array changes the light phase with 0 or π alternatively. In this case, the light focuses at the edge of the output circle and couples to the edge waveguide. Two focus points with equal intensity appear for the two different diffraction orders. Although one of the two focus points can couple to an output waveguide, light scattered at the other causes inherent loss of 3 dB for the edge wavelength of the FSR.

11.4.2. POLARIZATION DEPENDENCE OF SEMICONDUCTOR AWG

After the light signal transmits through the fiber, its polarization is uncertain and unstable. In this context, elimination of polarization dependence in optical systems and devices is important in order to avoid signal degradation due to polarization fluctuation. In AWG optical filters, the polarization dependence of the peak wavelength is the most important issue for WDM applications.

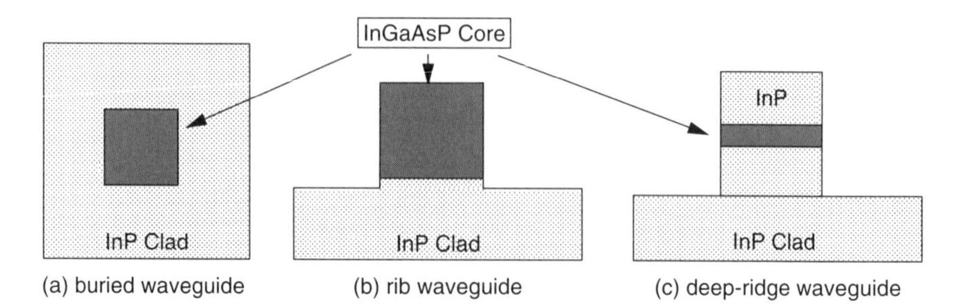

Fig. 11.10 Some examples of waveguide structures for the non-birefringent waveguide: (a) buried waveguide with a square core, (b) rib waveguide with a square-like core, and (c) deep-ridge waveguide with relatively narrow waveguide width.

As the polarization dependence in AWG arises from the polarization dependence of the effective refractive index in the waveguide, the most straightforward way to make a polarization-insensitive AWG is to use non-birefringent waveguides, which have an equal refractive index for any polarization states. Figure 11.10 shows three typical examples of non-birefringent waveguides. The buried square waveguide [9] (a) is widely used in silica-based and polymer-based AWGs. However, it is not suitable for semiconductor AWGs. A thick waveguide core is necessary in order to reduce bending loss, and it makes the fabrication and integration process difficult. Therefore, this waveguide is not usually used in semiconductor-based AWGs. The raised-stripe, or rib, waveguide [10] (b) also needs a thick (2 μm) waveguide core to eliminate the birefringence. In contrast, the polarization independent deep-ridge waveguide [11, 12] (c) can be either thick or thin. By selecting a thin (0.3–0.5-μm-thick) waveguide core, integration with other devices becomes easy because the spot size and thickness of the core layer are similar to those of common semiconductor laser diodes. In addition, the strong lateral confinement by air enables a small bending radius of less than 100 μm. This small bending radius enables us to fabricate extremely small AWGs. The waveguide structure is very simple, and only one epitaxial growth step and a single dry etching step are necessary in order to make the entire waveguide. However, strict control of the waveguide width is necessary in order to make the waveguide non-birefringent. Figure 11.11 shows the birefringence of the deep-ridge waveguide as a function of waveguide width. Although the birefringence has a strong dependence on the waveguide width, the slope of the birefringence becomes flatter as the composition wavelength of the core becomes shorter. When the composition wavelength of the InGaAsP core layer is 1.05 μm, the

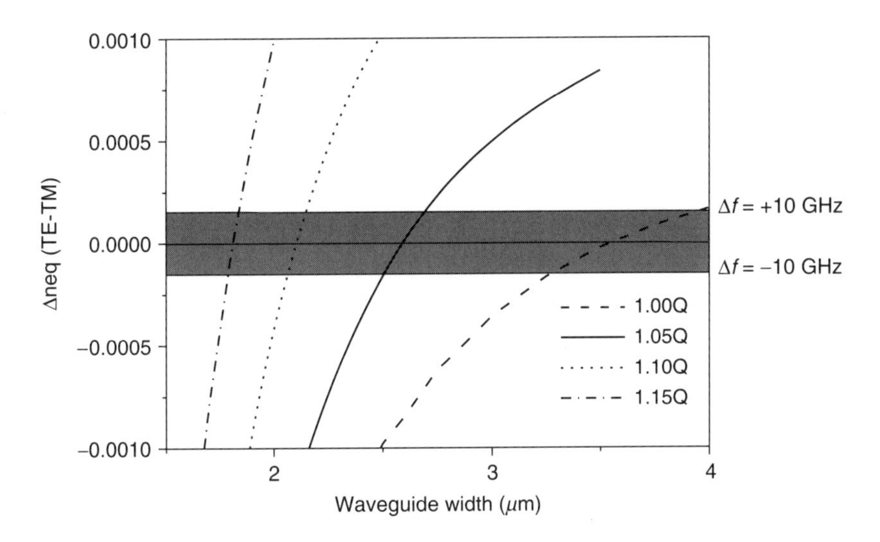

Fig. 11.11 Polarization dependence of the equivalent refractive index in a semiconductor deep-ridge waveguide as a function of the waveguide width. The four curves represent different compositions of the core material. Labels, such as 1.0 Q, mean core material with a band-gap wavelength of 1.0 μm. The vertical axis represents the polarization dependence of the peak frequency in an AWG fabricated with the waveguide. The shaded region represents the region of polarization dependence of less than 10 GHz.

fabrication tolerance is about $\pm 0.1\ \mu$m. Although the required accuracy for width control is still severe, recent progress in dry etching technology [13, 14] can possibly control the width with good reproducibility.

Although using non-birefringent waveguides is the simplest strategy, this is very difficult in some applications of the AWG. Integration with other optical devices requires a waveguide structure compatible with the device in terms of the optical properties and the processing. This makes application of non-birefringent waveguides difficult. In this case, we can use compensation technique to cancel out the birefringence of the waveguides by inserting either compensating waveguides [15, 16] or a half-wavelength plate [17] to cancel the birefringence of the waveguide. The latter method is widely used in the silica-based AWGs, but is difficult to use in semiconductor devices. The compensation of the birefringence in a semiconductor AWG mostly relies on the former method.

Figure 11.12 [16] shows the structure of a semiconductor AWG with birefringence compensation. A buried rib heterostructure waveguide is used in the AWG because of the compatibility with BH devices. The arrayed waveguide in the device consists of two different regions. Although both

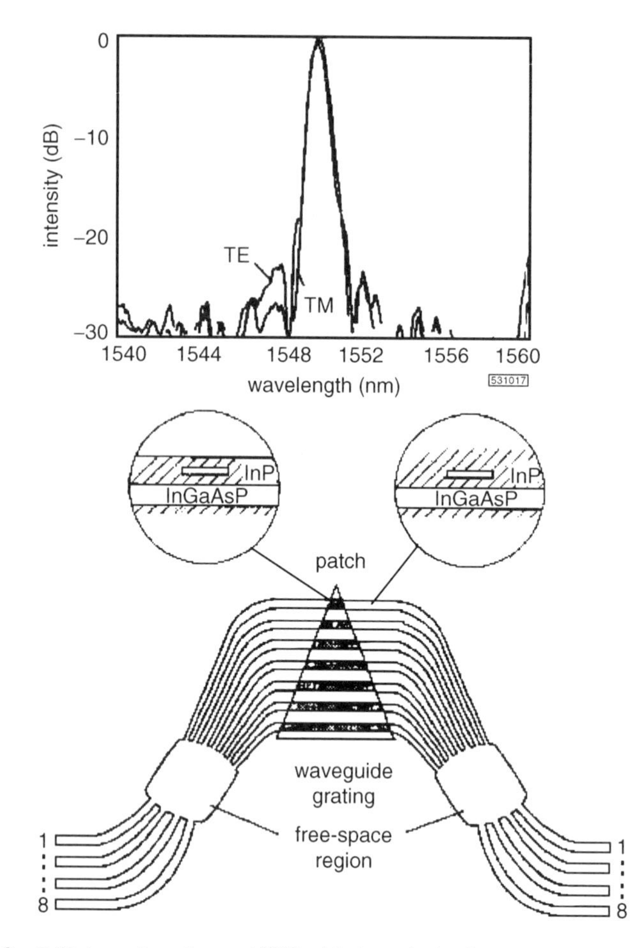

Fig. 11.12 [16] A semiconductor AWG with the polarization-compensation structure. A waveguide in the array consists of two different serially connected waveguides. Drawings in the middle show waveguide structures in two regions. The upper figure shows transmission spectra for TE and TM polarizations.

regions have the same waveguide structure, the upper clad of the waveguide is etched to 0.2 μm in a region with a triangular shape. The thinner clad stresses the optical field in the vertical direction and consequently enhances the birefringence in this region.

As waveguides in the array contain two different waveguide structures, the center wavelength of the AWG is given by an extension of Eq. (11.8) as

$$\lambda_0^m = (\Delta L_1 \cdot n_{g_1} + \Delta L_2 \cdot n_{g_2})/m, \tag{11.11}$$

with the equivalent refractive index $n_{g1(2)}$ in the two different regions.

If the refractive index n^{TE} for TE polarization is different from n^{TM} for TM polarization, the center wavelength of the AWG shows polarization dependence, and

$$\Delta\lambda_0^m \equiv \lambda_0^{m\mathrm{TE}} + \lambda_0^{m\mathrm{TM}} = (\Delta L_1 \cdot \delta n_{\mathrm{g}_1} + \Delta L_2 \cdot \delta n_{\mathrm{g}_2})/m, \quad (11.12)$$

with $\delta n = n^{\mathrm{TE}} - n^{\mathrm{TM}}$. As most semiconductor waveguides have a larger refractive index for TE than for TM, δn is always positive. The path length difference $\Delta L_{1(2)}$, however, can be positive or negative. This implies that the polarization dependence of the center wavelength can be canceled out if the length of the two waveguides is designed so that

$$\Delta L_1 \cdot \delta n_{\mathrm{g}_1} + \Delta L_2 \cdot \delta n_{\mathrm{g}_2} = 0. \quad (11.13)$$

Under this condition, center wavelength is given by

$$\lambda_0^m = \frac{\Delta L_1 \cdot (n_{\mathrm{g}_1} - n_{\mathrm{g}_2} \cdot \delta n_{\mathrm{g}_2}/\delta n_{\mathrm{g}_2})}{m}. \quad (11.14)$$

The solution simultaneously satisfying Eqs. (11.13) and (11.14) gives a design for a polarization-independent AWG at the desired center wavelength λ_0. The difference in the birefringence, i.e., $\delta n_{\mathrm{g}_1} \neq \delta n_{\mathrm{g}_2}$, or the difference in the refractive index, i.e., $n_{\mathrm{g}_1} \neq n_{\mathrm{g}_2}$, give a non-trivial solution to the equations. The waveguide in the thick clad region has a birefringence of 0.23%, implying that the conventional AWG fabricated by this waveguide has a peak wavelength difference of 3.5 nm, that is, 0.23% of 1.5 μm, for TE and TM polarization. This polarization dependence can be canceled out by the enlarged birefringence in the thin clad region of 0.54%, if the path-length difference is satisfied, $\Delta L_2 = 0.43\Delta L_1$ $(= 0.23/0.54\Delta L_1)$.

In principle, this technique can be applied for any kind of waveguide, a powerful option for making a polarization-insensitive AWG when use of the non-birefringent waveguide is difficult.

11.4.3. CHARACTERISTICS OF SEMICONDUCTOR AWGs

In a well-designed AWG, channel cross-talk is mainly determined by the phase error in the arrayed waveguides; if there is a fluctuation of the phase at the arrayed waveguides, the focused light at the second slab waveguide cannot be accurately focused on the output channels. Therefore, reduction of the phase error becomes very important. In general, the phase error becomes large as the refractive index difference between the core and cladding becomes large. This means that semiconductor-based waveguides usually

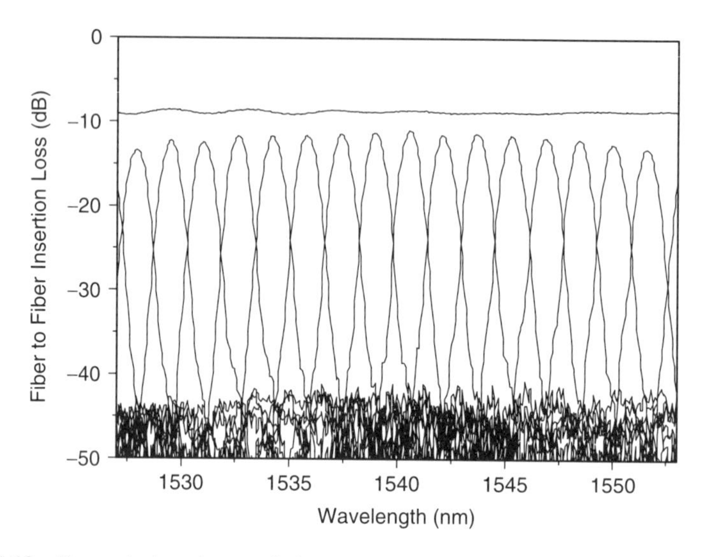

Fig. 11.13 Transmission characteristics of a 16-ch 200-GHz AWG as a function of the light wavelength. Spectra for 16 output ports overlap. The flat line at the top of the graph represents the transmission loss of the reference waveguide. The loss for the reference is mainly determined by coupling between the waveguide and the fibers.

are at a disadvantage as far as the amount of cross-talk is concerned. However, we can decrease the cross-talk by optimizing the waveguide structure and using high-quality fabrication techniques. Figure 11.13 shows the characteristic of a semiconductor-based AWG with 16 channels and 200-GHz channel spacing. The channel cross-talk is lower than −30 dB for all channels. The channel cross-talk values are comparable to those of silica-glass-based devices (−30 to −40 dB).

To enlarge transmission capacity in WDM systems, both a further reduction of channel spacing and a further increase of channel number are desirable. The reduction of channel spacing means a longer optical path and a larger device size in AWGs, which leads to fabrication difficulties and may increase channel cross-talk. Although the silica-base AWGs have achieved more than 100 [18] channels and an extremely small channel spacing of 10 GHz [19, 20], the phase error mentioned before, as well as limited wafer size, may limit the number of channels and channel spacing in the semiconductor AWGs. Recent fabrication technologies, however, enable us to make relatively large AWGs. Semiconductor-based AWGs with 64 channels with a channel spacing as small as 50-GHz have been fabricated [21]. Figure 11.14 shows a photograph of the 64-channel AWG. The device contains 64 input and output waveguides and an arrayed waveguide

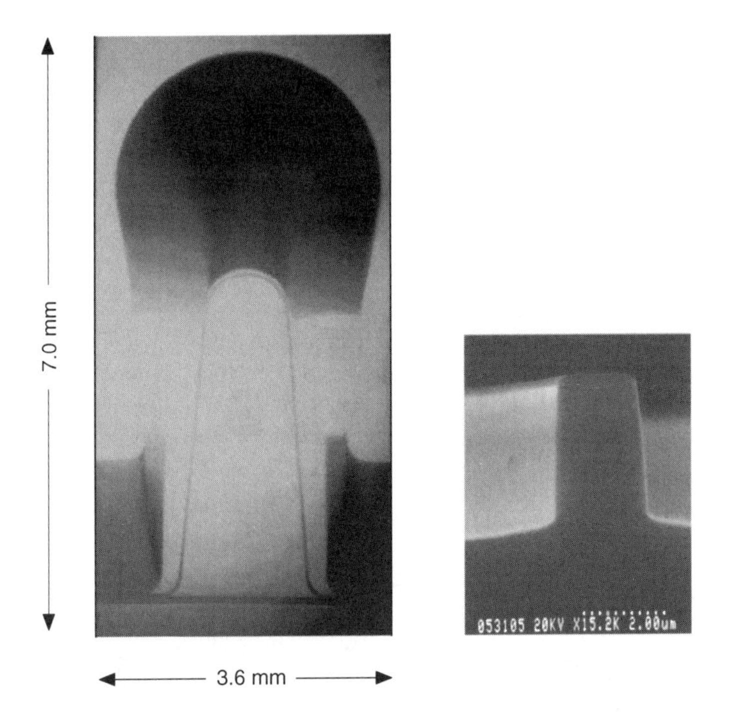

Fig. 11.14 [21] Photograph of a 64-ch 50-GHz channel spacing AWG. The AWG contains 64 input and output waveguides and an array with 232 waveguides in a 3.6 × 7.0 mm chip. The enlarged photograph on the right shows the intersection of a waveguide in the AWG.

with 232 waveguides on a small chip of 3.6 × 7.0 mm [2]. On the right is a photograph of a waveguide that was fabricated by using a Br_2-N_2 reactive ion beam etching. As can be seen, the sidewall is very smooth and almost perpendicular to the substrate, which are very important factors for obtaining uniform and low-loss waveguides and consequently making AWGs with a large number of channels and high resolution. Figure 11.15 shows transmission characteristics of a 64-ch 50-GHz spacing AWG as a function of the light wavelength. The bottom figure shows overlapped spectra for 64 output ports. The peak intensities of each channel are nearly uniform, and the cross-talk is less than −20 dB for all the channels, as shown in the bottom figure. The cross-talk is over −25 dB around the central channels. The insertion loss of the device is about −14.4 dB at the center output port and 2 dB lower for the outermost port. This high insertion loss is mainly due to the coupling loss at the fiber and the Fresnel reflections of the two uncoated facets. The estimated on-chip loss is about 5 dB. The top figure shows a transmission spectrum at the center for TE and TM polarization. The device exhibits polarization-insensitive characteristics.

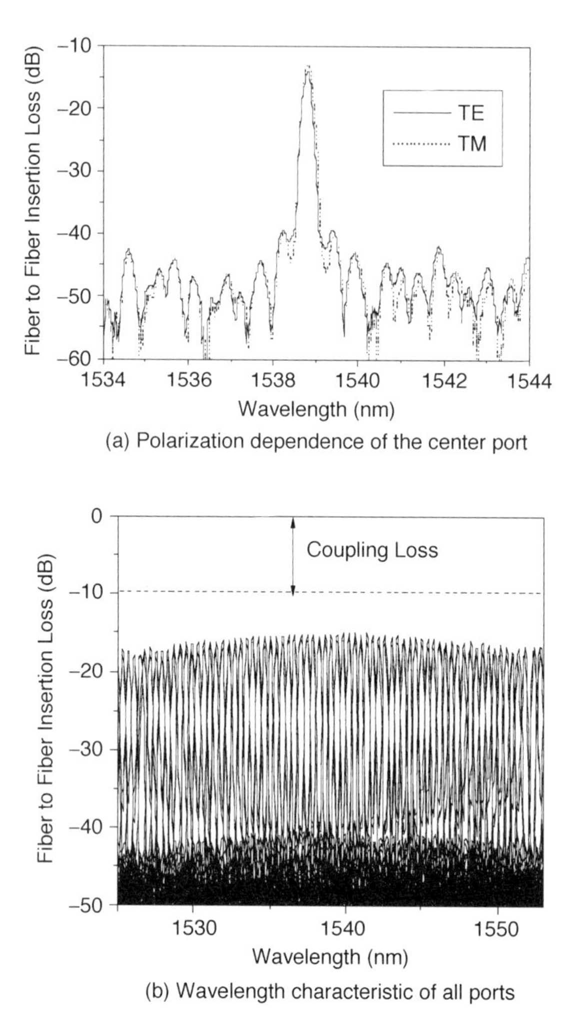

Fig. 11.15 Transmission characteristics of a 64-ch 50-GHz spacing AWG as a function of the light wavelength. The lower figure shows overlapped spectra for the 64 output ports. The upper figure shows a transmission spectrum at the center for the TE and the TM polarizations.

11.4.4. INSERTION LOSS OF SEMICONDUCTOR AWGs

The insertion loss is a serious problem in semiconductor-based AWGs when compared with silica-based AWGs. The loss comprises propagation loss, Fresnel loss (reflection at the semiconductor and air interface), coupling loss with fibers, and excess loss. Although the propagation loss per centimeter of a semiconductor waveguide is relatively large (0.5–a few dB/cm), a small device and consequent shorter waveguide partially compensate for the large propagation loss per unit length. The propagation loss, therefore, is not a significant issue at the present scale of AWGs. Although the Fresnel reflection at the as-cleaved facet of semiconductor waveguides causes a significant loss of 1.5 dB per facet, an anti-reflection coating can suppress it to a negligible level. Consequently, there remain two significant origins of loss in the semiconductor AWG: the coupling at the slab array interface and the coupling between a semiconductor waveguide and an optical fiber

The excess loss of the AWG is caused by inconsistencies in the field distribution of the array waveguides and slab waveguide. At the slab array interface, the expanded field in the slab waveguide couples to much narrower fields in many waveguides of the array. The coupling efficiency, therefore, is given by a sum of the overlap integral,

$$\eta = \sum_{i}^{N} \left(\int f(x) g_i(x) dx \right)^2, \tag{11.15}$$

where $f(x)$ and $g_i(x)$ are the optical field distributions in the slab and in the i-th waveguide of the array, respectively. The strong confinement in a deep-ridge waveguide gives us a simple approximation given by [21]

$$\eta \approx \frac{8W}{\pi^2(W+G)}. \tag{11.16}$$

where W is the waveguide width and G is the gap between waveguides. Figure 11.16 shows the excess loss of an AWG caused due to coupling at the slab array interfaces. Please note that the excess loss of AWG is twice the coupling loss because there are two interfaces, one at the input and the other at the output slab waveguides. Because scattering of light illuminating the gap between waveguides causes significant loss, reduction of the gap width significantly reduces the coupling loss. Excess loss as low as 2.8 dB has been obtained by making a very narrow gap of 0.3 μm [12]. The inset in the figure is a photograph of the etching shape at the junction between the deep-ridge waveguides and slab waveguide, where we can see very narrow gaps of about 0.3 μm are well formed.

Equation (11.16) predicts a remaining loss of $8/\pi^2$, ~1.8 dB, even for an infinitesimal gap width. This loss is inherent in deep-ridge waveguides because the strong confinement implies that the optical field is almost zero

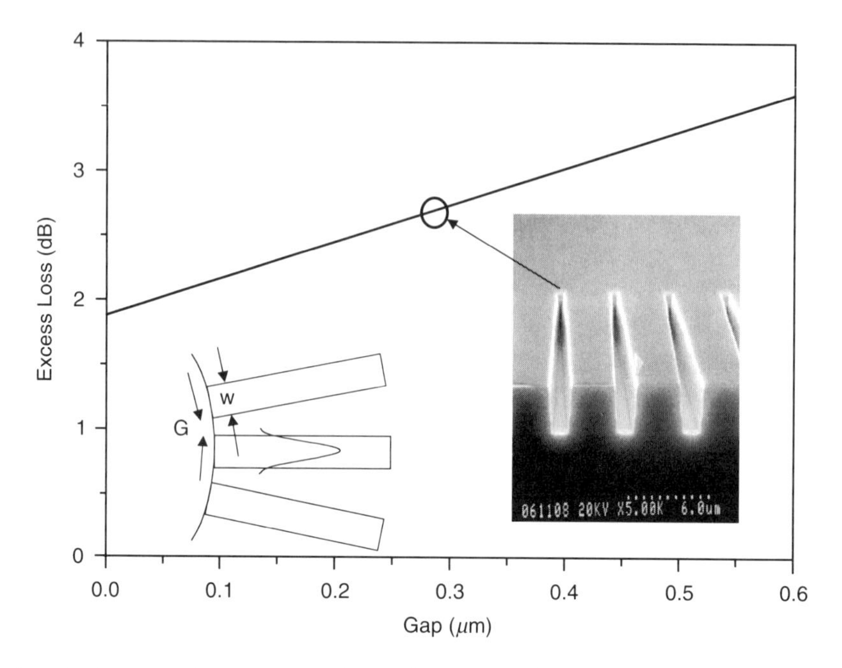

Fig. 11.16 Excess loss in semiconductor AWGs caused by the coupling loss between the slab and the arrayed waveguides as a function of the gap between adjacent waveguides in the array. The photograph shows the slab array interface in a fabricated device. The waveguide gap is as narrow as 0.3 μm, which reduces the excess loss to less than 3 dB.

at the edge of waveguide and that light illuminating the edge region cannot couple to any waveguides. A weak confinement waveguide, where the edge region has significant optical field, is therefore needed at the interface to remove this inherent loss. Introducing a shallow-ridge waveguide at the slab-array interface in an AWG has resulted in excess loss values as low as 0.4 dB [22, 23].

The other origin of the large loss in semiconductor AWG is the coupling between optical fibers and the waveguides, because a semiconductor waveguide has a very small spot size compared with that of a fiber. Figure 11.17 shows coupling loss between a deep-ridge waveguide and flat-ended and lensed fiber as a function of waveguide width. A large loss of over 4 dB/facet is estimated when a flat-ended fiber is used. On the other hand, a lensed fiber can drastically decrease coupling loss. However, the coupling tolerance becomes small, and stability problems arise in the coupling, plus mounting costs increase. A significant improvement can be obtained for the coupling loss and tolerance if we integrate the spot-size converter with the AWG. Although various types of spot-size converter have been reported,

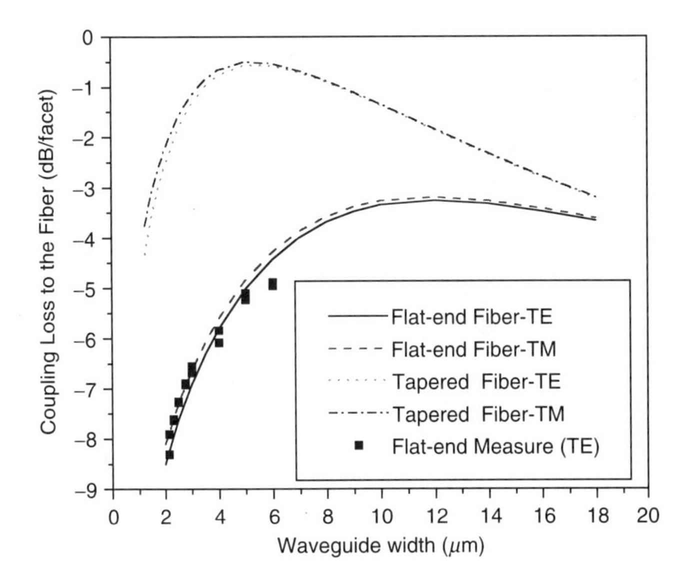

Fig. 11.17 Coupling loss between an optical fiber and a semiconductor deep-ridge waveguide as a function of waveguide width. Lines in the figure show calculated coupling losses from fiber to fiber, which are twice as large as the coupling loss from fiber and the waveguide. Coupling losses are shown for flat-end fibers and for tapered fibers with TE and TM polarizations. Experimental data are shown by squares.

most of them are optimized for integration with buried heterostrucure laser diodes. It is therefore difficult to integrate it with an AWG with high index difference waveguides, such as deep-ridge waveguides.

Recently, a few publications report a spot-size converter optimized for integration with AWGs. Figure 11.18 [24] shows a photograph of the spot-size converter-integrated AWG. As shown in the right figure, it consists of two different waveguide structures. At the spot-size converter section, the laterally tapered buried waveguide structure is used for effective spot-size conversion. At the AWG section, the deep-ridge waveguide structure is used for compactness and low loss. The converter section is 200-μm long, and it is relatively small compared with the total chip size.

Figure 11.19 [25] shows the wavelength characteristics of the device, where insertion loss between fiber-to-fiber was measured with flat-ended fiber on the input and output of the AWG. Insertion loss as small as 7 dB is obtained at the central channel without any lenses. The transmission peaks of the output channels are nearly uniform. By integrating the spot-size converter, the 1-dB-down coupling tolerance is enlarged to $\pm 2\ \mu$m, which is three times larger than in the lensed-fiber case. A fiber coupling loss of

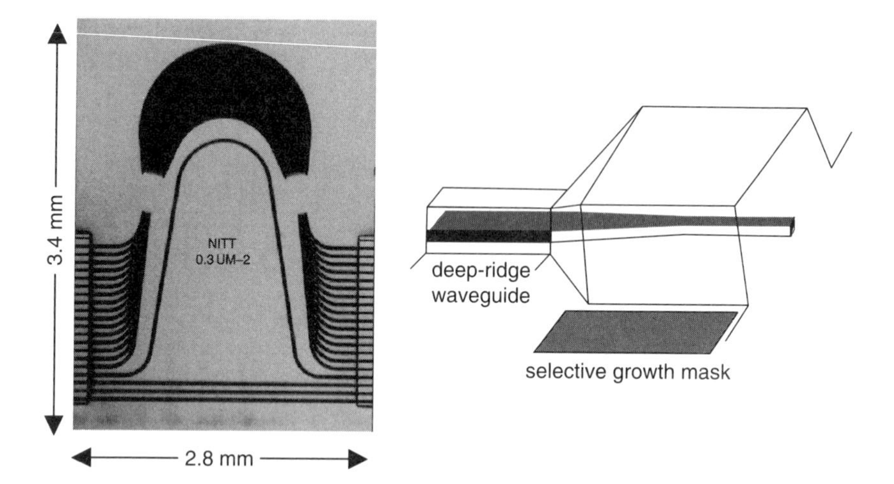

Fig. 11.18 Semiconductor AWG integrated with the spot-size converter. On the left is a photograph of the device. The right figure shows the structure of the spot-size converter.

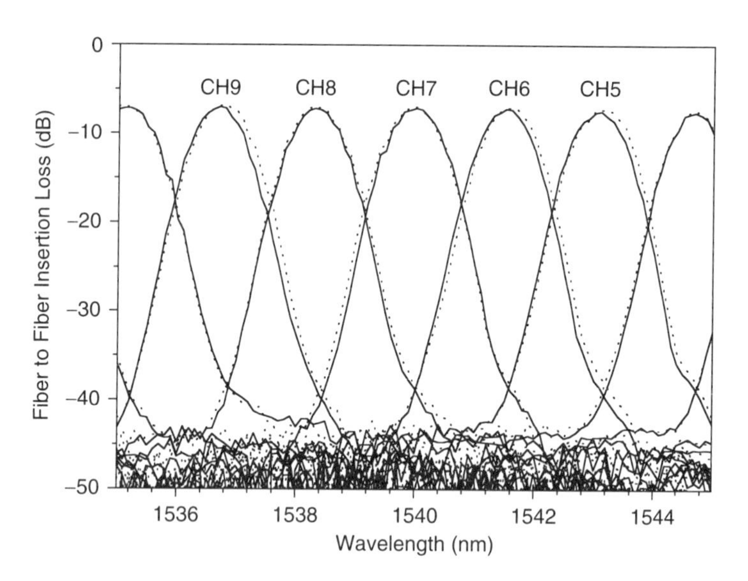

Fig. 11.19 [26] Transmission characteristics of an AWG integrated with a spot-size converter as a function of the light wavelength. Flat-ended fibers are coupled to the input and the output port at the center of the AWG. The vertical axis represents fiber-to-fiber insertion loss.

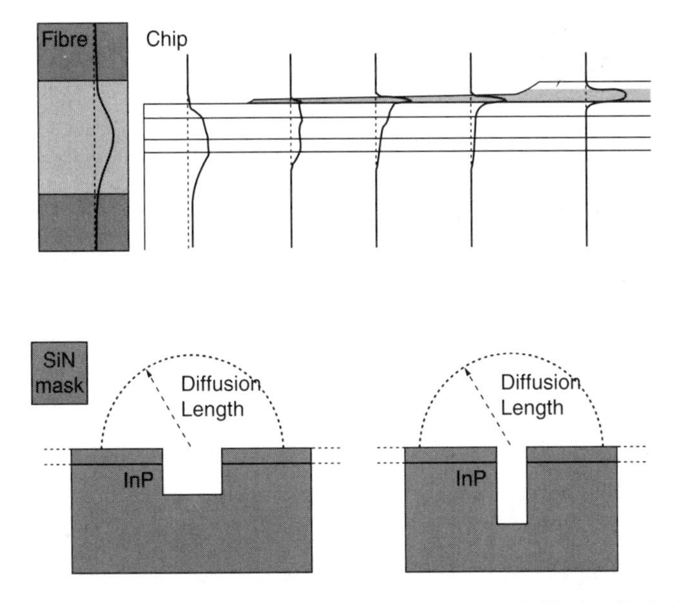

Fig. 11.20 [26] A spot-size converter with a vertical taper. Diffusion limited etching controls the thickness of the waveguide core locally (lower figure).

less than 2 dB/facet was measured, and further improvement is possible by optimizing the device parameters.

Figure 11.20 shows another example of a spot-size converter with the vertical taper. The waveguide has a multi-layer structure, forming two possible core layers that can confine the optical field vertically. In the waveguide region, light is confined in the relatively thick upper core layer. In the spot-size converter, the thickness of the upper core layer is tapered to zero. This forces the light into the lower core layer, which has a weak optical confinement and consequently a large optical spot size. Although the vertical taper has the same effect on the optical spot size, it requires a much more difficult fabrication process. The diffusion-limited etch shown in the lower part of the figure is used to make the vertical taper structure. The process is based on Br_2:CH_3OH etching at room temperature [26]. The etch rate in this condition is limited by the diffusion of the reactant material. If the etch region defined by a mask is smaller than the diffusion length, the volume of material etched out is constant in a unit time under the diffusion-limited condition. This implies that the etch rate depends on the mask width, so that etch rate is faster for a narrow etching region. Using this technique, a vertical taper can be successfully formed by changing the mask width for the etch process.

11.4.5. A TEMPERATURE-INSENSITIVE SEMICONDUCTOR AWG

The temperature dependence of wavelength in optical devices is a serious problem in building high-density wavelength division multiplexed (WDM) optical networks. The center wavelength in semiconductor optical filters has a typical temperature dependence of about 1 Å/°C. This coefficient is too high for these optical devices to be used without temperature control in WDM systems. Thus, a temperature-insensitive optical device is required. Compensation studies are now being carried out in dielectric material systems, such as athermal waveguides with a polymer [27] and temperature-insensitive fiber Bragg gratings with some packaging techniques [28]. Although such approaches are not suitable for semiconductor devices, reduction of the temperature sensitivity of semiconductor waveguides is still possible by using two different waveguides with different temperature sensitivities [29]. To cancel out the effect of the change in the optical path length, the same technique used for polarization compensation, described in Section 11.4.2, is applied to the temperature sensitivity problem. Let us consider an AWG that has two different waveguides in the array as shown in Fig. 11.21. All waveguides consist of two waveguides that have different values of temperature dependence of the refractive index dn/dT. Differentiating both sides of Eq. (11.11) with respect to temperature, T, we obtain the temperature dependence of the peak wavelength $d\lambda/dT$ as

$$\frac{d}{dT}(n_1 \cdot \Delta L_1 + n_2 \cdot \Delta L_2) = m \cdot \frac{d\lambda}{dT}. \tag{11.17}$$

Because the $d\Delta L/dT$ terms are more than one order of magnitude smaller than the dn/dT terms, we can neglect these terms. We then obtain

$$\frac{dn_1}{dT}\Delta L_1 + \frac{dn_2}{dT}\Delta L_2 = m \cdot \frac{d\lambda}{dT}. \tag{11.18}$$

Setting this expression equal to zero, we obtain a temperature-insensitive design of AWG, as follows:

$$\frac{\Delta L_1}{\Delta L_2} = -\frac{dn_2/dT}{dn_1/dT}. \tag{11.19}$$

The device shown in Fig. 11.21 is an 8×8 temperature-insensitive AWG designed and fabricated according to this concept. Here, 1.3-μm and 1.1-μm quaternary composition cores were used for the waveguide with high and low dn/dT values, respectively. The physical path length difference in the adjacent arrayed waveguides is 20 μm for obtaining a free

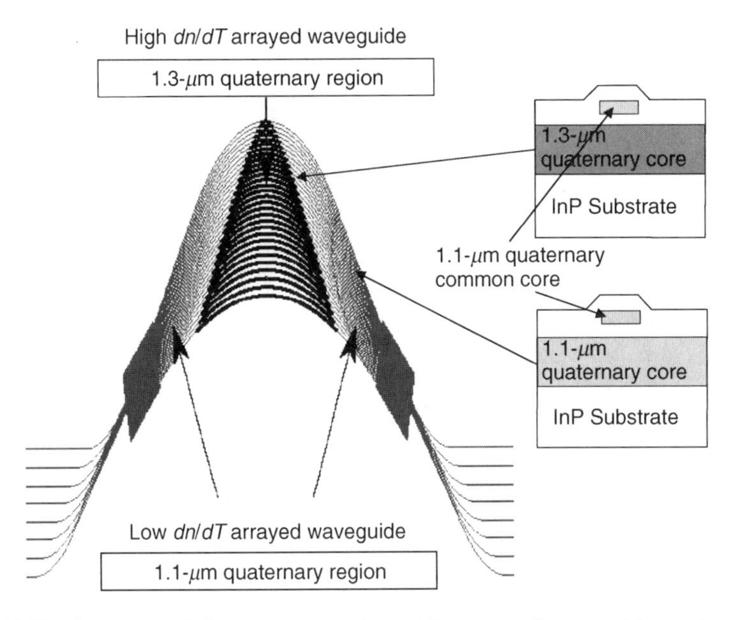

Fig. 11.21 Structure of the temperature-insensitive arrayed waveguide grating. The arrayed waveguide in the array consists of two different serially connected waveguides. On the left are the waveguide structures in the device. The waveguides have a buried rib structure with a striped core loaded on the slab core region. Although the striped core is common in both waveguides, compositions of the slab waveguides are different.

spectral range (FSR) of about 3.2 THz. Each waveguide has a buried-rib structure with different core materials. To make a junction between the 1.3- and 1.1-μm quaternary composition cores, 0.3-μm-thick core regions are butt-joined. Then, a 0.1-μm-thick intermediate InP layer is grown on the core layer. A common 1.1-μm quaternary composition rib, which is 0.23-μm thick and 1.5-μm wide, is laid on the InP layer. A Cl_2 RIE system was used to form the rib. Finally, the rib is buried with a 1.5-μm-thick InP layer. All of the layers are undoped epitaxial layers and are grown by low-pressure MOCVD. The observed free spectral range is 32.3 nm and peak spacing is 3.8 nm.

In Fig. 11.22 we can compare temperature dependence of a transmission peak in a conventional AWG on the left and in the temperature-insensitive AWG on the right. As the temperature increases, the spectrum of the conventional AWG shifts toward longer wavelengths. However, in the case of temperature-insensitive AWG, it remains unchanged even for a large temperature change between 25 and 75 degrees.

Figure 11.23 shows the temperature dependence of the peak wavelength for the eight different output ports in the temperature-insensitive AWG.

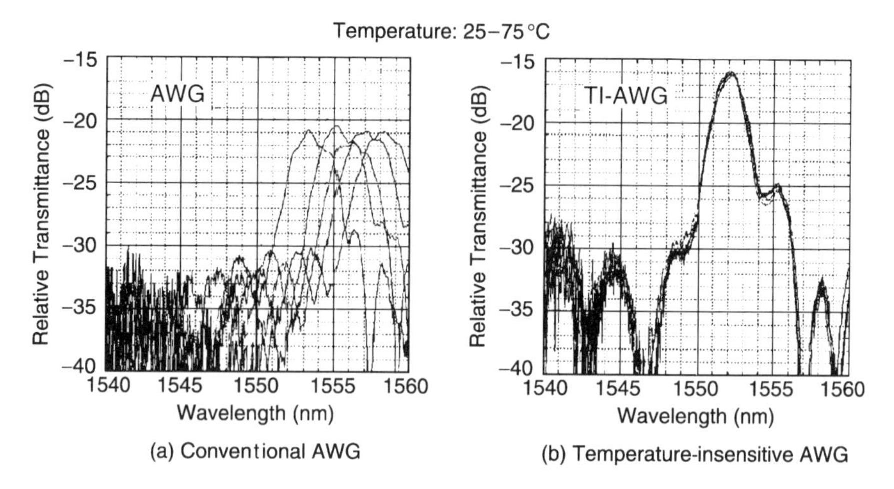

Fig. 11.22 Temperature dependence of transmission peak in a temperature-insensitive AWG (right) and a conventional semiconductor AWG (left). Each figure shows six overlapped spectra for different temperatures from 25 to 75 degrees centigrade with increments of 10 degrees.

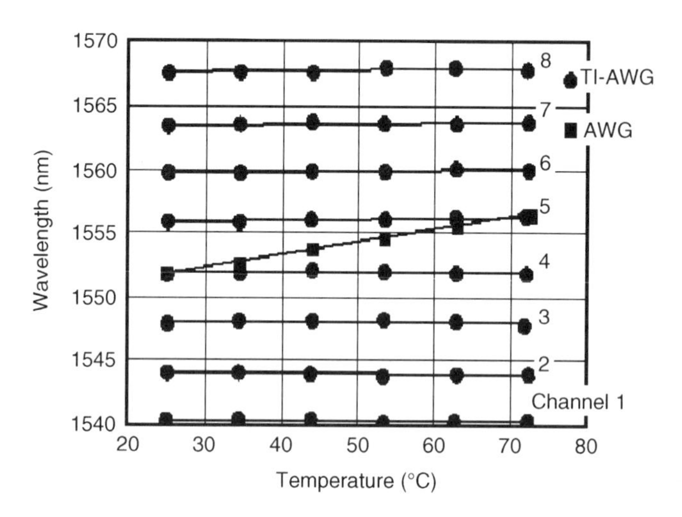

Fig. 11.23 Temperature sensitivities of peak wavelengths of the temperature-insensitive AWG. The eight lines represent peaks for different output ports. Temperature sensitivity of a conventional AWG is also shown as a reference.

Although the temperature variation of the device is large, the peak wavelengths stay almost perfectly constant for all eight output ports. The wavelength change is less than 0.1 Å/°C that is an order of magnitude smaller than that of conventional AWGs on InP substrates, as also indicated in the figure.

11.5. Multi-mode Interference (MMI) Couplers

Optical couplers, which split and/or combine optical signals, are important building blocks of optical integrated circuits. Although directional couplers have been used as optical couplers, they have some drawbacks in fabrication, especially for semiconductor devices. A directional coupler consists of two parallel waveguides coupled to each other through the overlapped evanescent field in their cladding layer. The field decreases rapidly with distance from the core layer in semiconductor waveguides, where the difference of the refractive index between core and clad is large. Therefore, severe control of the width and separation of two waveguides is required. In this context, a multi-mode interference (MMI) coupler, which has a much simpler structure and wider fabrication tolerance than the directional coupler, is a very attractive alternative. Although the basic concept of the MMI coupler was published in the 1970s [30, 31], the actual implementation in optical circuits has had to wait for the recent re-invention in the early 1990s [32].

Figure 11.24(a) shows the basic structure of a MMI coupler with two input and output ports. There are a lot of variations of the MMI structure, which changes according to the number of input and output waveguides. In an MMI coupler, the input and output waveguides are connected by a multi-mode waveguide. The light coming from an input waveguide couples to guided modes in the multi-mode waveguides at the interface of the input and the multi-mode waveguides. Figure 11.24(b) shows light propagation in the multi-mode waveguide. The optical field in the multi-mode waveguide can be expressed as a superposition of all propagating guided modes and can be written as

$$\varphi(y, z) = \sum_{i=0}^{n-1} c_i \phi_i(y) e^{-j\beta_i z}, \tag{11.20}$$

where $\phi_i(x)$ is the i-th mode profile, β_i the propagation constant of the mode, and c_i the expansion coefficient determined by the overlap integral

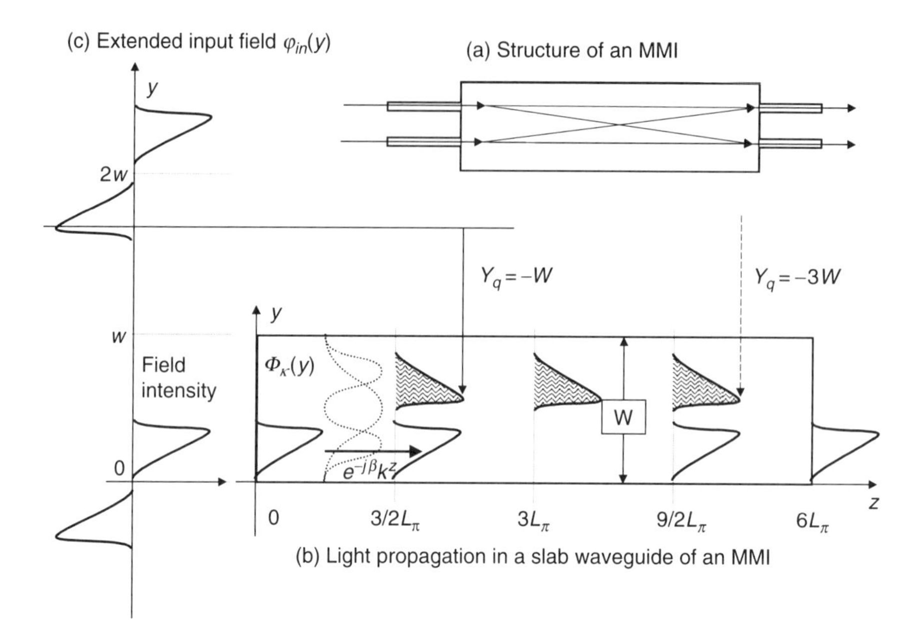

Fig. 11.24 Structure and operation principle of the multi-mode interference (MMI) coupler: (a) Structure of multi-mode interference (MMI) couplers containing two input and output ports. The input and output waveguides are connected by a relatively wide multi-mode waveguide that supports many different lateral modes. (b) Light propagation in the multi-mode waveguide in the MMI coupler. The input field couples to modes in the waveguide at the input side. The light in each mode propagates in the waveguide with its own propagation constant Φ_k and they interfere with each other at the output side of the waveguide. (c) The extended input field introduced to analyze the light propagation in the multi-mode waveguide. The field is completely the same as the input field within the waveguide, $0 < y < W$, but is different outside the waveguide, $y > W$ or $y < 0$, where the field contains an inverted image of the input field and periodical extension along the y-axis with a period of $2W$.

of the input field and the mode profile.

$$c_i = \frac{\int \varphi(y, 0)\phi_i(x)dy}{\int (\phi_i(\gamma))^2 dy} \tag{11.21}$$

If the optical confinement is strong enough, the optical field at the core–clad interface almost completely vanishes. Assume that the field at the interface is completely zero, then the mode profile $\phi_i(x)$ is then simply written as

$$\phi_i(y) = a_i \sin(\gamma_i y), \tag{11.22}$$

with the normalization factor a_i for the i-th mode, and

$$\gamma_i = (i+1)\frac{\pi}{W}. \tag{11.23}$$

The propagation constant β_i in the refractive index n_c is then given by

$$\beta_i = \sqrt{(n_c k_0)^2 - \gamma_i^2} \approx n_c k_0 - \frac{\pi(i+1)^2\lambda}{4n_c W^2}. \tag{11.24}$$

Defining the lowest order beat length as

$$L_\pi = \frac{\pi}{\beta_1 - \beta_0} \frac{4n_c W^2}{3\lambda}, \tag{11.25}$$

the propagation constant of the k-th mode is written as

$$\beta_i \approx \beta_0 - \frac{i(i+2)\pi}{3L_\pi}, \tag{11.26}$$

with the lowest order mode propagation constant, β_0,

$$\beta_0 \approx n_c k_0 - \frac{\pi}{3L_\pi}. \tag{11.27}$$

Substituting Eq. (11.27) into Eq. (11.20), we obtain

$$\varphi(y, z) = \sum_{i=0}^{n-1} c_i \phi_i(y) e^{-j\beta_0 z} e^{-j\frac{i(i+2)\pi}{3L_\pi} z}. \tag{11.28}$$

The first exponential term simply describes the phase change due to propagation in the slab region. The second exponential term, on the other hand, determines the relative phase between different modes and consequently determines the characteristics of the MMI. The regular i-dependence of the propagation constant leads the self-imaging properties due to the interference. The equations imply that there are two interesting points on z where the propagating light makes a focus.

At $z = 6L_\pi$, the second exponential term in Eq. (11.28) is unity for all possible modes, as the phase term is a multiple of 2π. Therefore, the optical field at this point is the same as the input field $\varphi(\gamma, 0)$. This self-imaging reconstructs the input field at this point.

At $z = 3L_\pi$, the second exponential term is unity for all even modes but negative unity for all odd modes. The even and the odd modes represent the symmetric and asymmetric components of the field profile, respectively. This means the image at this point is an inverted image of the input field to the center of the waveguide.

The general self-imaging properties with multiple focuses have been investigated in the reference [33]. The analysis predicts that multiple images appear at a length of L_{pN};

$$L_{pN} = 3L_{\pi} p/N, \tag{11.29}$$

where p and N are positive integers with no common divisor. Let us introduce the extended input field φ_{in} given by

$$\varphi_{\text{in}}(y) = \sum_{i=0}^{\infty} c_i \phi_i(y), \tag{11.30}$$

which is the input field re-defined by reverse Fourier transform of the Fourier components of the input field defined in Eq. (11.21). Note that this field is the same as the input field within the waveguide, $0 < y < W$, but is different outside the waveguide, $y > W$ or $y < 0$. Although the actual input field has no optical field outside the waveguide, the extended one has a periodical field that contains an inverted image of the input field and periodical extension along the y-axis with a period of $2W$. Using this extended input field, the field at L_{pN} is expressed by

$$\varphi(y, L_{pN}) = \sum_{q=0}^{N-1} \varphi_{\text{in}}(y - y_q)e^{j\Omega_q}, \tag{11.31}$$

with

$$y_q = p(2q - N)W/N \tag{11.32}$$

and

$$\Omega_q = p(N - q)q\pi/N. \tag{11.33}$$

Equation (11.31) expresses that the optical field at L_{pN} is a sum of N images that are the images of the extended input field successively sifted by $-2W/N$ in the y direction. The optical field at $z = 3/2L_{\pi}(9/2L_{\pi})$ is depicted in Fig. 11.24(a), where we can see two folded images containing two images of the extended input field: One is non-shifted and the other one is shifted by $y_q = -W$ $(-3W)$. The actual MMI coupler contains N waveguides on the input and the output sides of the multimode waveguide with a length of L_{pN}. As the input field is created by an input waveguide, the multimode waveguide with length L_{pN} distributes light from an input waveguide to N focusing points at the output. If the output waveguides are allocated on these focusing points, the light from an input waveguide is distributed to N output waveguides. Reciprocally, light in an output waveguide contains light from N input waveguides. Thus, this MMI coupler can be applied to $N \times N$ couplers if the input and output waveguides are properly aligned.

Although an $N \times N$ coupler requires a minimum length of $L_{1N} = 3L_\pi/N$, a simple power splitter or combiner, that is a $1 \times N$ and $N \times 1$ coupler, can be fabricated with a much shorter length. Let us consider a power splitter with an input waveguide at the center of the MMI. The symmetry of the structure implies that the Fourier expansion in Eq. (11.20) only contains even order terms. This significantly simplifies the relation, and reduces MMI length by a factor of four. Figure 11.25 shows the simulated

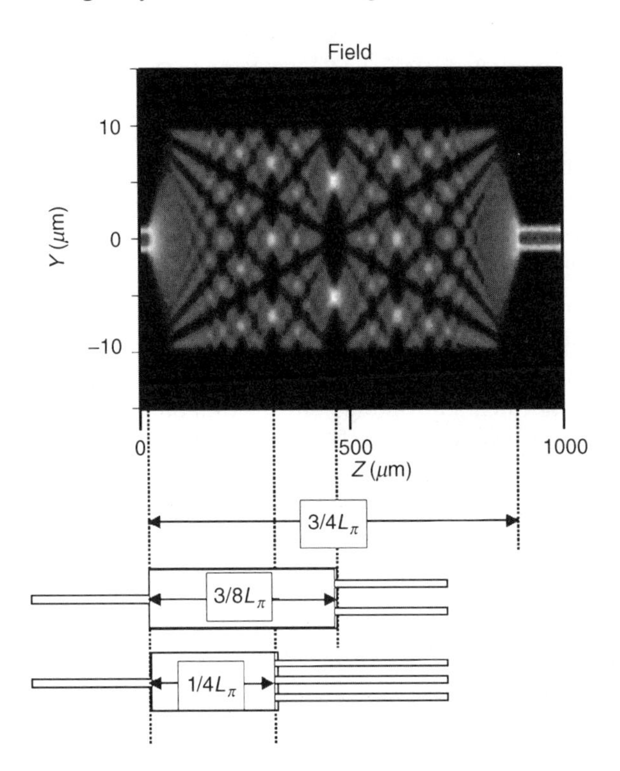

Fig. 11.25 Simulated light propagation in an multi-mode interference (MMI) coupler calculated by the beam propagation method (BPM). The device contains one input port at the center of the MMI waveguide. The symmetry of the device reduces device length by a factor of four. The simulated device (top) has a length of $1.5L_{pN}$ or forming a focus point on the output side. The light in the MMI waveguide forms several focusing points at the intermediate positions. The device with the intermediate length therefore works as a $1 \times N$ coupler (lower figures). (*See color plate*).

light propagation in an MMI device with a single-input waveguide at the center. The beam propagation method (BPM) was used for the calculation. The MMI waveguide has a width of 20 μm and an equivalent refractive index of 3.22, and the corresponding beat length L_p is 1100 μm. The length is set at $3L_\pi/4 = 830\ \mu$m. The optical field forms a single focusing point at the output waveguide, and N of focusing points appear at $z = 3L_\pi/4N$. If the output waveguides are aligned at these focusing points, a $1 \times N$ power splitter can be fabricated with the MMI length of $L_N = 3L_\pi/4N$. The MMI is four times shorter than the $N \times N$ coupler.

11.6. Monolithic Integration of Semiconductor Devices

Although active integrated devices have been used practically, application of semiconductor monolithic integration to passive devices at more than just a primitive level has only recently become possible, thanks to recent progress in fabrication technologies and device design. Two significant developments, the AWG and the MMI coupler, have significantly accelerated development of monolithic circuits or monolithic devices. AWGs, by providing the filtering function for multiplexing and demultiplexing signals, have been instrumental in expanding the application of semiconductor devices to WDM systems. Device integration is a very important issue for WDM systems because the systems tend to use a large number of optical devices. The significant advantage of the semiconductor-based AWG over AWGs based on other materials is its capability of monolithic integration with other semiconductor devices. Semiconductor monolithic technologies are now being applied in various devices for WDM systems, mostly with AWGs.

11.6.1. MONOLITHIC INTEGRATION OF SEMICONDUCTOR AWGs AND PHOTO-DETECTORS

Integration with a photodiode is relatively easy, but the advantage of the integration is clear; therefore, multi-wavelength AWG receiver [34–37] and WDM monitoring devices [38] have been fabricated at many laboratories.

Figure 11.26 shows an example of the semiconductor AWG integrated with photo-detectors. As shown in the right bottom figure, the device contains AWG connected to 16 photodiodes. The input signal, which contains 16 different wavelengths, is demultiplexed by the AWG and then each signal

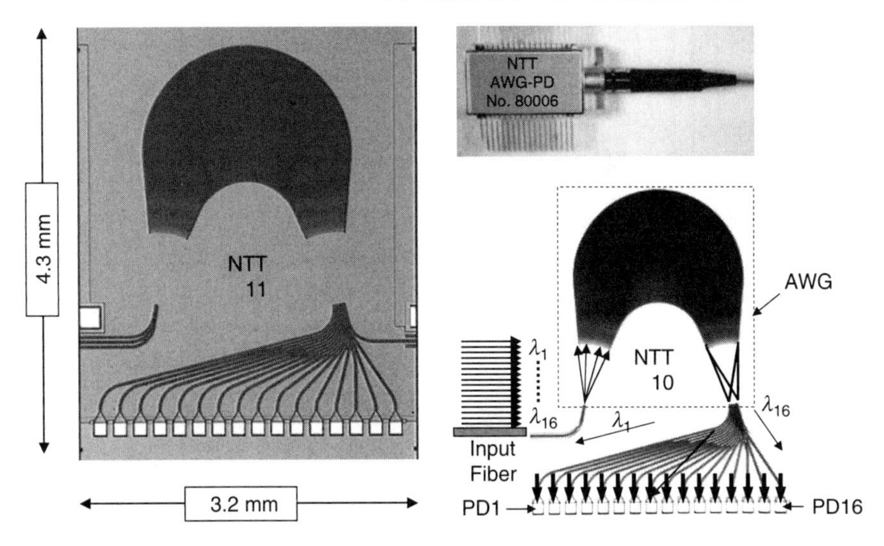

Fig. 11.26 An AWG integrated with photodiodes. On the left is a photograph of the chip, and on the top right a photograph of the packaged device. The drawing explains the operation of the device. (*See color plate*).

is detected by corresponding photo-detectors. On the left is a photograph of the chip. A small bending radius of 250 μm makes the chip as small as $3.4 \times 4.4\,\text{mm}^2$. The device has 16 photo-detectors for electrical monitoring and one optical output port for easy characterization of the chips. The AWG has 114 waveguides in the grating arm. A 2.45-μm-wide and 4.0-μm-deep ridge waveguide structure is used for obtaining polarization-insensitive characteristics. The path length difference ΔL is 27.25 μm, which gives a free spectral range (FSR) of 3.2 THz and the corresponding order of 56 at around 1.55 μm.

Figure 11.27 summarizes the fabrication process for the AWG integrated with a photo-detector. The device structure is formed in a sequential procedure comprising the crystal growths and the etchings formations. The first growth forms the layer structure for both the waveguide and photo-detectors. The waveguide layer consists of a 0.5-μm-thick n-InGaAsP ($n = 5 \times 10^{17}\,\text{cm}^{-3}$) core with band-gap wavelength of 1.05 μm. The photodiode layers on the waveguide layer consist of a 0.1-μm-thick undoped-InP spacer layer, a 0.3-μm-thick undoped InGaAs absorption layer, a 1.0-μm-thick p-InP cladding layer, and a 0.3-μm-thick p-InGaAs contact layer. This waveguide structure has a relatively large evanescent field in the cladding layer and enables us to obtain efficient evanescent coupling to the absorption layer.

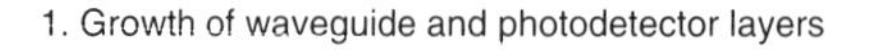

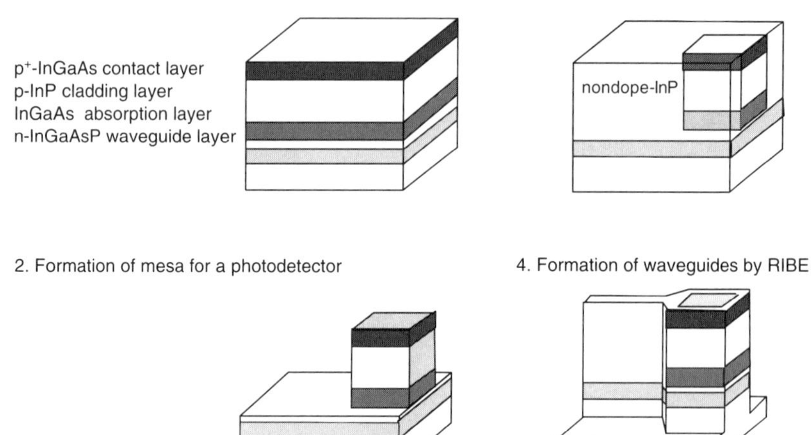

Fig. 11.27 Fabrication steps for an AWG integrated with photodiodes. 1. Crystal growth of waveguide and photo-detector layers. 2. Formation of mesa for photo-detectors by wet etching. 3. Crystal regrowth of non-doping layer for cladding. 4. Formation of waveguides by reactive ion-beam etching (RIBE).

After the first-growth, photo-detecter layers in the passive waveguide region are etched off to form mesas for a photo-detector. Then, a 1.5-μm-thick non-doped InP cladding layer is regrown on the core layer to form a cladding layer in the passive waveguide region. The regrowth of the non-doped cladding is essential in order to remove strong absorption of p-type materials. Finally, both waveguide and photodiode regions are etched simultaneously by Br_2-N_2-reactive ion-beam etching (RIBE). The photodiode is $180 \times 180\ \mu m^2$ and the electrical isolation between the photodiodes is more than ten mega-ohms. The 3.4 mm $\times$ 4.4 mm^2 chip is cleaved from the wafer and mounted on a Cu-W heat sink. Finally, the chip is packaged in a butterfly package with a Peltier cooler and thermister for temperature control. A two-lens system couples light from the input fiber into the device. The packaged device shown on the top right, is nearly the same size as the conventional laser diode module.

Figure 11.28 shows observed photo response of the 16 detectors integrated with the AWG as a function of the input wavelength. Each response curve shows the response for different detectors, PD1 to PD16. The response curves for all detectors show a clear peak. The peak wavelength difference between neighboring detectors is about 0.8 nm, which corresponds

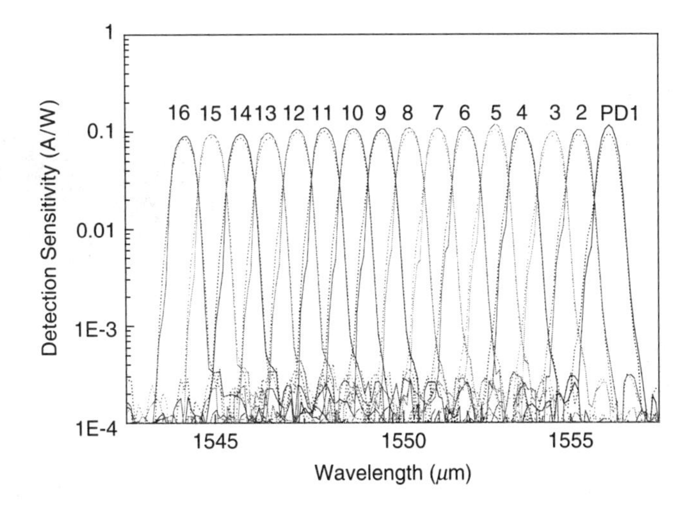

Fig. 11.28 Spectrum response of the efficiencies in an AWG integrated with photo-detectors. The device is designed for equally spaced WDM systems with channel separation of 100 GHz. In the figure, the 16 response curves for different detectors overlap. (*See color plate*).

to the 100-GHz channel spacing. A channel cross-talk less -20 dB has been obtained for all 16 channels.

A significant advantage of an AWG filter is its flexibility of design for wavelength allocation. As spectrum in wavelength is converted to spatial distribution on the output slab waveguide in AWGs, allocation of the output waveguide determines peak allocation in the wavelength response [39]. Figure 11.29 shows the spectral response of an AWG integrated with photo-detectors designed for the unequally spaced WDM system [40]. The unequally spaced WDM system is used for WDM transmission in dispersion-shifted fiber (DSF) to suppress four wave mixing (FWM) cross-talk. Design of the arrayed waveguide is almost the same as for the equally spaced AWG shown in Fig. 11.26, but allocation of the output waveguides is specially designed to obtain unequally spaced peaks in the wavelength. According to the AWG parameters, the ratio of the spatial movement of the focus point Δx in the second slab to the frequency change Δf is given by $\Delta x/\Delta f = 0.033\,\mu$m/GHz. The unequally spaced AWG is realized by allocating the output waveguide with spacing determined with the wavelength separation between channels and the parameter $\Delta x/\Delta f$. The right figure is an enlarged view of the interface of the second slab and the output waveguides, where we can see the unequal spatial allocation of the output waveguides. Due to this spatial allocation, the transmission spectrum of

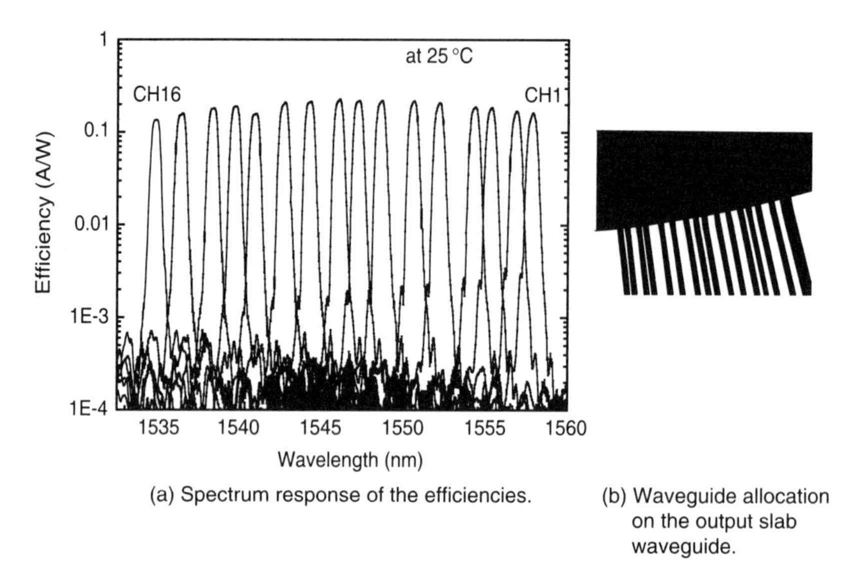

Fig. 11.29 An AWG integrated with photo-detectors designed for unequally spaced WDM systems. (a) Spectrum response of the efficiencies. (b) Waveguide allocation on the output slab waveguide.

the device contains transmission peaks unequally spaced in wavelength as shown in the left figure. The wavelength of each peak corresponds to the channel wavelength in an unequal-channel-spacing WDM system.

11.6.2. MONOLITHIC INTEGRATION OF AWG AND SEMICONDUCTOR OPTICAL AMPLIFIERS

Integration with semiconductor optical amplifiers (SOAs) is also an attractive target. The optical gain provided by SOAs may compensate the relatively large insertion loss in semiconductor monolithic circuits and enable us to make larger-scale optical circuits. The SOA is also attractive as an optical gate or switch because fast and low-crosstalk optical gating is possible by only on/off control of the currents. The device shown in Fig. 11.30, a WDM channel selector [41, 42], is a good example of the integration of a SOA. WDM channel selectors, which contain two AWGs connected by SOAs, will be key components in advanced WDM-based photonic network systems. Fast and low-cross-talk operations are expected because arbitrary demultiplexed signals can be selected by using only an on/off control of the SOA currents.

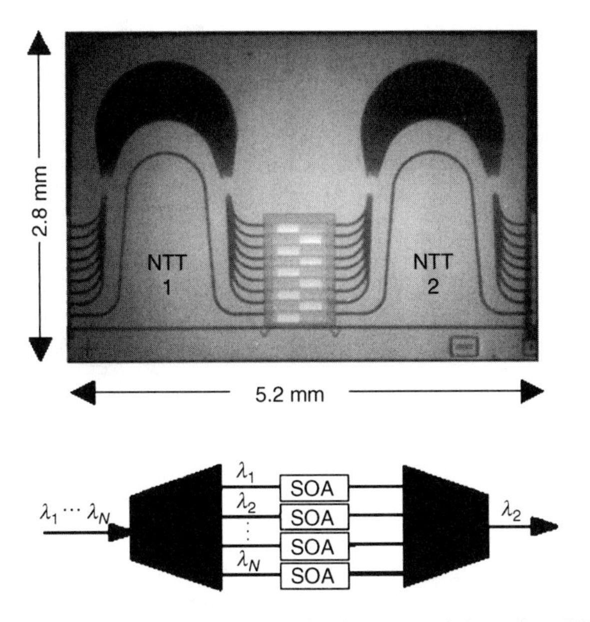

Fig. 11.30 Photograph of the WDM channel selector consisting of an AWG integrated with SOAs. The drawing at the bottom shows the schematic structure of the device.

WDM channel selectors are composed of numerous waveguides and components; therefore, reduction of device size is an issue in achieving low-cost mass production. The device in the figure uses a hybrid waveguide structure to make it compact and reliable; a deep-ridge structure with a small bending radius is used for passive waveguides, and a buried waveguide structure, whose reliability has been confirmed in conventional lasers, is used in the SOA regions

The device consists of two AWGs and an 8-ch. SOA array. Each AWG has 8 input and 8 output waveguides and an array of 28 waveguides. The free spectral range and the channel spacing are 1.6 THz and 200 GHz, respectively. Eight 600-μm-long SOAs are located between the two AWGs. The spacing in the SOA array is 125 μm. Total device size is 5.2×3.6 mm^2. WDM signals, which are fed into one of the input waveguides, are demultiplexed by the first AWG, and some of the demultiplexed signals are selected by adjusting the current of the 8 SOA gates. Then, the selected signals are multiplexed into one output waveguide in the second AWG.

The deep-ridge structure is used in most of the passive waveguide regions. The waveguide layer is a 0.5-μm-thick InGaAsP ($\lambda_g = 1.05$ μm). A non-doped InP buffer layer is inserted above the waveguide layer to prevent

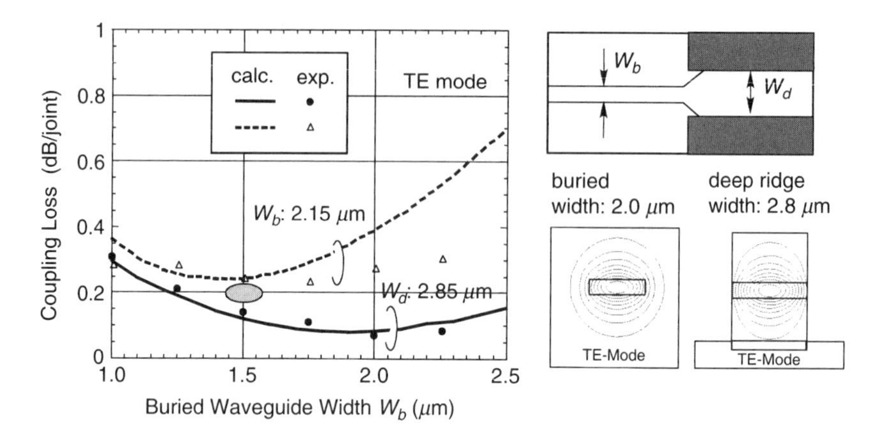

Fig. 11.31 Coupling efficiency between deep-ridge and buried waveguides for the structure shown at the right. On the bottom right are the optical field distributions in the deep-ridge and the buried waveguides.

absorption loss in the p-type InP. The 2.5-μm-wide, 4.5-μm-deep ridge waveguide is designed to obtain low polarization dependence [12, 43]. Because light is strongly confined by air against the lateral direction, a small bending radius is allowed in the waveguide layout, without excess loss. A minimum bending radius of 250 μm was chosen, which makes the AWGs as small as $2.2 \times 3.6\,\text{mm}^2$. On the other hand, the buried waveguide structure is used in the SOA regions. The active layer of the SOA is a 0.3-μm-thick InGaAsP ($\lambda_g = 1.55\,\mu\text{m}$). The buried waveguide is the same as that conventionally used in laser diodes. The injection current is almost completely confined to the active waveguide by the current blocking layers. This structure should provide high reliability because the active layer is buried in InP.

The optical coupling between the two waveguides was calculated by the finite difference method (FDM), and the results are shown in Fig. 11.31 [44]. Even though the structures are quite different, we can make field distributions of about the same size by choosing appropriate sizes for the waveguides. Therefore, the calculated coupling loss at the interface is as small as 0.2 dB. The losses were experimentally estimated from the excess loss of test waveguides with many interfaces, below 0.3 dB in these designs.

The fabrication process is summarized in Fig. 11.32. First, the active layer is deposited on an n-InP wafer. After part of it is removed by wet chemical etching, the passive layer is butt-jointed using MOVPE regrowth. Then, an undoped InP buffer layer is grown on the passive layer to prevent

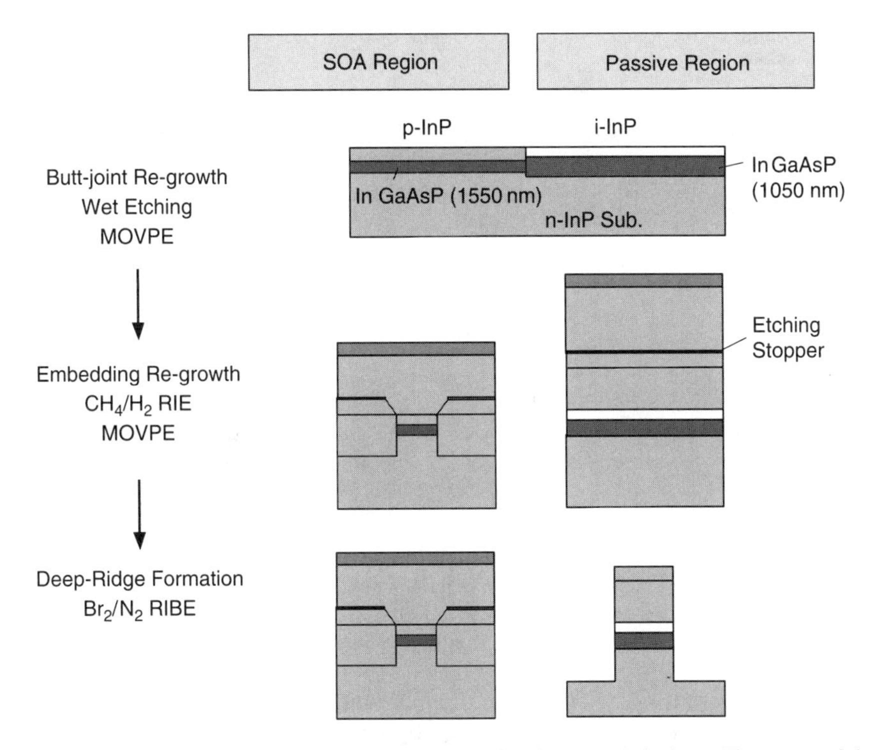

Fig. 11.32 Fabrication process for the AWG–SOA integrated devices. The sequential crystal growths and etchings form different waveguide structures for the semiconductor optical amplifiers (SOAs) and the passive waveguides.

absorption loss in the p-type cladding layer. Next, the buried waveguide is made by reactive ion etching with CH_4/H_2 gases, and two-step regrowth. Then, the embedding layers are also regrown on the flat passive region, where a quaternary etching stopper is inserted in the layers. Next, part of the cladding layer on the passive region is removed using the etching stopper, and Br_2/N_2 reactive ion-beam etching [13] forms the deep ridge. Finally, electrodes are formed in the same manner as in laser fabrication.

The filtering characteristics of the WDM channel selector are shown in Fig. 11.33. WDM signals containing 8 different wavelengths from 1546.0 to 1557.2 nm with a spacing of 1.6 nm (200 GHz) were used for input. The power of each channel was adjusted to $\sim$$-8$ dBm, as shown in the top left of the figure. The WDM signals were passed through the device with a pair of lensed fibers, and the filtered light was measured using an optical spectrum analyzer. The right figure shows the spectra of the filtered light, when one channel was selected by the SOA gate control.

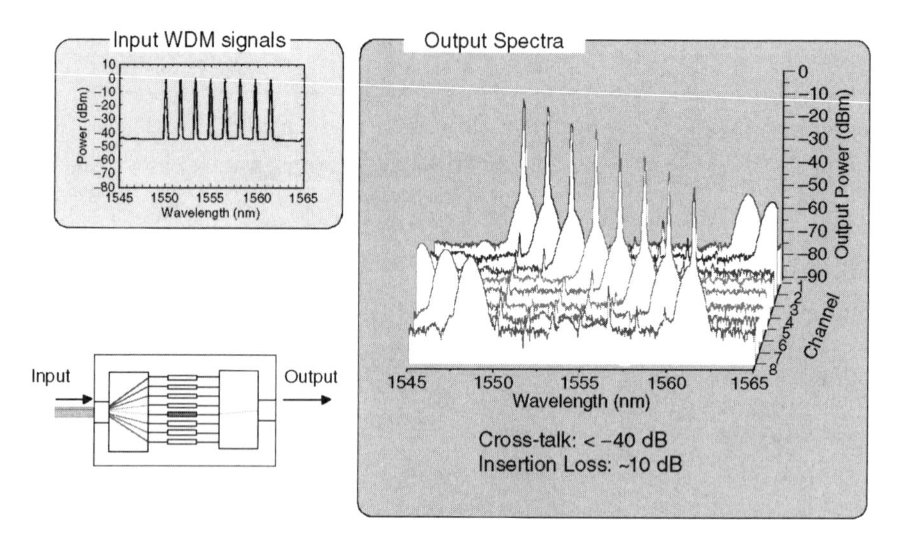

Fig. 11.33 Channel selection of WDM signal in the AWG–SOA integrated device. The input light contains eight different wavelengths as shown at the top left. An injection current excites one of the eight optical amplifiers (left bottom). The right figure shows the eight different spectra of the output light when a different amplifier is excited.

The SOA current was 200 mA in the ON state. All 8 channels were well selected with a low cross-talk of less than −40 dB, where the cross-talk is defined as the ratio between a selected signal and an unselected signal. The amplified spontaneous emission (ASE) passed through the second AWG filter is observed at ∼30 dB lower level than the signal, where the resolution of the used spectrum analyzer is 0.1 nm. Figure 11.34 shows the fiber-to-fiber transmittance for each channel as a function of the SOA current. The power of the input light is −12.4 dB, and the input was TE-polarized light. Zero-insertion-loss operations are obtained, except with Channel 8. High extinction ratios of more than 50 dB are also obtained due to large absorption in the SOA. Transmission characteristics of the AWG filter, which was cleaved from the monolithic device, were also measured. The AWG filter has low polarization sensitivity. The transmission peak difference between TE and TM input is ∼0.1 nm in wavelength and 0.4 dB in transmittance. However, the SOA has a polarization-dependent gain of 2–3 dB at a SOA current of 150 mA. On-chip loss without coupling for an AWG is estimated to be 10–12 dB, where the loss of the out-side port is 2 dB higher than that of the central port. The coupling loss between the fiber and the device is ∼3 dB. Therefore, the total loss is ∼26–30 dB [$= 2 \times (10\text{–}12 + 3)$]. Because the SOA almost completely compensates for the loss, fiber-to-fiber insertion losses are less than 1 dB at a SOA current of 150 mA.

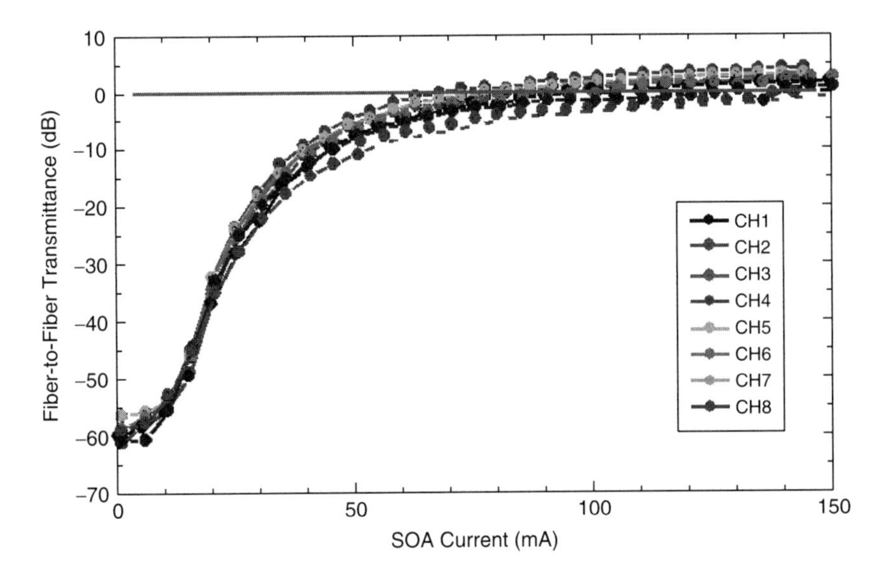

Fig. 11.34 Fiber-to-fiber gain of the AWG–SOA integrated device as a function of the injection current to an optical amplifier. Lensed fibers are coupled to the input and the output waveguides of the device. The fiber-to-fiber gain was measured for eight different wavelength channels with the corresponding amplifier excited. (*See color plate*).

The gain peak of the SOA is located at a slightly shorter wavelength. This makes the gain relatively small for Channel 8, which is the longer wavelength. However, when the SOA current is 220 mA, zero-insertion-loss operations for all the channels are achieved. The non-doped InP buffer layer in this device is 0.6-μm thick. A waveguide loss of 15 dB/cm was estimated from the test waveguides. This value is still larger than the loss of the non-doped waveguide, 2–3 dB/cm. Therefore, further improvements of transmittance are possible if the waveguide loss is reduced.

A similar structure consisting of an AWG and optical amplifiers has been also applied to make semiconductor lasers [45, 46]. These compact and reliable devices promise significant cost savings in WDM communication networks.

11.6.3. MONOLITHIC INTEGRATION OF AWG AND OPTICAL SWITCH

Optical cross-connects (OXCs) are important network elements in WDM systems, which exchange optical signals in different fibers at a node point. Figure 11.35 shows the schematic structure of an OXC, where the system

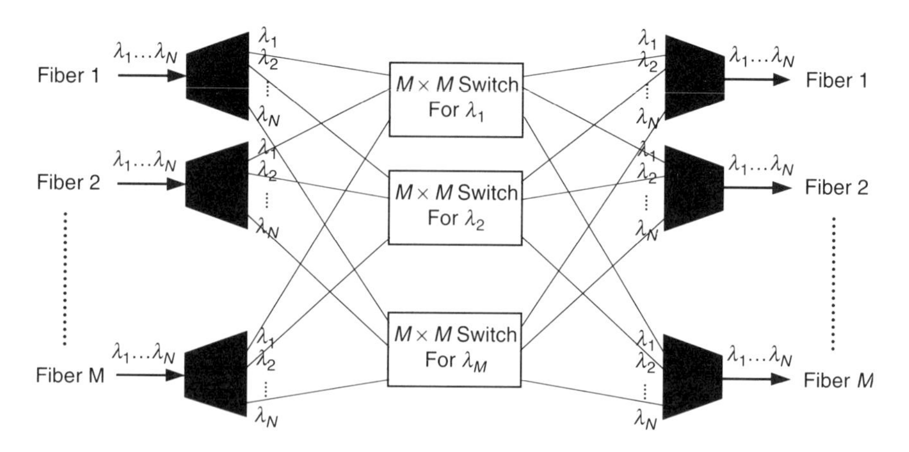

Fig. 11.35 Schematic structure of the optical cross-connect (OXC), where the system contains M input and output fibers and each fiber carries light with N different wavelengths. The input lights are independently demultiplexed by $1 \times N$ demultiplexers. Then demultiplexed lights from different fibers but with the same wavelength are concentrated to $M \times M$ matrix switches for selection of the proper destinations. The lights with different wavelength but for the same fiber are multiplexed again for the output.

contains M input and output fibers and each fiber carries light with N different wavelengths. The input lights are independently demultiplexed by $1 \times N$ demultiplexers. Then demultiplexed lights from different fibers but with the same wavelength are concentrated to $M \times M$ matrix switches for selection of the proper destinations. The lights with different wavelength but for the same fiber are multiplexed again for the output. Although the OXC can be implemented by connecting optical devices, such as filters and optical switches, complicated structure requires optical integration for practical implementation. Figure 11.36 shows a monolithically integrated WDM cross-connect device consisting of two 2×8 AWG multiplexers connected with four dilated electrooptic space switches [47, 48]. A folded design allows the use of one 2×8 AWG twice, for 1×4 demultiplexing and 4×1 multiplexing. Thus, the device equivalently contains four AWGs. Each dilated switch contains four Mach–Zehnder interferometer (MZI) switches operated by applying an electric field to one of the arms of the interferometer. The advantage of using a revere bias instead of current injection is the negligible power consumption and potentially very fast switching. Although most voltage-driven devices have strong polarization dependence, the orientation-sensitive Pockels' effect, which occurs only

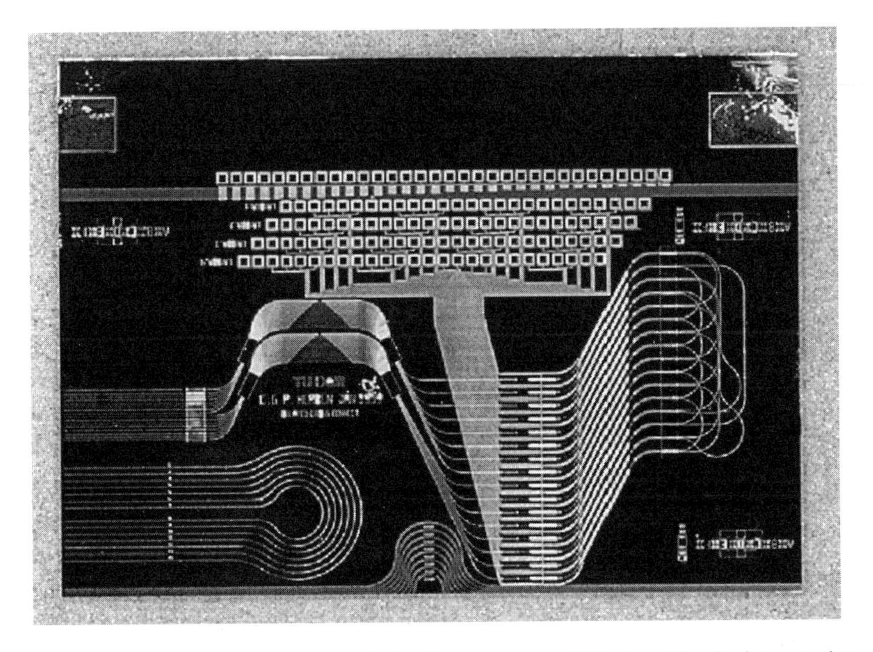

Fig. 11.36 Photograph of the monolithically integrated OXC chip. The device contains two AWGs and optical switches.

for TE, enables us to make polarization-insensitive switches. The phase shifters of the switches are aligned at an angle of 28 degrees relative to the [011]-crystallographic orientation to obtain polarization-independent operation [49]. The switches are aligned in such a way that the total number of crossings is minimized. All possible paths have an equal number of crossings to avoid nonuniformity in the response due to crossing losses. The device is designed for $4 \times 400\,\text{GHz}$ (3.2 nm). Total device size is $11 \times 6.5\,\text{mm}$.

Figure 11.37 shows the wavelength response measured in the monolithically integrated OXC device. Each curve represents output to the cross and the bar port observed with TE- and TM-polarized input light. Switches are set up so that the second wavelength channel is switched to cross-port in Fig. 11.37(a) or the fourth wavelength channel is switched to cross in Fig. 11.37(b). The device has total chip loss of 15–17 dB. The response curves indicate that signals are successfully switched between the cross and the bar state with crosstalk better than −20 dB.

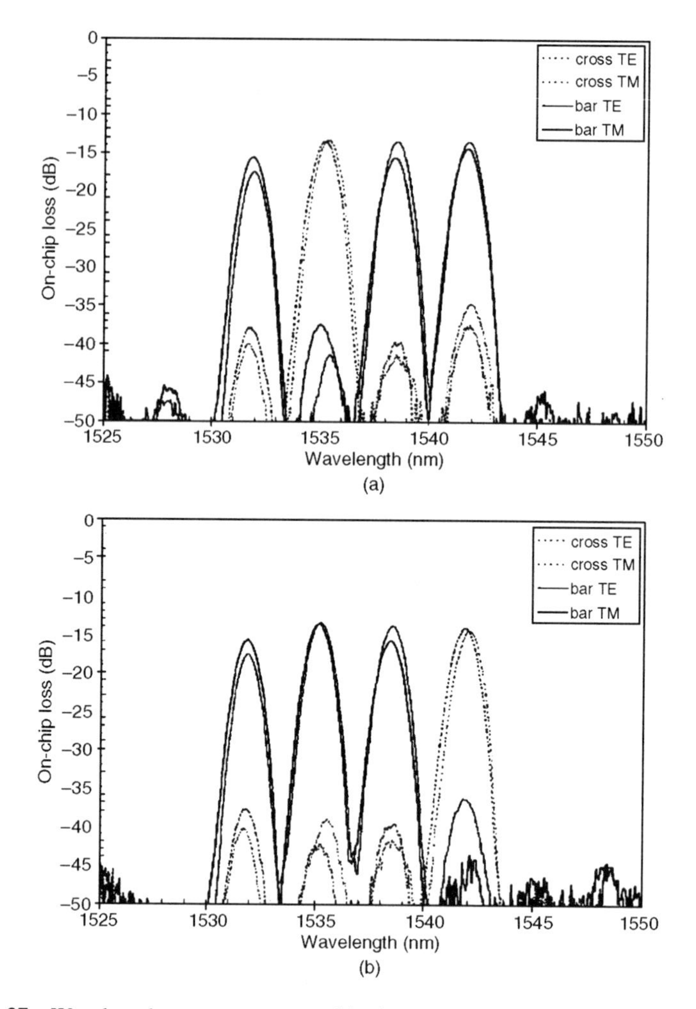

Fig. 11.37 Wavelength response measured in the monolithically integrated OXC device. Each curve represents the output to the cross and the bar port with TE and TM polarizations. (a) Second wavelength channel is switched to cross. (b) Fourth wavelength channel is switched to cross.

11.7. Summary

The explosive growth of Internet traffic makes the development of integrated photonic devices an important issue in building future photonic systems. Monolithic integration on semiconductor substrates is an attractive technology because it can include semiconductor devices, which are indispensable in optical systems. Although semiconductor-laser-based technologies have already been applied to make active devices, application to passive device had been limited to a primitive stage until the recent progress made in fabrication technologies and device designing. These technologies enable us to extend application of semiconductor materials to passive optical devices and consequently to complicated monolithic circuits containing both passive and active devices. These compact and reliable devices promise sucessful development for the next generation in WDM communication networks.

References

1. R. E. Nahory, M. A. Pollack, W. D. Johnston Jr., and R. L. Barnes, Appl. Phys. Lett., 33 (1978) 659.
2. R. L. Moon, G. A. Antypas, and L. W. James, J. Electron. Mater., 3 (1974) 635.
3. K. Utaka, Y. Suematsu, K. Kobayashi, and H. Kawanishi, "GaInAsP/InP integrated twin-guide lasers with first-order distributed Bragg reflectors at 1.3 μm wavelength," Japan J. Appl. Phys., 19:2 (1980) L137–L140.
4. F. N. Timofeev, P. Bayvel, J. E. Midwinter, and M. N. Sokolskii, "High-performance, free-space ruled concave grating demultiplexer," Electron. Lett., 31:17 (1996) 1466–1467.
5. E. S. Koteles, J. J. He, B. Lamontagne, A. Delage, L. Erickson, G. Champion, and M. Davies, "Recent advances in InP-based waveguide grating demultiplexers," Optical Fiber Commuun. Conf. 1998, Technical Digest (1998) 82–83.
6. V. Mizrahi, T. Erdogan, D. J. DiGiovanni, P. J. Lemaire, W. M. MacDonald, S. G. Kosinski, S. Cabot, and J. E. Sipe, "Four channel fibre grating demultiplexer," Electron. Lett., 30:10 (1994) 780–781.
7. M. K. Smit, "New focusing and dispersive planar component based on an optical phased array," Electron. Lett., 24:7 (1988) 385–386.
8. H. Tanobe, Y. Kondo, Y. Kadota, K. Okamoto, and Y. Yoshikuni, "Temperature insensitive arrayed waveguide gratings on InP substrates," IEEE Photon. Technol. Lett., 10:2 (1998) 235–237.
9. J. B. D. Soole, M. R. Amersfoort, H. P. LeBlanc, N. C. Andreadakis, A. Rajhel, C. Caneau, M. A. Koza, R. Bhat, C. Youtsey, and I. Adesida, "Polarisation-independent InP arrayed waveguide filter using square cross-section waveguides," Electron. Lett., 32:4 (1996) 323–324.

10. B. H. Verbeek, A. A. M. Staring, E. J. Jansen, R. van Roijen, J. J. M. Binsma, T. van Dongen, M. R. Amersfoort, C. van Damm, and M. K. Smit, "Large bandwidth polarization independent and compact 8-channel PHASAR demultiplexer/filter," Proceedings in OFC/IOOC'94, San Jose, CA PD13-1 (Feb. 1994) 63–66.
11. H. Bisessir, B. Martin, R. Mestric, and F. Gaborit, "Small-size, polarisation independent phased-array demultiplexers on InP," Electron. Lett., 31:24 (1995) 2118–2120.
12. M. Kohtoku, H. Sanjoh, S. Oku, Y. Kadota, and Y. Yoshikuni, "Polarization independent InP arrayed waveguide grating filter using deep ridge waveguide structure," CLEO/Pacific Rim'97 (1997) 284–285.
13. S. Oku, Y. Shibata, and K. Ochiai, "Controlled beam dry etching of InP by using Br_2-N_2 gas," J. Electron. Mater., 25 (1996) 585–591.
14. Y. S. Oei, L. H. Spiekman, F. H. Groen, I. Moerman, E. G. Metaal, and J. W. Pedersen, "Novel RIE-process for high quality InP-based waveguide structures," Proc. 7th Eur. Conf. On Int. Opt. (ECIO'95), Delft, The Netherlands (1995) 205–208.
15. C. G. M. Vreeburg, C. G. P. Herben, X. J. M. Leijtens, M. K. Smit, F. H. Groen, J. J. G. M. van der Tol, and P. Demeester, "A Low-loss 16-channel polarization dispersion-compensated PHASAR demultiplexer," IEEE Photon. Technol. Lett., 10:3 (1998) 382–384.
16. M. Zirngibl, C. H. Joyner, and P. C. Chou, "Polarisation compensated waveguide grating router on InP," Electron. Lett., 31 (1995) 1662–1664.
17. Y. Inoue, Y. Ohmori, M. Kawachi, S. Ando, T. Sawada, and H. Takahashi, "Polarization mode convertor with polyimide half waveplate in silica-based planar lightwave circuits," IEEE Photonics Technol. Lett., 6:5 (1994) 626–628.
18. K. Okamoto, K. Syuto, H. Takahashi, and Y. Ohmori, "Fabrication of 128-channel arrayed-waveguide grating multiplexer with 25 GHz channel spacing," Electron. Lett., 32:16 (1996) 1474–1476.
19. H. Yamada, K. Takada, and S. Mitachi, "Crosstalk reduction in a 10-GHz spacing arrayed-waveguide grating by phase-error compensation," J. Lightwave Technol., 16:3 (Mar. 1998) 364–371.
20. H. Yamada, K. Takada, Y. Inoue, Y. Ohmori, and S. Mitachi, "Statically-phase-compensated 10 GHz-spaced arrayed-waveguide grating," Electron. Lett., 32:17 (1996) 1580–1582.
21. M. Kohtoku, H. Sanjoh, S. Oku, Y. Kadota, Y. Yoshikuni, and Y. Shibata, "InP-based 64-channel arrayed waveguide grating with 50 GHz channel spacing and up to −20 dB crosstalk," Electron. Lett., 33:21 (1997) 1786–1787.
22. Y. C. Zhu, F. H. Groen, D. H. P. Maat, Y. S. Oei, J. Romijn, and I. Moerman, "A compact phasar with low central channel loss," ECIO'99 ThA6 (1999).

23. C. G. P. Herben, X. J. M. Leijtens, F. H. Groen, and M. K. Smit, "Low-loss and compact phased array demultiplexer using a double etch process," ECIO'99 ThA4 (1999).
24. M. Kohtoku, S. Oku, Y. Kadota, Y. Shibata, and Y. Yoshikuni, "Spotsize converters integrated with deep ridge waveguide structure," Electron. Lett., 34:25 (1998) 2403–2404.
25. J. Stulemeijer, A. F. Bakker, I. Moerman, F. H. Groen, and M. K. Smit, "InP-based spotsize converter for integration with switching devices," IEEE Photonics Technol. Lett., 11 (1999) 81–83.
26. T. Brenner and H. Melchior, "Local etch-rate control of masked InP/InGaAsP by diffusion-limited etching," J. Electrochem. Soc., 141:7 (1994) 1954–1956.
27. Y. Kokubun, S. Yoneda, and H. Tanaka, "Temperature-independent narrow-band optical filter at 1.3 μm wavelength by an athermal waveguide," Electron Lett., 32 (1996) 1998–2000.
28. T. E. Hammon, J. Bulman, F. Ouellette, and S. B. Poole, "A temperature compensated optical fiber Bragg grating band rejection filter and wavelength reference," OECC'96, 18C1-2 (1996) 350–351.
29. H. Tanobe, Y. Kondo, Y. Kadota, H. Yasaka, and Y. Yoshikuni, "A temperature insensitive InGaAsP/InP optical filter," IEEE Photon. Tech. Lett., 8 (1996) 1489–1491.
30. O. Bryngdahl, "Image formation using self-imaging technique," J. Opt. Soc. A., 63.
31. R. Ulrich, "Image formation by phase coincidences in optical waveguides," Optical Communications, 134:3 (1975) 259–264.
32. L. B. Soldano, F. B. Veerman, M. K. Smit, B. H. Verbeek, A. H. Dubost, and E. C. M. Pennings, "Planar monomode optical couplers based on multimode interference effects," J. Lightwave Technol., 10:12 (1992) 1843–1850.
33. M. Bachmann, P. A. Besse, and H. Melchior, "General self-imaging properties in N $\times$ N multimode interference couplers including phase relations," J. Appl. Phys., 33 (1994) 3905–3911.
34. M. Zirngibl, C. H. Joyner, and L. W. Stulz, "WDM receiver by monolithic integration of an optical preamplifier, waveguide grating router and photodiode array," Electron. Lett., 31:7 (1995) 581–582.
35. C. A. M. Steenbergen, C. van Dam, A. Looijen, C. G. P. Herben, M. de Kok, M. K. Smit, J. W. Pedersen, I. Moerman, R. G. F. Baets, and B. H. Verbeek, "Compact low loss 8 $\times$ 10 GHz polarisation independent WDM receiver," 22nd European Conference on Optical Communication, ECOC'96, Proceedings Vol. 1 (1996) 129–132.
36. M. R. Amersfoort, J. B. D. Soole, H. P. LeBlanc, N. C. Andreadakis, A. Rajhel, and C. Caneau, "8 $\times$ 2 nm polarization-independent WDM detector based on compact arrayed waveguide demultiplexer," IPR'95 PD-3 (1995).

37. A. A. M. Staring, C. van Dam, J. J. M. Binsma, E. J. Jansen, A. J. M. Verboven, L. J. C. Vroomen, J. F. de Vries, M. K. Smit, and B. H. Verbeek, "Packaged PHASAR-based wavelength demultiplexer with integrated detectors," 11th International Conference on Integrated Optics and Optical Fibre Communications, and 23rd European Conference on Optical Communications, Vol. 3 (1997) 75–78.
38. M. Kohtoku, S. Oku, Y. Kadota, Y. Shibata, J. Kikuchi, and Y. Yoshikuni, "Unequally spaced 16 channel semiconductor AWG WDM monitor module with low crosstalk (<−24 dB) characteristics," ECIO'99 ThA2 (1999).
39. K. Okamoto, M. Ishii, Y. Hibino, Y. Ohmori, and H. Toba, "Fabrication of unequal channel spacing arrayed-waveguide grating multiplexer modules," Electron. Lett., 31:17 (17th August, 1995) 1464–1466.
40. F. Forghieri, R. W. Tkach, A. R. Chraplyvy, and D. Marcuse, "Reduction of four-wave mixing crosstalk in WDM systems using unequally spaced channels," IEEE Photon. Technol. Lett., 6 (1994) 754–756.
41. M. Zirngibl, C. H. Joyner, and B. Grance, "Digitally tunable channel dropping filter/equalizer based on waveguide grating router and optical amplifier integration," IEEE Photon. Technol. Lett., 6 (1994) 513–515.
42. M. K. Smit and C. van Dam, "PHASAR-based WDM-deices; Principles, design and applications," IEEE J. Select. Topics in Quantum Electron., 2 (1996) 236–250.
43. M. Kohtoku, H. Sanjoh, S. Oku, Y. Kadota, and Y. Yoshikuni, "Polarization independent semiconductor arrayed waveguide gratings using a deep ridge waveguide structure," IEICE Trans. Electron., E81-C (1998) 1195–1204.
44. H. Ishii, H. Sanjoh, M. Kohtoku, S. Oku, Y. Kadota, Y. Yoshikuni, Y. Kondo, and K. Kishi, "Monolithically integrated WDM channel selectors on InP substrates," ECOC'98, Vol. 1, TuD06 (1998) 329–330.
45. S. Menezo, A. Talneau, S. Grosmaire, F. Gaborit, F. Delorme, and S. Slempkes, "10λ monolithic selectable source on InP integrating a DBR laser array and a PHASAR," ECIO'99 ThF2 (1999).
46. R. Monnard, A. K. Srivastava, C. R. Doerr, R.-J. Essiambre, C. H. Joyner, L. W. Stulz, M. Zirngibl, Y. Sun, J. W. Sulhoff, J. I. Zyskind, and C. Wolf, "Demonstration of a 16 × 10 Gb/s long-haul transmitter with 50-GHz channel spacing using two multifrequency lasers," 24th European Conference on Optical Communication (ECOC'98), Proceedings, Vol. 1 (1998) 193–194.
47. C. G. P. Herben, D. H. P. Maat, X. J. M. Leijtens, M. R. Leys, Y. S. Oei, and M. K. Smit, "Polarization independent dilated WDM cross-connect on InP," IEEE Photonics Technol. Lett., 11:12 (1999) 1599–1561.

48. C. R. Doerr, C. H. Joyner, L. W. Stulz, and R. Monnard, "Wavelength-division multiplexing cross connect in InP," IEEE Photon. Technol. Lett., 10:1 (1998) 117–119.
49. R. Krähenbühl, R. Kyburz, W. Vogt, M. Bachmann, T. Brenner, E. Gini, and H. Melchior, "Low-loss polarization insensitive InP/InGaAsP optical space switches for fiber optical communications," IEEE Photon. Technol. Lett., 8 (May 1996) 632–634.

Index

N

P

Q

R

S

X

Y

Z